Microbial Bioremediation: Processes, Techniques and Applications

Microbial Bioremediation: Processes, Techniques and Applications

Edited by Sansa Gilbert

☐ SYRAWOOD
PUBLISHING HOUSE

New York

Published by Syrawood Publishing House,
750 Third Avenue, 9th Floor,
New York, NY 10017, USA
www.syrawoodpublishinghouse.com

Microbial Bioremediation: Processes, Techniques and Applications
Edited by Sansa Gilbert

International Standard Book Number: 978-1-68286-714-3 (Hardback)

Cataloging-in-Publication Data

Microbial bioremediation : processes, techniques and applications / edited by Sansa Gilbert.
    p. cm.
Includes bibliographical references and index.
ISBN 978-1-68286-714-3
1. Bioremediation. 2. Microbiological chemistry. 3. Microbial biotechnology. I. Gilbert, Sansa.
TD192.5 .M53 2019
628.5--dc23

TABLE OF CONTENTS

Preface..IX

Chapter 1   **Production of functionalized oligo-isoprenoids by enzymatic cleavage of rubber**........................1
            Wolf Röther, Jakob Birke, Stephanie Grond, Jose Manuel Beltran and Dieter Jendrossek

Chapter 2   **Engineering the xylose-catabolizing Dahms pathway for production of poly
            (D-lactate-*co*-glycolate) and poly(D-lactate-*co*-glycolate-*co*-D-2-hydroxybutyrate)
            in *Escherichia coli***..9
            So Young Choi, Won Jun Kim, Seung Jung Yu, Si Jae Park, Sung Gap Im and
            Sang Yup Lee

Chapter 3   **Advances and bottlenecks in microbial hydrogen production**...21
            Alan J. Stephen, Sophie A. Archer, Rafael L. Orozco and Lynne E. Macaskie

Chapter 4   **_Mycobacterium smegmatis_ is a suitable cell factory for the production of
            steroidic synthons**...29
            Beatriz Galán, Iria Uhía, Esther García-Fernández, Igor Martínez, Esther Bahíllo,
            Juan L. de la Fuente, José L. Barredo, Lorena Fernández-Cabezón and
            José L. García

Chapter 5   **Intracellular metabolite profiling of _Saccharomyces cerevisiae_ evolved under
            furfural**...42
            Young Hoon Jung, Sooah Kim, Jungwoo Yang, Jin-Ho Seo and Kyoung Heon Kim

Chapter 6   **Tailor-made PAT platform for safe syngas fermentations in batch, fed-batch
            and chemostat mode with _Rhodospirillum rubrum_**...52
            Stephanie Karmann, Stéphanie Follonier, Daniel Egger, Dirk Hebel, Sven Panke
            and Manfred Zinn

Chapter 7   **Comparative studies of the composition of bacterial microbiota associated
            with the ruminal content, ruminal epithelium and in the faeces of lactating
            dairy cows**..63
            Jun-hua Liu, Meng-ling Zhang, Rui-yang Zhang, Wei-yun Zhu and
            Sheng-yong Mao

Chapter 8   **Microbial protein: future sustainable food supply route with low
            environmental footprint**..74
            Silvio Matassa, Nico Boon, Ilje Pikaar and Willy Verstraete

Chapter 9   **Towards the development of multifunctional molecular indicators combining
            soil biogeochemical and microbiological variables to predict the ecological
            integrity of silvicultural practices**..82
            Vincent Peck, Liliana Quiza, Jean-Philippe Buffet, Mondher Khdhiri,
            Audrey-Anne Durand, Alain Paquette, Nelson Thiffault, Christian Messier,
            Nadyre Beaulieu, Claude Guertin and Philippe Constant

Chapter 10   **The sequence capture by hybridization: a new approach for revealing the potential of mono-aromatic hydrocarbons bioattenuation in a deep oligotrophic aquifer**.. 96
Magali Ranchou-Peyruse, Cyrielle Gasc, Marion Guignard, Thomas Aüllo, David Dequidt, Pierre Peyret and Anthony Ranchou-Peyruse

Chapter 11   **Microbial synthesis of a novel terpolyester P(LA-*co*-3HB-*co*-3HP) from low-cost substrates**.. 107
Yilin Ren, Dechuan Meng, Linping Wu, Jinchun Chen, Qiong Wu and Guo-Qiang Chen

Chapter 12   **Syngas obtained by microwave pyrolysis of household wastes as feedstock for polyhydroxyalkanoate production in *Rhodospirillum rubrum*** 117
Olga Revelles, Daniel Beneroso, J. Angel Menéndez, Ana Arenillas, J. Luis García and M. Auxiliadora Prieto

Chapter 13   **Water-, pH- and temperature relations of germination for the extreme xerophiles *Xeromyces bisporus* (FRR 0025), *Aspergillus penicillioides* (JH06THJ) and *Eurotium halophilicum* (FRR 2471)** 123
Andrew Stevenson, Philip G. Hamill, Jan Dijksterhuis and John E. Hallsworth

Chapter 14   **Engineering a bzd cassette for the anaerobic bioconversion of aromatic compounds**... 134
María Teresa Zamarro, María J. L. Barragán, Manuel Carmona, José Luis García and Eduardo Díaz

Chapter 15   **Revealing the combined effects of lactulose and probiotic enterococci on the swine faecal microbiota using 454 pyrosequencing** 142
Jong Pyo Chae, Edward Alain B. Pajarillo, Ju Kyoung Oh, Heebal Kim and Dae-Kyung Kang

Chapter 16   **Molecular optimization of rabies virus glycoprotein expression in *Pichia pastoris***.. 152
Safa Ben Azoun, Aicha Eya Belhaj, Rebecca Göngrich, Brigitte Gasser and Héla Kallel

Chapter 17   **Synergistic chemo-enzymatic hydrolysis of poly (ethylene terephthalate) from textile waste** 166
Felice Quartinello, Simona Vajnhandl, Julija Volmajer Valh, Thomas J. Farmer, Bojana Vončina, Alexandra Lobnik, Enrique Herrero Acero, Alessandro Pellis and Georg M. Guebitz

Chapter 18   **Statistical test for tolerability of effects of an antifungal biocontrol strain on fungal communities in three arable soils**.. 173
Kai Antweiler, Susanne Schreiter, Jens Keilwagen, Petr Baldrian, Siegfried Kropf, Kornelia Smalla, Rita Grosch and Holger Heuer

Chapter 19   **Single-cell genomics based on Raman sorting reveals novel carotenoid-
containing bacteria in the Red Sea**...189
Yizhi Song, Anne-Kristin Kaster, John Vollmers, Yanqing Song,
Paul A. Davison, Martinique Frentrup, Gail M. Preston, Ian P. Thompson,
J. Colin Murrell, Huabing Yin, C. Neil Hunter and Wei E. Huang

Permissions

List of Contributors

Index

PREFACE

Microbial bioremediation is a process that uses microbes for the decay of contaminated media like soil, water, etc. It is a sustainable and cost-effective process. Bioremediation involves stimulating the growth of microorganisms with the addition of microbial cultures to the polluted media which then break down organic substances through oxidation-reduction reactions. Some popularly used technologies in this field include phytoremediation, mycoremediation, bioventing, bioaugmentation, etc. This book covers advanced concepts and theories related to bioremediation. It also elucidates new techniques and their applications in a multidisciplinary manner. The book is appropriate for students seeking detailed information in this area as well as for experts. It will help the readers in keeping pace with the rapid changes in this field.

The information shared in this book is based on empirical researches made by veterans in this field of study. The elaborative information provided in this book will help the readers further their scope of knowledge leading to advancements in this field.

Finally, I would like to thank my fellow researchers who gave constructive feedback and my family members who supported me at every step of my research.

Editor

Production of functionalized oligo-isoprenoids by enzymatic cleavage of rubber

Wolf Röther,[1] Jakob Birke,[1] Stephanie Grond,[2] Jose Manuel Beltran[2] and Dieter Jendrossek[1,*]

[1]Institute of Microbiology, University of Stuttgart, Stuttgart, Germany.

[2]Institute of Organic Chemistry, Eberhard Karls Universität Tübingen, Tübingen, Germany.

Summary

In this study, we show the proof of concept for the production of defined oligo-isoprenoids with terminal functional groups that can be used as starting materials for various purposes including the synthesis of isoprenoid-based plastics. To this end, we used three types of rubber oxygenases for the enzymatic cleavage of rubber [poly(cis-1,4-isoprene)]. Two enzymes, rubber oxygenase $RoxA_{Xsp}$ and rubber oxygenase $RoxB_{Xsp}$, originate from *Xanthomonas* sp. 35Y; the third rubber oxygenase, latex-clearing protein (Lcp_{K30}), is derived from Gram-positive rubber degraders such as *Streptomyces* sp. K30. Emulsions of polyisoprene (latex) were treated with $RoxA_{Xsp}$, $RoxB_{Xsp}$, Lcp_{K30} or with combinations of the three proteins. The cleavage products were purified by solvent extraction and FPLC separation. All products had the same general structure with terminal functions ($CHO-CH_2$- and -CH_2-COCH_3) but differed in the number of intact isoprene units in between. The composition and *m/z* values of oligo-isoprenoid products were determined by HPLC-MS analysis. Our results provide a method for the preparation of reactive oligo-isoprenoids that can likely be used to convert polyisoprene latex or rubber waste materials into value-added molecules, biofuels, polyurethanes or other polymers.

*For correspondence. E-mail dieter.jendrossek@imb.uni-stuttgart.de;

Funding Information
This work was supported by the Graduiertenkolleg GRK1708 to the University of Tübingen and by a grant of the Deutsche Forschungsgemeinschaft to D.J.

Introduction

Natural rubber has been produced in huge amounts for more than a century by cultivating the rubber tree (*Hevea brasiliensis*), and the material is used for a variety of applications, as an example for the production of rubbers, tyres, sealings, latex gloves and many other items. The main component of rubber is the hydrocarbon poly(cis-1,4-isoprene). For most of today's applications of rubber, an important material property is the molecular weight of the polymer that – when high – gives rise to superior material properties that are necessary for example for the production of tyres. However, no attention has been given so far to the use of rubber for the biotechnological preparation of low molecular fine chemicals (Förster-Fromme and Jendrossek, 2010; Kamm, 2014; Akhlaghi *et al.*, 2015; Schrader and Bohlmann, 2015). In this contribution, we describe the proof of concept for the use of rubber oxygenases to cleave polyisoprene-containing (waste) materials to low molecular products and to produce functionalized oligo-isoprenoids with defined structure. The generated products can be used either directly as biofuels or value-added materials which can be obtained by conversion of oligo-isoprenoids to new products such as polyurethanes and related isoprene-containing polymers.

Only two major types of rubber-cleaving enzymes have been described so far. One is the rubber oxygenase RoxA that was first isolated from *Xanthomonas* sp. 35Y (Tsuchii and Takeda, 1990; Braaz *et al.*, 2004) and has been found only in Gram-negative rubber-degrading bacteria (Birke *et al.*, 2013). The genome sequence of *Xanthomonas* sp. 35Y has been determined (Sharma, V., Siedenburg, G., Birke, J., Mobeen, F., Jendrossek, D., Srivastava, T.P. unpubl. data). RoxA of *Xanthomonas* sp. 35Y ($RoxA_{Xsp}$) is a *c*-type dihaem dioxygenase ($\approx$70 kDa, Fig. 1A) and cleaves poly(cis-1,4-isoprene) into 2-oxo-4,8-dimethyl-trideca-4,8-diene-1-al (ODTD), a C_{15} compound with a terminal keto and aldehyde group as the main product (Fig. 1B) (Braaz *et al.*,

2005; Schmitt *et al.*, 2010). The structure of RoxA$_{Xsp}$ has been solved (Seidel *et al.*, 2013), and molecular insights in the active site of RoxA$_{Xsp}$ as well as the cleavage mechanism have been obtained by the construction and biochemical characterization of RoxA$_{Xsp}$ muteins (Birke *et al.*, 2012).

The second type of rubber oxygenase is a protein designated as latex-clearing protein (Lcp) (Rose *et al.*, 2005; Hiessl *et al.*, 2012; Yikmis and Steinbüchel, 2012). Lcps ($\approx$40 kDa, Fig. 1A) are widespread in or even specific for Gram-positive rubber-degrading bacteria, such as *Streptomyces* sp. K30 (Lcp$_{K30}$) (Rose *et al.*, 2005), *Gordonia polyisoprenivorans*, *Gordonia westfalica* (Arenskötter *et al.*, 2001; Bröker *et al.*, 2008), and were recently isolated from *Gordonia polyisoprenivorans* VH2 (Hiessl *et al.*, 2014) *Streptomyces sp.* K30 (Birke *et al.*, 2015; Röther *et al.*, 2016) and from *Rhodococcus rhodochrous* RPK1 (Watcharakul *et al.*, 2016). The amino acid sequences of RoxAs and Lcps are not related although both enzymes catalyse the oxidative cleavage of the double bonds in poly(*cis*-1,4-isoprene) and both cleave polyisoprene to products with terminal keto and aldehyde groups (Fig. 1B). In contrast to RoxAs that cleave rubber to only one major end-product (ODTD), Lcps produce a mixture of oligo-isoprenoids (C$_{20}$, C$_{25}$, C$_{30}$ and higher oligo-isoprenoids, Fig. 2) (Ibrahim *et al.*, 2006; Birke and Jendrossek, 2014). Lcps are *b*-type cytochromes and share a common domain of unknown function (DUF2236) (Hiessl *et al.*, 2014; Birke *et al.*, 2015). Recently, the importance of several strictly conserved residues within the DUF2236 domain for stability and activity was determined (Röther *et al.*, 2016).

Very recently, a third type of rubber oxygenase, RoxB, was discovered (Birke *et al.*, 2017). The coding sequence is provided under the accession No KY

498024. RoxB$_{Xsp}$ was identified as a RoxA$_{Xsp}$ homologue in *Xanthomonas* sp. 35Y and shared some properties with RoxAs: RoxB$_{Xsp}$ is also a *c*-type dihaem protein with an apparent molecular weight of around 70 kDa), but it has only a low sequence similarity to RoxA$_{Xsp}$ (38%). However, RoxB$_{Xsp}$ differs from RoxAs in cleaving polyisoprene to a mixture of oligo-isoprenoids (C$_{20}$, C$_{25}$, C$_{30}$ and higher oligo-isoprenoids, Fig. 2B). This has previously been described only for Lcps. Therefore, RoxB$_{Xsp}$ combines properties of RoxAs and Lcps (Birke *et al.*, 2017). RoxB is related in amino acid sequence to the *latA* gene product of *Rhizobacter gummiphilus* (83%) (Kasai *et al.*, 2017). The *latA* gene was recently discovered to code for a protein that is responsible for the cleavage of polyisoprene in *R. gummiphilus*. However, no information on the properties of the expressed LatA protein is yet available.

Recombinant overexpression of rubber oxygenases

Despite the fact that all so far described rubber oxygenases must be post-translationary modified to incorporate the haem cofactor, overexpression of highly active rubber oxygenases is surprisingly easy: RoxA$_{Xsp}$ can be expressed extracellularly in quantities of $\approx$15 mg l^{-1} from recombinant *Xanthomonas* sp. 35Y strains which harbour a *roxA*$_{Xsp}$ gene on the chromosome under the control of an rhamnose-inducible promoter (Hambsch *et al.*, 2010; Birke *et al.*, 2012). We assume that the amount of produced rubber oxygenase can be increased by a combination of medium optimization, inducer concentration and the time point of addition and harvest. Lcps have been successfully overexpressed intracellularly in recombinant *E. coli* using conventional induction by rhamnose (Birke *et al.*, 2015; Watcharakul *et al.*, 2016) or via autoinduction (Andler and Steinbüchel, 2017). Secretion of the

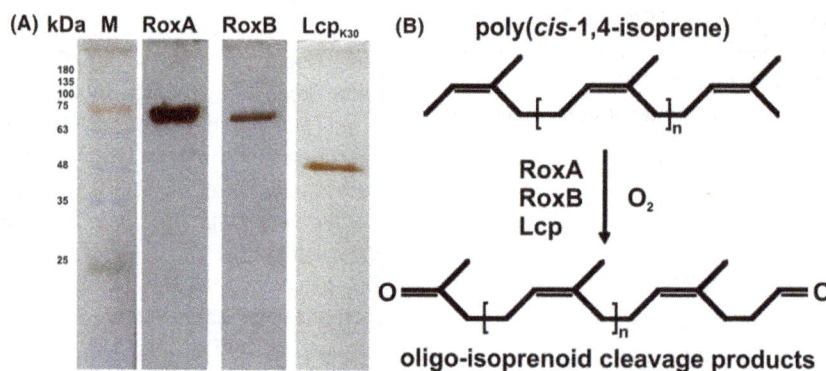

Fig. 1. SDS-PAGE of purified rubber oxygenases. RoxA$_{Xsp}$ and RoxB$_{Xsp}$ were purified from filter-concentrated supernatants of L-rhamnose/LB-medium-grown Δ*roxA Xanthomonas* sp. 35Y cells with genome-integrated *roxA*$_{Xsp}$ or *roxB*$_{Xsp}$ gene respectively. Lcp$_{K30}$ was purified from soluble French-press extracts of L-rhamnose/LB-medium-grown *E. coli* (p4782.1::*strep-lcp*$_{K30}$) via Strep-Tactin HC gravity flow affinity chromatography.
A. Purified proteins were separated by SDS-PAGE and stained with silver. A molecular mass standard (M) with kDa values indicated is shown.
B. Oxidative cleavage of rubber. Poly(*cis*-1,4-isoprene) ($100 < n < \approx 10\,000$) is oxidatively cleaved by rubber oxygenases to oligo-isoprenoids with terminal keto- and aldehyde groups. The methanol-soluble products differ in the number of intact isoprene units (*n*) with $1 \leq n < \approx 12$.

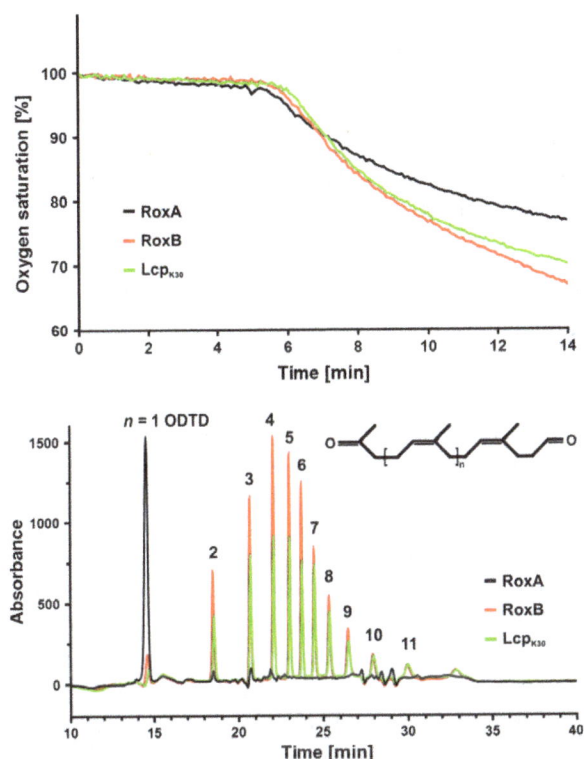

Fig. 2. Activities and product analysis of rubber oxygenases. Activities of purified rubber oxygenases (Lcp_{K30}, $RoxA_{Xsp}$ and $RoxB_{Xsp}$) were determined by following the consumption of dissolved oxygen at 37°C in a Oxy4 V2 apparatus, Presens, Regensburg, Germany, as described recently (Röther et al., 2017) (top). 4 μg each of Lcp_{K30}, $RoxA_{Xsp}$ or $RoxB_{Xsp}$ was added to 1 ml of an emulsion of polyisoprene latex in potassium phosphate buffer (100 mM, pH 7) at ≈5.5 min. The initial slopes correspond to specific activities of 6.2, 2.6, 6.4 U mg^{-1} for Lcp_{K30}, $RoxA_{Xsp}$ or $RoxB_{Xsp}$ respectively. One unit corresponds to the consumption of one molecule of dioxygen per minute. The products of polyisoprene cleavage were determined by HPLC-based analysis of the ethylacetate-extracted cleavage products (bottom). For Lcp_{K30} and $RoxB_{Xsp}$, a typical pattern of oligo-isoprenoids varying in the number of subunits (n = 2–11) was observed. For RoxA, 12-oxo-4,8-dimethyltrideca-4,8-diene-1-al (ODTD, n = 1) was detected as the major cleavage product.

mature Lcps via the TAT secretion pathway in *E. coli* (Yikmis et al., 2008) or *Bacillus subtilis* (van Dijl and Hecker, 2013) should be possible. However, the secretion pathways used for RoxA and RoxB proteins have not yet been identified. If pure proteins are necessary, tagged versions of Lcps can be purified in high yields using a one step affinity chromatography (≈ 15 mg Lcp_{K30} l^{-1} culture for Strep-tagged Lcp). The tag also offers the opportunity for enzyme immobilization. Furthermore, over-production of haem containing rubber oxygenases might be limited by the intracellular availability of the cofactor. An increase in the efficiency of haem biosynthesis, e g., by the expression of gamma-aminolevulinic acid synthase and gamma-aminolevulinic acid dehydratase could be used to overcome this limitation (Doss and Philipp-Dormston, 1975).

Purification of rubber oxygenases

We purified each one representative of the three types of rubber oxygenase ($RoxA_{Xsp}$, $RoxB_{Xsp}$ and Lcp_{K30}, Fig. 1) and used the purified proteins alone or in combination for the production of oligo-isoprenoids from polyisoprene latex. Produced oligo-isoprenoids were purified by HPLC and FPLC, and the identity of the isolated products was confirmed by ESI-MS analysis.

Untagged $RoxA_{Xsp}$ and $RoxB_{Xsp}$ were purified from the culture fluid of recombinant ΔroxA *Xanthomonas* sp. 35Y strains which harboured either the $roxA_{Xsp}$ or the $roxB_{Xsp}$ gene integrated into the chromosome under the control of an L-rhamnose-inducible promoter using a two-step purification procedure as described recently (Birke et al., 2012, 2017). Lcp_{K30} was expressed intracellularly in form of an N-terminal Strep-tagged protein and was purified from recombinant *E. coli* as described previously (Röther et al., 2016). Fig. 1A shows that all three proteins were of high purity and activity determinations confirmed high specific activities of 2.6 U mg^{-1} ($RoxA_{Xsp}$), 6.2 U mg^{-1} (Lcp_{K30}) and 6.4 U mg^{-1} ($RoxB_{Xsp}$) at 37°C for the three purified rubber oxygenases (Fig. 2 top). HPLC analysis of the solvent-extracted products confirmed the cleavage of polyisoprene to ODTD (C_{15} oligo-isoprenoid) as major product by $RoxA_{Xsp}$ and the formation of a mixture of C_{20} and higher oligo-isoprenoids in case of $RoxB_{Xsp}$ and Lcp_{K30} (Fig. 2 bottom). ODTD was present only in minor amounts in the products obtained from $RoxB_{Xsp}$ and Lcp_{K30}.

The finding of only one cleavage product (C_{15} oligo-isoprenoid ODTD) for the $RoxA_{Xsp}$-catalysed reaction and the identification of multiple cleavage products (C_{20} and higher oligo-isoprenoids) in case of the $RoxB_{Xsp}$- or Lcp_{K30}-cleaved polyisoprene suggested that $RoxA_{Xsp}$ on the one side and $RoxB_{Xsp}$ and Lcp_{K30} on the other side employ different cleavage mechanisms. We assume that $RoxA_{Xsp}$ has a 'molecular ruler' and uses an *exo*-type mechanism to cleave the polyisoprene chain (Seidel et al., 2013). This explains the formation of only one main cleavage product of a defined length (ODTD). In contrast, in case of $RoxB_{Xsp}$ and Lcp_{K30}, the formation of multiple products of different length suggests that these rubber oxygenases do not have such a molecular ruler and cleave the polyisoprene chain randomly in an *endo*-type mechanism resulting in the observed mixture of oligo-isoprenoids of different lengths.

Synergistic effect of RoxB and of Lcp on polyisoprene cleavage by RoxA

The generation of oligo-isoprenoids by *endo*-cleavage of polyisoprene molecules (with $RoxB_{Xsp}$ or Lcp_{K30})

increases the number of free polyisoprene chains. A higher concentration of polyisoprenoid ends should enhance the efficiency of polyisoprene cleavage by rubber oxygenases with an endo-type cleavage such as $RoxA_{Xsp}$. We therefore determined whether the amount of ODTD produced by RoxA could be increased by the presence of trace amounts of $RoxB_{Xsp}$ or Lcp_{K30}. The presence of 0.2 µg ml^{-1} purified $RoxB_{Xsp}$ or Lcp_{K30} in the assay mixture did not lead to the formation of substantial amounts of ODTD (factor being < 0.02 relative to 1.0 by 2 µg of $RoxA_{Xsp}$, Fig. 3). However, when combined, 2 µg ml^{-1} $RoxA_{Xsp}$ and 0.2 µg ml^{-1} purified $RoxB_{Xsp}$ or Lcp_{K30} increased the amount of produced ODTD by a factor of 1.4 or 1.5, respectively, in comparison with the values obtained with 2 µg $RoxA_{Xsp}$ or Lcp_{K30} alone (Fig. 3). Furthermore, the synergistic effect was investigated with respect to a kinetic effect enhancing the speed of the cleavage reaction, representing a major factor to be considered upon industrial employment of the reaction. To this end, the oxygen consumption rates by Lcp_{K30} (0.4 µg) and $RoxA_{Xsp}$ (4 µg) alone were determined, combined (added) in silico and were then compared to an experiment in which both enzymes were simultaneously present. As evident from Fig. 4, the simultaneous presence of low amounts of Lcp_{K30} increased the specific oxygen consumption by a factor of 1.4 (2.6 U mg^{-1}) relative to the in silico combined oxygen consumption rates (1.8 U mg^{-1}). These results also showed that the presence of terminal aldehyde and keto groups did not inhibit the cleavage of these oligo-isoprenoids to ODTD by $RoxA_{Xsp}$. Furthermore, the efficiency of rubber degradation was enhanced when each

an endo- and exo-type rubber oxygenase were simultaneously present. These data provide a plausible explanation for the presence of the roxA and roxB gene in Xanthomonas sp. 35Y due to a synergistic effect; in the presence of both gene products, ODTD is the only observed cleavage product for the facilitated uptake into the cells and use as a source of carbon and energy.

Separation and purification of oligo-isoprenoids

As shown in Fig 2B, the cleavage of polyisoprene by $RoxB_{Xsp}$ or by Lcp_{K30} yielded a mixture of oligo-isoprenoids (C_{20} and higher oligo-isoprenoids). For the application of these compounds as fine chemicals or as building blocks for (polymer) plastic synthesis in organic chemistry, the preparation of large amounts of pure oligo-isoprenoids is preferable. To demonstrate the isolation of isoprenoids at a higher scale, we increased the volume of polyisoprene latex and replaced the HPLC-based separation of oligo-isoprenoids by an FPLC separation system because FPLC systems can be up-scaled more easily than HPLC-based separations. As a proof of principle, we treated 1 litre of 5% (wt/vol) polyisoprene latex in 100 mM potassium phosphate buffer, pH 7 with 4 mg of purified Lcp_{K30} and incubated the assay mixture for 24 h at room temperature while stirring at 200 rpm. The produced oligo-isoprenoids were solvent-extracted with 100 ml ethylacetate. The solvent was evaporated, and the products (≈100 mg) were dissolved in 5 ml methanol. 200 µl of the dissolved products was then applied to a PEP RPC HR5/5-FPLC column that had been equilibrated with 50% methanol: water and eluted

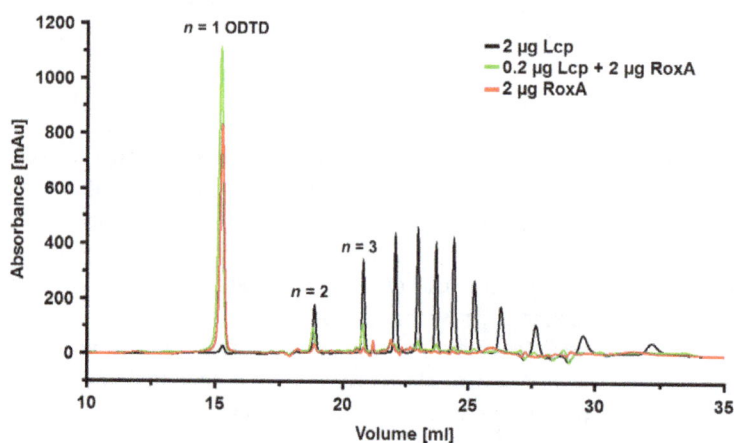

Enzyme combination	Relative amount of ODTD
Lcp 0.2 µg	<0.01
RoxB 0.2 µg	<0.02
Lcp 2 µg	0.03
RoxB 2 µg	0.15
RoxA 2 µg	1.00
RoxA 4 µg	1.70
Lcp 2 µg + RoxA 2 µg	2.11
RoxB 2 µg + RoxA 2 µg	2.15
Lcp 0.2 µg + RoxA 2 µg	1.40
RoxB 0.2 µg + RoxA 2 µg	1.52

Fig. 3. Synergistic effect during rubber cleavage. Polyisoprene latex was cleaved by different amounts and combinations of rubber oxygenases as indicated (left). The amounts of produced ODTD (Table on the right) were determined from the ODTD areas in HPLC chromatograms (exemplary shown in the image on the left). The ODTD-specific area obtained for 2 µg of $RoxA_{Xsp}$ was set as 1.0. The addition of only 0.2 µg Lcp increased ODTD formation by 2 µg of $RoxA_{Xsp}$ by a factor of 1.4 and only trace amounts of higher oligo-isoprenoids (n = 2 and 3; n indicates the number of intact isoprene units, see structure shown in Fig. 1B) were determined. A similar effect with 1.5-fold higher amount of produced ODTD was observed for a combination of 2 µg $RoxA_{Xsp}$ and 0.2 µg $RoxB_{Xsp}$.

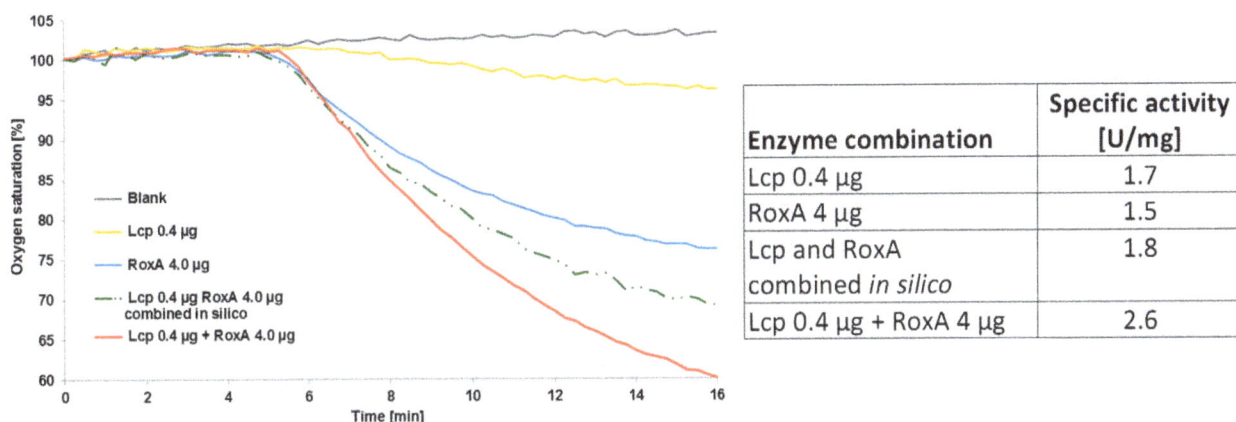

Enzyme combination	Specific activity [U/mg]
Lcp 0.4 µg	1.7
RoxA 4 µg	1.5
Lcp and RoxA combined *in silico*	1.8
Lcp 0.4 µg + RoxA 4 µg	2.6

Fig. 4. Synergistic effect of the presence of Lcp_{K30} on the specific activity of $RoxA_{Xsp}$. The oxygen consumption rates of 0.4 µg Lcp_{K30}, of 4 µg of $RoxA_{Xsp}$ and of a mixture of 0.4 µg Lcp_{K30} and of 4 µg of $RoxA_{Xsp}$ were recorded. The values for 0.4 µg Lcp_{K30} and 4 µg $RoxA_{Xsp}$ were combined *in silico* and the slope of the resulting curve was calculated to determine a theoretical specific activity. Note that the specific activities of the reaction in the presence of both enzymes were 1.4-fold higher (2.6 U mg^{-1}) compared to the *in silico* combined values of the two reactions with the single enzymes (1.8 U mg^{-1}).

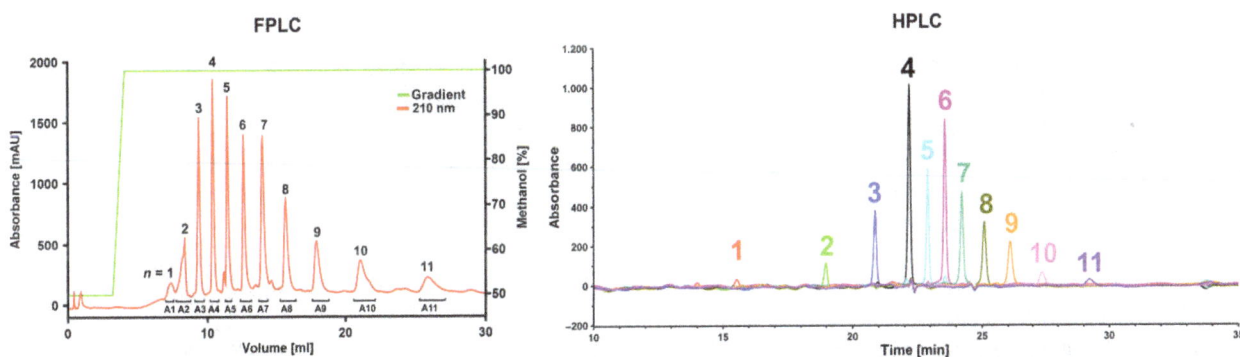

Fig. 5. Separation of oligo-isoprenoid mixtures by FPLC, HPLC and HPLC-MS. 200 µl oligo-isoprenoid solution in methanol (prepared by digestion of polyisoprene with Lcp_{K30} as described in the main text) was applied to a reversed-phase FPLC column (Pep RPC HR 5/5, 1 ml bed volume) and separated by a step gradient from 50% water: methanol to 100% methanol (green line) (left image). Absorption at 210 nm (red line) was used to fractionate peaks representing different oligo-isoprenoids (A1-A11, corresponding to n = 1–11). Aliquots of each separately collected fraction (A1 to A11) were applied to analysis via HPLC. An overlay image consisting of all eleven HPLC chromatograms is shown on the right. The superposition of the chromatograms shows the high resolution power of the used FPLC column. The masses (m/z values) of each isolated compound were confirmed by HPLC-ESI-MS and are provided in Table 1.

by the application of an increasing step gradient to 100% methanol at a constant flow rate of 1.5 ml min^{-1}. Peaks were automatically fractioned ($\approx$2 ml per peak) by monitoring the absorbance at 210 nm. As shown in Fig. 5 left, the same eleven individual peaks were identified that had been detected on the analytical HPLC column (Fig. 2). The compound of each of the separated peaks was collected individually, concentrated by evaporation and dissolved in 100 µl of methanol. When each of the isolated compounds was separately run on the analytical HPLC column, the successful isolation of each oligo-isoprenoid was demonstrated by the appearance of one homogeneous peak (Fig. 5, right). The m/z values of the isolated oligo-isoprenoids were determined by HPLC-MS and were in agreement with the structural

formulas and the theoretical values for the individual oligo-isoprenoids (Table 1).

Conclusions and outlook

Polyisoprene in form of natural rubber latex is a cheap bulk compound and is available in the ton-scale. Cleavage of polyisoprene by rubber oxygenases and separation of produced oligo-isoprenoids is fairly possible. In this study, eleven oligo-isoprenoids of the 'ODTD-family' with one to eleven central isoprene units (n) between the terminal aldehyde and keto functional groups could be separately prepared. The highest yields were obtained for ODTD ($RoxA_{Xsp}$ alone) and for the C_{30} to C_{50} compounds (Lcp_{K30} or $RoxB_{Xsp}$ alone). Purification of oligo-

Table 1. Oligo-isoprenoids produced by enzymatic cleavage of polyisoprene using purified rubber oxygenases.

No of intact isoprene units [n]	Elemental formula	m/z	m/z [M+H]$^+$	m/z [M+Na]$^+$	m/z [M+Na+CH$_3$OH]$^+$	FPLC peak area [%]
1	$C_{15}H_{24}O_2$ (ODTD) 12-oxo-4,8-dimethyl-trideca-4,8-diene-1-al	236.178	237.185	259.167	291.193	2.9
2	$C_{20}H_{32}O_2$	304.240	305.248	327.230	359.256	6.8
3	$C_{25}H_{40}O_2$	372.303	373.310	395.292	427.318	9.6
4	$C_{30}H_{48}O_2$	440.365	441.373	463.355	495.381	10.5
5	$C_{35}H_{56}O_2$	508.428	509.435	531.417	563.444	10.3
6	$C_{40}H_{64}O_2$	576.491	577.498	599.480	631.506	10.4
7	$C_{45}H_{72}O_2$	644.553	645.561	667.543	699.569	12.4
8	$C_{50}H_{80}O_2$	712.616	713.623	735.605	767.631	10.1
9	$C_{55}H_{88}O_2$	780.678	781.686	803.668	835.694	8.7
10	$C_{60}H_{96}O_2$	848.741	849.748	871.730	903.757	9.5
11	$C_{65}H_{104}O_2$	916.804	917.811	939.793	971.819	9.0

Polyisoprene latex was treated with purified rubber oxygenase (Lcp$_{K30}$), and cleavage products were extracted with ethylacetate and dissolved in methanol. Products were analysed by HPLC-ESI-MS analysis before and after purification of individual peaks by FPLC (Fig. 4). For each compound the theoretical m/z values and the values corresponding to the protonated ([M+H]$^+$), the sodium ion adduct ([M+Na]$^+$) and for the sodium ion+ methanol adduct forms ([M+Na+CH$_3$OH]$^+$) are indicated. The relative amounts (in %) of each prepared oligo-isoprenoid are also provided.

isoprenoids by FPLC can be easily up-scaled for the mass production of oligo-isoprenoids. The use of tyres and other materials containing vulcanized rubbers as substrates for enzymatic degradation by different rubber oxygenases is also possible; however, the presence of sulfur bridges and other components complicates the efficiency of enzymatic cleavage of vulcanized rubber waste and therefore limit – at present – the use of rubber oxygenases to the cleavage of unprocessed natural rubber latex. Mechanical, chemical and/or physical pre-treatments of rubber wastes (e.g. grinding, solvent extraction, desulphurization) might help to make processed rubber wastes also accessible for enzymatic cleavage. Isoprenoids derived from rubber can be used for the production of fragrances, hormones and pharmaceuticals, creating interest in cheap synthesis pathways see (Förster-Fromme and Jendrossek, 2010; Schewe et al., 2015). Furthermore, they can be also used in chemical or enzymatic cyclization reactions (Siedenburg et al., 2012, 2013) for the production of cyclic compounds or can be used as biofuels (Mewalal et al., 2017). This study provides purified, reactive oligo-isoprenoids that can likely be used to convert rubber waste, e.g., from tires into precursors for the synthesis of value-added compounds. The reactivity of the aldehydes might be directly used to form covalent bonds with other molecules (e.g. with amines). Alternatively, the keto groups of the oligo-isoprenoids can be chemically or enzymatically reduced to the corresponding mono- or di-alcohols. The reduction in the C$_{15}$ compound ODTD to the corresponding alcohol by enzymatic reduction has been previously demonstrated (Braaz et al., 2005). Enzymatic generation of isoprenoid-diols can help to provide precursors for the production of polymers from sustainably produced monomers, e.g., for the production of polyurethanes and might be an alternative to chemical methods for the conversion of polyisoprenes to polyurethanes (Anancharoenwong, 2011).

Acknowledgements

We thank Weber and Schaer (Hamburg) for providing polyisoprene and IBA Life sciences (Göttingen) for providing Strep-Tactin columns and cooperation during up-scaling of rubber oxygenase purification. This work was supported by the Graduiertenkolleg GRK1708 to the University of Tübingen and by a grant of the Deutsche Forschungsgemeinschaft to D.J.

Conflict of Interest

None declared.

References

Akhlaghi, S., Gedde, U.W., Hedenqvist, M.S., Brana, M.T.C., and Bellander, M. (2015) Deterioration of automotive rubbers in liquid biofuels: a review. *Renew Sustain Energy Rev* **43**: 1238–1248.

Anancharoenwong, E., 2011. Synthesis and characterization of cis-1, 4-polyisoprene-based polyurethane coatings; study of their adhesive properties on metal surface. Université du Maine, English. NNT: 2011LEMA1009.

Andler, R., and Steinbüchel, A. (2017) A simple, rapid and cost-effective process for production of latex clearing protein to produce oligopolyisoprene molecules. *J Biotechnol* **241**: 184–192.

Arenskötter, M., Baumeister, D., Berekaa, M.M., Pötter, G., Kroppenstedt, R.M., Linos, A., and Steinbüchel, A. (2001)

Taxonomic characterization of two rubber degrading bacteria belonging to the species *Gordonia polyisoprenivorans* and analysis of hyper variable regions of 16S rDNA sequences. *FEMS Microbiol Lett* **205:** 277–282.

Birke, J., and Jendrossek, D. (2014) Rubber oxygenase and latex clearing protein cleave rubber to different products and use different cleavage mechanisms. *Appl Environ Microbiol* **80:** 5012–5020.

Birke, J., Hambsch, N., Schmitt, G., Altenbuchner, J., and Jendrossek, D. (2012) Phe317 is essential for rubber oxygenase RoxA activity. *Appl Environ Microbiol* **78:** 7876–7883.

Birke, J., Röther, W., Schmitt, G., and Jendrossek, D. (2013) Functional identification of rubber oxygenase (RoxA) in soil and marine myxobacteria. *Appl Environ Microbiol* **79:** 6391–6399.

Birke, J., Röther, W., and Jendrossek, D. (2015) Latex clearing protein (Lcp) of *Streptomyces* sp. strain K30 Is a b-type cytochrome and differs from rubber oxygenase A (RoxA) in its biophysical properties. *Appl Environ Microbiol* **81:** 3793–3799.

Birke, J., Röther, W. and Jendrossek, D. (2017) RoxB is a novel type of rubber oxygenase that combines properties of rubber oxygenase RoxA and latex clearing protein (Lcp). *Appl Environ Microbiol* **83,** in press, https://doi.org/10.1128/aem.00721-17

Braaz, R., Fischer, P., and Jendrossek, D. (2004) Novel type of heme-dependent oxygenase catalyzes oxidative cleavage of rubber (poly-*cis*-1,4-isoprene). *Appl Environ Microbiol* **70:** 7388–7395.

Braaz, R., Armbruster, W., and Jendrossek, D. (2005) Heme-dependent rubber oxygenase RoxA of *Xanthomonas* sp. cleaves the carbon backbone of poly(cis-1,4-Isoprene) by a dioxygenase mechanism. *Appl Environ Microbiol* **71:** 2473–2478.

Bröker, D., Dietz, D., Arenskötter, M., and Steinbüchel, A. (2008) The genomes of the non-clearing-zone-forming and natural-rubber- degrading species *Gordonia polyisoprenivorans* and *Gordonia westfalica* harbor genes expressing Lcp activity in *Streptomyces* strains. *Appl Environ Microbiol* **74:** 2288–2297.

van Dijl, J.M., and Hecker, M. (2013) *Bacillus subtilis*: from soil bacterium to super-secreting cell factory. *Microb Cell Fact* **12:** 3.

Doss, M., and Philipp-Dormston, W.K. (1975) Over-production of porphyrins and heme in heterotrophic bacteria. *Z Naturforsch C Biosci* **30:** 425–426.

Förster-Fromme, K., and Jendrossek, D. (2010) Catabolism of citronellol and related acyclic terpenoids in pseudomonads. *Appl Microbiol Biotechnol* **87:** 859–869.

Hambsch, N., Schmitt, G., and Jendrossek, D. (2010) Development of a homologous expression system for rubber oxygenase RoxA from *Xanthomonas* sp. *J Appl Microbiol* **109:** 1067–1075.

Hiessl, S., Schuldes, J., Thuermer, A., Halbsguth, T., Broeker, D., Angelov, A., *et al.* (2012) Involvement of two latex-clearing proteins during rubber degradation and insights into the subsequent degradation pathway revealed by the genome sequence of *Gordonia polyisoprenivorans* strain VH2. *Appl Environ Microbiol* **78:** 2874–2887.

Hiessl, S., Boese, D., Oetermann, S., Eggers, J., Pietruszka, J., and Steinbüchel, A. (2014) Latex clearing protein-an oxygenase cleaving Poly(cis-1,4-Isoprene) rubber at the cis double bonds. *Appl Environ Microbiol* **80:** 5231–5240.

Ibrahim, E., Arenskötter, M., Luftmann, H., and Steinbüchel, A. (2006) Identification of poly(cis-1,4-isoprene) degradation intermediates during growth of moderately thermophilic actinomycetes on rubber and cloning of a functional lcp homologue from *Nocardia farcinica* strain E1. *Appl Environ Microbiol* **72:** 3375–3382.

Kasai, D., Imai, S., Asano, S., Tabata, M., Iijima, S., Kamimura, N., *et al.* (2017) Identification of natural rubber degradation gene in *Rhizobacter gummiphilus* NS21. *Biosci Biotechnol Biochem* **81:** 614–620.

Mewalal, R., Rai, D.K., Kainer, D., Chen, F., Külheim, C., Peter, G.F., and Tuskan, G.A. (2017) Plant-derived terpenes: a feedstock for specialty biofuels. *Trends Biotechnol* **35:** 227–240.

Rose, K., Tenberge, K.B., and Steinbüchel, A. (2005) Identification and characterization of genes from *Streptomyces* sp strain K30 responsible for clear zone formation on natural rubber latex and poly(*cis*-1,4-isoprene) rubber degradation. *Biomacromolecules* **6:** 180–188.

Röther, W., Austen, S., Birke, J., and Jendrossek, D. (2016) Molecular insights in the cleavage of rubber by the latex-clearing-protein (Lcp) of *Streptomyces* sp. strain K30. *Appl Environ Microbiol* **82:** 6593–6602.

Röther, W., Birke, J., and Jendrossek, D. (2017) Assays for the detection of rubber oxygenase activities. *Bio-protocol* **7:** 1–14.

Schewe, H., Mirata, M.A., and Schrader, J. (2015) Bioprocess engineering for microbial synthesis and conversion of isoprenoids. *Adv Biochem Eng Biotechnol* **148:** 251–286.

Schmitt, G., Seiffert, G., Kroneck, P.M.H., Braaz, R. and Jendrossek, D.., 2010. Spectroscopic properties of rubber oxygenase RoxA from *Xanthomonas* sp., a new type of dihaem dioxygenase. *Microbiology (Reading, Engl.)* **156:** 2537–2548.

Schrader, J. and Bohlmann, J., 2015. *Biotechnology of Isoprenoids.* Cham: Springer. https://doi.org/10.1007/978-3-319-20107-8

Seidel, J., Schmitt, G., Hoffmann, M., Jendrossek, D., and Einsle, O. (2013) Structure of the processive rubber oxygenase RoxA from *Xanthomonas* sp. *Proc Natl Acad Sci USA* **110:** 13833–13838.

Siedenburg, G., Jendrossek, D., Breuer, M., Juhl, B., Pleiss, J., Seitz, M., *et al.* (2012) Activation-independent cyclization of monoterpenoids. *Appl Environ Microbiol* **78:** 1055–1062.

Siedenburg, G., Breuer, M., and Jendrossek, D. (2013) Prokaryotic squalene-hopene cyclases can be converted to citronellal cyclases by single amino acid exchange. *Appl Microbiol Biotechnol* **97:** 1571–1580.

Tsuchii, A., and Takeda, K. (1990) Rubber-degrading enzyme from a bacterial culture. *Appl Environ Microbiol* **56:** 269–274.

Vickers, C.E., Behrendorff, J.B.Y.H., Bongers, M., Brennan, T.C.R., Bruschi, M., and Nielsen, L.K. (2015) Production of industrially relevant isoprenoid compounds in engineered microbes. *Microbiol Monogr* **26:** 303–334.

Watcharakul, S., Röther, W., Birke, J., Umsakul, K., Hodgson, B., and Jendrossek, D. (2016) Biochemical and spectroscopic characterization of purified Latex Clearing Protein (Lcp) from newly isolated rubber degrading *Rhodococcus rhodochrous* strain RPK1 reveals novel properties of Lcp. *BMC Microbiol* **16:** 92.

Yikmis, M., and Steinbüchel, A. (2012) Historical and recent achievements in the field of microbial degradation of natural and synthetic rubber. *Appl Environ Microbiol* **78:** 4543–4551.

Yikmis, M., Arenskoetter, M., Rose, K., Lange, N., Wernsmann, H., Wiefel, L., and Steinbüchel, A. (2008) Secretion and transcriptional regulation of the latex-clearing protein, Lcp, by the rubber-degrading bacterium *Streptomyces* sp strain K30. *Appl Environ Microbiol* **74:** 5373–5382.

Engineering the xylose-catabolizing Dahms pathway for production of poly(D-lactate-*co*-glycolate) and poly(D-lactate-*co*-glycolate-*co*-D-2-hydroxybutyrate) in *Escherichia coli*

So Young Choi,[1] Won Jun Kim,[1] Seung Jung Yu,[2] Si Jae Park,[3] Sung Gap Im[2] and Sang Yup Lee[1,*]

[1]*Metabolic and Biomolecular Engineering National Research Laboratory, Department of Chemical and Biomolecular Engineering (BK21 Plus Program), BioProcess Engineering Research Center, and KAIST Institute (KI) for the BioCentury, Korea Advanced Institute of Science and Technology (KAIST), 291 Daehak-ro, Yuseong-gu, Daejeon 34141, Korea.*
[2]*Department of Chemical and Biomolecular Engineering (BK21 Plus Program), KAIST, 291 Daehak-ro, Yuseong-gu, Daejeon 34141, Korea.*
[3]*Department of Chemical Engineering and Materials Science, Ewha Womans University, 52 Ewhayeodae-gil, Seodaemun-gu, Seoul 03760, Korea.*

Summary

Poly(lactate-*co*-glycolate), PLGA, is a representative synthetic biopolymer widely used in medical applications. Recently, we reported one-step direct fermentative production of PLGA and its copolymers by metabolically engineered *Escherichia coli* from xylose and glucose. In this study, we report development of metabolically engineered *E. coli* strains for the production of PLGA and poly(D-lactate-*co*-glycolate-*co*-D-2-hydroxybutyrate) having various monomer compositions from xylose as a sole carbon source. To achieve this, the metabolic flux towards Dahms pathway was modulated using five different synthetic promoters for the expression of *Caulobacter crescentus* XylBC. Further metabolic engineering to concentrate the metabolic flux towards D-lactate and glycolate resulted in production of PLGA and poly(D-lactate-*co*-glycolate-*co*-D-2-hydroxybutyrate) with various monomer fractions from xylose. The engineered *E. coli* strains produced polymers containing 8.8–60.9 mol% of glycolate up to 6.93 g l^{-1} by fed-batch cultivation in a chemically defined medium containing xylose. Finally, the biocompatibility of poly(D-lactate-*co*-glycolate-*co*-D-2-hydroxybutyrate) was confirmed by live/dead assay using human mesenchymal stem cells.

*For correspondence. E-mail leesy@kaist.ac.kr

Funding information
Ministry of Science, ICT and Future Planning (NRF-2012M1A2A20 26556).

Introduction

Poly(lactate-*co*-glycolate), PLGA, is a random copolymer of lactic and glycolic acids and is approved by Food and Drug Administration and European Medicine Agency for various biomedical and therapeutic applications, such as surgical sutures, prosthetic devices, drug delivery and tissue engineering due to its biocompatibility and biodegradability with controlled degradation characteristics (Makadia and Siegel, 2011; Danhier *et al.*, 2012). Currently, it is synthesized by random ring-opening copolymerization of lactide and glycolide, the cyclic dimers of lactic and glycolic acids, respectively, using metal catalysts (Wu and Wang, 2001). Such chemical methods are rather complicated as they involve the production and purification of monomers, synthesis of pre-polymer and removal of residual monomers and catalysts.

As a more environmentally friendly method, we recently reported production of PLGA and various D-lactate and glycolate containing polyhydroxyalkanoates (PHAs) by metabolically engineered *Escherichia coli* strains from renewable resources (Choi *et al.*, 2016). In the above study, the xylose metabolizing Dahms pathway of *Caulobacter crescentus* was established in *E. coli* by introducing the xylose dehydrogenase (XylB) and xylonolactonase (XylC) of *C. crescentus*. It resulted in production of glycolate from xylose, but also resulted in significant growth retardation; in this study, the acid names, D-lactate and glycolate, are generally used unless they are free acids excreted into the medium (D-lactic acid and glycolic acid).

Based on the *in silico* simulations, *E. coli* was engineered to utilize glucose and xylose simultaneously by inactivating glucose phosphotransferase system. This co-utilizing *E. coli* strain showed recovered cell growth with enhanced glycolate production. Further metabolic engineering was performed to more efficiently produce D-lactate and glycolate, which were then converted to D-lactyl and glycolyl-CoAs by evolved propionyl-CoA transferase (Pct540) followed by polymerization to PLGA by evolved PHA synthase (PhaC1437). The engineered *E. coli* produced PLGA to 1.95 g l^{-1} with a polymer content of 36.2 wt% (Choi *et al.*, 2016). It was also interesting to note that poly(D-lactate-*co*-glycolate-*co*-D-2-hydroxybutyrate) [poly(D-LA-*co*-GA-*co*-D-2HB)] was produced when there was no manipulation of *ilvA* or supplementation of L-isoleucine. Thus, we became interested in examining whether this terpolyester is also biocompatible like PLGA. In this study, we focused on the Dahms pathway utilizing xylose to improve the production of PLGA and poly(D-LA-*co*-GA-*co*-D-2HB). Effects of employing five different synthetic promoters to express XylBC$_{ccs}$ of the Dahms pathway on cell growth, polymer production and monomer fractions were examined. Using the constructed engineered strains, fed-batch cultures were performed to compare their performance with respect to cell growth and polymer production. Also, biocompatibility of poly(D-LA-*co*-GA-*co*-D-2HB) was examined for the first time by Live/Dead assay using human mesenchymal stem cells (hMSCs).

Results and discussion

Effects of XylBC$_{ccs}$ expression levels on cell growth and metabolites production

To produce GA *in vivo*, the heterologous Dahms pathway was established by introducing *C. crescentus* XylBC$_{ccs}$ in *E. coli* XL1-Blue strain using pTacxylBC plasmid as previously described (Choi *et al.*, 2016; Table 1). When *E. coli* XL1-Blue harbouring pTacxylBC was cultivated using xylose as a sole carbon source, 0.60 g l^{-1} of glycolic acid was produced. As before (Choi *et al.*, 2016), significant cell growth retardation was observed and the final cell density (OD$_{600}$) was 1.09, which corresponds to only 39.7% of that obtained with the control strain *E. coli* XL1-Blue harbouring an empty vector pTac15k (Figs 1 and 2 and Table 1).

To further investigate the relationship between cell growth and the Dahms pathway flux, *in silico* simulation was performed (Fig. S1). When XylBC$_{ccs}$ were expressed, *E. coli* was able to catabolize xylose via two metabolic pathways: the heterologous Dahms pathway and the native *E. coli* xylose catabolic pathway via the pentose phosphate pathway (Fig. 1). Therefore, after addition of the XylBC$_{ccs}$ reactions into the *E. coli* iJO1336 model, the maximum growth rate was examined by varying the relative ratio of utilizing Dahms pathway

Table 1. All bacterial strains and plasmids used in this study.

Strains and plasmids	Relevant characteristics[a]	Reference or Source
Strains		
XL1-Blue	*recA1 endA1 gyrA96 thi-1 hsdR17 supE44 relA1 lac* [F′ *proAB lacI*^{q}*ZΔM15* Tn*10* (Tet^R)]	Stratagene[b]
X15	XL1-Blue Δ*pflB* Δ*frdB* Δ*adhE* Δ*poxB*	This study
X15I	X15 P*ldhA::Ptrc*	This study
X15Id	X15 P*ldhA::Ptrc* Δ*dld*	This study
X15Ida	X15 P*ldhA::Ptrc* Δ*dld* Δ*ackpta*	This study
X17Id	X15 P*ldhA::Ptrc* Δ*dld* Δ*aceB* Δ*glcDEFGB*	This study
Plasmids		
pPs619C1437Pct540	pBluescriptII KS(+) derivative; *R. eutropha* PHA biosynthesis operon promoter of the, *phaC1*$_{Ps6-19}$ variant (*phaC1437*; E130D, S325T, S477G, Q481K), *pct*$_{Cp}$ variant (*pct540*; V193A, Silent mutations: T78C, T669C, A1125G, T1158C), transcriptional terminator of the *R. eutropha* PHA biosynthesis operon; Ap^R	Yang *et al.* (2011)
pTac15k	pACYC177 derivative; *tac* promoter; Km^R	Laboratory stock
pMIoxC	lox66-cat-lox71 cassette, Cm^R Ap^R	Kim *et al.* (2008)
pKD46	Ap^R, λ-Red recombinase under arabinose inducible araBAD promoter, temperature sensitive origin	Datsenko and Wanner (2000)
pJW168	Cre recombinase under IPTG inducible lacUV5 promoter, temperature sensitive origin, Ap^R	Palmeros *et al.* (2000)
pMtrc9	pMIoxC derivative, *trc* promoter downstream of lox66-cat-lox71 cassette, Ap^R	Laboratory stock
pTacxylBC	pTac15k derivative; *tac* promoter, *C. crescentus xylBC* genes; Km^R	Choi *et al.* (2016)
pTacxylBC_xylAB	pTac15k derivative; *tac* promoter, *C. crescentus xylBC* genes, *Escherichia coli* W3110 *xylAB* genes; Km^R	This study
pP1xylBC	pTac15k derivative; BBa_J23100 promoter, *C. crescentus xylBC* genes; Km^R	This study
pP2xylBC	pTac15k derivative; BBa_J23101 promoter, *C. crescentus xylBC* genes; Km^R	This study
pP3xylBC	pTac15k derivative; BBa_J23118 promoter, *C. crescentus xylBC* genes; Km^R	This study
pP4xylBC	pTac15k derivative; BBa_J23105 promoter, *C. crescentus xylBC* genes; Km^R	This study
pP5xylBC	pTac15k derivative; BBa_J23117 promoter, *C. crescentus xylBC* genes; Km^R	This study

a. Ap, ampicillin; Km, Kanamycin; R, resistance.
b. Stratagene Cloning System, La Jolla CA, USA.

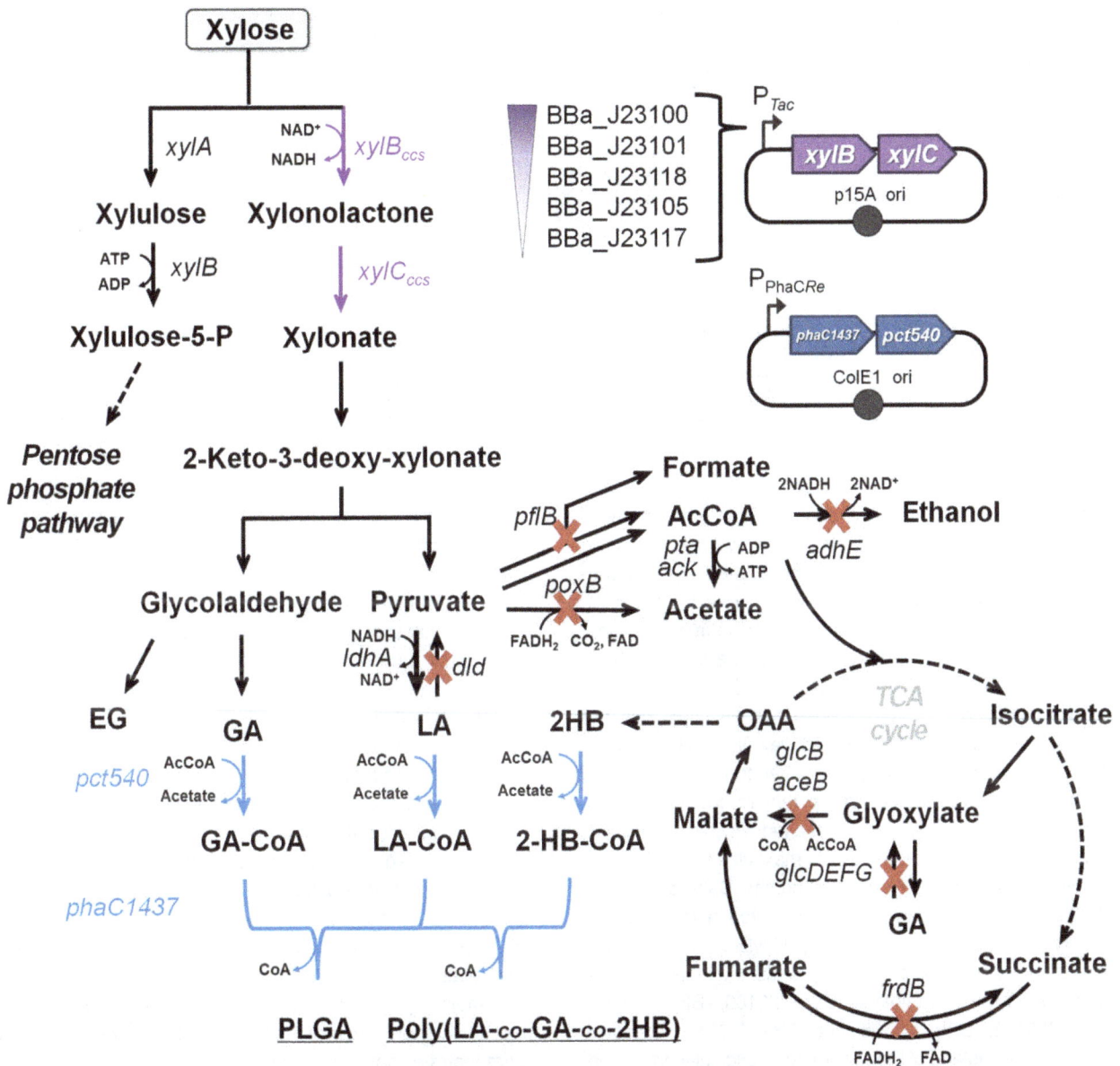

Fig. 1. Metabolic engineering of *Escherichia coli* for the production of PLGA and its copolymers from xylose. The overall strategies for the production of PLGA and its copolymers are shown. The native pathways in *E. coli* are shown in black arrow. Coloured arrows represent heterologous pathways introduced. The inactivated metabolic pathways are indicated by red X, and strengthened metabolic pathway by replacement of native promoter is shown in bold arrow. More than one conversion steps of metabolic pathways are simplified using dotted arrows. The genes shown are as follows: *xylA*, xylose isomerase; *xylB*, xylulokinase; *xylB*$_{ccs}$, xylose dehydrogenase; *xylC*$_{ccs}$, xylonolactonase; *ldhA*, D-lactate dehydrogenase; *pflB*, pyruvate formate lyase; *poxB*, pyruvate oxidase; *adhE*, acetaldehyde/alcohol dehydrogenase; *frdB*, fumarate reductase; *pta*, phosphotransacetylase; *ack*, acetate kinase; *aceB*, malate synthase A; *glcB*, malate synthase G; *glcDEFG*, glycolate oxidase; *pct540*, evolved propionyl-CoA transferase; *phaC1437*, evolved PHA synthase. Metabolites shown are as follows: Xylulose-5-P, xylulose-5-phosphate; AcCoA, acetyl-CoA; LA, D-lactate; GA, glycolate; EG, ethylene glyool; 2HB, D-2-hydroxybutyrate; GA-CoA, glycolyl-CoA; LA-CoA, D-lactyl-CoA; 2-HB-CoA, 2-D-hydroxybutyryl-CoA; OAA, oxaloacetate.

and native xylose catabolic pathway in the presence of xylose as a sole carbon source (Fig. S1). The result showed that there was an inversely proportional relationship between the Dahms pathway flux and the maximum growth rate (Fig. S1). Based on the experimental and simulation results, it was concluded that the Dahms

pathway affected cell growth more negatively compared with the native *E. coli* xylose catabolic pathway.

Thus, it was attempted to strengthen the metabolic flux through the native *E. coli* xylose catabolic pathway to enhance cell growth. In *E. coli*, xylose is converted by xylose isomerase (XylA) to xylulose, which is then

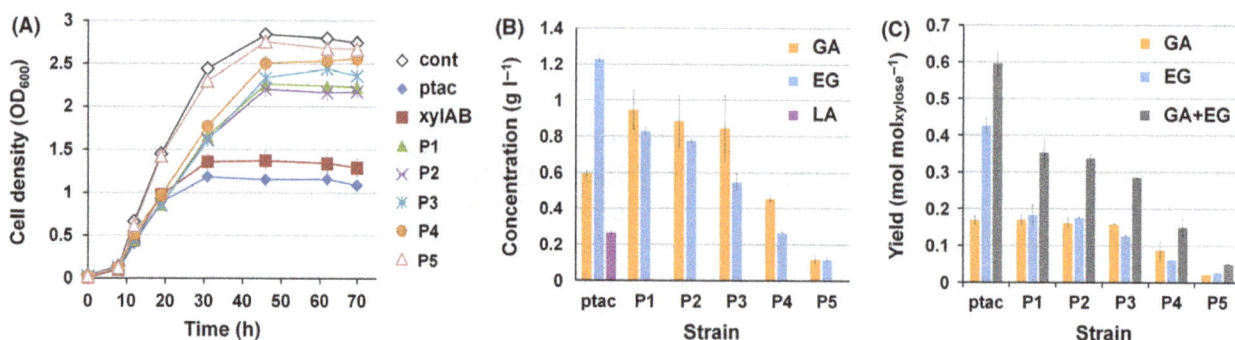

Fig. 2. Flask cultures of *Escherichia coli* XL1-blue expressing XylBC*ccs* using different promoters.
A. The time profiles of cell density (OD$_{600}$).
B. Production of glycolic acid and ethylene glycol, and (C) metabolites yield (mol mol$_{xylose}$$^{-1}$) at the end of cultivation (70 h). GA, EG and LA indicate glycolic acid, ethylene glycol and D-lactic acid, respectively. The strains used were XL1-Blue harbouring pTac15k (cont), pTacxylBC (ptac), pTacxylBC_xylAB (xylAB) and pP1xylBC-pP5xylBC (P1–P5), respectively. Error bars represent the SD values obtained in triplicate experiments.

phosphorylated to xylulose-5-phosphate by xylulose kinase (XylB). Xylulose-5-phosphate is further metabolized via pentose phosphate pathway (Fig. 1). The *E. coli* XylA and XylB were overexpressed together with XylBC*ccs* using plasmid pTacxylBC_xylAB (Table 1). However, this resulted in only 1.19-fold increase in final cell density, which was 47% of control strain *E. coli* XL1-Blue harbouring an empty vector pTac15k (Fig. 2A).

As enhancing the native xylose pathway flux was not effective, reducing the Dahms pathway flux itself was chosen as an alternative strategy. However, too much reduction in the Dahms pathway flux can potentially reduce GA production and consequently decreased PLGA production. Thus, the optimal metabolic flux to maximize the PLGA production was examined by adjusting the Dahms pathway flux. To achieve this, we constructed five plasmids pP1xylBC, pP2xylBC, pP3xylBC, pP4xylBC and pP5xylBC that allow expression of XylBC*ccs* under five different synthetic Anderson promoters, BBa_J23100, BBa_J23101, BBa_J23118, BBa_J23105 and BBa_J23117 (Table 1), respectively, instead of the *tac* promoter originally employed. These constitutive promoters, BBa_J23100 (P1 hereafter for convenience), BBa_J23101 (P2), BBa_J23118 (P3), BBa_J23105 (P4), and BBa_J23117 (P5), have relative promoter strengths of 1, 0.7, 0.56, 0.24 and 0.06, respectively (http://parts.igem.org/).

When *E. coli* XL1-Blue transformed with each XylBC*ccs* expression plasmid was cultivated, the OD$_{600}$ profile showed that the cell concentration was increased as the promoter strength decreased (Fig. 2A). Even when XylBC*ccs* were expressed under P1 promoter, the strongest one among the five synthetic promoters, cell density was restored to 81.2% of the control strain (*E. coli* XL1-Blue harbouring an empty vector pTac15k). When the weakest P5 promoter was employed, cell density reached was 97.4% of that obtained with the control strain (Fig. 2A).

As shown in Fig. 2B, production of GA and ethylene glycol, which are both Dahms pathway metabolites, was found to be highly dependent on the strength of the promoter employed, and showed decreasing tendency with weaker promoter (except for the *tac* promoter). To compare the relative extent of Dahms pathway utilization in each strain, the molar yields GA and ethylene glycol on xylose were calculated (Fig. 2C). The sum of the yields of GA and ethylene glycol clearly showed a decreasing tendency with decrease in promoter strength (Fig. 2C). These results confirm that the Dahms pathway flux could be effectively modulated by controlling expression levels of XylBC*ccs* using the synthetic promoters having various strengths. Also, use of the top three strong promoters (P1, P2 and P3) resulted in glycolic acid production to 0.96, 0.88 and 0.85 g l^{-1}, which are higher than that (0.60 g l^{-1}) obtained using the *tac* promoter. Although the relative flux towards Dahms pathway was reduced by employing a weaker promoter, enhanced cell growth led to higher level of GA production. These results demonstrate that modulating the Dahms pathway flux is effective for improving glycolic acid production.

Metabolic engineering for D-lactate production

Escherichia coli XL1-Blue expressing XylBC*ccs* using five different synthetic promoters produced 0.11–0.95 g l^{-1} of glycolic acid and negligible amount of D-lactic acid at the end of cultivation (70 h) (Fig. 2B). Because the reported PLGA producers (Choi *et al.*, 2016) were developed based on a *ptsG*-deleted *E. coli* strain, we newly constructed *E. coli* strains to efficiently produce D-lactate using the *E. coli* XL1-Blue strain as a base strain, while employing the effective engineering targets found from the previous study. *E. coli* was engineered by knocking out the *poxB* (encoding pyruvate oxidase), *pflB* (encoding pyruvate formate lyase), *frdB* (encoding fumarate reductase) and *adhE* (encoding acetaldehyde/alcohol dehydrogenase) genes in the chromosome of XL1-Blue (Fig. 1). Then, the native promoter of the *ldhA* gene (encoding D-lactate dehydrogenase) was replaced with the strong *trc* promoter and the *dld* (encoding D-lactate dehydrogenase) gene was deleted to construct X15ld strain (Fig. 1 and Table 1).

Into the X15ld strain, the above plasmids expressing XylBC$_{ccs}$ under different promoters (*tac* or P1-P5) were introduced. Flask cultures of these engineered X15ld strains produced up to 1.81 g l^{-1} of D-lactic acid (Fig. 3A). Use of the weakest P5 promoter resulted in the highest D-lactic acid titre of 1.81 g l^{-1}, while the relatively mid-strength P3 and P4 promoters resulted in lower D-lactic acid production of 0.87 and 0.98 g l^{-1}, respectively. As cell growth was enhanced, acetic acid accumulation was also significantly increased (Fig. 3A). To reduce acetic acid production, the *ack-pta* genes (encoding acetate kinase and phosphotransacetylase) were deleted from X15ld strain to make X15lda strain. Removal of *ack-pta* genes effectively eliminated acetic acid formation and significantly increased D-lactic acid production up to 5.72 g l^{-1}, which is 3.1-fold higher than that obtained with X15ld strain employing P5 promoter (Fig. 3B).

Biosynthesis of PLGA and poly(D-LA-co-GA-co-D-2HB)

For the biosynthesis of PLGA and poly(D-LA-*co*-GA-*co*-D-2HB), engineered *C. propionicum* propionyl-CoA transferase (Pct540) and engineered *Pseudomonas* sp. MBEL 6–19 PHA synthase (PhaC1437) were chosen as these two enzymes allowed successful production of PLGA and poly(D-LA-*co*-GA-*co*-D-3-hydroxybutyrate) in our previous study (Choi *et al.*, 2016); plasmid pPs619C1437Pct540 was employed for expressing Pct540 and PhaC1437. The X15ld and X15lda strains

harbouring pPs619C1437Pct540 and different XylBC$_{ccs}$ expression vectors were cultured. Flask cultures of these engineered strains produced PLA, PLGA and poly(D-LA-*co*-GA-*co*-D-2HB) having various monomer compositions (Fig. 4 and Fig. S2). When XylBC$_{ccs}$ was expressed under the *tac* promoter in X15ld, only small amount of PLA was produced (Fig. 4A). The X15ld strains expressing XylBC$_{ccs}$ under P1-P5 promoters produced poly(D-LA-*co*-GA-*co*-D-2HB)s. As the promoter strength decreased from P1 to P5, the GA mole fractions were decreased from 24.4 mol% to 1.4 mol%, while the polymer contents were increased from 12.3 wt% to 21.4 wt% (Fig. 4A). The X15lda produced polymers having higher D-LA fractions due to the increased D-LA flux as described above (Fig. S2). However, the polymer contents obtained were significantly lower than those (~21.4 wt%) obtained with X15ld; the highest polymer content obtained was 7.3 wt% with X15lda strain harbouring pP5xylBC and pPs619C1437Pct540. The GA mole fractions were also decreased except for the *tac* promoter case in which the increase of D-LA production resulted in enhanced PLGA production. The GA mole fractions were only 5.1 and 2.1 mol% when P1 and P2 promoters were used, respectively. When P3, P4 and P5 promoters were employed, only poly(D-LA-*co*-D-2HB) was produced without GA incorporation (Fig. S2). As X15lda strain was not effective for PLGA production, X15ld strain was used for further study.

The produced polymers contained D-2HB up to 7.0 mol% (Fig. 4A). This D-2HB incorporation has been previously reported to be originated from 2-ketobutyrate, which is an intermediate of the endogenous L-isoleucine biosynthetic pathway in *E. coli* (Choi *et al.*, 2016; Yang *et al.*, 2016). To prevent D-2HB incorporation into polymers, threonine dehydratase (IlvA), an enzyme responsible for 2-ketobutyrate formation, was inactivated by feeding 2 mM L-isoleucine in culture medium as IlvA is allosterically inhibited by L-isoleucine (Umbarger, 1956).

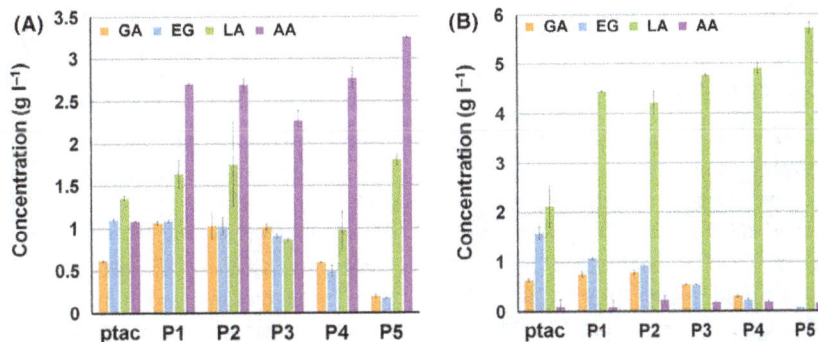

Fig. 3. Production of glycolic and D-lactic acids and major by-products.
Flask cultures of (A) X15ld strain and (B) X15lda strain harbouring different XylBC$_{ccs}$ expression vectors. GA, EG, LA and AA indicate glycolic acid, ethylene glycol, D-lactic acid and acetic acid, respectively. Error bars represent the SD values obtained in triplicate experiments.

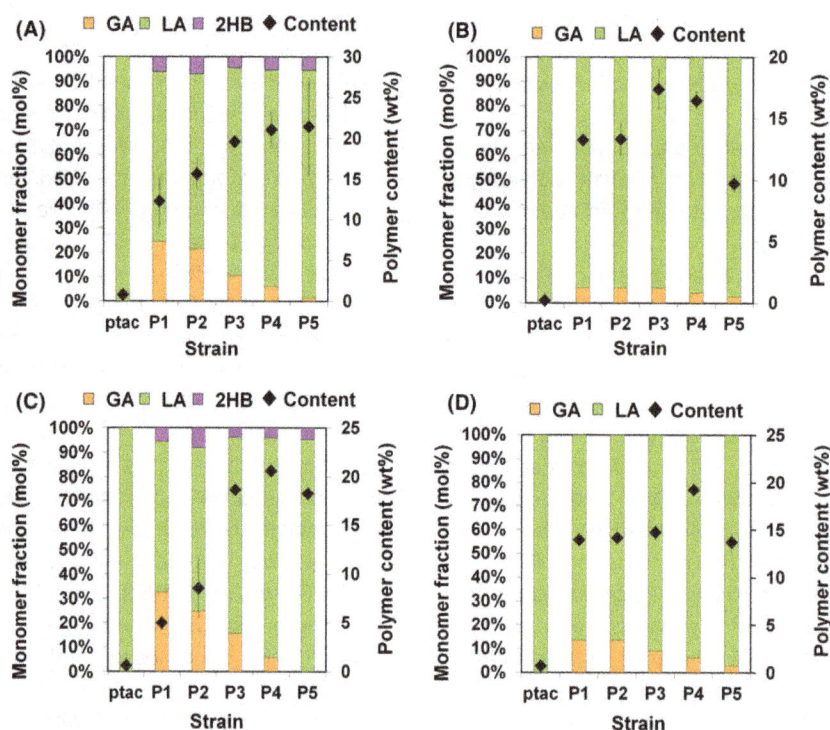

Fig. 4. Polymer contents and compositions.
Polymers produced by X15ld harbouring different XylBC$_{ccs}$ expression vectors and pPs619C1437Pct540 in medium (A) alone or (B) with 2 mM L-isoleucine supplementation. Polymers produced by X17ld harbouring different XylBC$_{ccs}$ expression vectors and pPs619C1437Pct540 in medium (C) alone or (D) with 2 mM L-isoleucine supplementation. GA, LA and 2HB indicate glycolic acid, D-lactic acid and D-2-hydroxybutyric acid, respectively. The polymer contents are represented by diamonds. Error bars represent the SD values obtained in triplicate experiments.

L-Isoleucine supplementation effectively eliminated D-2HB incorporation and resulted in production of PLGA (Fig. 4B). However, most of the strains produced polymers with reduced polymer content and GA mole fraction (~ 6.2 mol%) (Fig. 4B). Although previously verified, we analysed an example polymer produced by X15ld harbouring pPs619C1437Pct540 and pP4xylBC, poly (88.5 mol% D-LA-co-6.1 mol% GA-co-5.4 mol% D-2HB), by 1D (^{1}H and ^{13}C) NMR spectroscopy (Fig. S3). The results clearly confirmed again that poly(D-LA-co-GA-co-D-2HB) was produced in the engineered *E. coli*.

Enhanced PLGA production by fed-batch cultivation

To examine whether PLGA production can be further enhanced and the polymer compositions can be varied, fed-batch cultures of the X15ld harbouring pPs619C1437Pct540 and different XylBC$_{ccs}$ expression vectors were performed. Cell concentrations (dry cell weight, DCW, g l^{-1}) reached were 1.95, 4.98, 6.43, 13.91, 16.52 and 22.2 g DCW l^{-1} when the *tac*, P1, P2, P3, P4 and P5 promoters were employed, respectively (Fig. 5A and Fig. S4). As can be seen from these fed-batch fermentation results, higher cell concentration

could be reached when weaker promoter was employed for the expression of XylBC$_{ccs}$; the strain expressing XylBC$_{ccs}$ under the weakest P5 promoter showed 11.4-fold higher cell concentration than that obtained with the strain expressing XylBC$_{ccs}$ under the *tac* promoter. These results again show the negative relationship between cell growth and Dahms pathway flux modulated through adjusting the promoter strength.

Among the six engineered strains examined, X15ld harbouring pP3xylBC and pPs619C1437Pct540 resulted in the highest glycolic acid titre of 16.49 g l^{-1} (Fig. 5A and Fig. S4). When the promoters *tac*, P1, P2, P3, P4 and P5 were employed, glycolic acid titre obtained was 2.28, 5.69, 6.88 16.49, 12.46 and 7.51 g l^{-1}, respectively. Glycolic acid production increased as the promoter strength decreased up to P3 promoter and then started to decrease with further reduced promoter strength. So, as observed in flask cultures, enhanced cell growth by employing weak promoters positively affected GA titre. However, use of too weak promoter resulted in overall lower GA titre due to too much decreased Dahms pathway flux. Thus, it is important to optimize the key pathway flux (e.g. Dahms pathway) affecting cell growth for enhanced GA production. Ethylene glycol production also showed similar patterns (Fig. 5A and

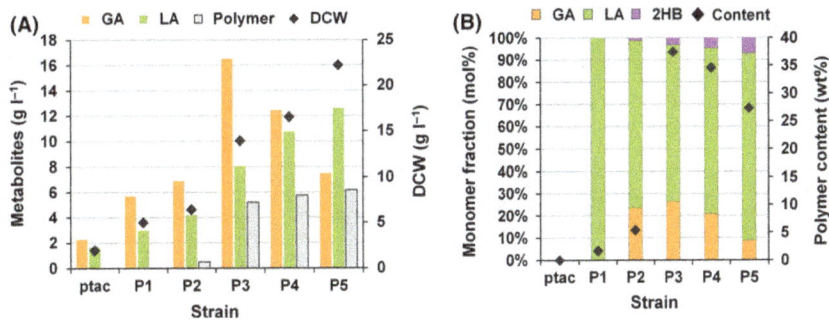

Fig. 5. Fed-batch cultures of X15ld expressing XylBC$_{ccs}$, PhaC1437 and Pct540.
(A). Concentrations of dry cell weight (DCW) and metabolites (D-lactic and glycolic acids and polymer), and (B) polymer contents and monomer compositions at the end of fed-batch cultures. The strains used were X15ld harbouring pPs619C1437Pct540 and different XylBC$_{ccs}$ expression vectors: pTacxylBC (ptac) and pP1xylBC-pP5xylBC (P1-P5). GA, LA and 2HB indicate glycolic acid, D-lactic acid and D-2-hydroxybutyric acid, respectively.

Fig. S4). Ethylene glycol titre was highest with P1 promoter (11.69 g l^{-1}) and became lower as the promoter strength was further decreased. In the case of D-lactic and acetic acids, the production titres were inversely proportional to the promoter strength (Fig. 5A and Fig. S4). As promoter strength decreased from *tac* to P5, production of D-lactic acid and acetic acid increased from 1.66 to 12.64 g l^{-1} and 1.70 to 10.23 g l^{-1}, respectively. (Fig. 5A and Fig. S4).

When the *tac* and P1 promoters were employed for XylBC$_{ccs}$ expression, only negligible amount of polymer was produced due to insufficient D-LA and GA production (Fig. 5A and Fig. S4). Use of P2 promoter resulted in production of only 5.3 wt% of poly(75.2 mol% D-LA-*co*-23.5 mol% GA-*co*-1.3 mol% D-2HB) at the end of fed-batch culture (Fig. 5). When P3 promoter was used, poly (70.5 mol% D-LA-*co*-26.3 mol% GA-*co*-3.2 mol% D-2HB) was produced to the highest polymer content of 37.4 wt %, which is significantly higher than that (19.6 wt%) obtained by flask culture (Fig. 5B). Also, the highest GA mole fraction of 26.3 mol% could be obtained. When the weakest P5 promoter was used, poly(84.2 mol% D-LA-*co*-8.8 mol% GA-*co*-7.0 mol% D-2HB) was produced to the polymer content of 27.3 wt%. Although the polymer content was lower than those obtained with the strains employing P3 and P4 promoters, the higher cell concentration led to the highest polymer titre of 6.1 g l^{-1} (Fig. 5). The polymer mole fractions were not much changed during the fed-batch culture (Fig. S4).

Construction of X17ld strain to further increase the glycolate fraction

As the monomer fraction of PLGA is an important characteristic to modulate the polymer degradation rate for various applications, further metabolic engineering was performed to enhance GA production. Because GA can be converted into glyoxylate by the native GA oxidase (GlcDEFG), the corresponding *glcDEFG* genes were deleted in the chromosome of X15ld. Also, the *glcB* and *aceB* genes encoding malate synthases were deleted to concentrate glyoxylate because glyoxylate can be converted into GA by endogenous glyoxylate reductase (YcdW); this strain was named X17ld (Fig. 1). Engineered X17ld strains harbouring pPs619C1437Pct540 and different XylBC$_{ccs}$ expression vectors were constructed and cultured in flasks to compare strain performances. The X17ld strains expressing XylBC$_{ccs}$ under P1, P2 and P3 promoters produced poly(D-LA-*co*-GA-*co*-D-2HB) with higher GA mole fractions compared to those obtained with the similarly engineered X15ld strains (Fig. 4A and C). The GA mole fraction decreased with decreasing promoter strength except for the *tac* promoter. When P1 promoter was employed, polymer having 32.7 mol% GA was produced to the polymer content of 5.0 wt% (Fig. 4C). When 2 mM of L-isoleucine was supplemented into the culture medium, the engineered X17ld strains produced PLGA containing up to 13.6 mol % of GA (for P1 promoter) that were also higher than those (up to 6.2 mol% of GA) obtained by employing similarly engineered X15ld strains (Fig. 4B and D).

Next, fed-batch cultures of these engineered X17ld strains were performed. As the engineered strain expressing XylBC$_{ccs}$ under *tac* promoter (the X17ld harbouring pTacxylBC and pPs619C1437Pct540) could not grow well, the results of this fed-batch culture are not included. As shown in Fig. 6A and Fig. S5, the engineered X17ld strains also showed similar tendencies in cell concentration and metabolites production with those observed with similarly engineered X15ld strains. As promoter strength decreased, the concentrations of cell, D-lactic acid and acetic acid were increased. The highest glycolic acid and ethylene glycol concentrations were 8.81 and 6.7 g l^{-1}, when the X17ld employing P3 promoter was used (Fig. 6A and Fig. S5).

In the fed-batch cultures, the GA mole fraction of produced polymers could be further increased

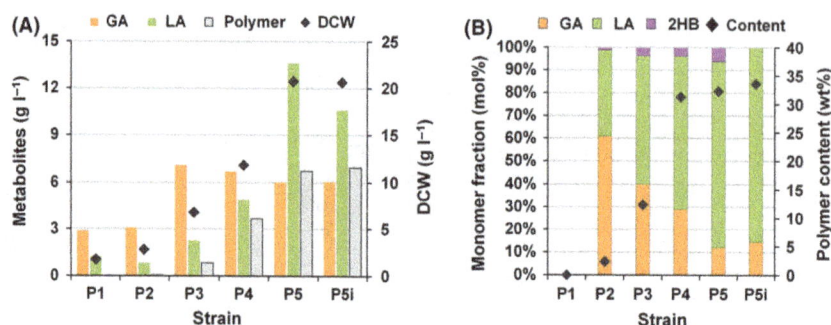

Fig. 6. Fed-batch cultures of X17ld expressing XylBC$_{ccs}$, PhaC1437 and Pct540.
(A). Concentrations of dry cell weight (DCW) and metabolites (D-lactic and glycolic acids and polymer), and (B) polymer contents and monomer compositions at the end of fed-batch cultures. The strains used were X17ld harbouring pPs619C1437Pct540 and different XylBC$_{ccs}$ expression vectors: pP1xylBC-pP5xylBC (P1–P5). P5i indicates that X17ld strain harbouring pPs619C1437Pct540 and pP5xylBC was cultured in L-isoleucine-supplemented medium. GA, LA and 2HB indicate glycolic acid, D-lactic acid and D-2-hydroxybutyric acid, respectively.

(Fig. 6B). When P2 and P3 promoters used, GA mole fractions were 60.9 and 40.1 mol%, respectively (Fig. 6B). Similarly to the results of fed-batch cultures of engineered X15ld strains, the highest polymer titre (6.70 g l^{-1}) was obtained when the weakest P5 promoter was employed (Fig. 6A). The engineered X17ld strain employing P5 promoter produced poly(81.7 mol% D-LA-co-12.1 mol% GA-co-6.2 mol% 2HB) to the polymer content of 32.3 wt% (Fig. 6B). To produce PLGA without incorporation of D-2HB, L-isoleucine was supplemented into the culture medium during fed-batch cultures of X17ld harbouring pPs619C1437Pct540 and pP5xylBC. As a result, 6.93 g l^{-1} of poly(85.4 mol% D-LA-co-14.6 mol% GA) could be successfully produced to a polymer content of 33.4 wt% (Fig. 6 and Fig. S5).

Molecular weights, thermal properties and biocompatibility of poly(D-LA-co-GA-co-D-2HB)

As the biocompatibility of poly(D-LA-co-GA-co-D-2HB) is not known, we investigated it with two representative polymers, poly(70.4 mol% D-LA-co-26.4 mol% GA-co-3.2 mol% D-2HB) and poly(84.2 mol% D-LA-co-8.8 mol% GA-co-7.0 mol% D-2HB) produced by fed-batch cultures. Before doing so, these polymers were characterized with respect to molecular weights by gel permeation chromatography (GPC) and thermal properties by differential

scanning calorimetry (DSC). Previously, several studies also reported for the synthesis of 2HB containing polymers such as homo poly(D-2HB) and copolymers including poly(D-LA-co-D-2HB) and characterization of molecular weight, biodegradation, thermal and mechanical properties (Park *et al.*, 2012, 2013; Matsumoto *et al.*, 2013; Chae *et al.*, 2016).

The number average (M_n) and weight average (M_w) molecular weights of 26.4 mol% GA containing polymer were 10,128 and 13,304 Da, respectively, with the polydispersity index of 1.314 (Table 2). The 8.8 mol% GA containing polymer had molecular weights of 16,612 (M_n) and 26,308 (M_w) Da with the polydispersity index of 1.584. These molecular weights are similar to those of PLGAs produced by engineered *E. coli* strains we previously reported (Choi *et al.*, 2016). Also, they are comparable to those of PLGAs suitable for use in making microsphere for drug delivery carrier and scaffolds for tissue engineering (Makino *et al.*, 2004; Campos *et al.*, 2014; Xie *et al.*, 2015). The thermal properties of the two polymers were measured by DSC analysis (Table 2). They exhibited glass transition temperature at 45.0 and 49.9 °C, respectively, and no melting temperatures, indicating that the produced polymers are amorphous. These properties are also similar to those of chemically synthesized PLGA (Park, 1995).

Table 2. Polymer characteristics of two sample poly(D-LA-co-GA-co-D-2HB) produced by engineered *Escherichia coli*.

Monomer fraction[a] (mol%)			Molecular weight[b]			Thermal properties[c]		
LA	GA	2HB	M_n (Da)	M_w (Da)	M_w M_n^{-1}	T_g	T_m	ΔC_p
70.4	26.4	3.2	10,128	13,304	1.314	45.0	–	0.773
84.2	8.8	7.0	16,612	26,308	1.584	49.9	–	0.638

a. LA, GA and 2HB indicate D-lactate, glycolate and D-2-hydroxybutyrate, respectively.
b. M_n, number average molecular weight; M_w, weight average molecular weight; M_w M_n^{-1}, polydispersity index.
c. T_g, glass transition temperature (°C); T_m, melting temperature (°C); ΔC_p, heat capacity (J g^{-1} K^{-1}).

Next, we examined the biocompatibility of one of these D-2HB containing PLGA polymers, poly(84.2 mol% D-LA-co-8.8 mol% GA-co-7 mol% D-2HB). To evaluate the polymer cytotoxicity, the live/dead assay of human mesenchymal stem cell (hMSC) was performed. The hMSCs were seeded on cover glass (spin coated with each polymer) and cultivated for 4 days. The poly(84.2 mol% D-LA-co-8.8 mol% GA-co-7 mol% D-2HB) exhibited high cell viability similar to that of commercial PLGA (Fig. 7 and Fig. S6). This result demonstrated that poly(D-LA-co-GA-co-D-2HB) containing D-2HB as an additional monomer is biocompatible.

Conclusion

We recently reported one-step direct fermentative production of PLGA, poly(D-LA-co-GA-co-D-2HB), and other PLGA copolymers by metabolically engineered *E. coli* strains (Choi *et al.*, 2016). As the introduction of heterologous Dahms pathway producing GA negatively affected cell growth, the metabolic flux towards Dahms pathway was modulated using five synthetic promoters of different strengths. Through further metabolic engineering described above, PLGA and poly(D-LA-co-GA-co-D-2HB) with various monomer compositions could be more efficiently produced. Also, the GA mole fraction in poly(D-LA-co-GA-co-D-2HB) could be modulated between 8.8 and 60.9 mol% of GA. The polymer titre could also be increased up to 6.93 g l^{-1} in the case of poly(85.4 mol% D-LA-co-14.6 mol% GA). As poly(D-LA-co-GA-co-D-2HB) has been less characterized, their molecular weights, thermal properties and biocompatibility were examined. Poly(D-LA-co-GA-co-D-2HB) was found to be biocompatible and thus can be similarly used for medical applications like PLGA.

This microbial production system can be further improved with respect to polymer titre, yield and productivity by adapting various systems metabolic engineering strategies (Lee *et al.*, 2011, 2012; Choi *et al.*, 2016; Lee and Kim, 2015; Song and Lee, 2015). One of the bottlenecks is less optimal PHA synthase activity and specificity. As the crystal structure of PHA synthase from *Ralstonia eutropha* has been recently resolved (Wittenborn *et al.*, 2016; Kim *et al.*, 2017a,b), much better and rational engineering of PHA synthase based on the structural information will improve the bio-based production of PLGA and various polymers.

Experimental procedures

Bacterial strains and plasmids

All bacterial strains, plasmids and primers used in this study are listed in Table 1 and Table S1. *E. coli* XL1-Blue (Stratagene Cloning Systems, La Jolla, CA, USA) was used for general gene cloning studies. *E. coli* XL1-Blue and its mutants were used as host strains for the synthesis of PLGA and PLGA copolymers. All DNA manipulations were performed following standard procedures (Sambrook and Russell, 2001). PCR was performed with the C1000 Thermal Cycler (Bio-Rad, Hercules, CA, USA).

Plasmid pPs619C1437Pct540, which expresses the *Pseudomonas* sp. MBEL 6–19 PHA synthase containing quadruple mutations of E130D, S325T, S477G, and Q481K (PhaC1437) and the *Clostridium propionicum* propionyl-CoA transferase mutant containing V193A and four silent nucleotide mutations of T78C, T669C, A1125G, and T1158C (Pct540) under the *Ralstonia eutropha* PHA biosynthesis operon promoter, has been previously described (Table 1; Yang *et al.*, 2011). Plasmid pTacxylBC expressing *Caulobacter crescentus* XylBC$_{ccs}$ (xylose dehydrogenase and xylonolactonase) has also been reported (Table 1; Choi *et al.*, 2016).

To construct pTacxylBC_xylAB plasmid, *E. coli* xylAB genes encoding xylose isomerase and xylulose kinase were amplified from the chromosomal DNA of *E. coli* XL1-Blue by PCR using the primers xylAB_F and xylAB_R (Table S1). For expression of the XylAB under *tac* promoter, the *tac* promoter region was amplified from the pTac15k using the primers tac_xylAB_F and tac_xylAB_R. The *xylAB*, *tac* and linear DNA of pTacxylBC digested by PstI and SphI were assembled into pTacxylBC_xylAB (Table 1 and Table S1).

To express XylBC$_{ccs}$ with different strength, the five Anderson promoters (BBa_J23100, BBa_J23101, BBa_J23118, BBa_J23105 and BBa_J23117) were used (http://parts.igem.org). To construct pP1xylBC plasmid,

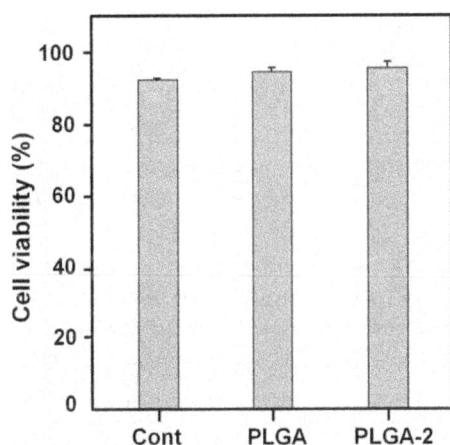

Fig. 7. Live/dead assay.
Percentage of live human mesenchymal stem cells (hMSCs) on each substrate: cover glass (Cont), PLGA-coated glass (PLGA) or poly(D-LA-co-GA-co-D-2HB)-coated glass (PLGA-2). Error bars represent the SD values obtained in triplicate experiments.

pTacxylBC plasmid was amplified by inverse PCR using the primer 100_F and 100_R to replace the original *tac* promoter with the BBa_J23100 promoter. The construction of pP2xylBC-pP5xylBC plasmids was performed in the same manner (Table 1 and Table S1).

Deletion of chromosomal genes and promoter replacement

Deletion of *E. coli* chromosomal genes was achieved using one-step inactivation method as previously described (Datsenko and Wanner, 2000). The primers and detailed methods are described in previous paper (Choi *et al.*, 2016). Briefly, the linear DNA fragment having homologous region with the target gene was prepared by PCR using pMloxC (Kim *et al.*, 2008). The PCR fragments were introduced by electroporation into the recombinant *E. coli* harbouring pKD46 to express λ recombinase (Datsenko and Wanner, 2000). The mutants in which double homologous recombination occurred were selected, and then, the chloramphenicol resistance gene was eliminated by introducing pJW168 encoding the *cre* recombinase in presence of 1 mM isopropyl β-D-1-thiogalactopyranoside (IPTG; Sigma-Aldrich, St. Louis, MO, USA; Palmeros *et al.*, 2000). For the replacements of chromosomal *ldhA* promoter, plasmid pMtrc9 having *trc* promoter at the downstream of lox66-cat-lox71 cassette was used in place of pMloxC.

Cultivation and metabolites analysis

Escherichia coli was routinely cultured at 37 °C in Luria-Bertani (LB) medium containing 10 g l^{-1} tryptone, 5 g l^{-1} yeast extract and 10 g l^{-1} NaCl. For the production, recombinant *E. coli* strains were cultured in MR medium supplemented with 100 mM 3-(N-morpholino) propansulfonic acid (MOPS) in 30 °C in a rotary shaker at 250 rpm. The MR medium (pH 7.0) contains (per litre) 6.67 g KH_2PO_4, 4 g $(NH_4)_2HPO_4$, 0.8 g $MgSO_4 \cdot 7H_2O$, 0.8 g citric acid and 5 ml trace metal solution. The trace metal solution contains (per litre of 0.5 M HCl) 10 g $FeSO_4 \cdot 7H_2O$, 2 g $CaCl_2$, 2.2 g $ZnSO_4 \cdot 7H_2O$, 0.5 g $MnSO_4 \cdot 4H_2O$, 1 g $CuSO_4 \cdot 5H_2O$, 0.1 g $(NH_4)_6Mo_7O_{24} \cdot 4H_2O$ and 0.02 g $Na_2B_4O_7 \cdot 10H_2O$. Xylose and $MgSO_4 \cdot 7H_2O$ were sterilized separately. Seed cultures were prepared in 25-ml test tubes containing 5-ml LB medium at 30 °C overnight in a rotary shaker at 250 rpm. One ml of overnight culture was used to inoculate 250-ml flask containing 100 ml of 100 mM MOPS added MR medium containing 10 mg l^{-1} of thiamine supplemented with 20 g l^{-1} of xylose. Ampicillin (Ap, 50 μg ml^{-1}) and kanamycin (Km, 30 μg ml^{-1}) were added to the medium depending on the resistance marker of the employed plasmid. One mM of IPTG (Sigma

Aldrich) was added at the beginning of the flask cultivation. To inactivate IlvA, 2 mM of L-isoleucine was added to the culture medium.

Fed-batch cultures carried out in a 6.6L Bioreactor (Bioflo 3000, New Brunswick Scientific, Edison, NJ, USA) containing 100 mM MOPS added MR medium supplemented with 20 g l^{-1} of xylose, and 10 mg l^{-1} of thiamine. The dissolved oxygen concentration (DOC) was maintained above 10% of air saturation by changing the agitation speed. The pH was controlled at 7.0 by ammonia solution. The following feeding solution was used: 700 g l^{-1} xylose plus 0.8 g $MgSO_4 \cdot 7H_2O$ and 10 mg thiamine. The feeding solution was manually fed into the fermentor when the xylose concentration dropped to below 5 g l^{-1} to increase it to about 20–25 g l^{-1}. For L-isoleucine-supplemented culture, the L-isoleucine was manually fed into the fermentor when the L-isoleucine concentration dropped to below 1 mM to increase it to about 2 mM.

The carbon sources (xylose) and metabolites (acetate, glycolate, D-lactate, ethylene glycol, formate, succinate) were measured by high-performance liquid chromatography (HPLC) (1515 isocratic HPLC pump; Waters, Milford, MA, USA) equipped with refractive index detector (2414; Waters). A MetaCarb 87H column (Agilent, Palo Alto, CA, USA) was eluted isocratically with 0.01 N H_2SO_4 at 25 °C at flow rate of 0.5 ml min^{-1}. The amino acid (L-isoleucine) was measured as previously described (Shin *et al.*, 2016). For the detection of amino compounds, the supernatant of the culture was reacted with *o*-phthalaldehyde and the derivatized amino acids were analysed by Agilent 1100 HPLC (Agilent). Cell growth was monitored by measuring the absorbance at 600 nm (OD_{600}) using an Ultrospec 3000 spectrophotometer (Amersham Biosciences, Uppsala, Sweden).

Polymer analysis

To analyse the contents and monomer compositions of polymers produced by engineered *E. coli*, the samples were prepared as previously described and were analysed by gas chromatography (GC; Braunegg *et al.*, 1978; Choi *et al.*, 2016). The cells were washed with distilled water and then lyophilized. The polymers in lyophilized cells were converted into corresponding hydroxymethyl esters by acid-catalysed methanolysis. The resulting methyl esters were measured by GC (Agilent 6890N; Agilent) equipped with Agilent 7683 automatic injector, flame ionization detector and a fused silica capillary column (ATTM-Wax, 30 m, ID 0.53 mm, film thickness 1.20 m; Alltech, Deerfield, IL, USA). Polymer content is defined as the weight percentage of polymer concentration to dry cell concentration (wt% of dry cell weight).

Polymers were extracted from the cells by the chloroform extraction (Choi and Lee, 1999). The harvested cells were washed with distilled water and lyophilized. The lyophilized cells were located in chloroform refluxed Soxhlet apparatus (Corning, Lowell, MA, USA) overnight. The chloroform solution was concentrated and then precipitated by addition of ice cold methanol. The structure, molecular weight and thermal properties of the polymers were determined by nuclear magnetic resonance (NMR) spectroscopy, gel permeation chromatography (GPC) and differential scanning calorimetry (DSC), respectively, as previously described (Jung and Lee, 2011).

Biocompatibility test

The polymers were dissolved in chloroform at 2 w v^{-1}%. The solution of each polymer was dropped on 18-mm cover glass and spin coated at 1200 g for 1 min. The spin-coated cover glass was kept at room temperature for overnight to allow evaporation of chloroform. The cover glass and spin-coated cover glass were washed with 1X Dulbecco's phosphate-buffered saline for 3 times and cured by ultraviolet light for 30 min for sterilization. The human mesenchymal stem cells (Lonza, Muenchensteinerstrasse 38, CH-4002 Basel, Switzerland) were seeded onto the sterilized cover glasses and polymer films at a seeding density of 2.5 × 10^4 cells cm^{-2} and cultured for 4 days with Minimum Essential Media alpha (α-MEM, Gibco, Grand Island, NY, USA) supplemented with 17% foetal bovine serum (FBS, Welgene, Gyeongsan-si, Gyeongsangbuk-do, Republic of Korea), 100 units ml^{-1} penicillin (Gibco) and 100 μg ml^{-1} streptomycin (Gibco) in 5% CO_2 and 37 °C incubator. The medium is exchanged every 2 days.

A live/dead assay is performed using LIVE/DEAD® Viability/Cytotoxicity Assay Kit for mammalian cells (Invitrogen, Carlsbad, CA, USA). The kit reagents, Calcein AM and ethidium homodimer-1 (EthD-1), are diluted in 1X Dulbecco's phosphate-buffered saline (DPBS) according to the kit instructions. The cells cultured on cover glasses and PLGA films were gently washed with 1X DPBS for 3 times, and 1 ml of diluted Calcein AM and EthD-1 solution was added to the cover glasses and PLGA films for 45 min at room temperature. The live cells were stained with Calcein AM, and dead cells were stained with EthD-1. After staining, cells were gently washed with 1X DPBS for 3 times again and imaged with a fluorescence microscope (Ti-U; Nikon, Minato-ku, Tokyo, Japan). The percentage of live and dead cells was counted with IMAGEJ software (https://imagej.nih.gov/ij/) with three independent experiments.

Acknowledgements

This work was supported by the Technology Development Program to Solve Climate Changes (Systems Metabolic Engineering for Biorefineries) from the Ministry of Science, ICT and Future Planning (MSIP) through the National Research Foundation (NRF) of Korea (NRF-2012M1A2A2026556 and NRF-2012M1A2A2026557).

Conflict of interest

The authors declare that they have conflict of interest as the related technologies are patented and under commercialization discussion.

References

Braunegg, G., Sonnleitner, B.Y., and Lafferty, R.M. (1978) A rapid gas chromatographic method for the determination of poly-β-hydroxybutyric acid in microbial biomass. *Appl Microbiol Biotechnol* **6:** 29–37.

Campos, D.M., Gritsch, K., Salles, V., Attik, G.N., and Grosgogeat, B. (2014) Surface entrapment of fibronectin on electrospun PLGA scaffolds for periodontal tissue engineering. *Biores Open Access* **3:** 117–126.

Chae, C.G., Kim, Y.J., Lee, S.J., Oh, Y.H., Yang, J.E., Joo, J.C., *et al.* (2016) Biosynthesis of poly(2-hydroxybutyrate-co-lactate) in metabolically engineered *Escherichia coli*. *Biotechnol Bioproc Eng* **21:** 169–174.

Choi, J., and Lee, S.Y. (1999) Efficient and economical recovery of poly(3-hydroxybutyrate) from recombinant *Escherichia coli* by simple digestion with chemicals. *Biotechnol Bioeng* **62:** 546–553.

Choi, K.R., Shin, J.H., Cho, J.S., Yang, D., and Lee, S.Y. (2016) Systems metabolic engineering of *Escherichia coli*. *EcoSal Plus* **7:** 1–56. https://doi.org/10.1128/ecosalplus. ESP-0010-2015.

Choi, S.Y., Park, S.J., Kim, W.J., Yang, J.E., Lee, H., Shin, J., *et al.* (2016) One-step fermentative production of poly (lactate-*co*-glycolate) from carbohydrates in *Escherichia coli. Nat Biotechnol* **34:** 435–440.

Danhier, F., Ansorena, E., Silva, J.M., Coco, R., Le Breton, A., and Préat, V. (2012) PLGA-based nanoparticles: an overview of biomedical applications. *J Control Release* **161:** 505–522.

Datsenko, K.A., and Wanner, B.L. (2000) One-step inactivation of chromosomal genes in *Escherichia coli* K-12 using PCR products. *Proc Natl Acad Sci USA* **97:** 6640–6645.

Jung, Y.K., and Lee, S.Y. (2011) Efficient production of polylactic acid and its copolymers by metabolically engineered *Escherichia coli*. *J Biotechnol* **151:** 94–101.

Kim, J.M., Lee, K.H., and Lee, S.Y. (2008) Development of a markerless gene knock-out system for *Mannheimia succiniciproducens* using a temperature-sensitive plasmid. *FEMS Microbiol Lett* **278:** 78–85.

Kim, J.E., Kim, Y.-J., Choi, S.Y., Lee, S.Y., and Kim, K.-J. (2017a) Crystal structure of *Ralstonia eutropha* polyhydroxyalkanoate synthase C-terminal domain and reaction mechanisms. *Biotechnol J* **12:** 1–12.

Kim, Y.-J., Choi, S.Y., Kim, J.E., Jin, K.S., Lee, S.Y., and Kim, K.-J. (2017b) Structure and function of the N-terminal domain of *Ralstonia eutropha* polyhydroxyalkanoate synthase, and the proposed structure and mechanisms of the whole enzyme. *Biotechnol J* **12:** 1–12.

Lee, S.Y., and Kim, H.U. (2015) Systems strategies for developing industrial microbial strains. *Nature Biotechnol* **33:** 1061–1072.

Lee, J.W., Kim, T.Y., Jang, Y.S., Choi, S., and Lee, S.Y. (2011) Systems metabolic engineering for chemicals and materials. *Trends Biotechnol* **29:** 370–378.

Lee, J.W., Na, D., Park, J.M., Lee, J., Choi, S., and Lee, S.Y. (2012) Systems metabolic engineering of microorganisms for natural and non-natural chemicals. *Nature Chem Biol* **8:** 536–546.

Makadia, H.K., and Siegel, S.J. (2011) Poly lactic-co-glycolic acid (PLGA) as biodegradable controlled drug delivery carrier. *Polymers* **3:** 1377–1397.

Makino, K., Nakajima, T., Shikamura, M., Ito, F., Ando, S., Kochi, C., *et al.* (2004) Efficient intracellular delivery of rifampicin to alveolar macrophages using rifampicin-loaded PLGA microspheres: effects of molecular weight and composition of PLGA on release of rifampicin. *Colloids Surf B Biointerfaces* **36:** 35–42.

Matsumoto, K.I., Terai, S., Ishiyama, A., Sun, J., Kabe, T., Song, Y., *et al.* (2013) One-pot microbial production, mechanical properties, and enzymatic degradation of isotactic P [(R)-2-hydroxybutyrate] and its copolymer with (R)-lactate. *Biomacromol* **14:** 1913–1918.

Palmeros, B., Wild, J., Szyblalski, W., Borgne, S.L., Hermandez-Chavez, G., Gosset, G., *et al.* (2000) A family of removable cassettes designed to obtain antibiotic-resistance-free genomic modifications of *Escherichia coli* and other bacteria. *Gene* **247:** 255–264.

Park, T.G. (1995) Degradation of poly(lactic-co-glycolic acid) microspheres: effect of copolymer composition. *Biomaterials* **16:** 1123–1130.

Park, S.J., Lee, T.W., Lim, S.-C., Kim, T.W., Lee, H., Kim, M.K., *et al.* (2012) Biosynthesis of polyhydroxyalkanoates containing 2-hydroxybutyrate from unrelated carbon source by metabolically engineered *Escherichia coli*. *Appl Microbiol Biotechnol* **93:** 273–283.

Park, S.J., Kang, K.H., Lee, H., Park, A.R., Yang, J.E., Oh, Y.H., *et al.* (2013) Propionyl-CoA dependent biosynthesis of 2-hydroxybutyrate containing polyhydroxyalkanoates in metabolically engineered *Escherichia coli*. *J Biotechnol* **165:** 93–98.

Sambrook, J., and Russell, D.W. (2001) *Molecular Cloning: A Laboratory Manual*, 3rd edn. Cold Spring Harbor, NY: Cold Spring Harbor Lab Press.

Shin, J.H., Park, S.H., Oh, Y.H., Choi, J.W., Lee, M.H., Cho, J.S., *et al.* (2016) Metabolic engineering of *Corynebacterium glutamicum* for enhanced production of 5-aminovaleric acid. *Microb Cell Fact* **15:** 174.

Song, C.W., and Lee, S.Y. (2015) Combining rational metabolic engineering and flux optimization strategies for efficient production of fumaric acid. *Appl Microbiol Biotechnol* **99:** 8455–8464.

Umbarger, H.E. (1956) Evidence for a negative-feedback mechanism in the biosynthesis of isoleucine. *Science* **123:** 848.

Wittenborn, E.C., Jost, M., Wei, Y., Stubbe, J., and Drennan, C.L. (2016) Structure of the catalytic domain of the class I polyhydroxybutyrate synthase from *Cupriavidus necator*. *J Biol Chem* **291:** 25264–25277.

Wu, X.S., and Wang, N. (2001) Synthesis, characterization, biodegradation, and drug delivery application of biodegradable lactic/glycolic acid polymers. Part II: biodegradation. *J Biomater Sci Polym Ed* **12:** 21–34.

Xie, X., Lin, W., Xing, C., Yang, Y., Chi, Q., Zhang, H., *et al.* (2015) In vitro and in vivo evaluations of PLGA microspheres containing nalmefene. *PLoS ONE* **10:** e0125953.

Yang, T.H., Jung, Y.K., Kang, H.O., Kim, T.W., Park, S.J., and Lee, S.Y. (2011) Tailor-made type II *Pseudomonas* PHA synthases and their use for the biosynthesis of polylactic acid and its copolymer in recombinant *Escherichia coli*. *Appl Microbiol Biotechnol* **90:** 603–614.

Yang, J.E., Kim, J.W., Oh, Y.H., Choi, S.Y., Lee, H., Park, A.-R., *et al.* (2016) Biosynthesis of poly(2-hydroxyisovalerate-co-lactate) by metabolically engineered *Escherichia coli*. *Biotechnol J* **11:** 1572–1585.

3

Advances and bottlenecks in microbial hydrogen production

Alan J. Stephen,[1] Sophie A. Archer,[1]
Rafael L. Orozco[2] and Lynne E. Macaskie[2,*]
*Schools of [1]Chemical Engineering, [2]Biosciences,
University of Birmingham, Edgbaston, Birmingham, B15
2TT, UK.*

Summary

Biological production of hydrogen is poised to become a significant player in the future energy mix. This review highlights recent advances and bottlenecks in various approaches to biohydrogen processes, often in concert with management of organic wastes or waste CO_2. Some key bottlenecks are highlighted in terms of the overall energy balance of the process and highlighting the need for economic and environmental life cycle analyses with regard also to socio-economic and geographical issues.

Introduction

Hydrogen provides a CO_2-free sustainable alternative to fossil fuels. A pioneering global initiative, the 'Hydrogen Council', comprising thirteen leading energy, transport and related industries, intends to increase investment in the hydrogen and fuel cell sectors (currently €1.4 Bn year^{-1}) to stimulate hydrogen as a key part of the future energy mix via new policies and schemes (Anon, 2017).

Hydrogen is currently obtained mainly by steam reforming of hydrocarbons, releasing multiple greenhouse gas emissions (DOE, 2013). Hence, new H_2 production methods are required such as biological production (bio-H_2; Dincer and Acar, 2015). H biotechnologies are maturing towards benchmarking against established clean energy from electrolysis of water, solar photovoltaics and wind farms. Biohydrogen can be made fermentatively from wastes, providing a simultaneous method of organic waste management (Chang *et al.*,

*For correspondence. E-mail L.E.Macaskie@bham.ac.uk

Funding Information
Natural Environment Research Council (NE/L014076/1).

2011). This short review highlights progress and bottlenecks of bio-H_2 towards a sustainable development goal to ensure access to affordable, reliable, sustainable and modern energy for all. Biohydrogen has been reviewed in comparison with other hydrogen production processes (Nikolaidis and Poullikkas, 2017).

Biohydrogen embraces any H_2 production involving biological material (Mohan and Pandey, 2013). The energy source can be solar or can come from conversion of fixed carbon substrates (or both, in various combinations). An approach to CO_2-end of pipe treatment (e.g. from flue gas from fossil fuel combustion or carbon-neutral fermentation of biomass) is to grow algae on waste CO_2. Algal biohydrogen production is well-described, but O_2 from algal oxygenic photosynthesis inhibits the hydrogenase that makes H_2. A key study (Kubas *et al.*, 2017) will open the way to developing O_2-resistant hydrogenase. Emerging technology uses cyanobacteria (blue-green algae) that make H_2 via hydrogenase and also nitrogenase; their O_2-sensitivity is managed by temporal separation of photosynthetic O_2 evolution and nitrogenase action, and by compartmentalization into microanaerobic heterocysts (Tiwari and Pandey, 2012). Despite a note that cyanobacterial biohydrogen is probably uneconomic (Singh *et al.*, 2016), an environmental life cycle analysis (LCA) has shown for the first time that cyanobacterial bio-H_2 has the potential to be a competitor to desulfurized natural gas; the associated environmental impact of producing and extracting each gas, including use in a solid oxide fuel cell, was calculated and simulated respectively using the LCA software SIMAPRO (Archer *et al.*, 2017). This research used published data from a raceway growth system (James *et al.*, 2009). However, at latitudes above ~40°N, the generally low incident solar energy makes stand-alone photobiological H_2 systems seasonal and uneconomic without some form of process intensification. Boosting light delivery (e.g. LEDs, quantum dots) can be effective, but these may risk photopigment saturation and inhibition; this approach may be questionable economically and would be best addressed by a life cycle analysis. In sunny countries, light is plentiful, but in this case, 'delivering cold' is needed to extend crop product and food life; cooling is energy-demanding and a global challenge (Strahan, 2017).

Another challenge is organic materials from agri-food and municipal wastes, which must be managed to avoid landfilling which yields methane, a potent greenhouse gas. Current practices use anaerobic digestion (AD) with biogas – methane used for power. We review some options for combining waste treatments with bio-H_2 technology as possibly the best approach to tackling effectively these dual socio-economic problems; stand-alone biohydrogen is possibly uneconomic, but this awaits a life cycle analysis, currently in progress.

Biohydrogen production from waste: fermentation strategies for sustainable 'waste to hydrogen energy'

Fermentation is the disposal of excess metabolic reductant (NADH) onto organic compounds in the absence of alternative electron acceptors such as O_2 and NO_3^- (Guo et al., 2010). The mixed-acid fermentation ('dark fermentation') pathway of the paradigm Escherichia coli (Fig. 1A) is simple, has high rates of H_2 production but has limitations (Saratale et al., 2013; Fig. 1A inset). Hexose sugars can stoichiometrically deliver 12 mol H_2 mol hexose^{-1}.

The mixed-acid fermentation, while irreversible, is thermodynamically limited to 2–4 mol H_2 mol hexose^{-1} (Hallenbeck, 2012). The 'NADH pathway' of some microorganisms (Hallenbeck, 2012, 2017) can deliver a higher H yield, but is reversible under a positive H_2 partial pressure, which is required for with a downstream H fuel cell. Thermophilic bacteria have advantages but require input of heat energy. Hence, the focus has been mainly on mesophilic bacteria (Balachandar et al., 2013).

Most mixed-acid fermentations follow a similar schematic: the cell forms reduced metabolic end-products: organic acids (including toxic formate) and alcohol (Fig. 1A). Up to 2 mol H_2 mol^{-1} hexose (Hallenbeck and Ghosh, 2009) is produced via the activity of formate hydrogen lyase (which splits formate to $H_2 + CO_2$), that is < 20% of the theoretical maximum H_2. Sustained bio-H_2 production is limited by end-product (ethanol) toxicity and acidification of the medium by accumulating organic acids (Redwood, 2007).

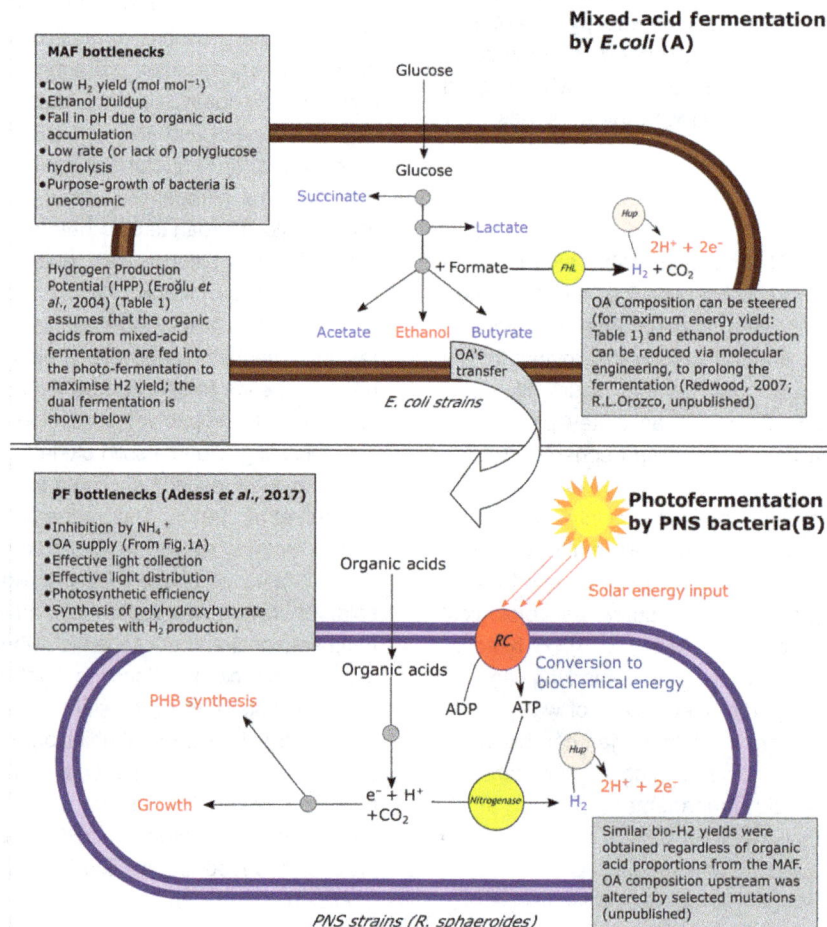

Fig. 1. Mixed-acid fermentation (MAF) of E. coli (A) and use of purple non-sulfur bacteria (B) in photofermentation (PF) of organic acids (OAs) into H_2. The organic acids are taken up by (e.g.) R. sphaeroides, and reducing power is generated as NADH (not shown). This reducing power can either be used for polyhydroxybutyrate synthesis or growth to maintain cellular redox or alternatively can be used for H_2 production under light when growth is restricted by limitation of N or P source. Italicized bottlenecks are those overcome by use of the dual system (see text).

The organic acids provide a means to overcome the thermodynamic limitation via their use in a coupled photofermentation reactor (Redwood et al., 2012a,b; Hallenbeck, 2013, 2017) via electrodialysis (Fig 2). If organic acid mixtures are fed to purple non-sulfur bacteria (e.g. *Rhodobacter sphaeroides*), the off-gas (typically > 90% H) is suitable for direct use in fuel cells (Nakada et al., 1999). This anoxygenic photofermentative H_2 process (Fig. 1B) requires input of light energy (to help overcome the thermodynamic barrier in converting organic acids into H_2 (Hallenbeck, 2013)). Nitrogen-deficient conditions are essential; in purple non-sulfur bacteria, H_2 biogenesis is a side reaction of nitrogenase, which normally fixes N_2 and is downregulated in the presence of fixed nitrogen. Utilizable organic acids also feed a competing pathway to make polyhydroxybutyrate which detracts from the H_2 yield (Fig. 1B). Redwood et al. (2012a,b) incorporated an electrodialysis step to concentrate the organic acids (by ~eightfold) and link the mixed-acid and photofermentation steps (Fig. 2). Electrodialysis separates anions (negatively charged organic acids in the dark fermentation medium), removing them and also preventing the transfer of inhibitory NH_4^+ into the photofermentation medium. This continuous dual fermentation process combines high H_2 production rates and yield (Redwood et al., 2012b); the electrical energy demand of electrodialysis is counterbalanced, in part, by a third H_2 stream from electrolysis of water.

Redwood (2007) calculated the break-even current efficiency to quantify the role played by specific organic acids (Table 1). Butyrate is the most attractive organic acid for electrodialysis with the lowest break-even current efficiency at 13% (Table 1). Butyrate is a neglected organic acid product from *E. coli* which can predominate under some conditions (Redwood, 2007; R.L. Orozco unpublished). Using this example (Figs 1 and 2), the energy balance for bio-H_2 (via fermentation of food

Table 1. Properties of organic acids relevant to their separation from spent medium by electrodialysis.

Organic acid	Carbons	Valence	pK_a	HPP mol^{-1}	BCE (%)
Butyrate	4	1	4.81	10	13
Lactate	3	1	3.86	6	21.6
Formate	1	1	3.75	2	N/A
Acetate	2	1	4.76	7	32.5
Succinate	4	2	4.19, 5.57	7	27.1

The break-even current efficiency (BCE: (energy expended/energy gained) × 100)) was calculated for individual organic acids. The lower the BCE, the less energy required to transport the organic acid. The electrical energy required for organic acid transport via electrodialysis relates to the number of charges and number of carbons; butyrate (4 carbons, 1 charge) is the most favourable and also has the highest proportion of charged butyrate (c.f. butyric acid) according to the pK_a. HPP is hydrogen production potential of the dual system as defined by Eroğlu et al. (2004).

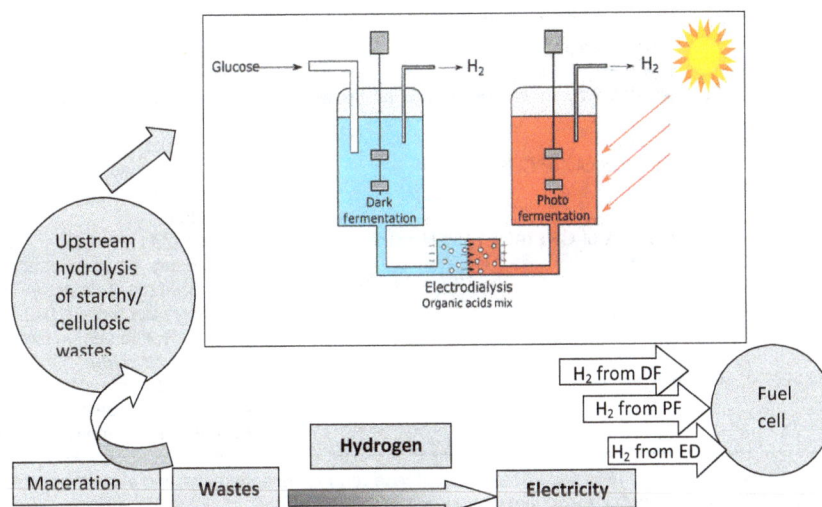

Fig. 2. System for energy delivery from wastes via biohydrogen A fusion of chemical and biochemical engineering for conversion of waste into electricity via integrated biohydrogen technology. Electrodialysis (ED) separates the organic acid (OA) products from the mixed-acid fermentation of (e.g.) *E. coli* (formate is converted to H_2 + CO_2 via formate hydrogen lyase). OAs pass from the dark fermentation medium to the photofermentation, typically being concentrated by ~eightfold via electrodialysis for dilution into the photofermentation vessel. Alcohol is not removed by ED; this would require a catalytic oxidation stage to give the corresponding organic acid; this has been achieved via using Au(0) nanoparticle catalyst made on *E. coli* cells (Deplanche et al., 2007). Two bio-H_2 streams are formed from the combined dark- and photofermentations, with a third H_2 stream from electrolysis of water. The maximum H_2 yield from the mixed-acid fermentation is 2 mol sugar^{-1}; hence, the dark fermentation can be viewed as a generator of OAs rather than as the primary H supply. A schematic of upstream waste conversion into sugar feed is shown (see text), and downstream use of hydrogen in a fuel cell for electricity production. Note that bio-H_2 is free of catalyst poisons, which extends fuel cell life. Not all wastes (e.g. sugary fruits, bakery products) require extensive upstream treatment. The main box is the biotechnology; the grey flow sheet is the chemical engineering required to realize the positive energy balance. Both are equally important.

waste) exceeded that from anaerobic digestion, wind and solar power, even without factoring in the additional electrochemically made H_2. (Redwood *et al.*, 2012b). Although ~half of the organic acid is available (anionic) at the pH of the fermentation (according to the pK_a values: Table 1), the electrodialysis chamber itself is alkaline due to OH^- release.

Two key findings are salient. First, the role of the dark fermentation is more important as a supply of organic acids into the photofermentation than for its bio-H_2 *per se*. Second, recent work (R.L. Orozco and A.J. Stephen, unpublished) showed that the H_2 yield in the photofermentation was largely independent of the actual organic acid proportions in the feed from the mixed-acid fermentation and was optimal at ~40 mM organic acids. Hence, any source of organic acids could be potentially used from a dual system or, indeed, in a stand-alone photofermentation.

Bacterial photofermentation

Purple non-sulfur photosynthetic bacteria produce H_2 from a variety of organic substrates including organic acids (Lazaro *et al.*, 2012), sugars (Keskin and

Table 2. Some approaches to increase photofermentation H productivity (Reviewed by Adessi *et al.*, 2017).

Approach/Rationale	Outcomes/comments	References
'Black box' mathematical relationships between input and output streams Box-behnken statistical design/methods	Permits multivariable analysis: measures cause and effect; hence can be empirical SCE (glycerol) > doubled (*R.palustris*)	Abo-Hashesh *et al.* (2013), Show and Lee (2013) and Ghosh *et al.* (2012a,b,c)
Modelling metabolic fluxes	Guided interventions: success using lactate but not malate or acetate	Golomysova *et al.* (2010) and Hädicke *et al.* (2011)
Deletion of polyhydroxybutyrate synthesis pathway	Increased H_2 yield (by 1.5-fold c.f. wild type)	Kim *et al.* (2011)
Reducing pigment concentration	Allows greater light penetration[a]	Ma *et al.* (2012)
Use of quantum dots to 'upgrade' light	Doubled photosynthetic efficiency	M.D. Redwood, unpublished[b]

SCE, substrate conversion efficiency.
a. 27% increase in H_2 yield was obtained.
b. Collaborative study with Photon Science Institute, University of Manchester: M.D. Redwood, L.E. Macaskie and D.J. Binks, unpublished work. But note: current commercial quantum dots would be grossly uneconomic at scale.

Table 3. Options for delivery of bio-H_2 into power, all via electro-photofermentation (Figs 1 and 2; M.D. Redwood, R.L.Orozco and L.E. Macaskie, unpublished work)[a].

Feedstock (upstream)	Power (downstream)	Comments
Fermentation of food wastes	Fuel cell electricity[b] or combined heat and power[c]	Food wastes (FW) required (tonnages). Anaerobic digestion (AD) has monopoly on FW. Bio-H_2 can power a fuel cell directly.
Fermentation of cellulosic wastes	Fuel cell electricity or CHP	Comminution/maceration energy demand adversely affects overall energy balance[e]. Upstream hydrolysis is required.
OAs obtained from anaerobic digestion (AD)	'Hythane': mix of CH_4 (AD) + bio-H_2; CHP	AD interrupted at acetogenesis stage; organic acids diverted into a bolt-on photofermentation. Overall AD residence time is reduced. This increases process complexity but gives a higher energy output. Gas is compatible with current infrastructure. Scenario 1: 20% more power[d]. Scenario 2: 70% more power[d]
OAs used directly from wastes (e.g. wastewaters) or CHP	Fuel cell electricity	Organic acid waste streams (tonnage scale) are (e.g.) vinasse (from bioethanol production) and municipal wastewater treatment plants (see text).

a. Calculations were made independently of incentivization schemes as these tend to be ephemeral and skew the longer term picture. Likewise, increasing/decreasing feed-in tariffs would complicate economic assessments.
b. Fuel cell technology is still emergent at large scale, and FCs fail prematurely (see Rabis *et al.*, 2012).
c. Combined heat and power (CHP: well-established technology). In this scenario, the methane stream from anaerobic digestion can be supplemented with photofermentatively derived H_2 to make 'hythane' for CHP.
d. Scenario 1: diversion of 10% of the organic acids into photofermentation and use of hythane in CHP. Scenario 2: diversion of 80% of the organic acids into photofermentation and use of AD-methane in CHP plus use of the photofermentation H_2 in a fuel cell would give 70% more power (R.L. Orozco, unpublished). The proportion of flow diverted from the acetogenesis step of anaerobic digestion (via electroseparation) could be simply ramped in response to incident light intensity to feed the photofermentation; at night the flow would pass to the methanogenic reactor as normal. By combining the two processes, the residence time in the system would also be reduced as compared to traditional anaerobic digestion due to reduced flow entering the methanogenesis reactor daily.
e. Using *Miscanthus* as an example, the energy demand of comminution to 4 mm particles is 184 kJ kg dry matter^{-1}; energy from H is 10 kJ l^{-1} (at 1 atm and 125°C); that from the dark fermentation was only 110 kJ kg cellulose; hydrolysate; hence the PF (~4 times the H_2 as the dark fermentation) is key to a positive energy balance from complex substrates.

Hallenbeck, 2012) and industrial and agricultural efflu-ents (Saratale *et al.*, 2013), with high H$_2$ yields from acetic, butyric and lactic acids (Hallenbeck, 2013). Bac-teria used include *Rhodobacter sphaeroides* (Han *et al.*, 2013), *R. rubrum* (Zürrer and Bachofen, 1979), *R. palus-tris* (Oh *et al.*, 2004; Xiaobing, 2012) and *R. capsulatus* (Zhang *et al.*, 2016); despite some differences, they all follow a similar general scheme (Fig. 1B), metabolizing organic acids to reduce NAD$^+$ to the cellular reductant NADH (Oh *et al.*, 2013). Excess reductant must be dissi-pated to reoxidize NADH and maintain cellular redox bal-ance. This is achieved via cellular growth, channelling of carbon into cellular reserves (synthesis of polyhydroxy-butyrate) or via H$_2$ production under nitrogen-deficient conditions, via nitrogenase, which produces H$_2$ as an electron sink for excess reducing power (as with cyanobacteria: above). Nitrogenase normally fixes N$_2$ into NH$_3$ under light (to supply the large energy demand

of N-fixation, via ATP). Without N, the enzyme uses the reductant and ATP to produce H$_2$ (2H$^+$ + 2e^- + 4 ATP $\Rightarrow$ H$_2$ + 4ADP + Pi). NADH is not a sufficiently strong reductant for this reaction; it is 'upgraded' to the stronger reductant ferredoxin via the input of energy, which is supplied by light through the action of the photosynthetic apparatus, via reverse electron transport. This apparatus also produces the ATP required for nitrogenase action (Hallenbeck, 2011). Various papers have studied the role of light (e.g. Uyar *et al.*, 2007; Nath, 2009), showing that optimum light conversion efficiency occurs at light inten-sities much lower than light saturation points; e.g. Uyar *et al.* (2007) showed light saturation for *R. sphaeroides* at 270 W m^{-2} but similar substrate conversion efficiency could be achieved at light intensities as low 88 W m^{-2}. Furthermore, optimum light intensities can be species specific; e.g. *R. sphaeroides* and *R. palustris* under simi-lar conditions (Light intensity = 2500 Lux) had substrate

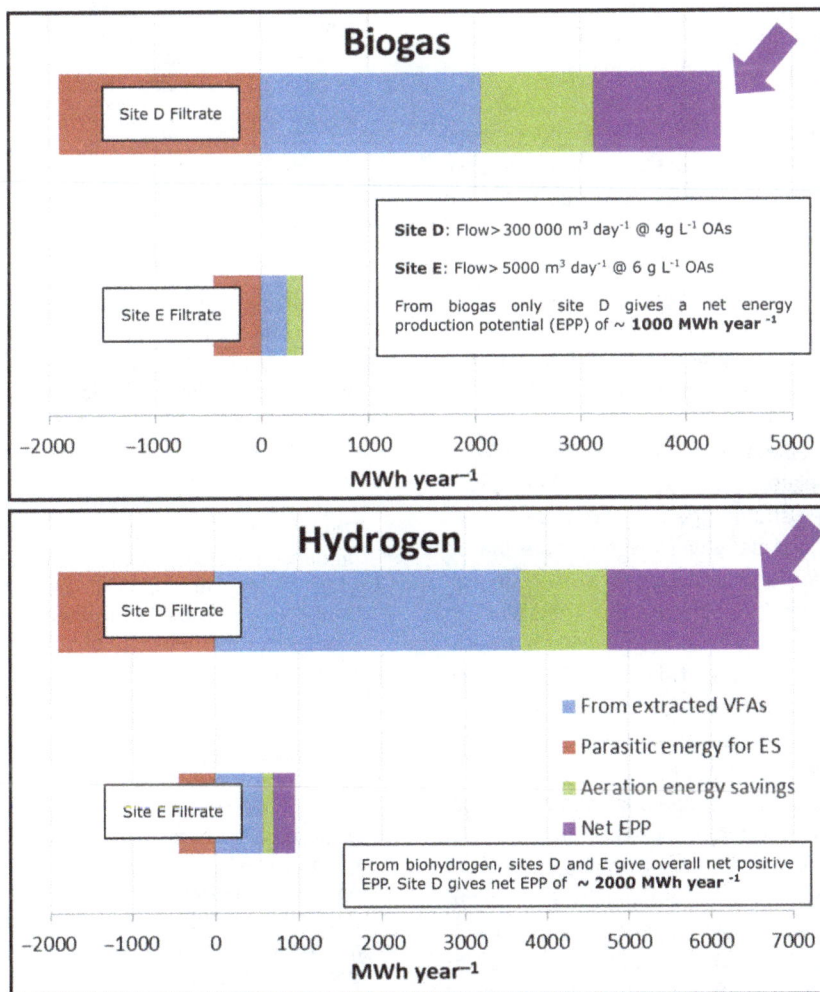

Fig. 3. Energy production potential (EPP) from use of real wastewater organic acids in a stand-alone photofermentation (real test data using *R. sphaeroides*: R.L.Orozco, I. Mikheenko and L.E. Macaskie, unpublished). As an organic acids liquid stream is used directly, the upstream dark fermentation is not required, and there is no sacrificial energy demand for maceration.

conversion efficiencies of 60–70% and 47% respectively (Han et al., 2013; Oh et al., 2013).

Hallenbeck and Liu (2015) reviewed advances in the field, highlighting various approaches to improve substrate conversion efficiency (Table 2), while recent publications provide an up-to-date overview of recent developments for photobiological biohydrogen technologies (Adessi et al., 2017; Hallenbeck, 2017).

Towards an economically competitive biohydrogen process from waste

Table 3 summarizes various options for a biohydrogen process. In the UK, food wastes at scale are generally centralized and 'committed' by agreements into anaerobic digestion and a 'bolt-on' addition into existing anaerobic digestion and combined heat and power (CHP) processes is one option as there is insufficient waste available for a realistic stand-alone bio-H_2 process (unpublished survey; Sustainable Resource Solutions Ltd). Agricultural wastes are currently unattractive due to high energy demands of comminution/maceration and upstream hydrolysis. A survey of wastes has indicated that vinasse (from bioethanol production) and in-process streams from UK Utility companies contain sufficient organic acids to warrant trialling for data into a full life cycle analysis.

The organic acid content of a typical vinasse waste is > 40 g l^{-1} (Ryznar-Luty et al., 2008; Esapaña-Gamboa et al., 2012); the high concentration of betaine (trimethylglycine, a zwitterionic osmoprotectant; 20 g l^{-1}) is not potentially problematic because at the low pH of vinasse (pH 3–4), it would be protonated (i.e. inaccessible to the anion transfer in electroseparation). Moreover, betaine was reported to stimulate nitrogenase activity, but it was not used as a nitrogen source (Igeňo et al., 1995).

Selected UK utility company wastewaters were trialled as potential targets for hydrogen bioenergy following filtration to remove debris but with no other modifications (Fig. 3). The energy production potential from biohydrogen via photofermentation was twice that from biogas (Fig. 3). Hence, H_2 energy from organic acid wastes is a viable option for energy production by heavily populated, industrialized countries but may be limited seasonally by available natural sunlight. Stand-alone photobiological hydrogen production has major potential in solar-rich countries with the option to also treat wastes in areas of high population density. An environmental life cycle analysis has been developed for cyanobacterial bio-H_2 (Archer et al., 2017). The next step is to apply a similar LCA for various options with respect to geographical location, other socio-economic factors and the global increase in demand for cooling to safeguard food supplies for expanding populations.

Acknowledgements

AJS and SAA acknowledge with thanks studentships from the EPSRC Doctoral Training Centre 'Fuel Cells and their Fuels'. The work was supported in part by NERC (grant NE/L014076/1) to LEM.

Conflict of Interest

None declared.

References

Abo-Hashesh, M., Desaunay, N., and Hallenbeck, P.C. (2013) High yield single stage conversion of glucose to hydrogen by photofermentation with continuous cultures of Rhodobacter capsulatus JP91. Biores Technol 128: 513–517.

Adessi, A., Corneli, E., and Philippi, S. (2017) Photosynthetic purple nonsulfur bacteria in hydrogen production systems: new approaches in the use of well known and innovative substrates. In Modern Topics in the Phototrophic Prokaryotes. Hallenbeck, P.C. (ed.). Cham, Switzerland: Springer, pp. 321–350.

Anon (2017) URL http://www.hydrogeneurope.eu/wp-content/uploads/2017/01/170113-Hydrogen-Council-International-Press-Releases.pdf.

Archer, S.A., Murphy, R.J. and Steinberger-Wilckens, R.. (2017). Systematic review and life cycle analysis of biomass derived fuels for solid oxide fuel cells. In Proceedings: Fuel Cell and Hydrogen Technical Conference 2017. 31st May – 1st June 2017. Birmingham, UK.

Balachandar, G., Khanna, N., and Das, D. (2013) Biohydrogen production from organic wastes by dark fermentation. In Biohydrogen. Pandey, A., Chang, J.-S., Hallenbeck, P.C., and Larroche, C. (eds). Burlington, MA, USA: Elsevier, pp. 103–144.

Chang, A.C.C., Chang, H.-F., Lin, F.-J., Lin, K.-H., and Chen, C.-H. (2011) Biomass gasification for hydrogen production. Int J Hydrogen Energy 36: 14252–14260.

Deplanche, K., Attard, G.A., and Macaskie, L.E. (2007) Biorecovery of gold from jewellery waste by Escherichia coli and biomanufacture of active Au-nanomaterial. Adv Mater Res 20–21: 647–652.

Dincer, I., and Acar, C. (2015) Review and evaluation of hydrogen production methods for better sustainability. Int J Hydrogen Energy 40: 11094–11111.

DOE (2013). Report of the hydrogen production expert panel: a subcommittee of the hydrogen and fuel cell technical advisory committee. URL https://www.hydrogen.energy.gov/pdfs/hpep_report_2013.pdf.

Eroğlu, E., Gündüz, U., Yücel, M., Türker, L., and Eroğlu, I. (2004) Photobiological hydrogen production by using olive mill wastewater as a sole substrate source. Int J Hydrogen Energy 29: 163–171.

Esapaña-Gamboa, E., Mijangos-Cortēs, J.O., Hernández-Zárate, G., Maldonado, J.A.D., and Alzate-Gaviria, L.M. (2012) Methane production from hydrous ethanol using a modified UASB. Biotechnol Biofuels 5: 82–92.

Ghosh, D., Sobro, I.F., and Hallenbeck, P.C. (2012a) Optimization of the hydrogen yield from single-stage photofermentation of glucose by *Rhodobacter capsulatus* JP91 using response surface methodology. *Biores Technol* **123:** 199–206.

Ghosh, D., Sobro, I.F., and Hallenbeck, P.C. (2012b) Stoichiometric conversion of biodiesel derived crude glycerol to hydrogen: response surface methodology study of the effects of light intensity and crude glycerol and glutamate concentration. *Biores Technol* **106:** 154–160.

Ghosh, D., Tourigny, A., and Hallenbeck, P.C. (2012c) Near stoichiometric reforming of biodiesel derived crude glycerol to hydrogen by photofermentation. *Int J Hydrogen Energy* **37:** 2273–2277.

Golomysova, A., Gomelsky, M., and Ivanov, P.S. (2010) Flux balance analysis of photoheterotrophic growth of purple nonsulfur bacteria relevant to biohydrogen production. *Int J Hydrogen Energy* **35:** 12751–12760.

Guo, X.M., Trably, E., Latrille, E., Carrère, H. and Steyer, J.P. (2010) Hydrogen production from agricultural waste by dark fermentation: a review. *Int J Hydrogen Energy* **35,** 10660–10673.

Hädicke, O., Grammel, H., and Klamt, S. (2011) Metabolic network modeling of redox balancing and biohydrogen production in purple nonsulfur bacteria. *BMC Syst Biol* **5:** 150.

Hallenbeck, P.C. (2011) Microbial paths to renewable hydrogen production. *Biofuels* **2:** 285–302.

Hallenbeck, P.C. (2012) Fundamentals of dark hydrogen production: multiple pathways and enzymes. In *State of the Art and Progress in Biohydrogen*. Azbar, N., and Levin, D. (eds). Sharjah, United Arab Emirates: Bentham Science Publishers, pp. 94–111.

Hallenbeck, P.C. (2013) Photofermentative biohydrogen production. In *Biohydrogen*. Pandey, A., Chang, J.-S., Hallenbeck, P.C., and Larroche, C. (eds). Burlington, MA, USA: Elsevier, pp. 145–159.

Hallenbeck, P.C. (ed) (2017) *Modern Topics in the Phototrophic Prokaryotes*. Cham, Switzerland: Springer.

Hallenbeck, P.C., and Ghosh, D. (2009) Advances in fermentative biohydrogen production: the way forward? *Trends Biotechnol* **27:** 287–297.

Hallenbeck, P.C., and Liu, Y. (2015) Recent advances in hydrogen production by photosynthetic bacteria. *Int J Hydrogen Energy* **41:** 4446–4454.

Han, H., Jia, Q., Liu, B., Yang, H., and Shen, J. (2013) Fermentative hydrogen production from acetate using *Rhodobacter sphaeroides* RV. *Int J Hydrogen Energy* **38:** 10773–10778.

Igeño, M.I., Del Moral, C.G., Castillo, F., and Caballero, J. (1995) Halotolerenace of the phototrophic bacterium *Rhodobacter capsulatus* E1F1 is dependent on the nitrogen source. *Appl Environ Microbiol* **61:** 2970–2975.

James, B.D., Baum, G.N., Perez, J. and Baum, K.N. (2009) *Technoeconomic boundary analysis of biological pathways to hydrogen production*. NREL Technical Monitor: Ali Jalalzadeh-Azar, A. Subcontract No. AFH-8-88601-01. Subcontract Report. NREL/SR-560-46674. Golden, CO, USA: NREL, pp 1–207.

Keskin, T., and Hallenbeck, P.C. (2012) Hydrogen production from sugar industry wastes using single-stage photofermentation. *Biores Technol* **112:** 131–136.

Kim, M.S., Kim, D.H., Son, H.N., Ten, L.N., and Lee, J.K. (2011) Enhancing photo-fermentative hydrogen production by *Rhodobacter sphaeroides* KD131 and its PHB synthase deleted-mutant from acetate and butyrate. *Int J Hydrogen Energy* **36:** 13964–13971.

Kubas, A., Orain, C., De Sancho, D., Saujet, L., Sensi, M., Gauquelin, C., *et al.* (2017) Mechanism of O₂ diffusion and reduction in FeFe hydrogenases. *Nat Chem* **9:** 88–95.

Lazaro, C.Z., Vich, D.V., Hirasawa, J.S., and Varesche, M.B.A. (2012) Hydrogen production and consumption of organic acids by a phototrophic microbial consortium. *Int J Hydrogen Energy* **37:** 11691–11700.

Ma, C., Wang, X., Guo, L., Wu, X., and Yang, H. (2012) Enhanced photo-fermentative hydrogen production by *Rhodobacter capsulatus* with pigment content manipulation. *Biores Technol* **118:** 490–495.

Mohan, S.V., and Pandey, A. (2013) Biohydrogen production: an introduction. In *Biohydrogen*. Pandey, A., Chang, J.-S., Hallenbeck, P.C., and Larroche, C. (eds). Burlington, MA, USA: Elsevier, pp. 1–24.

Nakada, E., Nishikata, S., Asada, Y., and Miyake, J. (1999) Photosynthetic bacterial hydrogen production combined with a fuel cell. *Int J Hydrogen Energy* **24:** 1053–1057.

Nath, K. (2009) Effect of light intensity and initial pH during hydrogen production by an integrated dark and photofermentation process. *Int J Hydrogen Energy* **34:** 7497–7501.

Nikolaidis, P., and Poullikkas, A. (2017) A comparative overview of hydrogen production processes. *Renew Sustain Energy Rev* **67:** 597–611.

Oh, Y.-K., Seol, E.-H., Kim, M.-S., and Park, S. (2004) Photoproduction of hydrogen from acetate by a chemoheterotrophic bacterium *Rhodopseudomonas palustris* P4. *Int J Hydrogen Energy* **29:** 1115–1121.

Oh, Y.K., Raj, S.M., Jung, G.Y., and Park, S. (2013) Metabolic engineering of microorganisms for biohydrogen production. In *Biohydrogen*. Pandey, A., Chang, J.-S., Hallenbeck, P.C., and Larroche, C. (eds). Burlington, MA, USA: Elsevier, pp. 45–65.

Rabis, A., Rodriguez, P., and Schmidt, J.J. (2012) Electrocatalysis for polymer electrolyte fuel cells. Recent achievements and future challenges. *ACS Catal* **2:** 864–890.

Redwood, M.D.. (2007) Bio-hydrogen production and biomass-supported palladium catalyst for energy production and waste minimization. PhD Thesis. Burlington, MA, UK: University of Birmingham.

Redwood, M.D., Orozco, R.L., Majewski, A.J., and Macaskie, L.E. (2012a) Electro-extractive fermentation for efficient biohydrogen production. *Biores Technol* **107:** 166–174.

Redwood, M.D., Orozco, R.L., Majewski, A.J., and Macaskie, L.E. (2012b) An integrated biohydrogen refinery: synergy of photofermentation, extractive fermentation and hydrothermal hydrolysis of food wastes. *Biores Technol* **119:** 384–392.

Ryznar-Luty, A., Krzywonos, M., Obis, E., and Miskiewicz, T. (2008) Aerobic biodegradation of vinasse by a mixed culture of bacteria of the genus *Bacillus*. Optimization of temperature, pH and oxygenation state. *Pol J Environ Stud* **17:** 101–112.

Saratale, G.D., Saratale, R.G., and Chang, J.-S. (2013) Biohydrogen from renewable resources. In *Biohydrogen*.

Pandey, A., Chang, J.-S., Hallenbeck, P.C., and Larroche, C. (eds). Burlington, MA, USA: Elsevier, pp. 185–221.

Show, K.Y., and Lee, D.J. (2013) Biioreactor and bioprocess design for biohydrogen production. In *Biohydrogen*. Pandey, A., Chang, J.-S., Hallenbeck, P.C., and Larroche, C. (eds.). Burlington, MA, USA: Elsevier, pp. 317–337.

Singh, J.S., Kumar, A., Rai, A.N., and Singh, D.P. (2016) Cyanobacteria: a precious bio-resource in agriculture, ecosystem and environmental sustainability. *Front Microbiol* **7:** 1–19. article 529

Strahan, D.. (ed.) (2017) *Clean cold and the global goals*. URL www.birmingham.ac.uk/Documents/college-eps/energy/Publications/Clean-Cold-and-the-Global-Goals.pdf.

Tiwari, A., and Pandey, A. (2012) Cyanobacterial hydrogen- a step towards clean environment. *Int J Hydrogen Energy* **37:** 139–150.

Uyar, B., Eroglu, I., Yücel, M., Gündüz, U., and Türker, L. (2007) Effect of light intensity, wavelength and illumination protocol on hydrogen production in photobioreactors. *Int J Hydrogen Energy* **32:** 4670–4677.

Xiaobing, W. (2012) Enhanced photo-fermentative hydrogen production from different organic substrate using hupL inactivated *Rhodopseudomonas palustris*. *Afr J Microbiol Res* **6:** 5362–5370.

Zhang, Y., Yang, H., and Guo, L. (2016) Enhancing photo-fermentative hydrogen production performance of *Rhodobacter capsulatus* by disrupting methylmalonate-semialdehyde dehydrogenase gene. *Int J Hydrogen Energy* **41:** 190–197.

Zürrer, H., and Bachofen, R. (1979) Hydrogen production by the photosynthetic bacterium *Rhodospirillum rubrum*. *Appl Environ Microbiol* **37:** 789–793.

4

Mycobacterium smegmatis is a suitable cell factory for the production of steroidic synthons

Beatriz Galán,[1,#] Iria Uhía,[1,2,#] Esther
García-Fernández,[1] Igor Martínez,[1] Esther Bahíllo,[3]
Juan L. de la Fuente,[3] José L. Barredo,[3] Lorena
Fernández-Cabezón[1] and José L. García[1,*]

[1]Department of Environmental Biology, Centro de
Investigaciones Biológicas, Consejo Superior de
Investigaciones Científicas, Ramiro de Maeztu 9, 28040
Madrid, Spain.
[2]MRC Centre for Molecular Bacteriology and Infection,
Department of Medicine, Imperial College London,
London SW7 2AZ, UK.
[3]Department of Biotechnology, Gadea Biopharma,
Parque Tecnológico de León, Nicostrato Vela s/n, 24009
León, Spain.

Summary

A number of pharmaceutical steroid synthons are currently produced through the microbial side-chain
cleavage of natural sterols as an alternative to multi-
step chemical synthesis. Industrially, these synthons
have been usually produced through fermentative
processes using environmental isolated microorganisms or their conventional mutants. Mycobacterium
smegmatis mc^{2}155 is a model organism for tuberculosis studies which uses cholesterol as the sole carbon
and energy source for growth, as other mycobacterial
strains. Nevertheless, this property has not been
exploited for the industrial production of steroidic
synthons. Taking advantage of our knowledge on the
cholesterol degradation pathway of M. smegmatis
mc^{2}155 we have demonstrated that the MSMEG_6039
(kshB1) and MSMEG_5941 (kstD1) genes encoding a
reductase component of the 3-ketosteroid 9α-hydro-
xylase (KshAB) and a ketosteroid Δ^1-dehydrogenase
(KstD), respectively, are indispensable enzymes for
the central metabolism of cholesterol. Therefore, we
have constructed a MSMEG_6039 (kshB1) gene deletion mutant of M. smegmatis MS6039 that transforms
efficiently natural sterols (e.g. cholesterol and phytosterols) into 1,4-androstadiene-3,17-dione. In addition,
we have demonstrated that a double deletion mutant
M. smegmatis MS6039-5941 [ΔMSMEG_6039 (ΔkshB1)
and ΔMSMEG_5941 (ΔkstD1)] transforms natural sterols into 4-androstene-3,17-dione with high yields.
These findings suggest that the catabolism of cholesterol in M. smegmatis mc^{2}155 is easy to handle and
equally efficient for sterol transformation than other
industrial strains, paving the way for valuating this
strain as a suitable industrial cell factory to develop à
la carte metabolic engineering strategies for the
industrial production of pharmaceutical steroids.

Introduction

Androstenedione (4-androstene-3,17-dione; AD) and
androstadienedione (1,4-androstadiene-3,17-dione; ADD)
are key intermediates of microbial steroid metabolism
(García et al., 2012). These compounds belong to the
17-keto steroid family and are used as the starting materials for the preparation of different pharmaceutical steroids (Malaviya and Gomes, 2008; Donova and Egorova,
2012; García et al., 2012). These synthons can be produced by microbial side-chain cleavage of cholesterol or
phytosterols as an alternative to multi-step chemical synthesis based on digoxigenin, a steroid found exclusively
in the flowers and leaves of Digitalis plants, as a starting
material. Industrially, AD and ADD have been produced
through fermentative processes using wild microorganisms that have been subsequently modified and optimized by conventional mutagenic procedures (Donova
et al., 2005; Andor et al., 2006; Donova and Egorova,
2012; García et al., 2012). Mycobacterium spp. NRRL
3805B and 3683B capable of forming AD and ADD from
sterols, respectively, are examples of these mutants
used at industrial scale (Donova et al., 2005; Donova
and Egorova, 2012). A drawback of the AD and ADD
industrial production based on these wild strains is the
usual and concomitant accumulation of unwanted by-
products which hinder downstream processes. The use
of molecularly defined mutants has been envisioned to
avoid such drawbacks, but the lack of genetic data on
the microbial catabolism of steroids has hampered the

*For correspondence. E-mail: jlgarcia@cib.csic.es

#These authors contributed equally to this work.
Funding Information
The technical work of A. Valencia is greatly appreciated. We thank I.
Calvillo for assistance in the analytical GC/MS experiments. This
study was supported by grants from the Ministry of Science and
Innovation (BFU2006-15214-C03-01, BFU2009-11545-C03-03) and
Ministry of Economy and Competitiveness (BIO2012-39695-C02-01).

construction of genetically engineered mutants so far. Nevertheless, some attempts have been conducted to construct site-directed mutant strains of *Rhodococcus* (i.e. the best characterized cholesterol-degrading organism) to produce AD, ADD and 9α-hydroxy-4-androstene-3,17-dione (9OH-AD) from natural sterols (e.g. cholesterol or phytosterols), but these mutants have not been used at industrial scale yet (van der Geize *et al.*, 2000, 2001a,b, 2002a,b, 2008; Wilbrink *et al.*, 2011; Yeh *et al.*, 2014).

To create *à la carte* mutants able to produce AD and ADD from cholesterol or phytosterols we have tested *Mycobacterium smegmatis* mc²155 as a model strain

based on our current knowledge on sterol catabolism in this microorganism (Fig. 1) (García *et al.*, 2012). The 3-ketosteroid 9α-hydroxylase (Ksh) has been proposed as the key enzyme for ring-B opening, and therefore, the removal of this activity should render ADD as end-product. The Ksh enzymes of *Rhodococcus* and *Mycobacterium* have been characterized as two-component monooxygenases, composed of an oxygenase (KshA) and a ferredoxin reductase (KshB) (van der Geize *et al.*, 2002a; Capyk *et al.*, 2009, 2011; Petrusma *et al.*, 2009, 2012, 2014; Hu *et al.*, 2010; Bragin *et al.*, 2013; Penfield *et al.*, 2014). Therefore, theoretically, a deletion mutant of one of the Ksh encoding genes, i.e. *kshA* or *kshB*,

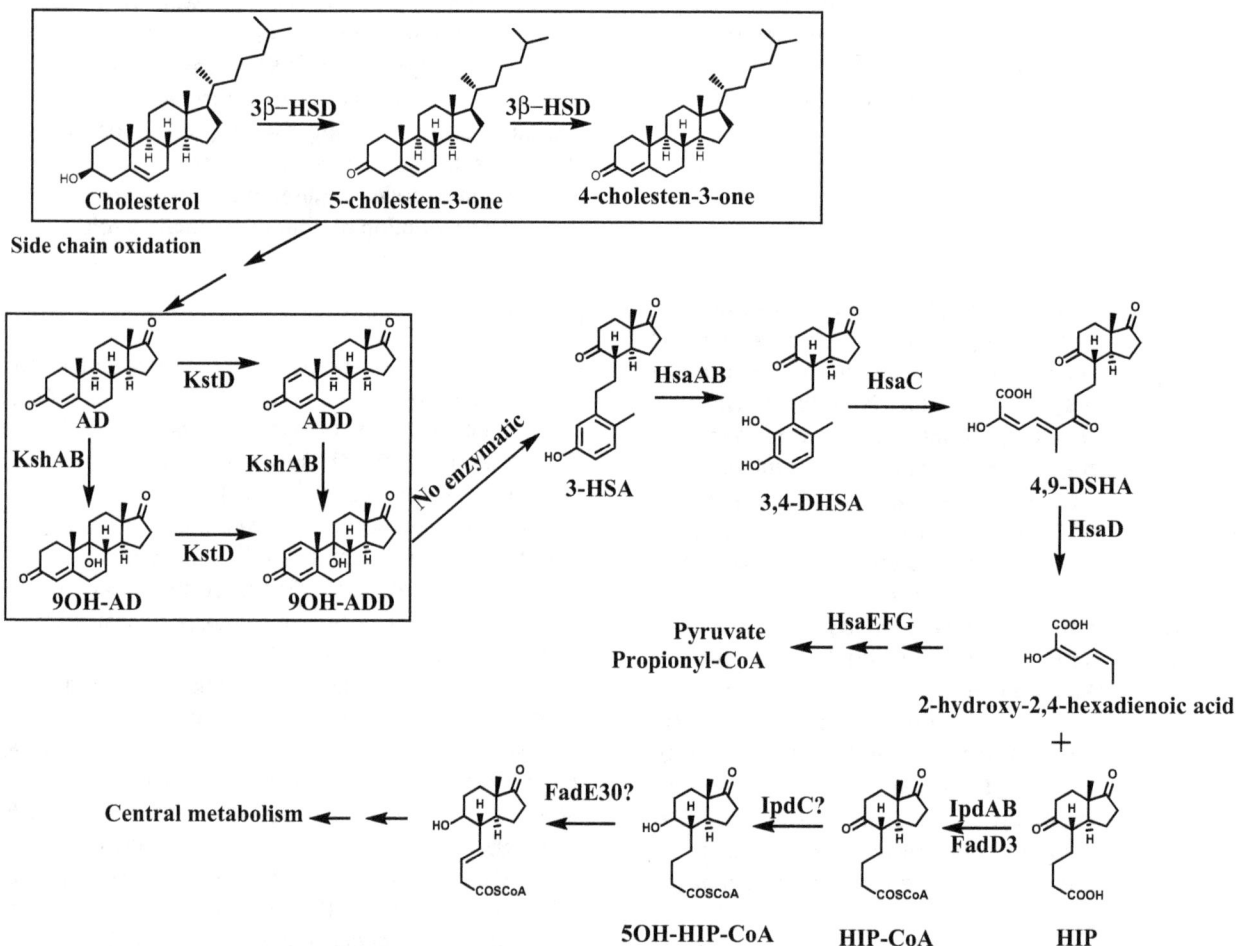

Fig. 1. Proposed pathway for cholesterol degradation in *Mycobacterium smegmatis*. Cholest-4-en-3-one or any of the subsequent metabolites from degradation of the side-chain up to (and including) AD may undergo a dehydrogenation reaction to introduce a double bond in the position 1, leading to compound cholest-1,4-diene-3-one in the case of cholest-4-en-3-one, or to the corresponding 1,2-dehydro derivatives for other molecules. The side-chain degradation of this compounds will be identical to that of the cholest-4-en-3-one to the common intermediate 9α-hydroxyandrosta-1,4-diene-3,17-dione (9OHAD). 3β-hydroxysteroid dehydrogenase (3β-HSD), 3-ketosteroid-Δ¹-dehydrogenase (KstD), 3-ketosteroid-9α-hydroxylase (KshAB), 3-hydroxy-9,10-secoandrosta-1,3,5(10)-trien-9,17-dione (3-HSA), 3,4-dihydroxy-9,10-secoandrosta-1,3,5(10)-trien-9,17-dione (3,4-HSA), 4,5,9,10-diseco-3-hydroxy-5,9,7-trioxoandrosta-1(10),2-diene-4-oic acid (4,9-DSHA), 3aα-H-4α(3′-propionic acid)-7aβ-methylhexahydro-1,5-indanedione (HIP), 3aα-H-4α(3′-propionic acid)-5aH-7β-methylhexahydro-1-indanone (5OH-HIP), 3-hydroxy-9,10-secoandrosten-1,3,5(10)-trien-9,17-dione dyoxigenase (HsaC), 4,5,9,10-diseco-3-hydroxy-5,9,7-trioxoandrosta-1(10),2-diene-4-oic acid hydroxylase (HsaD), 2-hydroxy-2,4-hexadienoic acid hydratase (HsaE), 4-hydroxy-2-hydroxy-2-ketovalerate aldolase (HsaF), acetaldehyde dehydrogenase (HsaG), HIP-CoA transferase (LpdAB and FadD3), FadE30 (Acyl-CoA dehydrogenase).

should accumulate ADD. On the other hand, 3-ketosteroid Δ^1-dehydrogenase (KstD) is the key enzyme that transforms AD into ADD and thus, strains lacking KstD should theoretically accumulate 9OH-AD. Finally, double mutants in Ksh and KstD should render AD as the main product for sterol degradation (Fig. 1) (García et al., 2012).

Nevertheless, these previous assumptions cannot be always easily confirmed by gene inactivation since cholesterol catabolic pathway usually presents functional redundancy for some steps of the pathway; this means that one catalytic step can be carried out by several homologous enzymes. Two KstD enzymes were reported in Rhodococcus erythropolis SQ1 (van der Geize et al., 2000) and accordingly, targeted disruption of only one of the kstD genes did not result on the accumulation of intermediates, but the strain lacking both dehydrogenase activities was able to convert AD into 9OH-AD stoichiometrically (van der Geize et al., 2002b). Similarly, Fernández de las Heras et al. (2012) have demonstrated the existence of three KstD activities in Rhodococcus ruber strain Chol-4 and showed that the triple KstD mutant was able to convert AD into 9OH-AD. However, these mutants were unable to accumulate 9OH-AD from cholesterol or phytosterols (van der Geize et al., 2001a; Fernández de las Heras et al., 2012). Brzostek et al. (2009) have identified six putative KstD enzymes within the M. smegmatis genome and a targeted disruption of one of them (KstD1) resulted in partial inactivation of the cholesterol degradation pathway and the consequent accumulation of AD. More recently, Wei et al. (2010) have identified and deleted a kstD gene (named $ksdD_M$) in Mycobacterium neoaurum NwIB-01, a strain isolated from a sterol contaminated soil that naturally accumulates ADD from phytosterols. Interestingly, the resulting deleted mutant (named NwIB-2) was able to accumulate AD from phytosterols, but ADD is still present in the culture medium, suggesting that this strain contains other KstD enzymes (Wei et al., 2010). Three homologues of KstD have been characterized in M. neoaurum ATCC 25795 and thus, single deletion of these genes failed to result in a stable and maximum accumulation of 9OH-AD due to residual KstD activities (Yao et al., 2014).

On the other hand, van der Geize et al. (2002a, 2008) have demonstrated that Ksh mutants of R. erythropolis SQ1 deleted in kshA or kshB genes can produce ADD from AD, but surprisingly they do not accumulate AD, ADD or other metabolites during sterol conversion. Interestingly, the degradation of phytosterols was not impaired in a kshA⁻ mutant and rates of degradation were comparable to those of the parent strain, suggesting that Rhodococcus has alternative enzymes to catabolize phytosterols. Moreover, the kshB⁻ mutant failed to

cleave the side-chain of sterols and, although phytosterols were oxidized to their stenone derivatives, they were not metabolized further, suggesting that kshB could be also involved in the degradation of sterol side-chain in Rhodococcus (van der Geize et al., 2002a). The situation can be more complex taking into account that some strains of Rhodococcus may contain up to five KshA homologous proteins, each displaying unique steroid induction patterns and substrate ranges, confirmed that the 9α-hydroxylation can take place at different steps of steroid oxidation (Petrusma et al., 2011).

On the basis of our knowledge on cholesterol metabolism in M. smegmatis (Uhía et al., 2012) we detected that this organism has several important differences with the sterol catabolism in Rhodococcus and we anticipated that these differences might render M. smegmatis as a suitable cell factory for the metabolic manipulation of this pathway. The results presented below demonstrate this hypothesis and show that M. smegmatis is an ideal cell factory to develop metabolic engineering strategies for the industrial production of AD, ADD or other steroid intermediates using natural sterols as feedstock.

Results

Identification of Ksh and KstD enzymes in M. smegmatis

In contrast to the frequent observation that the components of bacterial multicomponent oxygenases are encoded in a single operon, the kshA and kshB genes encoding Ksh activity in Rhodococcus (van der Geize et al., 2002a) or in M. tuberculosis (Cole et al., 1998) are located far from each other in the genome. Nevertheless, using the sequence of the annotated kshA and kshB genes from Rhodococcus (van der Geize et al., 2002a, 2008) and M. tuberculosis (Capyk et al., 2009) as probes, we have localized the corresponding orthologues in M. smegmatis (Table 1). This analysis revealed that there are at least two genes encoding putative oxygenase components, named kshA1 and kshA2, and two genes encoding putative reductase components, named kshB1 and kshB2. This means that M. smegmatis could produce theoretically four different Ksh hydroxylases, i.e. Ksh1 (KshA1B1), Ksh2 (KshA2B2), Ksh3 (KshA1B2), Ksh4 (KshA2B1). However, a detailed analysis of the upstream sequences of these genes showed that only the MSMEG_5925 (kshA1) and MSMEG_6039 (kshB1) genes are preceded by promoters inducible by cholesterol, having the consensus operator sequence for the binding of the KstR repressor, one of the regulators of cholesterol catabolism (named KstR regulon) (Kendall et al., 2007). Moreover, microarray expression experiments carried out on M. smegmatis indicate that kshA1 and kshB1 are induced 1.9-fold and 7.0-fold, respectively, in cholesterol

Table 1. *In silico* analysis of *M. smegmatis* mc[2] 155 genome.

Protein	Gene[a] *M. smegmatis* mc[2] 155	Protein (aa length)	Identity[b] (%)	KstR-regulon operator	Induction fold[c]
KshA	*MSMEG_5925*	383	67	Yes	7.00
	MSMEG_2870	386	53	No	1.00
KshB	*MSMEG_6039*	353	63	Yes	1.88
	MSMEG_2893	351	56	No	1.21
KstD	*MSMEG_5941*	566	65	Yes	13.00
	MSMEG_4864	587	39	No	-1.46
	MSMEG_2869	558	40	No	-1.17
	MSMEG_2867	522	31	No	-1.14
	MSMEG_4870	546	30	No	-1.35
	MSMEG_5835	547	17	No	1.14

a. Gene identifications correspond to the last annotation of *M. smegmatis* genome and are different to the annotations described by Brzostek *et al.* (2005).
b. Identities were established using the proteins of *R. jostii* RHA1 as reference.
c. Induction fold was calculated comparing gene expression in *M. smegmatis* cultured in cholesterol verus glycerol containing media (Uhía *et al.*, 2012).

with respect to glycerol (Table 1) (Uhía *et al.*, 2012). On the contrary, *kshA2* and *kshB2* are not induced by cholesterol under the tested conditions (Table 1) (Uhía *et al.*, 2012). These results strongly suggest that most probably only Ksh1 encoded by *kshA1B1* is involved in the metabolism of cholesterol in *M. smegmatis*.

Among the six putative *kstD* genes identified in the *M. smegmatis* genome by Brzostek *et al.* (2009) that might encode the KstD-like activity only the *MSMEG_5941* (*kstD1*) gene is controlled by a promoter containing the consensus KstR operator sequence and is differentially expressed in the presence of cholesterol (Table 1) (Uhía *et al.*, 2012). This observation suggests that in *M. smegmatis* most probably only *kstD1* was involved in the catabolism of cholesterol.

Therefore, based on these analyses we anticipated the hypothesis that in *M. smegmatis* a single deletion of *kshB1* (or *kshA1*) and a double deletion of *kshB1* (or *kshA1*) and *kstD1* might be sufficient to generate mutants able to accumulate ADD or AD, respectively, when cultured in the presence of sterols.

Construction of the ΔkshB1 M. smegmatis *mutant*

As mentioned above, to eliminate the two-component Ksh1 hydroxylase activity (KshA1B1) in *M. smegmatis*, we assumed that it was enough to delete the *MSMEG_6039* (*kshB1*) gene encoding the reductase component, thus we constructed the MS6039 mutant (*ΔkshB1*) (Fig. S1). Ksh1 activity can be eliminated either by deleting *kshB1* or *kshA1* (or both), but we decided to test the single deletion of *kshB1* instead of *kshA1*, because in the absence of the corresponding oxygenase component, the reductase component of multicomponent oxygenases usually acts as a gratuitous scavenger of reduction power, generating futile cycles that might reduce the efficiency of the biotransformation

process (Galán *et al.*, 2000; Blank *et al.*, 2010) and therefore, in this sense, it can be more convenient to suppress the reductase instead the oxygenase component. However, it is also true that other reductase enzymes present in the cell could replace their function. Nevertheless, we also decided to delete *kshB1* because the effect of the deletion of the *kshB* homologous gene has already been analysed in *Rhodococcus* (van der Geize *et al.*, 2002a,b) and we would like to compare the performance of *M. smegmatis* having the same gene deletion.

The engineered MS6039 mutant (*ΔkshB1*) was unable to efficiently grow in cholesterol or phytosterols as a sole carbon and energy source when compared with the wild-type strain (Fig. 2A). However, the mutant perfectly grows using glycerol as a carbon source (Fig. 2B). To confirm that the deletion of *kshB1* was responsible for the observed phenotype, we transformed the MS6039 mutant with the plasmid pMV6039 harbouring the *kshB1* gene. Figure 2D shows that the complemented MS6039 (pMV6039) strain recovered the capacity to grow in sterols as a sole carbon and energy source. These results confirmed that *kshB1* plays an essential role in the catabolism of sterols in *M. smegmatis*. In addition, considering that the MS6039 mutant is unable to efficiently grow in sterols even at long incubation periods, we can conclude that the KshB1 reductase activity of Ksh1 cannot be replaced by other mycobacterial reductases, i.e. KshB2 or other KshB-like reductases, in the tested conditions, either because they are not expressed in these conditions or because they cannot interact with the KshA1 oxygenase component.

These experiments also revealed that wild-type *M. smegmatis* is able to use efficiently different phytosterols as a carbon and energy source (Fig. 2C), and this is remarkable because the capacity of this strain to grow

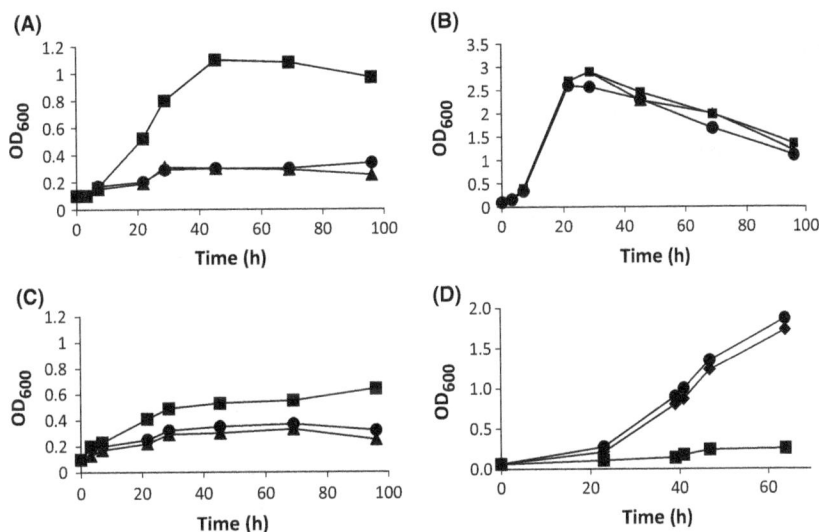

Fig 2. Growth curves of *M. smegmatis* mc²155, mutants and complemented strains cultured in shake flasks with different carbon sources. (A) Strains cultured with 0.4 g l⁻¹ of cholesterol (mc²155 (squares), MS6039 (circles) and MS6039-5941 (triangles)); (B) Strains cultured with 18 mM glycerol (mc²155 (squares), MS6039 (circles) and MS6039-5941 (triangles)); (C) Strains cultured with 0.4 g l⁻¹ of phytosterols (mc²155 (squares), MS6039 (circles) and MS6039-5941 (triangles)); (D) Gene complementation of MS6039 mutant strain cultured with 1.8 mM of cholesterol (mc²155 (pMV261) (diamonds), MS6039 (pMV261) (squares) and MS6039 (pMV6039) (circles)). The data reported are the averages of three different assays.

on a mixture of phytosterols has not been well documented in the literature so far.

On the other hand, the slight growth of MS6039 in the presence of sterols suggested that the mutant was able to degrade their side-chain and predicted a possible accumulation of metabolic intermediates in the culture medium. As expected, Figure 4 shows the accumulation of ADD in the culture medium of MS6039 mutant when cultured in shake flasks, whereas *M. smegmatis* wild-type strain completely mineralizes cholesterol (or phytosterols) and does not accumulate any intermediate in these culture conditions (data not shown). The ADD molar yield for the transformation of cholesterol or phytosterols was 91% for both feedstocks. Nevertheless, we have also detected the presence of small amounts of 22-hydroxy-23,24-bisnorchol-1,4-dien-3-one (1,4-HBC, 20-hydroxymethylpregna-1,4-dien-3-one) as a by-product (Fig. 4C and D).

In addition, we have tested the production of ADD from phytosterols in 5- and 20-l stirred jar bioreactors using higher concentrations of phystosterols (20 g l⁻¹) as substrate to study the industrial potential of MS6039 mutant strain. By carrying several replicates of these experiments in different operational conditions, we concluded that between 55% and 70% of added phytosterols are consumed and the molar yield to ADD varies from 67% to 80%, depending on the bioreactor scale. AD was also obtained in a yield from consumed phytosterols between 22% and 30%; other by-products, such as 1,4-HBC were detected as traces (Fig. S2). These

experiments carried out at high sterol concentrations showed a significant contamination of unconverted AD suggesting that KstD activity is a bottleneck for the complete transformation of phytosterols into ADD at industrial scale. This problem can be overcome by changing the operational conditions (e.g. preinoculation growth, inoculation conditions, bioreactor agitation configuration, aeration flow, composition of culture medium, pH control, substrate additions, etc.) in combination with the overexpression of *kstD* genes in MS6039 that lead to a significant increase in ADD/AD ratio (Gadea personal communication).

The results contrast with the data obtained with the equivalent *kshB⁻* mutant of *R. erythropolis* strain SQ1 (RG4 mutant), which only accumulates sitostenone from β-sitosterol, because apparently, the RG4 mutant does not degrade the side-chain of β-sitosterol. To explain this performance, it was proposed that the KshB reductase from *R. erythropolis* was not only involved in the production of 9OH-AD, but also in sterol side-chain removal (van der Geize *et al.*, 2002b). In *M. smegmatis*, the absence of KshB1 does not hinder the complete degradation of the sterol side-chain (see Discussion).

Construction of the ΔkshB1 and ΔkstD1 M. smegmatis double mutant

According to our previous genomic analysis, it should be possible to produce AD from sterols in *M. smegmatis* by eliminating the KstD activity in the MS6039 mutant.

Therefore, to test this hypothesis, we constructed the *M. smegmatis* MS6039-5941 double mutant (*ΔkstD1*, *ΔkshB1*) by deleting the *MSMEG_5941* (*kstD1*) gene in the MS6039 (*ΔkshB1*) mutant (Fig. S1). As expected, the MS6039-5941 (*ΔkshB1-ΔkstD1*) double mutant was unable to efficiently grow in cholesterol or phytosterols as the sole carbon and energy source, but the double mutant grows in glycerol at the same rate than the wild-type strain (Fig. 2B).

Accordingly to our predictions, when the MS6039-5941 double mutant was cultured in minimal media containing glycerol as carbon and energy source and cholesterol (or phytosterols) as feedstock, the culture yields a large accumulation of AD in shake flasks (Fig. 4A and B). The AD molar yields for sterol transformations were 90% and 84% when cholesterol or phytosterols were used as substrates respectively. Interestingly, in these culture conditions, small amounts of 22-hydroxy-23,24-bisnorchol-4-en-3-one (4-HBC, 20-hydroxymethylpregna-4-en-3-one) were accumulated during the biotransformation (Fig. 4C and D).

These results support the hypothesis that although *M. smegmatis* has six putative KstD dehydrogenases, the KstD1 enzyme encoded by the *MSMEG_5941* gene is the main KstD used for sterol catabolism in this bacterium. This observation contrasts with other cholesterol-degrading bacteria, where several KstD dehydrogenases are involved in the metabolism of sterols (van der Geize *et al.*, 2000, 2002b; Fernández de las Heras *et al.*, 2012; Yao *et al.*, 2014).

The MS6039-5941 double mutant was tested in 2-l jar bioreactor in the presence of 10 g l^{-1} of phytosterols. An almost complete transformation of phytosterols into AD (88–90%), 4-HBC (10–11%) and very small amounts of ADD and 1,4-HBC were detected in several replicates of this experiment (Fig. S3).

Interestingly, the accumulation of small amounts of ADD observed in the bioreactor experiments suggests that one of the other identified *kstD2*, *kstD3*, *kstD4*, *kstD5* or *kstD6* homologous genes (Table 1) is somewhat active in the mutant. Thus, to further improve the AD production yield, this residual KstD activity should be identified and eliminated. Moreover, in this sense, it will be also important to understand why the three last carbons at C-17 are not efficiently converted into propionyl-CoA, rendering 4-HBC and 1,4 HBC as by-products of the pathway.

Discussion

The first remarkable finding presented above is the experimental demonstration that *M. smegmatis* mc^{2}155 is able to efficiently metabolize mixtures of phytosterols as carbon and energy sources through the same catabolic pathway utilized to mineralize cholesterol (Fig. 1) (Uhía *et al.*, 2012). This finding is important because it supports the proposal of considering *M. smegmatis* as a useful cell factory for the industrial production of steroidic synthons from raw sterols. Surprisingly, in spite of the large accumulated knowledge on *M. smegmatis* mc^{2}155 biology, this organism has not been used as a cell factory for this industrial purpose so far. The only case reported in the literature is an antibiotic resistant mutant deposited as *M. smegmatis* VKPM Ac-1552 in the Russian National Collection of Industrial Microorganisms (VKPM) that appears to transform sterols into AD (Russian Federation Patent no. 2 126 837 (1999), reviewed in Donova (2007)).

The observation that the growth of *M. smegmatis* mc^{2}155 in sterols is impaired by a deletion of the *kshB1* gene (*MSMEG_6039*) suggests that KshB1 is the main reductase component of the two-component KshAB 9α-hydroxylase in the cholesterol degradative pathway of this organism. Apparently, this reductase activity is not redundant in *M. smegmatis* and thus, it cannot be replaced by similar enzymes, as it has been demonstrated for other key enzymes of the pathway (Uhía *et al.*, 2011). In addition, the *kshB1* gene does not appear to be critical for the degradation of the side-chain of sterols in *M. smegmatis*, as suggested in *Rhodococcus* (van der Geize *et al.*, 2002a), and therefore, its absence does not impair an efficient transformation of sterols into AD or ADD in the mutant strain.

On the other hand, the significant accumulation of AD in the MS6039-5941 (*ΔkshB1-ΔkstD1*) double mutant ascribes a fundamental role to the *kstD1* gene (*MSMEG_5941*) in the catabolism of cholesterol. Nevertheless, in this mutant, we have observed some enzyme redundancy because we were able to detected small amounts of ADD when high concentrations of phytosterols are transformed. Therefore, we can conclude that at least one of the other five putative KstD enzymes identified in *M. smegmatis* mc^{2}155 (Table 1) can also replace this function, but this alternative KstD activity is very low and does not appear to fulfil the relevant function observed in other cholesterol-degrading strains (van der Geize *et al.*, 2000, 2002b; Fernández de las Heras *et al.*, 2012; Yao *et al.*, 2014).

The results presented in this work suggest that the metabolism of sterols in *M. smegmatis* mc^{2}155 concerning to the two central/key enzymes investigated, i.e. KstD and Ksh, is less redundant than in *Rhodococcus*. Perhaps, the low redundancy of these central enzymes can explain why many strains currently used at industrial scale to transform phytosterols into AD, ADD or 9OH-AD are mycobacterial mutants obtained by conventional mutagenesis, i.e. because single mutations like those produced in this work might render a producer strain.

Genomic analyses of some of these industrial mutants have confirmed this hypothesis (data not shown).

Another interesting difference observed between the sterol catabolism of *Mycobacterium* and *Rhodococcus* is the presence of 4-HBC and 1,4-HBC alcohols as metabolic by-products in the cultures of our mutants (Figs 3C, D and 4C, D). It has been described that the cultures of equivalent *Rhodococcus* mutants only accumulate the corresponding acids (Wilbrink *et al.*, 2011; Yeh *et al.*, 2014). The RG32 mutant of *Rhodococcus rhodochrous* DSM43269, a mutant completely devoid of Ksh by inactivation of five *kshA* genes, was able to produce very small amounts of ADD from β-sitosterol (7% molar) but large amounts of 3-oxo-23,24-bisnorchola-1,4-dien-22-oic acid (1,4-BNC) (67% molar) (Wilbrink *et al.*, 2011). A *kshB⁻* mutant of *Rhodococcus equi* USA-18 was also able to produce ADD and 1,4-BNC from sterols in similar molar ratios (Yeh *et al.*, 2014).

In this regard, 4-HBC alcohol was detected many years ago together with AD during phytosterol transformations performed by the industrial strain of *Mycobacterium* sp. NRRL B-3805 (Marsheck *et al.*, 1972). This alcohol was postulated as a side reaction product but not as a physiological intermediate of the major pathway leading to production of C-17-ketonic products (Marsheck *et al.*, 1972). It has been proposed that 4-HBC and 1,4-HBC derive from 4-BNC and 1,4-BNC, respectively, by the action of a carboxyl-reductase by one or two consecutive enzymatic steps (Szentirmai, 1990; Xu *et al.*, 2016). Nevertheless, these carboxyl-reductases have not been identified in mycobacteria yet (Xu *et al.*, 2016). Assuming this hypothesis, these enzymes should not be active, or very low active, in *Rhodococcus*, since only the 4-BNC/1,4-BNC acids are detected in this organism (Yeh *et al.*, 2014).

The accumulation of these C-22 intermediates suggests that the elimination of the last isopropyl group of the sterol side-chain seems to be highly dependent of the 9α-hydroxylation by Ksh. We assume that in the absence of 9α-hydroxylation, the enzymes responsible for removing the last propionic acid of the side-chain at C-17, i.e. the postulated acyl-CoA-dehydrogenase, enoyl-CoA-hydratase and aldol-lyase enzymes (García *et al.*, 2012), do not work efficiently. This inefficiency can be caused because they are feedback inhibited by the accumulation of AD and/or ADD, or because the real

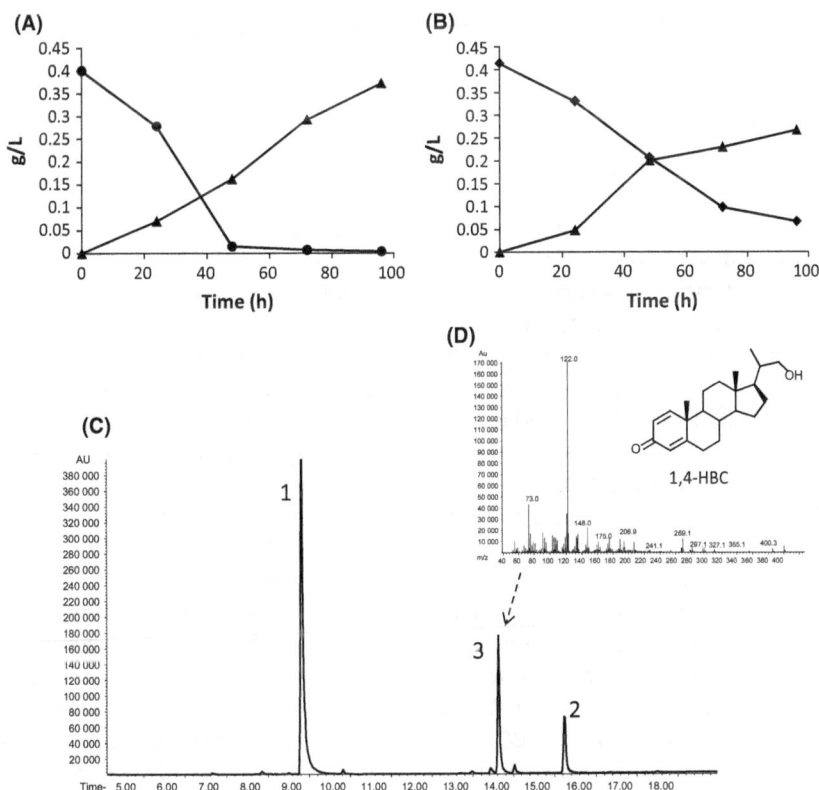

Fig 3. Production of ADD by the MS6039 mutant in shake flasks. ADD is represented by triangles. (A) 9 mM glycerol + 0.4 g l⁻¹ of cholesterol (circles) used as substrates; (B) 9 mM glycerol + 0.4 g l⁻¹ of phytosterols (diamonds) used as substrates; (C) Analysis by CG-MS of the products after 96 h of incubation on phytosterols. (1) ADD, (2) cholestenone (internal standard), and (3) 1,4-HBC. (D) Chemical structure and fragmentation pattern of 1,4-HBC. The data reported are the averages of three different assays.

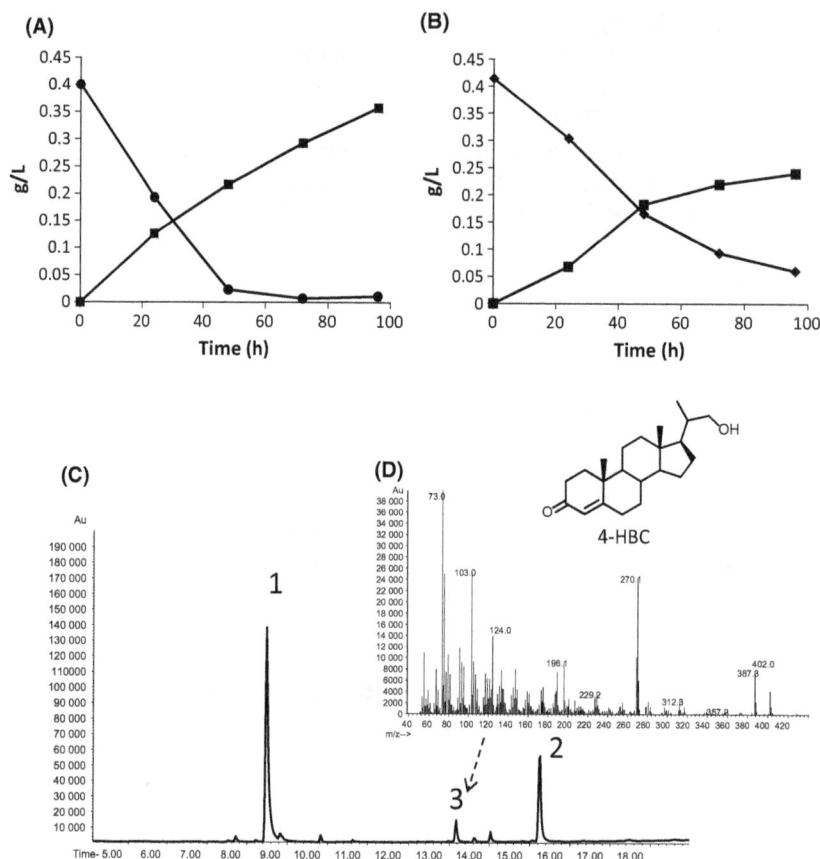

Fig 4. Production of AD by the MS6039-5941 mutant in shake flasks. AD is represented by squares. (A) 9 mM glycerol + 0.4 g l^{-1} of cholesterol (circles) used as substrates; (B) 9 mM glycerol + 0.4 g l^{-1} of phytosterols (diamonds) used as substrates; (C) Analysis by GC/MS of the products after 96 h of culture on phytosterols. (1) AD, (2) cholestenone (internal standard) and (3) 4-HBC. (D) Chemical structure and fragmentation pattern of 4-HBC. The data reported are the averages of three different assays.

substrates of these enzymes are the 9α-hydroxy derivatives of 1,4-BNC-CoA or 4-BNC-CoA. In this sense, it has been demonstrated that the best substrate of Ksh from *M. tuberculosis* is 4-BNC-CoA and 1,4-BNC-CoA (Capyk *et al.*, 2011), suggesting that 9α-hydroxylation occurs before the release of the last propionic acid at C-17. If the previous 9α-hydroxylation of the steroid is critical to be recognized as a substrate for some of the enzymes involved in the release of the side-chain, this might explain why in *Rhodococcus* the absence of KshB impairs the complete degradation of the sterol side-chain (van der Geize *et al.*, 2002a). Then, most probably the homologous enzymes of *Mycobacterium* are efficient enough on non-hydroxylated substrates and thus, the 9α-hydroxylation of sterols is not so critical to allow the complete side-chain degradation. In *M. neoaurum* ATCC 25795, the accumulation of 4-HBC and 1,4-HBC was only detected after the deletion of the *hsd4A* gene coding for a dual-functional enzyme with both 17β-hydroxysteroid dehydrogenase and β-hydroxyacyl-CoA dehydrogenase activities. However, these compounds are not the substrates of the Hsd4A *in vitro* (Xu *et al.*, 2016).

This means that some partial or complete blockage of the side-chain degradation might cause the accumulation of 4-HBC or 1,4-HBC as by-products. This blockage can be caused not only by specific mutations, but also by retro-inhibitions of the enzymes due to the accumulation of certain intermediates when the cells are cultured with high concentrations of sterols.

During the course of this work, Xu *et al.* (2015) have described a double (ΔkshA1, ΔkstD1) mutant of *Mycobacterium* sp. (apparently a derivative of *M. neoaurum* NwIB-02 (ΔkstD1) (Wei *et al.*, 2010) that accumulates AD, ADD, 1,4-HBC and 4-HBC at different concentrations depending of fermentation temperature when cultured on phytosterols. This result confirmed that mycobacteria accumulate C-22 alcohols instead of the corresponding C-22 acids. Interestingly, they have also demonstrated that the residual KstD activity is not active on 9OH-AD (Xu *et al.*, 2015) suggesting that a Δ^1-dehydrogenation by KstD precedes Ksh hydroxylation. Remarkably, in contrast with our MS6039-5941 (ΔkshB1-ΔkstD1) double mutant, this *M. neoaurum* mutant still retains other Ksh and KstD active isoenzymes, because this mutant completely metabolized high amounts of the sterol nucleus (39.8% by moles), which considerably reduces the final conversion yield of the process (Xu *et al.*, 2015). This result strongly reinforces our proposal that *M. smegmatis* can be considered as a good cell factory to produce steroid synthons by metabolic engineering.

The finding that MS6039 and MS6039-5941 mutants can produce large amounts of ADD and AD from natural sterols, respectively, constitutes an experimental demonstration that metabolic engineering can be implemented in a model mycobacterial system like *M. smegmatis* mc²155 to generate pharmaceutical steroid synthons at industrial scale using a rational metabolic engineering approach. In fact, these engineered strains are already competing at industrial scale, i.e. under industrial operational conditions, with the existing strains isolated from environmental sources that were modified for their industrial use long time ago by conventional mutagenic procedures. Thus, our results lay the foundations for the valuation of *M. smegmatis* as a useful tool to develop *à la carte* engineered strains as cell factories to transform with high efficiency natural sterols into valuable pharmaceutical steroids.

Experimental procedures

Chemicals

Cholest-4-en-3-one was purchased from Fluka (Steinheim, Germany). ADD and AD were purchased from TCI America. Chloroform and glycerol were purchased from Merck (Darmstardt, Germany). Cholesterol, N,O-bis(trimethylsilyl) trifluoroacetamide (BSTFA), gentamicin, pyridine, Tween 80 and tyloxapol were from Sigma (Steinheim, Germany). Oligonucleotides were from Sigma-Genosys.

Bacterial strains and culture conditions

The strains as well as the plasmids used in this work are listed in Table 2. *M. smegmatis* mc²155 and its mutant strains were grown at 37°C in an orbital shaker at 200 r.p.m. Middlebrook 7H9 broth medium (Difco) containing 10% albumin-dextrose-catalase supplement (Becton Dickinson, New Jersey, USA) and 0.05% Tween 80 was used as rich medium. 7H10 agar (Difco) plates supplemented with albumin-dextrose-catalase were also used for solid media. Middlebrook 7H9 broth medium (Difco) without albumin-dextrose-catalase supplements containing 18 mM glycerol was used as a minimal medium. The minimal culture media used for the production of AD or ADD contained a mixture of 9 mM glycerol and 0.4 g l^{-1} cholesterol, or a mixture of 9 mM glycerol and 0.4 g l^{-1} of phytosterols. Commercial phytosterols (provided by Gadea Biopharma, Spain) contained a mixture of different sterols (w/w percentage): brassicasterol (2.16%), stigmasterol (8.7%), campesterol (36.8%) and β-sitosterol (54.4%). Cholesterol and phytosterols were

Table 2. Bacterial strains and plasmids used in this study.

Strains or plasmids	Genotype and/or description	Source or reference
Strains		
Mycobacterium smegmatis		
mc²155	*ept-1*, mc²6 mutant efficient for electroporation	Snapper *et al.* (1990)
MS6039	*M. smegmatis* mc²155 Δ*MSMEG_6039*	This study
MS6039-5941	*M. smegmatis* mc²155 Δ*MSMEG_6039* Δ*MSMEG_5941*	This study
mc²155 (pMV261)	mc²155 strain harbouring plasmid pMV261	This study
MS6039 (pMV261)	MS6039 strain harbouring plasmid pMV261	This study
MS6039 (pMV6039)	MS6039 strain harbouring plasmid pMV6039	This study
Escherichia coli		
DH10B	F⁻, *mcrA*, Δ (*mrrhsdRMS-mcrBC*), Φ80d*lacZ*ΔM15, Δ*lacX74*, *deoR*, *recA1*, *araD139*, Δ(*ara-leu*)7697, *galU*, *galK*, λ⁻, *rpsL*, *endA1*, *nupG*	Invitrogen
Plasmids		
pJQ200x	Suicide vector used to perform allelic exchange mutagenesis in *Mycobacterium*, P15A *ori*,*sacB*, Gm^r, *xylE*	Jackson *et al.* (2001)
pJQ6039	pJQ200x containing one fragment upstream and another downstream of *MSMEG_6039* gene	This study
pJQ5941	pJQ200x containing one fragment upstream and another downstream of *MSMEG_5941* gene	This study
pGEM®-T Easy	*E. coli* cloning vector; Amp^r	Promega
pGEMT6039	pGEMT-Easy harbouring the *MSMEG_6039* gene encoding KshB1 from *M. smegmatis* mc² 155	This study
pMV261	*Mycobacterium/E. coli* shuttle vector with the kanamycin resistance *aph* and the promoter from the *hsp60* gene from *M. tuberculosis*	Stover *et al.* (1991)
pMV6039	pMV261 harbouring the *MSMEG_6039* gene encoding KshB1 from *M. smegmatis* mc² 155	This study

dissolved in 10% tyloxapol prior to its addition to the minimal medium when assayed in flasks. Due to the low solubility of cholesterol and phytosterols, stock solutions were warmed at 80°C in agitation, sonicated in a bath for 1 h and then autoclaved. Gentamicin (5 µg/ml) was used for selection of *M. smegmatis* mutant strains when appropriate.

Bioreactor experiments were performed using the culture media previously described (Herrington and Spassov, 2003) containing a vegetal oil to dissolve phytosterols and corn steep liquor as supplementary carbon and nitrogen sources. The bioconversion experiments were performed in stainless steel (20-l) or glass (2- and 5-l) jar bioreactors with efficient stirring. The entire process was performed at 37°C.

Escherichia coli DH10B strain was used as a host for cloning. It was grown in rich LB medium at 37°C in an orbital shaker at 200 r.p.m. LB agar plates were used for solid media. Gentamicin (10 µg ml^{-1}), ampicillin (100 µg ml^{-1}) or kanamycin (50 µg ml^{-1}), were used for plasmid selection and maintenance in this strain.

Gene deletions

The knock-out strains of *M. smegmatis* named MS6039 and MS6039-5941 were constructed by homologous recombination using the pJQ200x plasmid, a derivative of the suicide pJQ200 vector that does not replicate in *Mycobacterium* (Jackson *et al.*, 2001). The strategy consist, for each gene, in generating two fragments of ~700 bp, the first one containing the upstream region and few nucleotides of the 5′end of the gene and the second one containing the downstream region and few nucleotides of the 3′end of the gene, that are amplified by PCR using the oligonucleotides described in Table 3 and *M. smegmatis* genomic DNA as template (Fig. S1). *M. smegmatis* genomic DNA extraction was performed as described (Uhía *et al.*, 2011). The two fragments generated were digested with the corresponding enzymes and cloned into the plasmid pJQ200x using *E. coli* DH10B competent cells as described (Uhía *et al.*, 2011).

Plasmid DNA from *E. coli* DH10B recombinant strains was extracted using the High Pure Plasmid Purification Kit (Roche, Basel, Switzerland), according to the manufacturer's instructions. This procedure was performed for genes *MSMEG_6039* and *MSMEG_5941*, generating the pJQ6039 and pJQ5941 plasmids respectively. Plasmid pJQ6039 was electroporated into competent *M. smegmatis* mc^{2}155 to obtain strain MS6039. Plasmid pJQ5941 was electroporated into competent MS6039 cells to obtain MS6039-5941 strain. Single cross-overs were selected on 7H10 agar plates containing gentamicin and the presence of the *xylE* gene encoded in pJQ200x was confirmed by spreading catechol over the single colonies of electroporated *M. smegmatis*. The appearance of a yellow coloration indicates the presence of the *xylE* gene. Colonies were also contra-selected in 10% sucrose. A single colony was grown in 10 ml of 7H9 medium with 5 µg ml^{-1} gentamicin up to an optical density of 0.8–0.9 and 20 µl of a 1:100 dilution was plated onto 7H10 agar plates with 10% sucrose to select for double cross-overs. Potential double cross-overs (sucrose-resistant colonies) were screened for gentamicin sensitivity and the absence of the *xylE* gene. The mutant strains MS6039 and MS6039-5941 were analysed by PCR and DNA sequencing to confirm the deletions of *MSMEG_6039* and *MSMEG_5941* genes.

Construction of pMV6039 plasmid

To isolate the *kshB1* gene from *M. smegmatis* mc^2 155 genomic DNA was extracted and amplified by PCR using the primers MSMEG_6039F (CGGAATTCTGACCTAAG GAGGTGAATGTGACTGATGAGCCCCTGGG) and MS MEG_6039R (CGAAGCTTCTATTCGTCGTAGGTGAC TTCG). The amplified fragment of 1092 bp was cloned into the commercial plasmid pGEM®-T Easy to generate pGEMT6039 plasmid that was transformed in *E. coli* DH10B competent cells. The cloned fragment was further digested with *Eco*RI and *Hind*III to clone into pMV261, a shuttle plasmid that replicates in *E. coli* and *Mycobacterium* generating the plasmid pMV6039 that was

Table 3. Primers used in this study.

Primer	Sequence (5′–3′)^a	Use
MSMEG_6039 up F	ctagctcgagccagttgtgcacaccgatg	*MSMEG_6039* deletion (amplification of upstream region)
MSMEG_6039 up R	ctagactagtgcagcgtcaggaactggc	*MSMEG_6039* deletion (amplification of upstream region)
MSMEG_6039 down F	ctagactagtggtgcacatggagatcaacg	*MSMEG_6039* deletion (amplification of downstream region)
MSMEG_6039 down R	ctagtctagagtaccagtcgatcggtgtc	*MSMEG_6039* deletion (amplification of downstream region)
MSMEG_5941 up F	ctagactagtgatgttgcgaatgtcgatgtc	*MSMEG_5941* deletion (amplification of upstream region)
MSMEG_5941 up R	ctagtctagaccaccacaacgtcgtactcc	*MSMEG_5941* deletion (amplification of upstream region)
MSMEG_5941 down F	ctagtctagaccatgacattcggttacctgg	*MSMEG_5941* deletion (amplification of downstream region)
MSMEG_5941 down R	ctaggagctcgcaggagatctcgaaatcg	*MSMEG_5941* deletion (amplification of downstream region)

a. Restriction sites are underlined.

transformed into *E. coli* DH10B to generate the recombinant strain *E. coli* DH10B (pMV6039). Once the sequence of the pMV6039 was checked it was used to transform electrocompetent cells of *M. smegmatis* MS6039 generating the *M. smegmatis* MS6039 (pMV6039) recombinant strain.

GC/MS analyses

To perform GC/MS analysis, culture aliquots (0.2 ml) were extracted twice at various extents of incubation with an equal volume of chloroform. Previously to its extraction, 10 µl of a solution of 10 mM cholesterol (when phytosterols were used as substrate) or 10 mM cholestenone (when cholesterol was used as substrate) dissolved in chloroform were added to the aliquots as internal standards. The chloroform fraction was concentrated by evaporation and the trimethylsilyl ether derivatives were formed by reaction with 50 µl of BSTFA and 50 µl of pyridine and heating at 60°C for 45 min. Calibration standards were derivatized in the same way. The GC/MS analysis was carried out using an Agilent 7890A gas chromatograph coupled to an Agilent 5975C mass detector (Agilent Technologies, Palo Alto, CA, USA). Mass spectra were recorded in electron impact (EI) mode at 70 eV within the m/z range 50–550. The chromatograph was equipped with a 30 m × 0.25 mm i.d. capillary column (0.25_m film thickness) HP-5MS (5% diphenyl 95% dimethylpolysiloxane from Agilent Technologies). Working conditions in the sample were as follows: split ratio (20:1), injector temperature, 320°C; column temperature 240°C for 3 min, then heated to 320°C at 5°C min−1.For quantification of the peak area, the quantitative masses were 329 + 458 m/z for cholesterol; 343 + 384 m/z for cholestenone; 382 + 472 m/z for campesterol; 394 + 484 m/z for stigmasterol; 357 + 486 m/z for β-sitosterol; 244 + 286 m/z for AD and 122 + 284 m/z for ADD in selected ions of monitoring. EI mass spectra and retention data were used to assess the identity of compounds by comparing them with those of standards in the NIST Mass Spectral Database and commercial standards (NIST 2011).

High-performance liquid chromatography analyses

To perform high-performance liquid chromatography (HPLC) analysis, samples (1 g) were withdrawn and extracted with 10 mL of ethyl acetate during two hours in a magnetic stirrer. One aliquot was centrifuged 10 min at 10 000 r.p.m. and supernatant was diluted 1:10 in acetonitrile. Samples were filtered (0.2 µm pore size) prior to the chromatographic analysis. The HPLC analysis was carried out using a Waters liquid chromatograph with a PDA detector system. A Phenomenex

C18 column (Nucleosil C18, 100 Å, 250 × 4.6 mm, 5 µm particles) was employed and the temperature of the column was fixed in 50°C. Working conditions were as follows: the mobile phase was a mixture water: acetonitrile: acetic acid (48: 52: 0.1 v/v); flow rate was 1.1 mL min−1; the injection volumen was 10 µl. Peaks were monitorized at 240 nm and calibrations were performed using highly purified standards of each compound.

Conflict of interest

None declared.

References

Andor, A., Jekkel, A., Hopwood, D.A., Jeanplong, F., Ilkoy, E., Kónya, A., et al. (2006) Generation of useful insertionally blocked sterol degradation pathway mutants of fast-growing mycobacteria and cloning, characterization, and expression of the terminal oxygenase of the 3-ketosteroid 9α-hydroxylase in *Mycobacterium smegmatis* mc²155. *Appl Environ Microbiol* **72:** 6554–6559.

Blank, L.M., Ebert, B.E., Buehler, K., and Bühler, B. (2010) Redox biocatalysis and metabolism: molecular mechanism and metabolic network analysis. *Antioxid Redix Signal* **13:** 349–394.

Bragin, E.Y., Shtratnikov, V.Y., Dovbnya, D.V., Schelkunov, M.I., Pekov, Y.A., Malakho, S.G., et al. (2013) Comparative analysis of genes encoding key steroid core oxidation enzymes in fast-growing *Mycobacterium* spp. strains. *J Steroid Biochem Mol Biol* **138:** 41–53.

Brzostek, A., Sliwiński, T., Rumijowska-Galewicz, A., Korycka-Machała, M., and Dziadek, J. (2005) Identification and targeted disruption of the gene encoding the main 3-ketosteroid dehydrogenase in *Mycobacterium smegmatis*. *Microbiology* **151:** 2393–2402.

Brzostek, A., Pawelczyk, J., Rumijowska-Galewic, A., Dziadek, B., and Dziadek, J. (2009) *Mycobacterium tuberculosis* is able to accumulate and utilize cholesterol. *J Bacteriol* **191:** 6584–6591.

Capyk, J.K., D'Angelo, I., Strynadka, N.C., and Eltis, L.D. (2009) Characterization of 3-ketosteroid 9α-hydroxylase, a Rieske oxygenase in the cholesterol degradation pathway of *Mycobacterium tuberculosis*. *J BiolChem* **284:** 9937–9946.

Capyk, J.K., Casabon, I., Gruninger, R., Strynadka, N.C., and Eltis, L.D. (2011) Activity of 3-ketosteroid 9α-hydroxylase (KshAB) indicates cholesterol side chain and ring degradation occur simultaneously in *Mycobacterium tuberculosis*. *J Biol Chem* **286:** 40717–40724.

Cole, S.T., Brosch, R., Parkhill, J., Garnier, T., Churcher, C., Harris, D., et al. (1998) Deciphering the biology of *Mycobacterium tuberculosis* from the complete genome sequence. *Nature* **393:** 537–544. Erratum in: *Nature* **396:** 190.

Donova, M.V. (2007) Transformation of steroids by Actinobacteria: a review. *Appl Biochem Microbiol* **43:** 1–14.

Donova, M.V., and Egorova, O.V. (2012) Microbial steroid transformations: current state and prospects. *Appl Microbiol Biotechnol* **94:** 1423–1447.

Donova, M.V., Gulevskaya, S.A., Dovbnya, D.V., and Puntus, I.F. (2005) *Mycobacterium* sp. mutant strain producing 9α-hydroxyandrostenedione from sitosterol. *Appl Microbiol Biotechnol* **67:** 671–678.

Fernández de las Heras, L., van der Geize, R., Drzyzga, O., Perera, J., and Navarro-Llorens, J.M. (2012) Molecular characterization of three 3-ketosteroid-Δ(1)-dehydrogenase isoenzymes of *Rhodococcus ruber* strain Chol-4. *J Steroid Biochem Mol Biol* **132:** 271–281.

Galán, B., Díaz, E., Prieto, M.A., and García, J.L. (2000) Functional analysis of the small component of the 4-hydroxyphenylacetate 3-monooxygenase of *Escherichia coli* W: a prototype of a new Flavin:NAD(P)H reductase subfamily. *J Bacteriol* **182:** 627–636.

García, J.L., Uhía, I., and Galán, B. (2012) Catabolism and biotechnological applications of cholesterol degrading bacteria. *Microbial Biotechnol* **5:** 679–699.

van der Geize, R., Hessels, G.I., and Dijkhuizen, L. (2000) Targeted disruption of the *kstD* gene encoding a 3-ketosteroid Δ1-dehydrogenase isoenzyme of *Rhodococcus erythropolis* strain SQ1. *Appl Environ Microbiol* **66:** 2029–2036.

van der Geize, R., Hessels, G. and Dijkhuizen, L. (2001a) Preparing genetically modified organism unable to degrade the steroid nucleus, useful for producing corticosteroid intermediates, comprises inactivating the dehydrogenase genes. Patent WO 200131050.

van der Geize, R., Hessels, G.I., van Gerwen, R., van der Meijden, P., and Dijkhuizen, L. (2001b) Unmarked gene deletion mutagenesis of *kstD*, encoding 3-ketosteroid Δ1-dehydrogenase, in *Rhodococcus erythropolis* SQ1 using *sacB* as counter-selectable marker. *FEMS Microbiol Lett* **205:** 197–202.

van der Geize, R., Hessels, G.I., van Gerwen, R., van der Meijden, P., and Dijkhuizen, L. (2002a) Molecular and functional characterization of *kshA* and *kshB*, encoding two components of 3-ketosteroid 9α-hydroxylase, a class IA monooxygenase, in *Rhodococcus erythropolis* strain SQ1. *Mol Microbiol* **45:** 1007–1018.

van der Geize, R., Hessels, G.I., and Dijkhuizen, L. (2002b) Molecular and functional characterization of the *kstD2* gene of *Rhodococcus erythropolis* SQ1 encoding a second 3-ketosteroid Δ1-dehydrogenase isoenzyme. *Microbiology* **148:** 3285–3292.

van der Geize, R., Hessels, G.I., Nienhuis-Kuiper, M., and Dijkhuizen, L. (2008) Characterization of a second *Rhodococcus erythropolis* SQ1 3-ketosteroid 9α-hydroxylase activity comprising a terminal oxygenase homologue, KshA2, active with oxygenase-reductase component KshB. *Appl Environ Microbiol* **74:** 7197–7203.

Herrington, E.J. and Spassov, G. (2003) Process for fermentation of phytosterols to androstadienedione. Patent WO 2003/064674.

Hu, Y., van der Geize, R., Besra, G.S., Gurcha, S.S., Liu, A., Rohde, M., *et al.* (2010) 3-ketosteroid 9α-hydroxylase is an essential factor in the pathogenesis of *Mycobacterium tuberculosis*. *Mol Microbiol* **75:** 107–121.

Jackson, M., Camacho, L.R., Gicquel, B., and Guilhot, C. (2001) Gene replacement and transposon delivery using the negative selection marker *sacB*. In *Mycobacterium Tuberculosis Protocols*, Vol. **54,** Parish, T., and Stocker, N.G. (eds). Totowa, NJ: Humana Press, pp. 59–75.

Kendall, S.L., Withers, M., Soffair, C.N., Moreland, N.J., Gurcha, S., Sidders, B., *et al.* (2007) A highly conserved transcriptional repressor controls a large regulon involved in lipid degradation in *Mycobacterium smegmatis* and *Mycobacterium tuberculosis*. *Mol Microbiol* **65:** 684–699.

Malaviya, A., and Gomes, J. (2008) Androstenedione production by biotransformation of phytosterols. *Bioresour Technol* **99:** 6725–6737.

Marsheck, W., Kraychy, S., and Muir, R. (1972) Microbial degradation of sterols. *Appl Microbiol* **23:** 72–77.

Penfield, J.S., Worrall, L.J., Strynadka, N.C., and Eltis, L.D. (2014) Substrate specificities and conformational flexibility of 3-ketosteroid 9α-hydroxylases. *J Biol Chem* **289:** 25523–25536.

Petrusma, M., Dijkhuizen, L., and van der Geize, R. (2009) *Rhodococcus rhodochrous* DSM 43269 3-ketosteroid 9α-hydroxylase, a two-component iron-sulfur-containing monooxygenase with subtle steroid substrate specificity. *Appl Environ Microbiol* **75:** 5300–5307.

Petrusma, M., Hessels, G., Dijkhuizen, L., and van der Geize, R. (2011) Multiplicity of 3-ketosteroid-9α-hydroxylase enzymes in *Rhodococcus rhodochrous* DSM 43269 for specific degradation of different classes of steroids. *J Bacteriol* **193:** 3931–3940.

Petrusma, M., Dijkhuizen, L., and van der Geize, R. (2012) Structural features in the KshA terminal oxygenase protein that determine substrate preference of 3-ketosteroid 9α-hydroxylase enzymes. *J Bacteriol* **194:** 115–121.

Petrusma, M., van der Geize, R., and Dijkhuizen, L. (2014) 3-Ketosteroid 9α-hydroxylase enzymes: Rieske non-heme monooxygenases essential for bacterial steroid degradation. *Antonie Van Leeuwenhoek* **106:** 157–172.

Snapper, S.B., Melton, R.E., Mustafa, S., Kieser, T., and Jacobs, W.R. Jr (1990) Isolation and characterization of efficient plasmid transformation mutants of *Mycobacterium smegmatis*. *Mol Microbiol* **4:** 1911–1919.

Stover C.K., de la Cruz V.F., Fuerst T.R., Burlein J.E., Benson L.A., Bennett L.T., *et al.* (1991) New use of BCG for recombinant vaccines. *Nature* **351:** 456–460.

Szentirmai, A. (1990) Microbial physiology of sidechain degradation of sterols. *J Ind Microbiol* **6:** 101–115.

Uhía, I., Galán, B., Morales, V., and García, J.L. (2011) Initial step in the catabolism of cholesterol by *Mycobacterium smegmatis* mc2155. *Environ Microbiol* **13:** 943–959.

Uhía, I., Galán, B., Kendall, S.L., Stoker, N.G., and García, J.L. (2012) Cholesterol metabolism in *Mycobacterium smegmatis*. *Environ Microbiol Rep* **4:** 168–182.

Wei, W., Wang, F.Q., Fan, S.Y., and Wei, D.Z. (2010) Inactivation and augmentation of the primary 3-ketosteroid-Δ1-dehydrogenase in *Mycobacterium neoaurum* NwIB-01: biotransformation of soybean phytosterols to 4-androstene- 3,17-dione or 1,4-androstadiene-3,17-dione. *Appl Environ Microbiol* **76:** 4578–4582.

Wilbrink, M.H., Petrusma, M., Dijkhuizen, L., and van der Geize, R. (2011) FadD19 of *Rhodococcus rhodochrous*

DSM 43269, a steroid-Coenzyme A ligase essential for degradation of C-24 branched sterol side chains. *Microbiology* **158**: 3054–3062.

Xu, X.W., Gao, X.Q., Feng, J.X., Wang, X.D., and Wei, D.Z. (2015) Influence of temperature on nucleus degradation of 4-androstene-3, 17-dione in phytosterol biotransformation by *Mycobacterium* sp. *Lett Appl Microbiol* **61**: 63–68.

Xu, L.Q., Liu, Y.J., Yao, K., Liu, H.H., Tao, X.Y., Wang, F.Q., *et al.* (2016) Unraveling and engineering the production of 23,24-bisnorcholenic steroids in sterol metabolism. *Sci Rep* **6**: 21928.

Yao, K., Xu, L.Q., Wang, F.Q., and Wei, D.Z. (2014) Characterization and engineering of 3-ketosteroid-Δ^1-dehydrogenase and 3-ketosteroid-9α-hydroxylase in *Mycobacterium neoaurum* ATCC 25795 to produce 9α-hydroxy-4-androstene-3,17-dione through the catabolism of sterols. *Metab Eng* **24**: 181–191.

Yeh, C.H., Kuo, Y.S., Chang, C.M., Liu, W.H., Sheu, M.L., and Meng, M. (2014) Deletion of the gene encoding the reductase component of 3-ketosteroid 9α-hydroxylase in *Rhodococcus equi* USA-18 disrupts sterol catabolism, leading to the accumulation of 3-oxo-23,24-bisnorchola-1,4-dien-22-oic acid and 1,4-androstadiene-3,17-dione. *Microb Cell Fact* **13**: 130.

Intracellular metabolite profiling of *Saccharomyces cerevisiae* evolved under furfural

Young Hoon Jung,[1] **Sooah Kim,**[2] **Jungwoo Yang,**[2] **Jin-Ho Seo**[3] **and Kyoung Heon Kim**[2,*]

[1]*School of Food Science and Biotechnology, Kyungpook National University, Daegu 41566, South Korea.*
[2]*Department of Biotechnology, Graduate School, Korea University, Seoul 02841, South Korea.*
[3]*Department of Agricultural Biotechnology and Center for Food and Bioconvergence, Seoul National University, Seoul 08826, South Korea.*

Summary

Furfural, one of the most common inhibitors in pretreatment hydrolysates, reduces the cell growth and ethanol production of yeast. Evolutionary engineering has been used as a selection scheme to obtain yeast strains that exhibit furfural tolerance. However, the response of *Saccharomyces cerevisiae* to furfural at the metabolite level during evolution remains unknown. In this study, evolutionary engineering and metabolomic analyses were applied to determine the effects of furfural on yeasts and their metabolic response to continuous exposure to furfural. After 50 serial transfers of cultures in the presence of furfural, the evolved strains acquired the ability to stably manage its physiological status under the furfural stress. A total of 98 metabolites were identified, and their abundance profiles implied that yeast metabolism was globally regulated. Under the furfural stress, stress-protective molecules and cofactor-related mechanisms were mainly induced in the parental strain. However, during evolution under the furfural stress, *S. cerevisiae* underwent global metabolic allocations to quickly overcome the stress, particularly by maintaining higher levels of metabolites related to energy generation, cofactor regeneration and recovery from cellular damage. Mapping the mechanisms of furfural tolerance conferred by evolutionary engineering in the present study will be led to rational design of metabolically engineered yeasts.

*For correspondence. E-mail khekim@korea.ac.kr

Funding Information
Ministry of Science, ICT and Future Planning (2011-0031359).

Introduction

Lignocellulose is the most abundant and promising resource for producing fuels and bio-based chemicals. To efficiently produce fermentable sugars from lignocellulose, lignocellulose must be pre-treated because of its high recalcitrance. However, the generation of various degradation by-products, including 2-furaldehyde (furfural), 5-hydroxymethyl-2-furaldehyde, organic acids and phenolics, which negatively affect microbial metabolism during fermentation, is unavoidable (Liu, 2006; Almeida *et al.*, 2009; Jung *et al.*, 2014). It is because physico-chemical pretreatments are performed at extreme conditions such as high temperatures and/or extreme pH values. Furthermore, furfural, which is derived from pentose sugars, is known to be one of the most potent contributors to the toxicity of pretreatment hydrolysates for fermentative microorganisms (Heer and Sauer, 2008). Furfural significantly reduces cell proliferation and ethanol production either by inhibiting several enzymes that are essential to central metabolism, including dehydrogenases, or by damaging and blocking the synthesis of DNA, RNA, protein and cell wall (Zaldivar *et al.*, 1999; Modig *et al.*, 2002; Horváth *et al.*, 2003; Almeida *et al.*, 2009; Liu, 2011; Ask *et al.*, 2013; Wilson *et al.*, 2013). Fortunately, unless the furfural level is lethal, *Saccharomyces cerevisiae* can metabolize it into less toxic compounds such as furfuryl alcohol and furoic acid by consuming NAD(P)H at the beginning of fermentation (Liu *et al.*, 2004; 2005).

Various strategies to ameliorate furfural toxicity, including the physical and chemical detoxification of hydrolysates prior to fermentation, have been investigated (Palmqvist and Hahn-Hägerdal, 2000; Jung and Kim, 2014). However, because of the high cost of the detoxification processes, strategies to enhance the inherent resistance of microbes to furfural have been receiving much attention. Many efforts have been made to develop furfural-resistant fermentative strains. For example, the pentose phosphate pathway, γ-aminobutyric acid (GABA) shunt, cofactor interconversion, high osmolality glycerol signalling and DNA binding processes seem to be associated with growth improvement under furfural stress (Modig *et al.*, 2002; Gorsich *et al.*, 2006; Kim *et al.*, 2012; Wang *et al.*, 2013a; Glebes *et al.*, 2015). Several genes under furfural stress, which are involved in stress tolerance (e.g. dehydrogenases), cofactor balance (e.g. oxidoreductases and transhydrogenase) and other functions (e.g. sulfur assimilation and glucose phosphorylation), have been identified (Nilsson *et al.*, 2005; Liu, 2006; 2011; Heer *et al.*, 2009; Miller *et al.*, 2009; Yang *et al.*, 2012; Wilson *et al.*, 2013). The production of reactive oxygen species (ROS) and changes in energy status also influence cellular physiology of *S. cerevisiae* (Allen *et al.*, 2010; Ask *et al.*, 2013).

In recent years, evolutionary engineering of microbes, which relies on selective pressure towards an appropriate phenotype, has also been investigated. For example, through evolution in the presence of furfural, *S. cerevisiae* showed improvement in tolerance to furfural toxicity and in the ability to convert furfural to less toxic materials (Heer *et al.*, 2009; Liu *et al.*, 2009). The regulation of central carbon metabolism, redox balance, membrane fatty acids and amino acids in the presence of furfural has been investigated at the proteomic, lipidomic and metabolomic levels to determine the effects of these components on furfural tolerance in yeast (Lin *et al.*, 2009; Xia and Yuan, 2009; Ding *et al.*, 2011; Wang *et al.*, 2013b). In particular, through metabolic profiling of yeast adapted in lignocellulosic hydrolysates containing multiple inhibitors including furfural, alanine, GABA and glycerol have been suggested as the key metabolites (Wang *et al.*, 2013b). However, as microbial metabolism is tightly and globally regulated by a large number of intracellular metabolites following mass and energy conservation laws (Patil *et al.*, 2005), metabolic approaches for an individual compound need to be more thoroughly investigated.

Presently, there are insufficient evolutionary engineering studies regarding the strategies adopted by yeast for coping with furfural at the metabolite level. In this study, responses of both the parental strain and the evolved yeast for furfural were studied by analysing all the intracellular metabolites. First, an evolutionary engineering strategy in the presence of furfural stress was applied to *S. cerevisiae* to improve its tolerance for many generations. Second, the physiological basis of furfural resistance was explored by a match-up of the fermentation profiles of both the parental and the evolved strains. Finally, global profiles of the metabolites expressed in the parental and the evolved yeast were obtained using gas chromatography/time-of-flight mass spectrometry (GC/TOF MS) and compared. This study explored the metabolic perturbation patterns of *S. cerevisiae* both when the yeast encountered furfural by chance and when it was intentionally adapted to furfural.

Results and discussion

Evaluation of S. cerevisiae D₅A under furfural

To obtain evolved yeast strains which are tolerant to furfural, three different seed cultures of *S. cerevisiae* D₅A

(i.e. E_a, E_b and E_c) were cultivated independently in different tubes and transferred 50 times to fresh media containing 20 mM furfural. Cell concentrations and ethanol titres of the culture at each transfer were measured after 24 h of cultivation to monitor the evolutionary progress (Fig. 1). After approximately 5–10 transfers, cell growth and ethanol production in the presence of 20 mM furfural significantly increased. These rapid adaptation patterns have also been observed by other groups. For example, the long lag phases induced due to the presence of furfural were effectively shortened by evolution in two transfers under about 13.5 mM furfural (Wang *et al.*, 2013b) or by 20 transfers under 17 mM furfural (Heer and Sauer, 2008). To verify whether the phenotype of furfural tolerance in the evolved strains was maintained in the absence of furfural, the evolved strains (E_a, E_b and E_c) were transferred for approximately

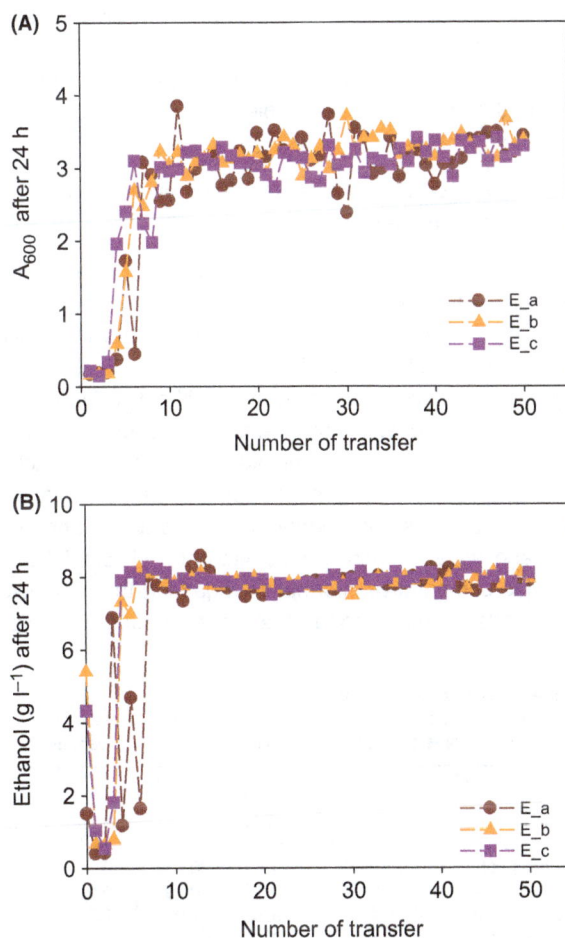

Fig. 1. Profiles of (A) cell growth (A₆₀₀, absorbance at 600 nm) and (B) ethanol production. For evolution, three seed cultures of *Saccharomyces cerevisiae* were independently grown in YPD medium containing 20 mM furfural at 30°C and 200 r.p.m. *Saccharomyces cerevisiae* D₅A was transferred after 24 h of cultivation under furfural during 50 transfers (~332 generations).

33 generations in YPD medium without furfural and was then recultivated with furfural (Fig. S1). When the evolved strains that were maintained in the absence of furfural were returned to furfural-containing media, its growth was still much higher than that of the parental strain. These results indicated a fixation of the phenotype in the evolved strains.

Next, we compared the cell growth behaviour of the evolved strains with that of the parental strain. In the presence of furfural (0–40 mM), the lag phase of S. cerevisiae D_5A increased from 2.5 to 45 h for the parental strain and from 3 to 28 h for the evolved strains as the furfural level increased from 0 to 30 mM (Fig. S2). This increase was probably due to furfural-induced inhibition of key enzymes in the glycolytic pathway (Palmqvist et al., 1999; Horváth et al., 2003; Wang et al., 2013b). In addition, under 20 mM furfural, compared with the parental strain, the evolved strains grew extremely fast (Fig. S3). Accordingly, around 20 g l^{-1} glucose and furfural were consumed within 18 h of fermentation by the evolved strains, and ethanol was produced at a rate of 0.3 g ethanol g^{-1} dry cell weight per hour to a maximum yield of 0.9 g g^{-1} after 18 h of fermentation (Table 1).

PCA of intracellular metabolites of the evolved strains versus the parental strains

Identifying the cellular metabolic reactions to environmental changes at the metabolite level is of great interest. In this study, six replicates of the parental S. cerevisiae strains and duplicates of the three evolved S. cerevisiae strains, which were grown with or without furfural, were collected at the early exponential phase for metabolite analysis. A total of 98 meaningful metabolites from different classes, including amines and phosphates, amino acids, fatty acids and phenolics, organic acids, and sugars and sugar alcohols, were identified (Table S1). To provide comparative information regarding the metabolomic differences among the four groups, principal

component analysis (PCA) was performed. The differences among the four groups were well explained by the PCA model, which showed an explained variation value (R^2X) of 0.94 and a predictive capability (Q^2) of 0.93. Although the first principal component (PC1) and the second principal component (PC2) showed less discrimination ($R^2X = 0.52$ and $Q^2 = 0.46$), PC1 and PC2 appeared to be the major variation factors induced by evolution and by furfural respectively (Fig. 2). Accordingly, on the basis of the differential distribution reflecting the importance of the original variables, it could be expected that metabolites with high loading in PC1 were related either to the glutathione and the thioredoxin reduction system (e.g. homoserine, cysteine, glutamate and 5'-deoxy-5'-methylthioadenosine) for relieving ROS accumulation or to the sugar metabolisms (e.g. glucose, galactose, fructose and mannose) to maintain energy balance. Conversely, cofactor-related metabolites (e.g. phenylacetate, salicylaldehyde and 3-hydroxypropionate) and amino acids (e.g. lysine, N-methylalanine, proline, glycine and threonine) were found to be relatively predominant in PC2 (Table S2). Several fatty acids, including palmitoleic acid, pentadecanoic acid and palmitic acid, did not contribute to the clear separation among the four groups. These results suggest that the intracellular fatty acid metabolism was not significantly affected by either the furfural stress or the evolutionary engineering.

Metabolic traits of the evolved strains under furfural stress

Biological interpretation of the identified metabolites is crucial for better understanding of the functional metabolism as a means of coping with furfural stress. In this study, after categorical annotation of the identified metabolites into suitable groups (Table S1), the value of

Table 1. Comparison of physiological values of the parental and evolved strains grown in YPD medium with or without 20 mM furfural. For the evolved strains, the mean values of E_a, E_b and E_c were used.

	Parental strain		Evolved strains	
Furfural concentration (mM)	0	20	0	20
Max. specific growth rate (h^{-1})	0.37	0.19	0.23	0.24
Cell dry weight at 48 h (g l^{-1})	3.2	2.0	2.9	2.7
Glucose depletion time (h)	9	30	15	18
Glucose consumption rate (g g^{-1} DCW h^{-1})	1.9	0.8	1.5	0.6
Furfural consumption rate (g g^{-1} DCW h^{-1})	NA	0.4	NA	0.3
Ethanol production rate (g g^{-1} DCW h^{-1})	0.9	0.3	0.7	0.3

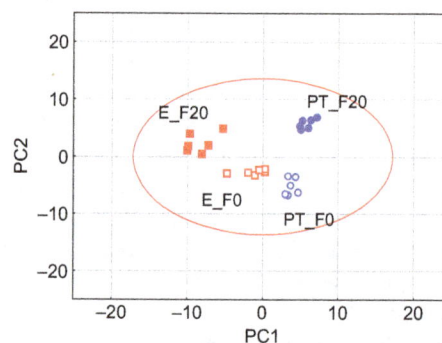

Fig. 2. Principal component analysis of identified metabolites in the parental and evolved strains. Both strains were grown in YPD medium with or without 20 mM furfural. PT_F0: the parental strain without furfural; PT_F20: the parental strain with 20 mM furfural; E_F0: the evolved strains without furfural; E_F20: the evolved strains with furfural. For the evolved strains, the mean values of E_a, E_b and E_c were used.

each metabolite was normalized by the sum of peak intensities of all the detected intracellular metabolites, which were analysed by GC/TOF MS, from each culture. Next, the average values of normalized data from both the parental and evolved strains grown under furfural stress were compared with the data from those grown without furfural to analyse the effect of furfural. Significant differences in the set of normalized data were evaluated ($p < 0.05$), and the variables without significant differences were considered to show similar expression levels, such as galactose in the parental strain and sucrose in the evolved strains, regardless of abundance changes. Relative comparison of the obtained fold changes was introduced to explain the metabolic fates caused by the evolutionary process.

The metabolic fates of parental and evolved S. cerevisiae grown with or without furfural stress were thoroughly investigated by selection procedures using stringency criteria (fold changes and p values). In this study, overall, the principal regulation mechanisms for coping with furfural toxicity differed markedly between the parental strain and the evolved strains. The parental strain tried to minimize primary metabolism and maximized the production of stress-related metabolites in response to furfural; the evolved strains, which was already habituated to the reduced environment, seemed to attempt to restore the anabolism suppressed by furfural. Specifically, we investigated carbohydrate metabolism, amino acid synthesis and cofactor-related pathways.

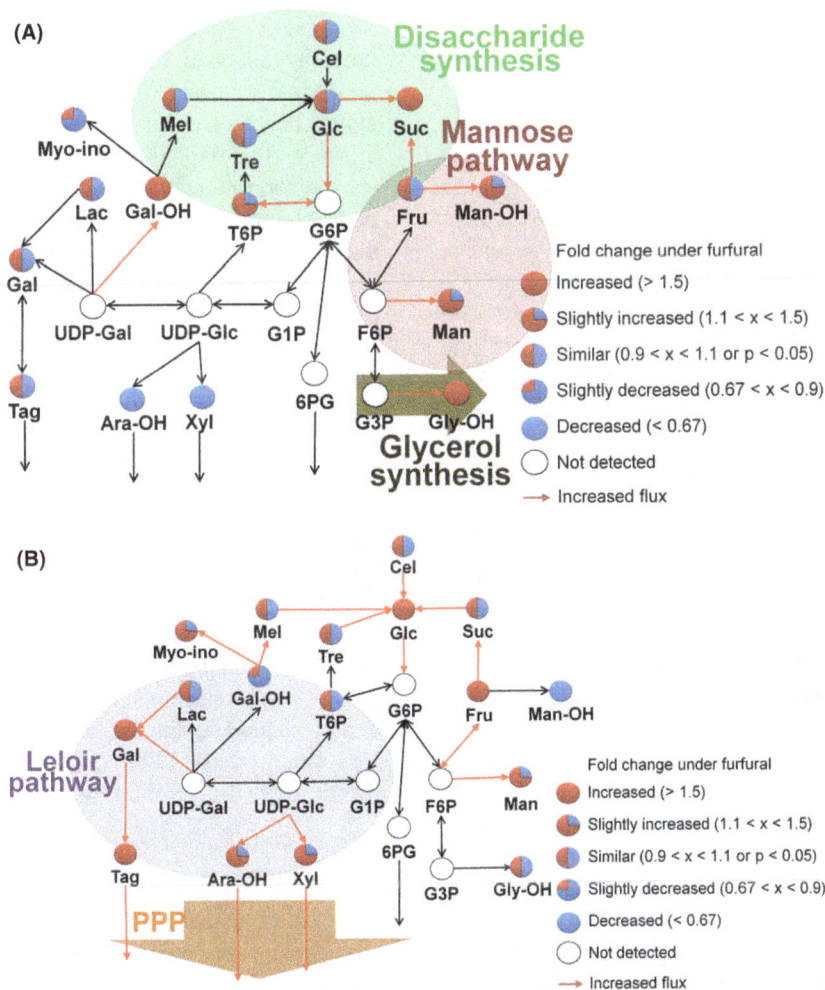

Fig. 3. The carbohydrate metabolic pathways in the (A) parental and (B) evolved strains. For the evolved strains, the mean values of E_a, E_b and E_c were used. The fold changes indicate the fold increases of metabolite abundances under the furfural stress in comparison with the metabolite abundances without furfural stress in the parental strain or the evolved strains. 6PG, 6-phosphogluconate; Ara-OH; arabinol; Cel, cellulose; Fru, fructose; F6P, fructose-6-phosphate; Gal, galactose; Gal-OH, galactinol; Glc, glucose; G1P, glucose-1-phosphate; G3P, glyceraldehyde-3-phosphate; G6P, glucose-6-phosphate; Gly-OH, glycerol; Lac, lactose; Man, mannose; Man-OH, mannitol; Mel, melibiose; Myo-ino, myo-inositol; Suc, sucrose; Tag, tagatose; Tre, trehalose; T6P, trehalose-6-phosphate; UDP-Gal, uridine diphosphate galactose; UDP-Glc, uridine diphosphate glucose; Xyl, xylose.

The central carbon metabolic pathway appeared to differ between the parental and evolved *S. cerevisiae* (Fig. 3). In the parental strain, sucrose, trehalose-6-phosphate, mannose, glycerol and others were higher in the furfural stress than those without furfural stress. As the presence of stress represses the expression of several enzymes in glycolysis, including aldehyde dehydrogenase, alcohol dehydrogenase and pyruvate dehydrogenase, a problem occurs in the generation of energy and building blocks (Cadière *et al.*, 2011). Thus, the higher abundance of glycerol may have acted as a protectant under the furfural stress (Wang *et al.*, 2013b). In addition, under stressful environments, yeasts must reduce ATP demands to recover from substrate-accelerated death caused by the imbalance between energy production and consumption. Thus, an ATP futile cycle through sugar phosphate and disaccharide synthesis would be induced to counteract the stress-induced ATP imbalance (Francois and Parrou, 2001; Jansen *et al.*, 2006). In this study, in the parental strain, the intracellular abundance of sucrose was higher as a safety valve under the furfural stress (Fig. 3A). On the other hand, in the evolved strains, monosaccharides such as glucose and fructose were higher, and various metabolites from the Leloir metabolism, including galactose, tagatose, xylose and arabitol, were significantly higher under the furfural stress (Fig. 3B). These results imply that in the evolved strains, the fluxes through the glycolytic and pentose phosphate pathways were recovered or intensified despite the furfural stress, probably to generate suitable amounts of energy, cofactors and other intermediate metabolites for the synthesis of aromatic amino acids and nucleotides.

With regard to the amino acid metabolism, in the parental strain, most of amino acids were lower in abundance under the furfural stress than those without furfural stress, possibly due to the shortage of energy resulting from the ATP futile cycle and due to the inhibition of primary metabolism under the furfural stress (Fig. 4A), as observed earlier in the carbohydrate metabolism (Fig. 3). In the evolved strains, the abundances of amino acids were significantly higher than those under the furfural stress (Fig. 4B), implying that glycolytic activity was restored over the course of evolution (Wang *et al.*, 2013b). Accordingly, in the evolved strains under the furfural stress, the abundances of the TCA cycle intermediates were maintained at the levels similar to those in the evolved strains without furfural (Fig. 4). This phenomenon on the TCA cycle intermediates was unlike those in the parental strain. In the evolved strains, the abundances of branched chain amino acids such as isoleucine, valine and leucine were also higher under the furfural stress than those without furfural stress (Fig. 4B), possibly either due to the source of acetyl-CoA or due to

the substrates for alanine synthesis, which provide energy efficiency via the alanine-GABA shunt and the alanine-glucose cycle (Wang *et al.*, 2013b). Meanwhile, in the evolved strains, the synthesis of glutamate and glutamine from ammonia in the central nitrogen metabolism was less active under the furfural stress (Fig. 4B). Probably, these reactions needed to decrease in order to save reduced cofactors as these reactions consume NAD(P)H.

Finally, changes in the metabolite abundances in the redox system were thoroughly investigated, as the consumption of NAD(P)H is necessary to metabolize furfural into less toxic compounds such as furfuryl alcohol and furoic acid (Liu *et al.*, 2004; 2005). Metabolism related to aromatic compounds originating from phenylalanine, including β-hydroxybutyrate, phenylacetate, phenyllactate, hydroxyphenylethanol, benzoate and salicylaldehyde, was significantly intensified (Fig. 5), which is probably affecting the enhanced flux to the pentose phosphate pathway (Park *et al.*, 2011). In the parental strain, due to lack of energy to cope with the furfural stress and damages to the protein, a stronger cofactor-regenerating mechanism with urea excretion was indispensable under the furfural stress (Fig. 5A). However, in the evolved strains, increases in metabolite abundances were observed in various metabolites as a correspondence mechanism under the furfural stress. In the evolved strains, along with aromatic compounds, the abundances of metabolites in the β-alanine cycle and several organic acids, including glycolate and glycerate, were higher possibly to secure NAD(P)H and/or acetyl-CoA availability under the furfural stress (Fig. 5B). In addition, to recover from the DNA or RNA damage caused by furfural, the thioredoxin cycle was well managed in the evolved strains under the furfural stress (Fig. 5B), probably for the restoration of nucleotides (Carmel-Harel and Storz, 2000; Shi *et al.*, 2011). The abundances of both putrescine and spermidine, which can act as protectants from abiotic stress or as substrates related to the protein synthesis initiation factor (Shimogori *et al.*, 1996; Gill and Tuteja, 2010), were higher under the furfural stress.

Conclusions

Significant metabolic rearrangements in response to furfural stress were revealed in *S. cerevisiae* by a combination of evolutionary engineering and metabolomics. The formation of stress-protective molecules, including glycerol and disaccharides, was important in maintaining the metabolic activity of the parental strain under furfural stress. Contrary to the *ad hoc* responses in the parental strain, the coping mechanisms in the evolved strains appeared to be strongly sessile throughout evolutionary

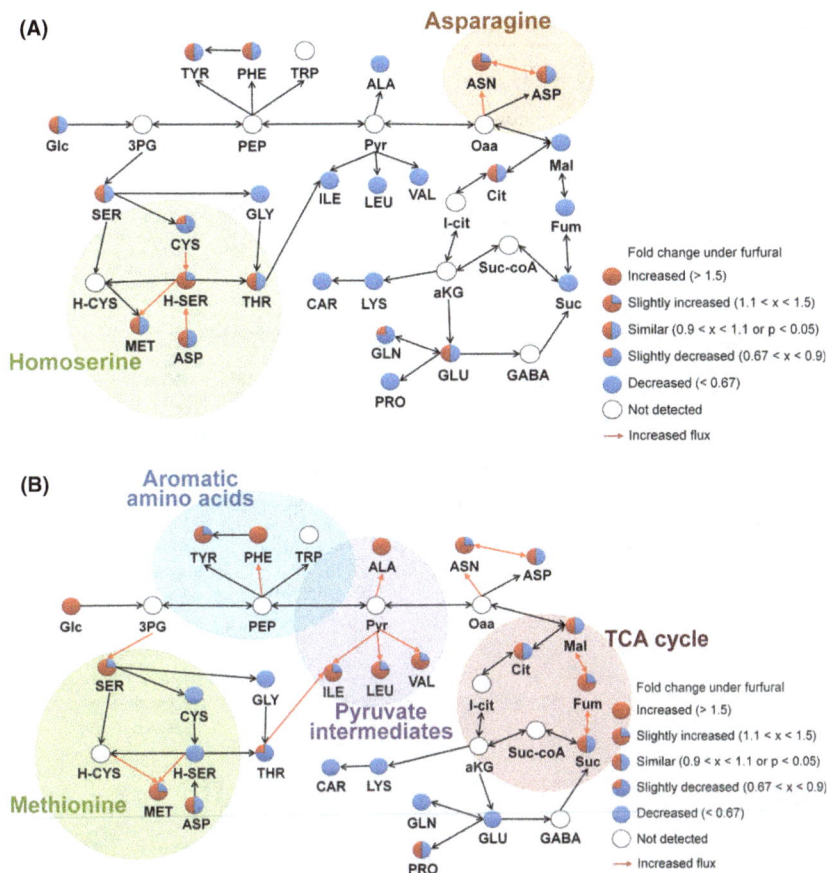

Fig. 4. The amino acid synthesis pathways in the (A) parental and (B) evolved strains. For the evolved strains, the mean values of E_a, E_b and E_c were used. The fold changes indicate the fold increases of metabolite abundances under the furfural stress in comparison with the metabolite abundances without furfural stress in the parental strain or the evolved strains. 3PG, 3-phosphoglycerate; aKG, α-keto glutarate; ALA, alanine; ASN, asparagine; ASP, aspartate; CAR, carnitine; Cit, citrate; CYS, cysteine; Fum, fumarate; GABA, γ-aminobutyric acid; Glc, glucose; GLN, glutamine; GLU, glutamate; GLY, glycine; H-CYS, homocysteine; H-SER, homoserine; I-cit, Isocitrate; ILE, isoleucine; LEU, leucine; LYS, lysine; Mal, malate; MET, methionine; Oaa, oxaloacetate; PEP, phosphoenolpyruvate; PHE, phenylalanine; PRO, proline; Pyr, pyruvate; SER, serine; Suc, succinate; Suc-coA, succinyl-CoA; THR, threonine; TRP, tryptophan; TYR, tyrosine; VAL, valine.

engineering. After rapid adaptation and physiological stabilization, we explored metabolism, which was remarkably strengthened by the improvement of glycolytic activity, salvation of spermidine and methionine and restoration of NAD(P)H pools. In conclusion, when the yeast recognizes the presence of furfural stress, they may globally regulate their metabolic status in advance in response to the furfural stress. The comparisons of defence mechanisms against furfural in the parental and evolved *S. cerevisiae* in this study provide new insights into the systems biology of yeast physiology.

Experimental procedures

Strain, media, culture conditions and evolution experiments

The parental strain *S. cerevisiae* D_5A (ATCC 200062) was used as a starting strain for the evolution

Fig. 5. The NAD(P)H pool metabolism in the (A) parental and (B) evolved strains. For the evolved strains, the mean values of E_a, E_b and E_c were used. The fold changes indicate the fold increases of metabolite abundances under the furfural stress in comparison with the metabolite abundances without furfural stress in the parental strain or the evolved strains. Metabolites written in red color are directly related to cofactor regulation. 2-4-HPE, 2-(4-hydroxyphenyl)ethanol; 3-HP, 3-hydroxypropionate; AcAce, acetoacetate; Ade, adenosine; AMP, adenosine-5'-monophosphate; ALA, alanine; APS, adenosine-5'-phosphosulfate; ASP, aspartate; Ben, benzoate; Beta-HB, β-hydroxybutyrate; CLA, cyano-L-alanine; CMP, cytidine-5'-monophosphate; CYS, cysteine; Eth, ethanolamine; GABA, γ-aminobutyric acid; GLU, glutamate; GLY, glycine; Glyce, glycerate; Glyco, glycolate; GOX, glyoxylate; GSH, glutathione; Gs-sG, glutathione-sulfur complex; Gua, guanine; H-CYS, homocysteine; Hpyr, hydroxypyridine; H-SER, homoserine; Ino, inosine; LEU, leucine; LYS, lysine; MET, methionine; MSX, methionine sulfoxide; MTA, 5'-deoxy-5'-methyl thioadenosine; NAD+, nicotinamide adenine dinucleotide +; N-mALA, N-methylalanine; Orn, ornithine; Oxa, oxalate; P2C, pyrrole-2-carboxylate; Pace, phenylacetate; PEP, phosphoenolpyruvate; PHE, phenylalanine; Plac, phenyllactate; PPP, pentose phosphate pathway; Pyr, pyruvate; Sac, saccharopine; Sal, salicylaldehyde; SER, serine; Spe, spermine; Sped, spermidine; Squ, squalene; Tere, terephthalate; Thr, threose; Thr-OH, threitol; Thy, thymine; TRP, tryptophan; Trx, thioredoxin; Ts-sT, thioredoxin-sulfur complex; TYR, tyrosine; Ura, uracil; Uri, uridine; VAL, valine; Xan, xanthine.

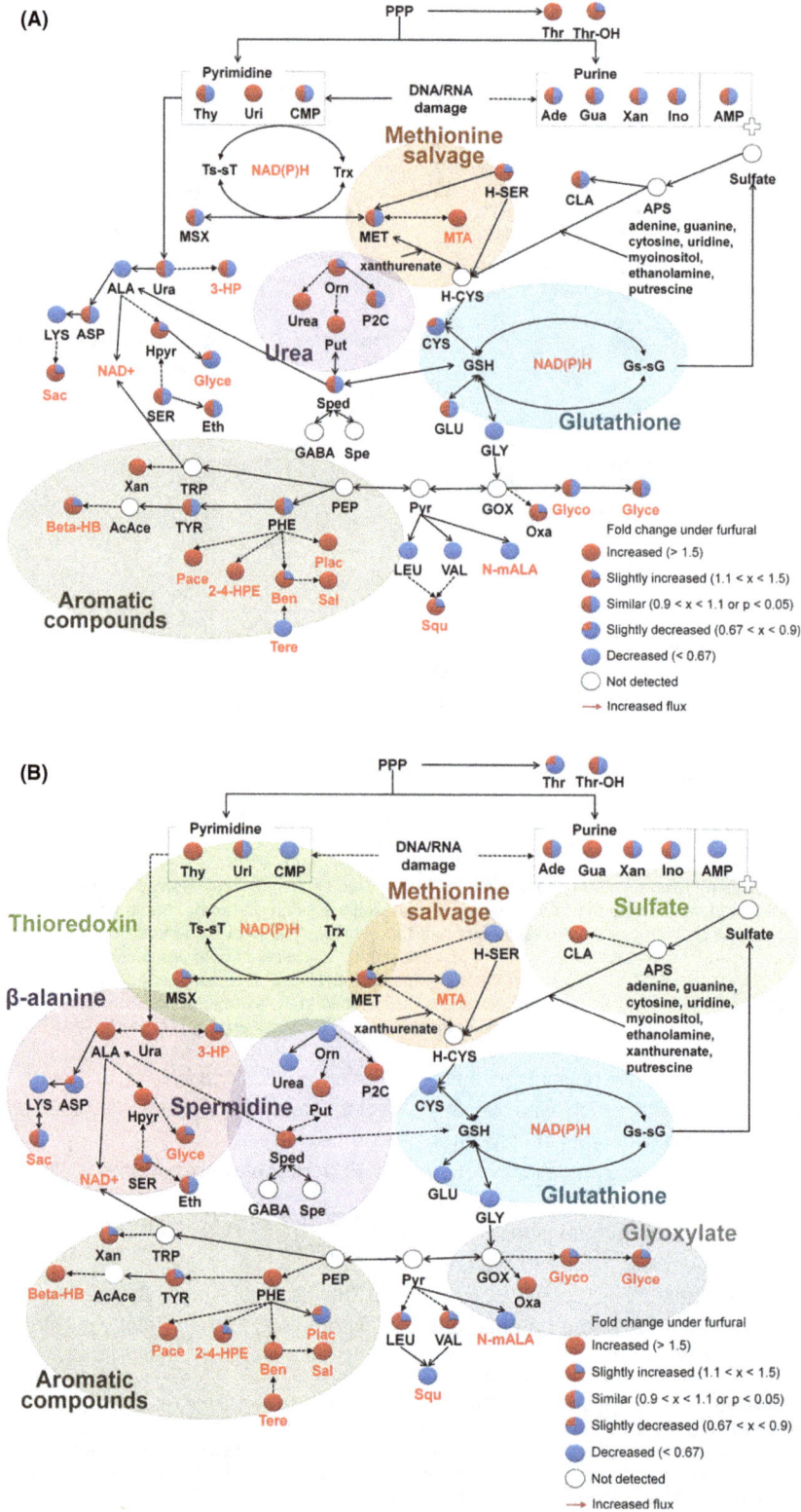

experiments. Three evolved phenotypes were independently generated in separate culture tubes through serial transfers. *Saccharomyces cerevisiae* was cultivated as a facultative anaerobe in 10 ml of YPD medium [1% (w/v)

yeast extract, 2% peptone and 2% glucose] containing 20 mM of furfural in a shaking flask at 30°C and 200 r.p.m. When the culture reached the late exponential phase, 1% (v/v) of cell cultures in each tube were transferred to a fresh medium containing 20 mM furfural independently. The cultivation was repeated under the same conditions for up to 50 transfers. The inoculation of 1% (v/v) of the culture into fresh medium in each transfer for repeated batch cultures was considered as $100\times$ dilution of the culture in each transfer. Based on this consideration, the propagation of cells during each culture was formulated as $2^n = 100$, in which n was solved for the number of generations in each culture. Therefore, the numbers of generations (n) for each transfer and 50 transfers were determined to be ~6.64 and ~332 generations respectively. Three evolved strains were isolated from each of the final cultures by streak-outs on YPD agar plates. For further experiments, biological duplications of all three evolved strains separately obtained by an independent evolutionary process were utilized.

Measurement of growth phenotype

To assess growth performance, the parental and evolved strains were cultivated in 100 ml of YPD medium containing various concentrations of furfural ranging from 0 to 40 mM in a shaking flask at 30°C and 200 r.p.m. Cell growth was measured as absorbance at 600 nm (A_{600}; Mark Microplate Spectrophotometer; Bio-Rad, Hercules, CA, USA). For the verification of phenotypic stability, the evolved strains were cultivated in medium without furfural for up to five transfers (~33 generations). The relative growth of the obtained cells was evaluated under furfural exposure.

For the analysis of extracellular metabolites, supernatants obtained after centrifugation at 13 000 r.p.m. for 5 min were filtered through 0.2 μm syringe filters prior to high performance liquid chromatography (HPLC; Agilent 1100; Agilent Technologies, Santa Clara, CA, USA) with a refractive index detector (G1362A; Agilent Technologies). HPLC was carried out on an Aminex HPX-87H column (H^+ form; Bio-Rad) operating at 65°C with 5 mM H_2SO_4 as a mobile phase and at a flow rate of 0.5 ml min^{-1} to measure the concentrations of glucose, ethanol, furfural and glycerol. All the analyses were conducted in duplicate. To determine the dry cell mass, 10 ml of culture broth was centrifuged at 13 000 r.p.m. for 5 min at 4°C and washed twice using phosphate-buffered saline (KH_2PO_4 0.24 g l^{-1}, KCl 0.2 g l^{-1}, NaCl 8 g l^{-1} and Na_2HPO_4 1.44 g l^{-1} at pH 7.4). The collected cell pellet was then dried using a speed vacuum concentrator (Labconco, Kansas City, MO, USA).

Sample preparation and intracellular metabolite analysis

Six replicates of the parental and evolved strains (i.e. biological duplications of all three evolved strains separately obtained from independent evolutionary processes) were prepared for metabolite analysis. Culture samples were collected at the early exponential phase, when the effect of furfural still remained. Fast filtration was carried out following the method described in a previous study (Kim *et al.*, 2013). In brief, within < 30 s, 1 ml of the collected sample was vacuum-filtered through a nylon membrane filter (0.45 μm pore size, 30 mm diameter; Whatman, Piscataway, NJ, USA), washed with 5 ml of distilled water at room temperature, rapidly mixed with 20 ml of acetonitrile/water (ACN) mixture (1:1, v/v) at −20°C and frozen in liquid nitrogen. After thawing on ice, the cell-loaded filters and solvent mixture were vortexed for 3 min for further extraction and centrifuged at 16 100 rcf for 5 min at 4°C. The supernatant (1 ml) was collected and vacuum-dried using a speed vacuum concentrator. The concentrate was then resuspended in 0.5 ml of ACN mixture to remove the lipids and wax and was dried again.

Prior to GC/TOF MS analysis, the dried metabolite concentrates were treated with a two-stage derivatization method including methoxyamination with 5 μl of 40 mg ml^{-1} methoxyamine hydrochloride in pyridine (Sigma-Aldrich) at 30°C for 90 min and silylation with 45 μl of *N*-methyl-*N*-trimethylsilyltrifluoroacetamide (Fluka, Buchs, Switzerland) at 37°C for 30 min. A mixture of fatty acid methyl esters was added to the derivatized metabolites as a retention index marker. GC/TOF MS analysis was performed using an Agilent 7890A GC (Agilent Technologies) coupled with a Pegasus HT TOF MS (LECO, St. Joseph, MI, USA). A 1 μl aliquot of the derivatized metabolite was injected into the GC in splitless mode and was separated on an RTX-5Sil MS column (30 m × 0.25 mm, 0.25 μm film thickness; Restek, Bellefonte, PA, USA) and an additional 10 m long integrated guard column with temperature programmed at 50°C for 1 min, followed by ramping to 330°C at 20°C min^{-1} and holding for 5 min. The ion source and transfer line temperatures were 250°C and 280°C, respectively, and the ions were generated by a 70 eV electron beam. The mass spectra of the metabolites were acquired in a range of 85–500 $m\ z^{-1}$ at an acquisition rate of 10 spectra s^{-1}.

Metabolite identification and statistical analysis

The spectra obtained by GC/TOF MS analysis were preprocessed using ChromaTOF software (ver. 3.34; LECO) and were then reprocessed by BinBase, an in-house programmed database built for metabolite identification (Skogerson *et al.*, 2011). After normalization of each culture by the total peak area, the data sets were analysed by STATISTICA (ver. 7.1; StatSoft, Tulsa, OK, USA) for PCA.

Acknowledgements

This work was financially supported by the Advanced Biomass R&D Center of Korea (2011-0031359) funded by the Korean Government (MSIP). Experiments were performed at the Institute of Biomedical Science and Food Safety at the Food Safety Hall, Korea University.

Conflict of interest

None declared.

References

Allen, S.A., Clark, W., McCaffery, J.M., Cai, Z., Lanctot, A., Slininger, P.J., et al. (2010) Furfural induces reactive oxygen species accumulation and cellular damage in Saccharomyces cerevisiae. Biotechnol Biofuels 3: 2.

Almeida, J.R.M., Bertilsson, M., Gorwa-Grauslund, M.F., Gorsich, S., and Lidén, G. (2009) Metabolic effects of furaldehydes and impacts on biotechnological processes. Appl Microbiol Biotechnol 82: 625–638.

Ask, M., Bettiga, M., Mapelli, V., and Olsson, L. (2013) The influence of HMF and furfural on redox-balance and energy-state of xylose-utilizing Saccharomyces cerevisiae. Biotechnol Biofuels 6: 22.

Cadière, A., Ortiz-Julien, A., Camarasa, C., and Dequin, S. (2011) Evolutionary engineered Saccharomyces cerevisiae wine yeast strains with increased in vivo flux through the pentose phosphate pathway. Metab Eng 13: 263–271.

Carmel-Harel, O., and Storz, G. (2000) Roles of the glutathione- and thioredoxin-dependent reduction systems in the Escherichia coli and Saccharomyces cerevisiae responses to oxidative stress. Annu Rev Microbiol 54: 439–461.

Ding, M.-Z., Wang, X., Yang, Y., and Yuan, Y.-J. (2011) Metabolomic study of interactive effects of phenol, furfural, and acetic acid on Saccharomyces cerevisiae. OMICS 15: 647–653.

Francois, J., and Parrou, J.L. (2001) Reserve carbohydrates metabolism in the yeast Saccharomyces cerevisiae. FEMS Microbiol Rev 25: 125–145.

Gill, S.S., and Tuteja, N. (2010) Polyamines and abiotic stress tolerance in plants. Plant Signal Behav 5: 26–33.

Glebes, T.Y., Sandoval, N.R., Gillis, J.H., and Gill, R.T. (2015) Comparison of genome-wide selection strategies to identify furfural tolerance genes in Escherichia coli. Biotechnol Bioeng 112: 129–140.

Gorsich, S.W., Dien, B.S., Nichols, N.N., Slininger, P.J., Liu, Z.L., and Skory, C.D. (2006) Tolerance to furfural-induced stress is associated with pentose phosphate pathway genes ZWF1, GND1, RPE1, and TKL1 in Saccharomyces cerevisiae. Appl Microbiol Biotechnol 71: 339–349.

Heer, D., and Sauer, U. (2008) Identification of furfural as a key toxin in lignocellulosic hydrolysates and evolution of a tolerant yeast strain. Microb Biotechnol 1: 497–506.

Heer, D., Heine, D., and Sauer, U. (2009) Resistance of Saccharomyces cerevisiae to high concentrations of furfural is based on NADPH-dependent reduction by at least two oxireductases. Appl Environ Microbiol 75: 7631–7638.

Horváth, I.S., Franzén, C.J., Taherzadeh, M.J., Niklasson, C., and Lidén, G. (2003) Effects of furfural on the respiratory metabolism of Saccharomyces cerevisiae in glucose-limited chemostats. Appl Environ Microbiol 69: 4076–4086.

Jansen, M.L.A., Krook, D.J.J., Graaf, K.D., van Dijken, J.P., Pronk, J.T., and de Winde, J.H. (2006) Physiological characterization and fed-batch production of an extracellular maltase of Schizosaccharomyces pombe CBS 356. FEMS Yeast Res 6: 888–901.

Jung, Y.H., Kim, K.H. (2014) Acidic pretreatment. In Pretreatment of Biomass: Processes and Technologies. Pandey, A., Negi, S., Binod, P. and Larroche, C. (eds). Amsterdam: Academic Press, pp. 27–50.

Jung, Y.H., Park, H.M., Kim, I.J., Park, Y.-C., Seo, J.-H. and Kim, K.H. (2014) One-pot pretreatment, saccharification and ethanol fermentation of lignocellulose based on acid-base mixture pretreatment. RSC Adv 4: 55318–55327.

Kim, H.-S., Kim, N.-R., Kim, W., and Choi, W. (2012) Insertion of transposon in the vicinity of SSK2 confers enhanced tolerance to furfural in Saccharomyces cerevisiae. Appl Microbiol Biotechnol 95: 531–540.

Kim, S., Lee, D.Y., Wohlgemuth, G., Park, H.S., Fiehn, O., and Kim, K.H. (2013) Evaluation and optimization of metabolome sample preparation methods for Saccharomyces cerevisiae. Anal Chem 85: 2169–2176.

Lin, F.-M., Qiao, B., and Yuan, Y.-J. (2009) Comparative proteomic analysis of tolerance and adaptation of ethanologenic Saccharomyces cerevisiae to furfural, a lignocellulosic inhibitory compound. Appl Environ Microbiol 75: 3765–3776.

Liu, Z.L. (2006) Genomic adaptation of ethanologenic yeast to biomass conversion inhibitors. Appl Microbiol Biotechnol 73: 27–36.

Liu, Z.L. (2011) Molecular mechanisms of yeast tolerance and in situ detoxification of lignocellulose hydrolysates. Appl Microbiol Biotechnol 90: 809–825.

Liu, Z.L., Slininger, P.J., Dien, B.S., Berhow, M.A., Kurtzman, C.P., and Gorsich, S.W. (2004) Adaptive response of yeasts to furfural and 5-hydroxymethylfurfural and new chemical evidence for HMF conversion to 2,5-bis-hydroxymethylfuran. J Ind Microbiol Biotechnol 31: 345–352.

Liu, Z.L., Slininger, P.J., and Gorsich, S.W. (2005) Enhanced biotransformation of furfural and hydroxymethylfurfural by newly developed ethanologenic yeast strains. Appl Biochem Biotechnol 121–124: 451–460.

Liu, Z.L., Ma, M., and Song M. (2009) Evolutionarily engineered ethanologenic yeast detoxifies lignocellulosic biomass conversion inhibitors by reprogrammed pathways. Mol Genet Genomics 282: 233–244.

Miller, E.N., Jarboe, L.R., Turner, P.C., Pharkya, P., Yomano, L.P., York, S.W., et al. (2009) Furfural inhibits growth by limiting sulfur assimilation in ethanologenic Escherichia coli strain LY180. Appl Environ Microbiol 75: 6132–6141.

Modig, T., Lidén, G., and Taherzadeh, M.J. (2002) Inhibition effects of furfural n alcohol dehydrogenase, aldehyde dehydrogenase and pyruvate dehydrogenase. *Biochem J* **363**: 769–776.

Nilsson, A., Gorwa-Grauslund, M.F., Hahn-Hägerdal, B., and Lidén, G. (2005) Cofactor dependence in furan reduction by *Saccharomyces cerevisiae* in fermentation of acid-hydrolyzed lignocellulose. *Appl Environ Microbiol* **71**: 7866–7871.

Palmqvist, E., and Hahn-Hägerdal, B. (2000) Fermentation of lignocellulosic hydrolysates. I: inhibition and detoxification. *Bioresour Technol* **74**: 17–24.

Palmqvist, E., Almeida, J.S., and Hahn-Hägerdal, B. (1999) Influence of furfural on anaerobic glycolytic kinetics of *Saccharomyces cerevisiae* in batch culture. *Biotechnol Bioeng* **62**: 447–454.

Park, S.-E., Koo, H.M., Park, Y.K., Park, S.M., Park, J.C., Lee, O.-K., *et al.* (2011) Expression of aldehyde dehydrogenase 6 reduces inhibitory effect of furan derivatives on cell growth and ethanol production in *Saccharomyces cerevisiae*. *Bioresour Technol* **102**: 6033–6038.

Patil, K.R., Rocha, I., Förster, J., and Nielsen, J. (2005) Evolutionary programming as a platform for *in silico* metabolic engineering. *BMC Bioinformatics* **6**: 308.

Shi, F., Li, Z., Sun, M., and Li, Y. (2011) Role of mitochondrial NADH kinase and NADPH supply in the respiratory chain activity of *Saccharomyces cerevisiae*. *Acta Biochim Biophys Sin* **43**: 989–995.

Shimogori, T., Kashiwagi, K., and Igarashi, K. (1996) Spermidine regulation of protein synthesis at the level of initiation complex formation of Met-tRNAi, mRNA and ribosomes. *Biochem Biophys Res Commun* **223**: 544–548.

Skogerson, K., Wohlgemuth, G., Barupal, D.K., and Fiehn, O. (2011) The volatile compound BinBase mass spectral database. *BMC Bioinformatics* **12**: 321.

Wang, X., Yomano, L.P., Lee, J.Y., York, S.W., Zheng, H., Mullinnix, M.T., *et al.* (2013a) Engineering furfural tolerance in *Escherichia coli* improves the fermentation of lignocellulosic sugars into renewable chemicals. *Proc Natl Acad Sci USA* **110**: 4021–4026.

Wang, X., Li, B.-Z., Ding, M.-Z., Zhang, W.-W., and Yuan, Y.-J. (2013b) Metabolomic analysis reveals key metabolites related to the rapid adaptation of *Saccharomyce cerevisiae* to multiple inhibitors of furfural, acetic acid, and phenol. *OMICS* **17**: 150–159.

Wilson, C.M., Yang, S., Rodriguez, M., Jr, Ma, Q., Johnson, C.M., Dice, L., *et al.* (2013) *Clostridium thermocellum* transcriptomic profiles after exposure to furfural or heat stress. *Biotechnol Biofuels* **6**: 131.

Xia, J.-M., and Yuan, Y.-J. (2009) Comparative lipidomics of four strains of *Saccharomyces cerevisiae* reveals different responses to furfural, phenol, and acetic acid. *J Agric Food Chem* **57**: 99–108.

Yang, J., Ding, M.-Z., Li, B.-Z., Liu, Z.L., Wang, X., and Yuan, Y.-J. (2012) Integrated phospholipidomics and transcriptomics analysis of *Saccharomyces cerevisiae* with enhanced tolerance to a mixture of acetic acid, furfural, and phenol. *OMICS* **16**: 374–386.

Zaldivar, J., Martinez, A., and Ingram, L.O. (1999) Effect of selected aldehydes on the growth and fermentation of ethanologenic *Escherichia coli*. *Biotechnol Bioeng* **65**: 24–33.

Tailor-made PAT platform for safe syngas fermentations in batch, fed-batch and chemostat mode with *Rhodospirillum rubrum*

Stephanie Karmann,[1,2] Stéphanie Follonier,[1]
Daniel Egger,[3] Dirk Hebel,[3] Sven Panke[2] and
Manfred Zinn[1,*]

[1]Institute of Life Technologies, University of Applied
Sciences and Arts Western Switzerland (HES-SO
Valais), Sion, Switzerland.
[2]Department of Biosystems Science and Engineering,
ETH Zurich (ETHZ), Basel, Switzerland.
[3]Infors AG, Bottmingen, Switzerland.

Summary

Recently, syngas has gained significant interest as renewable and sustainable feedstock, in particular for the biotechnological production of poly([*R*]-3-hydroxybutyrate) (PHB). PHB is a biodegradable, biocompatible polyester produced by some bacteria growing on the principal component of syngas, CO. However, working with syngas is challenging because of the CO toxicity and the explosion danger of H_2, another main component of syngas. In addition, the bioprocess control needs specific monitoring tools and analytical methods that differ from standard fermentations. Here, we present a syngas fermentation platform with a focus on safety installations and process analytical technology (PAT) that serves as a basis to assess the physiology of the PHB-producing bacterium *Rhodospirillum rubrum*. The platform includes (i) off-gas analysis with an online quadrupole mass spectrometer to measure CO consumption and production rates of H_2 and CO_2, (ii) an at-line flow cytometer to determine the total cell count and the intracellular PHB content and (iii) different online sensors, notably a redox sensor that is important to confirm that the culture conditions are suitable for the CO metabolization of *R. rubrum*. Furthermore, we present as first

applications of the platform a fed-batch and a chemostat process with *R. rubrum* for PHB production from syngas.

Introduction

Municipal solid waste (MSW) is a complex mixture of waste from households or public institutions. Although almost constant since 2004, the annual amount of MSW produced in the European Union is substantial, with over 5 tons per person in 2014 (Eurostat, 2014) and leading to a yearly amount of 66 million tons ending up in landfills (Eurostat, 2014).

One possible way to harvest this resource for a circular economy is pyrolysis, a method where organic material is thermochemically decomposed in the absence of oxygen (Velghe *et al.*, 2011). The gas fraction obtained from this process is called syngas, and it is composed of CO, H_2 and CO_2. It is of particular interest due to its potential as a cheap and renewable feedstock for (bio)-chemical processes. Furthermore, syngas does not interfere with the food chain and no agricultural land and water are needed for its production. It is also available as a direct waste product from the steel industry (Molitor *et al.*, 2016). Besides chemical catalysis (van de Loosdrecht and Niemantsverdriet, 2012; Mesters, 2016), also biological conversion is possible by some autotrophic bacteria that transform syngas into high-value products including H_2, methane, carboxylic acids (acetic and butyric acid), alcohols (ethanol, butanol) or esters (Munasinghe and Khanal, 2010; Bengelsdorf *et al.*, 2013). In contrast to chemical catalysis, bacteria are less sensitive to changes in the CO to H_2 ratio or traces of sulphur and chlorine in syngas (Mohammadi *et al.*, 2011).

Rhodospirillum rubrum, a Gram-negative, facultative photosynthetic, purple non-sulphur bacterium is a versatile CO-metabolizing bacterium. It can grow hetero- or autotrophically, aerobically as well as anaerobically. With the enzyme CO dehydrogenase (CODH) *R. rubrum* can oxidize CO with H_2O and produce CO_2 and H_2 (water-gas shift reaction) under anaerobic conditions at a redox potential below −300 mV (Heo *et al.*, 2001; Najafpour and Younesi, 2007; Younesi *et al.*, 2008). At the same time, *R. rubrum* can intracellularly accumulate carbon as poly([*R*]-3-hydroxybutyrate) (PHB) (Kerby *et al.*, 1992,

*For correspondence. E-mail manfred.zinn@hevs.ch

Funding information
Seventh Framework Programme, (Grant/Award Number: '311815').

1995; Do *et al.*, 2007). PHB is a biopolyester of the poly-hydroxyalkanoate (PHA) family that is particularly interesting because of its polypropylene-like properties, biodegradability and biocompatibility (Sudesh *et al.*, 2000; Zinn *et al.*, 2001). It has been shown that the PHB production on syngas reaches larger quantities when *R. rubrum* is cofed with acetate or malate. As an example, for cultures grown in serum vials with syngas as carbon source, the PHB content increased from 2 to 20 wt % when the acetate concentration in the medium was increased from 0 to 10 mM (Revelles *et al.*, 2016).

To date, PHB from syngas has not been produced at industrial scale yet because of insufficient PHB productivities. To be able to perform a systematic process optimization with *R. rubrum* and get a detailed understanding of the cell physiology, a tailor-made platform equipped with process analytical technology (PAT) is a necessity. PAT is part of the 'quality by design' guideline published by the American Food and Drug Administration (FDA). Its goal is to ensure the quality of a (pharmaceutical) product by a deep understanding of its production process (FDA, 2004). PAT includes real-time analytical monitoring, interpretation and feedback controls by a computer or the operator. For syngas fermentations, it is mainly desirable to follow the consumption of the substrate CO (dissolved in the culture broth and in the off-gas), the product formation (intracellular PHB and CO_2 and H_2 in the off-gas) and the biomass production.

Biomass quantification is a key parameter in process monitoring and control, it can be determined either at-line by measuring the optical density at 600 nm (OD_{600}), in-line with capacitance or optical probes (Kiviharju *et al.*, 2007), or as total cell count (TCC) with flow cytometry (FCM). FCM gained importance for PAT with the increasing availability of at-line methods and fully automated online systems (Abu-Absi *et al.*, 2003; Broger *et al.*, 2011; Hammes *et al.*, 2012). Furthermore, FCM can be used to quantify the intracellular PHB content in near real-time (Kacmar *et al.*, 2006; Lee *et al.*, 2013; Karmann *et al.*, 2016). A recently published staining protocol based on BODIPY493/503 and SYTO 62 for PHB-DNA-double staining ensures a complete recording of TCC including cells not containing PHB and the determination of the PHB content excluding false positives caused by cell debris or free PHB from lysed cells (Karmann *et al.*, 2016).

Typically, at-line or online gas chromatography (GC) is used to measure the syngas composition in the off-gas (Do *et al.*, 2007; Chipman, 2009; Revelles *et al.*, 2016). However, quadrupole mass spectrometry (QMS) is more suitable for online analysis of fermentation off-gases than GC as the measurement is faster (in the order of 1 min). Nevertheless, syngas is a complex mixture of gases and

difficult to quantify, in particular because N_2 and CO both have a molecular weight of 28. Therefore, a special analysis protocol is needed (Le *et al.*, 2015).

Few syngas fermentation platforms have been described in literature in the last years. Do *et al.* (2007) presented a fermentation platform including an *in situ* syngas production facility for the gasification of corn and equipped with tools to remove char, tar and ash from the gas before sparging it into a bioreactor, where *R. rubrum* was grown to produce PHB. The fermentation was followed with real-time off-gas analysis based on GC, but biomass and PHB quantifications were performed offline and, in case of PHB, involved a lengthy extraction procedure. Some years later, a new platform was developed in the same laboratory, this time including parallel 14 l bioreactors and a remote monitoring and control panel programmed with LabVIEW (Chipman, 2009). An alarm system was set up to send a text message to the operator in case of overpressure and if deviations in the gas mass flow rates occurred.

For syngas fermentations, safety aspects have to be considered as large amounts of H_2 and CO are continuously sparged into the bioreactor. H_2 can be hazardous if not handled properly because of a high flammability (Table 1). Strong ultraviolet radiation is generated when it burns and destructive explosions are likely to occur in case ignition happens in a confined space. Being colourless, odourless, tasteless and burning with an almost invisible flame, H_2 is difficult to detect. In addition, H_2 molecules are extremely small and light, which makes them prone to leak. One should also note that, although not corrosive, H_2 can alter the mechanical properties of metals such as steels, aluminium, titanium, nickel and their alloys ('hydrogen embrittlement') (Louthan *et al.*, 1972).

Like H_2, CO is colourless, odourless, tasteless and flammable (Table 1). It is, however, much heavier than

Table 1. Characteristics of H_2 and CO related to safety issues

	H_2	CO
Flammability range in air	4.0–75.0 vol%[a]	12.5–74.2 vol%[b]
Auto-ignition temperature in air	585°C[c]	605°C[b]
Specific gravity (compared to air at 20°C, 1 atm)	0.07[a]	0.97[b]
Permissible exposure limit (PEL)	n. a.	35 ppm[b]
Recommended exposure limit (REL)	n. a.	50 ppm[b]
Immediately dangerous to life or health value (IDLH)	n. a.	1200 ppm[b]

n. a. not available
a. Zabetakis (1965).
b. NIOSH (2016).
c. White *et al.* (2006).

H_2 and will mainly follow the air streams in case of leakage. The major hazard is its toxicity to humans, already in the ppm range. CO binds to haem proteins in human blood with a more than 200 times higher affinity than oxygen (Joels and Pugh, 1958), in particular to haemoglobin (Hb) thereby preventing oxygen transport. Exposure to 30–50 ppm CO will result in 5–8% COHb (of total Hb) and have adverse effects on healthy patients, while exposure to more than 600 ppm CO will increase the COHb up to 50%, the risk of death, caused by cardiac arrhythmia, is high (Dolan, 1985; ATSDR, 2012). As a result, exposure limits have been set by the safety institutions of most countries, for example by the U.S. National Institute for Occupational Safety and Health (NIOSH) and the Occupational Safety and Health Administration (Table 1).

In this article, we describe the syngas fermentation platform that we set up for the production of PHB by *R. rubrum* including (i) a series of safety measures for handling CO and H_2, and (ii) different PAT tools for monitoring the off-gas as well as the production of biomass and PHB during fed-batch and chemostat processes. This platform allowed the collection of particularly useful data for physiological studies that will build a basis to further improve the PHB productivity.

Results

Design of a safe laboratory platform

Because of the hazardousness of CO and H_2, an appropriate laboratory environment is required for syngas fermentations. This includes specific elements to (i) prevent gas leakages, (ii) detect gas leaks and raise an alarm in case they occur, and (iii) deal with gas leaks in case of emergencies so as to avoid any negative effects on human health and material damage.

An overview of the safety measures that we designed specifically for bioprocesses with two Labfors 5 benchtop bioreactors (3.6 and 13 l) is given in Fig. 1. Specifically, fume hoods ensure low levels of H_2 and CO in case of leakage, the gas cylinders are stored outside the building and safety shutoff electrovalves control the gas alimentation to the fume hoods. Detectors for CO and H_2 with both visual and sound alarms are used to warn the personnel in case of elevated concentrations. The first alarm thresholds are set to 100 ppm and 0.8 vol% for CO and H_2, respectively, and the second threshold to 150 ppm and 1.6 vol%. In case the second threshold is reached, the safety electrovalves are automatically closed, as do they if a power shortage occurs or if an emergency switch is triggered.

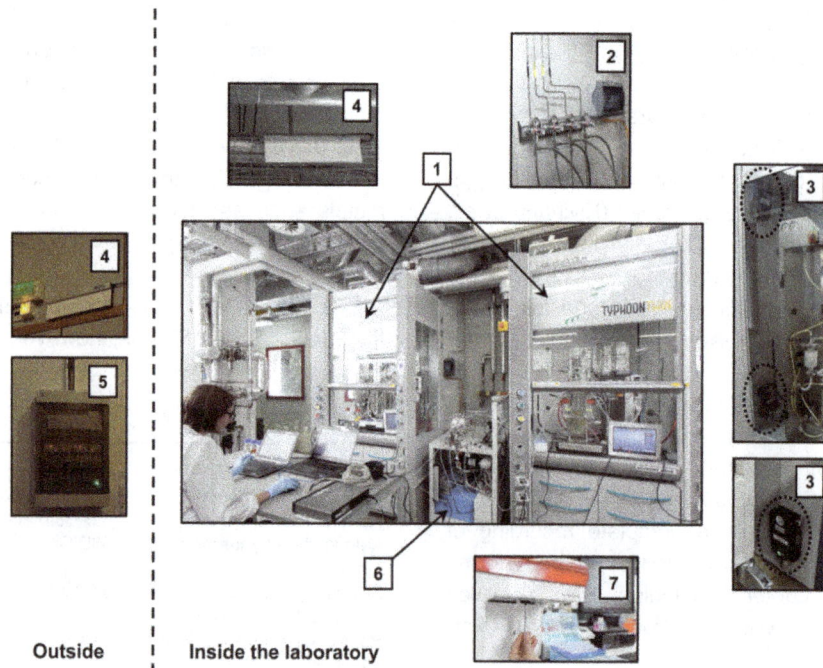

Outside | **Inside the laboratory**

Fig. 1. Safe PAT platform for syngas fermentations. Two fume hoods harbour each a bioreactor and ensure the rapid evacuation of H_2 and CO in case of leakage (1). Gas pipes are used for the transport of H_2, CO, CO_2 and N_2 from gas cylinders stored outside the building into the laboratory, and they are equipped with both safety shutoff electrovalves and manual pressure regulators (2). Detectors for H_2 and CO are located inside each fume hood in the upper and lower part, respectively, according to the gas density. Because of the higher risk of CO, one additional detector is placed for this gas in the vicinity of the fume hoods (3). Visual and sound alarms are used to warn the personnel both inside and outside the laboratory in case of a problem (4). The concentrations of H_2 and CO measured by the detectors are safely accessible on a control panel located outside the laboratory (5). A quadrupole mass spectrometer (QMS) continuously analyses the off-gases during a bioprocess (6), and the total cell count as well as the intracellular PHB content are measured at-line with a flow cytometer (7).

As an additional measure to prevent leakage, special care was taken to always use appropriate fittings including nuts, ferrules and tubing inserts for pipe and tubing connections. Furthermore, we selected as material for flexible tubings perfluoroalkoxy alkane (PFA), which exhibits similar properties as polytetrafluoroethylene and is resistant to high temperature and has a low gas permeability (Extrand and Monson, 2006).

Development of a continuous off-gas analysis method using mass spectrometry

We developed a fast and quantitative method to study the variation of H_2, CO, CO_2, N_2 and H_2O in the off-gas based on mass spectrometry. This technique is very powerful, yet quantification can be challenging in case of complex mixtures like syngas. Indeed, syngas analysis results in the overlap of the masses of several gases (e.g. both CO and N_2 have a mass to charge ratio m/z of 28) or fragments thereof (see Table S2, Supplementary data). The QMS software is able to distinguish the different gases by considering not only their molecular masses but also the masses of their fragments whose relative abundancy is fixed for given analysis settings. Nevertheless, accurate measurements can only be achieved if the analysis settings (voltage, electron emission, electron energy, dwell and settle times) are properly chosen and the system precisely calibrated (see Experimental procedures and Table S2). In addition, as the fragmentation pattern of the gases is critical for these analyses, we determined them experimentally under the same settings as for the off-gas analysis instead of relying on the software library. The analysis settings were optimized until the measurements of both pure gases and syngas mixtures of known composition reached satisfactory results, that is a relative error below 8% (arbitrary threshold). An example of such a verification test is given in Supplementary data Table S1, showing a relative error of the measurements < 5%.

Set-up of a syngas fermentation platform for batch, fed-batch and chemostat cultivations

We developed a versatile laboratory-scale fermentation platform for the optimization of PHB production from syngas with R. rubrum. The set-up for batch, fed-batch and chemostat cultivations is shown in Fig. 2. It was possible to use either a 3.6 or 13 l bioreactor as the vessels were easily interchangeable. A custom-made gas mixing station with five individual mass flow controllers (MFCs) enabled generating accurate mixtures of syngas. A liquid MFC allowed the precise feeding of a concentrated substrate (here acetic acid) for the fed-batch phase. The solution was delivered directly in the culture broth with a

metallic tube, which avoided irregular feeding by drop formation when applying very small flow rates. The bioreactors were equipped with standard sensors [pH and antifoam (AF)] and with a redox sensor providing online data. The off-gas was passed through a condenser to remove as much H_2O as possible before it reached the QMS. If necessary, part of the gas could be recycled using a gas-recycling loop. A balance was placed underneath the culture vessel to estimate the current culture volume that was necessary to calculate the feed or dilution rate (see Experimental procedures).

In case of chemostat cultivation, a second pump for larger flow rates continuously transferred medium (here modified RRNCO medium) into the bioreactor. To keep the volume constant during continuous cultivation, the culture broth was pumped out into the harvest container using a metallic tube installed at a level corresponding to a chosen volume (Fig. 2). All data were recorded by the same software and could be built into automated feedback-control loops.

Fed-batch fermentation with R. rubrum on syngas

The syngas fermentation platform was tested for R. rubrum in a fed-batch fermentation starting with a batch phase on modified RRNCO medium supplemented with 10 mM acetic acid and a continuous syngas aeration at a specific gassing rate of 0.1 l l^{-1} min^{-1}. The fed-batch phase was started after 43 h with a continuous feed of acetic acid (see Experimental procedures), required to trigger PHB accumulation (Fig. 3). During the batch phase, the cells grew exponentially at maximal specific growth rate μ_{max} of 0.1 h^{-1} (based on the OD_{600}) until CO limitation was reached. This limitation was indicated by the dissolved CO (DCO) concentration that reached non-detectable values (< 2 μM) 43 h after inoculation (Fig. 3). The redox potential, which has been previously shown to influence the activity of the CODH (Heo et al., 2001), followed a similar trend and dropped from −160 mV at the time of inoculation to −550 mV at the end of the batch phase. The CO consumption and the H_2 and CO_2 production rates determined from the off-gas measurements also reached a plateau at the end of the batch phase.

The PHB formation was followed by measuring the mean green fluorescence (mean FL1) of BODIPY 493/503 stained cells by FCM. Only a small amount of PHB, approximately 5%, was accumulated during the batch phase of the fermentation.

At the end of the fed-batch (after 184 h of cultivation), an OD_{600} of 23.5 and a cell dry weight (CDW) of 5.5 g l^{-1} were reached. The PHB content increased during the fed-batch phase and reached a final amount of almost 20 wt %. Interestingly, the TCC increased mainly during the first phase of the fed-batch (until 110 h) and stayed constant,

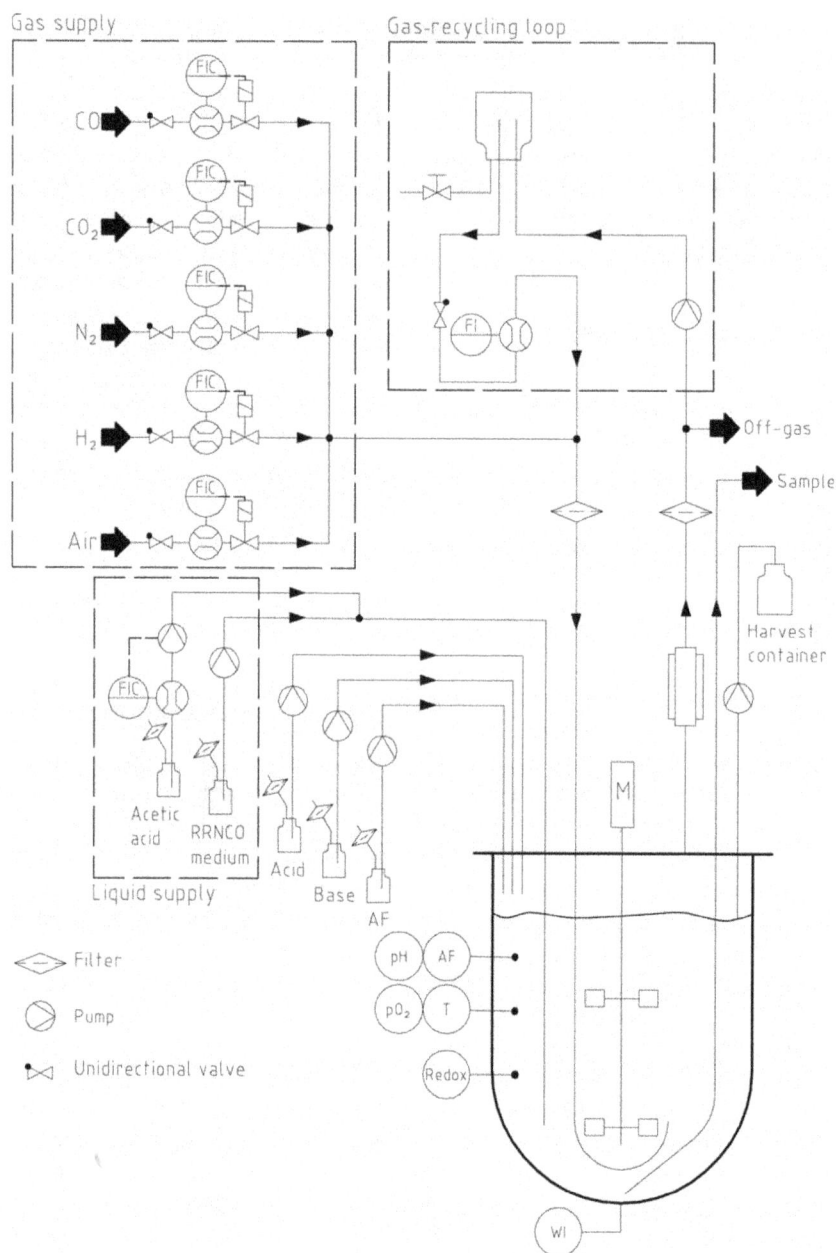

Fig. 2. Schematic layout of the bioreactor for fed-batch or chemostat cultivations on syngas. The gas supply is performed with five mass flow controllers (MFCs) shown as flow indicator controller (FIC) connected to different gas cylinders. The liquid supply consists of feed for acetic acid controlled with a liquid MFC and a feed for modified RRNCO medium. A gas-recycling loop can be activated to reuse off-gas with a high CO content. Online sensors are installed to measure and/or control pH, antifoam (AF), pO_2, temperature (T), redox potential (Redox) and a balance (WI). The off-gas passes a condenser and is then connected to the quadrupole mass spectrometer (QMS) for online off-gas analysis. *Only in use for continuous cultivations.

whereas the PHB content increased until the end. The redox potential remained constant at around −600 mV.

Chemostat cultivation with R. rubrum *on syngas*

A continuous bioprocess was performed starting with the same batch phase conditions as for the fed-batch described above. After 43 h, the continuous mode was started by constantly adding fresh modified RRNCO medium supplemented with 10 mM acetic acid at a dilution rate of $D = 0.02$ h^{-1}. Syngas was constantly sparged at 0.1 l l^{-1} min^{-1} into the medium. After two volume changes (50 h), all measured parameters reached a plateau indicating the steady state was achieved. Under these conditions, the OD_{600} was stable at 3.3 ± 0.04, the CDW at 0.67 ± 0.02 g l^{-1} and the

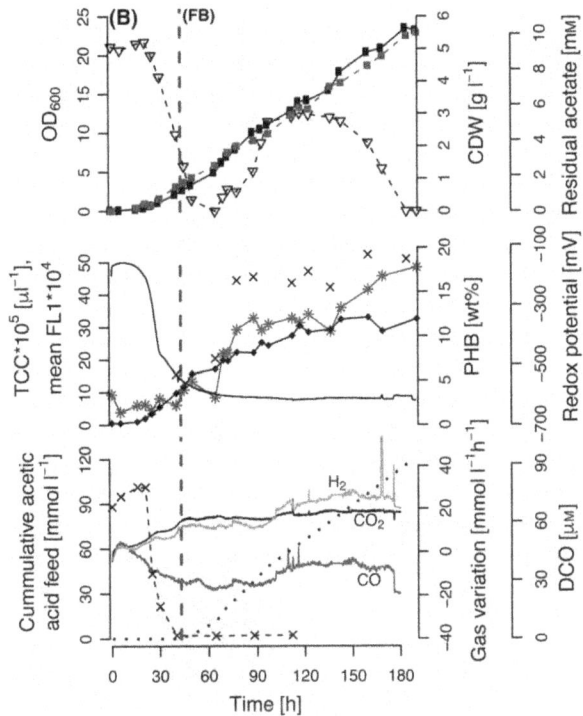

Fig. 3. Fed-batch fermentation with *Rhodospirillum rubrum* with continuous syngas aeration at 0.1 l l^{-1} min^{-1} and continuous feeding of acetic acid. The batch phase (B) and the fed-batch phase (FB) are separated by a grey dashed line. The upper panel shows the culture optical density at 600 nm (● OD$_{600}$), residual acetate (▽) and the cell dry weight (■ CDW). The middle panel displays the redox potential (-), intracellular PHB content in wt% determined by GC (×), as well as the PHB content (✳ mean FL 1) and the total cell count (♦ TCC) both measured by flow cytometry. The lower panel shows the cumulative acetic acid feed (···), the dissolved CO (× DCO) and the consumption and production rates of CO, H$_2$ and CO$_2$ as labelled in the plot.

Fig. 4. Continuous cultivation with *Rhodospirillum rubrum* on modified RRNCO medium containing 10 mM of acetate (*D* = 0.02 h^{-1}) and a continuous syngas supply at 0.1 l l^{-1} min^{-1}. The batch phase (B) and the chemostat phase (C) are separated by a grey dashed line. The upper panel contains biomass data with optical density at 600 nm (● OD$_{600}$), the cell dry weight (■ CDW) and the acetate concentration in the culture supernatant (▽). The middle panel shows the redox potential (–), the PHB content (✳ mean FL 1) and the total cell count (♦ TCC) measured by flow cytometry. The lower panel contains the concentration of dissolved CO (× DCO) and the CO, H$_2$ and CO$_2$ consumption and production rates as labelled in the plot.

intracellular PHB content at 4.7 wt%, as determined by FCM. Also, the CO consumption rate and H$_2$ and CO$_2$ production rates were constant. The DCO was below the detection limit of 2 µM, indicating a CO mass transfer limitation. The redox potential stayed constant at less than −600 mV even though the culture was constantly provided with (aerobic) modified RRNCO medium that had not been purged with N$_2$ or syngas (Fig. 4).

Discussion

To date, syngas fermentation has been performed mainly at small-scale in serum vials (Abubackar *et al.*, 2012; Revelles *et al.*, 2016). The advantage of this cultivation system is that it does not require a continuous syngas flow and therefore only limited safety measures are necessary. All manipulations can be simply accomplished in a fume hood. The gas tight vials can be incubated in regular incubators as no gas exchange with the environment is possible. However, a significant

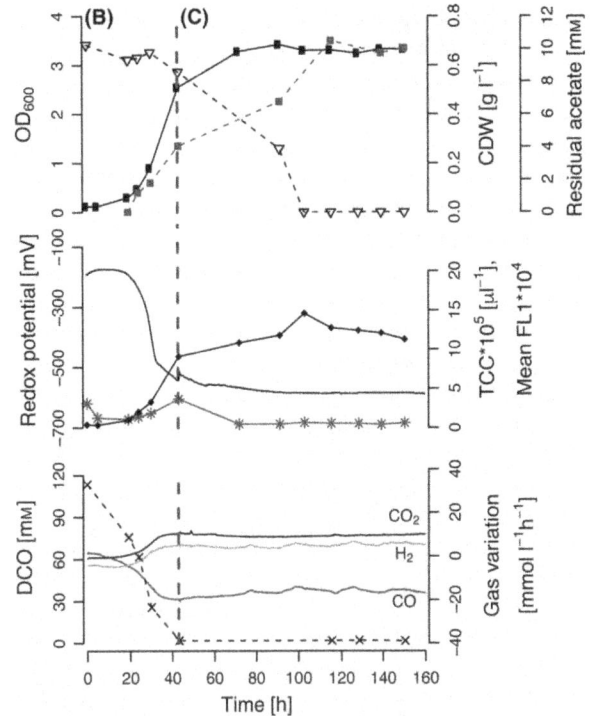

drawback of the serum vial cultivation technique is that only small growth rates of 0.02 h^{-1} and low final biomass concentration with an OD$_{600}$ around 1 can be reached (Revelles *et al.*, 2016). In addition, serum vials or shake flasks suffer from a low gas mass transfer and are rarely equipped with online sensors and consequently lead to a poor bioprocess understanding.

In this work, we set up a platform allowing to carry out syngas fermentation safely at larger scale while gaining a maximum of information about the bioprocess with specifically integrated PAT tools (redox sensor, QMS, FCM).

This platform was successfully applied to fed-batch and chemostat cultivations of *R. rubrum* on syngas. As expected, *R. rubrum* reached higher cell densities in both types of bioprocesses than with the cultivations in serum vials mentioned above. The CDW and PHB accumulation differed significantly between fed-batch and chemostat steady states (Figs 3 and 4). Due to the different amounts of total cosubstrate feeding, higher CDW and PHB values were found in the fed-batch fermentation. Furthermore, in the chemostat steady state the cells were growing exponentially and therefore had to invest a significant portion of carbon into biomass formation, whereas during the fed-batch cultivation, the cell number remained constant in the second half of the bioprocess (see TCC data in Fig. 3). Consequently, R. rubrum was able to use a bigger portion of carbon to synthesize PHB.

The online redox sensor confirmed that the redox potential in the medium was low enough (below −300 mV) to ensure the full activity of the CODH, the key enzyme of the CO metabolism of R. rubrum (Heo et al., 2001). A pO_2 sensor would not be sensitive enough to give this information as a pO_2 signal of 0.1% corresponds to a redox signal of up to −50 mV (Carius et al., 2013). The addition of a redox indicator such as resazurin to confirm anaerobic conditions was not required. Furthermore, no reducing agents like Na_2S or cysteine-HCl were necessary, even though aerobic acetic acid (fed-batch) or modified RRNCO medium (chemostat cultivation) were fed. This is important to mention as resazurin and cysteine-HCl have shown growth inhibiting effects on R. rubrum (Schultz and Weaver, 1982). Interestingly, the redox potential correlated very well with the DCO in the medium, especially during the batch phase (see Figs 3 and 4). This could potentially be implemented in the bioprocess control to rapidly detect CO limitation and replace the time-consuming at-line DCO quantification with myoglobin.

The CO content in the off-gas, determined with the QMS, stands in agreement with the DCO data that also reached a plateau after the switch to the fed-batch and the chemostat phase respectively (Figs 3 and 4). CO consumption and H_2 and CO_2 production rates increased in parallel. However, there is a shift in the off-gas variation during the fed-batch process at 112 h of cultivation that does not follow the theory where a smaller CO consumption rate is expected to lead to decreased H_2 and CO_2 production rates (Fig. 3). We realized that the off-gas filter was blocked due to too much humidity in the off-gas. This humid filter caused an overpressure in the bioreactor that presumably had an effect on the accuracy of the off-gas measurement. This problem can be solved by exchanging the off-gas condenser with a more powerful one. In general, the measurement of CO and CO_2 in the off-gas is necessary to establish carbon balances and calculate CO-based growth and PHB yields. Furthermore, the quantification of the H_2 production is also important as it can be sold as valuable by-product.

FCM is another flexible tool of the PAT platform. The PHB quantification by FCM needs much less time and sample volume than the GC method that requires 5–50 mg freeze-dried biomass, depending on the protocol (Brandl et al., 1988; Furrer et al., 2007). The PHB quantification by FCM has been shown to be at least as accurate as PHB quantification by GC (Karmann et al., 2016). In addition, during the fed-batch fermentation presented here (Fig. 3), the data obtained by the two quantification techniques correlated very well ($R^2 = 0.98$, supplementary data Fig. S2). However, care has to be taken at the very end of the process as stationary phase cells can have different staining properties than exponentially growing cells (Karmann et al., 2016), and for very small PHB contents (< 5 wt%) where the intercept of the correlation equation could introduce a small error (supplementary data Fig. S2).

In contrast to CDW, OD_{600} or measurements by in-line probes, the TCC determined by FCM provides information on whether the cells are dividing or whether the increasing biomass is due to intracellular PHB accumulation. As mentioned above, in Fig. 3 we can see clearly that the TCC is only increasing to a minor extent after 110 h of cultivation while the OD_{600} and CDW are still increasing continuously. This confirms that the latter measurements are influenced by PHB accumulation and thus do not provide accurate information about biomass in terms of the cell number.

Online sensors, especially temperature, pO_2 and pH sensors are part of the standard equipment of a bioreactor. In case of syngas fermentation, we showed the advantage of a redox sensor, an online QMS and an at-line FCM. All data from online measurements were collected by the same software and could be integrated in any feedback-control loop. For example, if the redox potential suggests a complete consumption of DCO, an automated increase in syngas flow rate or a higher agitation speed could be implemented to increase CO mass transfer. In contrast, if the QMS indicates a poor consumption of CO as it is the case at the beginning of the fermentation process, a pump in the gas-recycling loop could be switched on automatically and pump the CO-rich off-gas back into the culture vessel to avoid a waste of substrate (Fig. 2).

In conclusion, we presented here for the first time a safe laboratory-scale fermentation platform that is tailor-made for the production of PHB from syngas. We showed two applications of this PAT platform for syngas fermentation with R. rubrum, namely a fed-batch and a chemostat cultivation process. On one hand, with the fed-batch cultivation we showed that the PHB accumulation increased up to 20wt% when continuously feeding acetic acid as cosubstrate. On the other hand, chemostat cultivation is the technique of choice for process optimization as the influence of single cultivation parameters such as medium and syngas composition, pH, temperature and many more can be assessed individually in a very controlled and reproducible way. This is necessary for a better understanding of cell physiology to finally increase the PHB production from syngas. With only a few adaptations, the PAT platform for syngas fermentation can also be applied to biosynthesize a large variety of chemicals on other dangerous gas mixtures such as oxyhydrogen or methane. The QMS as well as the FCM are nearly unlimited in possible applications.

Experimental procedures

Safety installations

Two fume hoods (Typhoon twin, Labosystem, Rovellasca, Italy) were harbouring, each, a bioreactor and a cleaning- and sterilization-in-place unit (LabCIP; Infors AG, Bottmingen, Switzerland). Safety detectors for H_2 and CO (Models EC28 and CC28, respectively, GfG AG, Binz, Switzerland) were located inside each fume hood, and one detector for CO was located outside but in the vicinity of the fume hoods. The sound and visual alarms (inside and outside the laboratory) as well as the control panel outside the laboratory were acquired from GfG AG (Binz, Switzerland). Gas pipes for H_2, CO, CO_2 and N_2 (PanGas, Dagmersellen, Switzerland) were installed to transport the respective gases from cylinders stored outside the building into the laboratory and equipped with automatic shut-off electrovalves (Model 21X2KV; ODE S. r. l., Colico, Italy). The pipes were connected to semi-flexible PFA tubings (BGB Analytik AG, Boeckten, Switzerland) with appropriate fittings including nuts, ferrules and tubing inserts (Swagelok, Solon, United States).

Bacterial strain, growth conditions and media

Rhodospirillum rubrum S1(ATCC 11170), stored in 1.5 or 15 ml 16% glycerol stocks (OD_{600} = 3.33) at −80°C, was used for all experiments. The liquid medium for chemostat and fed-batch cultivations was a slightly modified version of the RRNCO medium (Kerby *et al.*, 1995). It contained per litre of deionized water: 250 mg of $MgSO_4 \cdot 7H_2O$, 132 mg of $CaCl_2 \cdot 2H_2O$, 1 g of NH_4Cl, 2.1 g of MOPS buffer, 1.0 g of yeast extract, 20 μM $NiSO_4$, 10 ml of a chelated iron-molybdenum solution (0.28 g of H_3BO_3, 2 g of Na_2EDTA, 0.4 g of $FeSO_4$ and 0.1 g of Na_2MoO_4 per litre of distilled water) and 2 μg of biotin. The medium was either sterile filtered (0.2 μm, Sartobran 300; Sartorius Stedim Biotech GmbH, Göttingen, Germany) or autoclaved except biotin and acetate which were always sterile filtered (0.22 μm filter; Sarstedt, Nümbrecht, Germany).

Precultures were grown in screw capped 1-l bottles (Müller + Krempel AG, Bülach, Switzerland) filled with 500 ml of sterile filtered modified RRNCO medium at pH 7 (0.22 μm filter; Sarstedt) containing 15 mM fructose. The precultures were inoculated directly with 12 ml glycerol stock to reach an OD_{600} of 0.08 and were then incubated at 30°C on an orbital shaker at 180 min^{-1} for 46–48 h. During this time, gas exchange with the aerobic environment was impossible and anaerobic conditions were reached after exhaustion of O_2 due to cell growth.

Bioreactor settings for fed-batch fermentation

The fed-batch cultivation was carried out in a 13-l benchtop bioreactor (Labfors 5; Infors AG) supervised by the supervisory control and data acquisition software Iris (Infors AG). The initial working volume was 10 l of autoclaved modified RRNCO medium supplemented with 10 mM acetate and 0.1 ml l^{-1} PPG2000 AF. The bioreactor was equipped with online sensors recording pH and redox potential (Hamilton Bonaduz AG, Bonaduz, Switzerland). The bioreactor was purged with syngas (25% CO, 25% H_2, 5% CO_2 and 45% N_2) at 0.1 l l^{-1} min^{-1} for at least one hour before inoculation to reach an OD_{600} of 0.1.

The pH in the culture broth was maintained at 7.0 ± 0.05 by automated addition of 4 M KOH or 3 M H_2SO_4. The temperature was kept constant at 30°C, and the culture was mixed with two six-blade Rushton impellers at 600 rpm. Syngas (25% CO, 25% H_2, 5% CO_2, 45% N_2) was sparged continuously into the culture through a sinter filter sparger at 0.1 l l^{-1} min^{-1}.

The batch phase was terminated after 43 h by starting the continuous feeding from a 2 M acetic acid stock solution with a pump controlled by a liquid MFC (mini Cori Flow M12; Bronkhorst Cori-Tech B.V., Ruurlo, the Netherlands). The feed rate was adjusted twice per day based on OD_{600} measurements as follows: As long as the acetate in the culture supernatant was consumed completely, it was calculated based on the empirically determined equation 0.17 mmol OD_{600}^{-1} l^{-1} h^{-1}. The culture volume was measured with a balance (Kern KMB-TM; Kern & Sohn GmbH) placed under the reactor. As soon as acetate accumulated in the medium, the feed rate was kept constant or reduced based on actual acetate consumption rates measured during the process.

Bioreactor settings for chemostat cultivation

The chemostat cultivation was carried out in a 3.6 l benchtop bioreactor (Labfors 5; Infors AG) with a constant working volume of 2 l. The batch phase was terminated after 43 h by starting the continuous feeding of sterile filtered modified RRNCO medium supplemented with 0.1 ml l^{-1} PPG2000 AF for a dilution rate $D = 0.02$ h^{-1}. Acetic acid was added separately at 0.2 g h^{-1} from a 2 M stock solution (final concentration of 10 mM in the total feed medium) (Fig. 2). The working volume was kept constant by continuously pumping culture through a harvest pipe installed at the volume of 2 l.

Samples were taken regularly during the bioprocesses. Analytics for biomass quantification, FCM and high pressure liquid chromatography (HPLC) and DCO were performed immediately. Pellets of biomass samples required for GC analysis were frozen at $-80°C$ and treated as indicated below.

Biomass quantification

Optical quantification of the biomass was performed spectrophotometrically at 600 nm (Libra S12; Biochrom, Cambridge, UK). For the measurement of the CDW, triplicates of 2 ml of cell culture (or suspension) were centrifuged in Eppendorf tubes (30,670 g, 4°C, 5 min), washed once with 1 ml 0.9% aqueous NaCl and then dried for 24 h in a 100°C oven. The tubes were then cooled down to room temperature in a desiccator and the CDW determined by subtracting the tare determined in advance.

FCM analysis for PHB and TCC determination

Fresh culture samples were analysed in duplicates with a BD Accuri C6 flow cytometer (FCM) (BD Bioscience, Erembodegern, Belgium) and a BODIPY 493/503-SYTO 62 double staining. The staining protocol published previously (Karmann et al., 2016) was adjusted slightly by increasing the incubation time to 30 min. This was due to the very slow growth rates especially at the end of the fed-batch process and during the phase of continuous cultivation. The influence of the growth phase (exponential or stationary) on the staining kinetics has been addressed recently by Karmann et al. (2016) because it is known that cell envelope properties change based on the growth phase (Wanner and Egli, 1990).

During the chemostat fermentation, PHB was only quantified by FCM. The PHB content in wt% was calculated based the correlation established during the fed-batch fermentation as shown in Supplementary data Fig. S2.

Determination of residual acetate by HPLC

The acetate concentration in the culture supernatant was determined by HPLC (1100 Series; Agilent, Santa Clara, USA) equipped with a 300 mm × 7.8 mm ion exchange column (Aminex® HPX-87H; Bio-Rad Laboratories, Hercules, USA). An isocratic elution was performed with 5 mM H_2SO_4 at a flow rate of 0.6 ml min^{-1} at 50°C for 30 min. Acetate was detected with a refractive index detector.

PHB quantification by GC

For the PHB quantification by GC, cell pellets from 30 to 50 ml culture (3,900 g, 4°C, 10 min) were harvested during the fed-batch fermentation and washed once with 0.9% aqueous NaCl, stored at $-80°C$ for at least 24 h and freeze-dried at $-80°C$ at a pressure of 0.25 mbar (Cryodos, Telstar, Terrassa, Spain). Methanolysis of 10–15 mg freeze-dried biomass was performed based on a previously published protocol with small modifications (Brandl et al., 1988): The biomass was mixed in a screw cap tube with 2 ml of methanol 15 vol% in H_2SO_4 and 2 ml of CH_2Cl_2 containing 2.5 g l^{-1} 2-ethyl-2-hydroxybutyric acid as internal standard. Purified bacterial PHB and 3-hydroxypentanoic acid (eNovation Chemicals, China) were used as external standards. The solution was incubated at 100°C for 3 h and then extracted in two steps with 2 ml dH_2O. The solvent phase was dried with anhydrous Na_2CO_3 and Na_2SO_4 and was subsequently analysed by GC with an Agilent 6890 GC-FID (G1530 A; Agilent) equipped with a BGD-Wax column (30 m·0.25 mm, film thickness 0.25 μm) (BGB Analytik, Böckten, Switzerland). The injection volume was 1 μl, the split ratio was 1:10 and the injection temperature was 250°C. The temperature was increased from 80°C to 240°C at a rate of 10°C min^{-1}, and 3 ml min^{-1} helium was used as carrier gas. A flame ionization detector at 285°C was used for detection.

DCO and off-gas quantification

The DCO was quantified based on an absorbance test using myoglobin as described earlier (Jones, 2007). The myoglobin solution was prepared without the dialysis step. Tests in our laboratory showed that this step did not improve the analysis.

The off-gas concentrations were quantified with a QIC Biostream quadrupole mass spectrometer and the data analysed using the QGA FT software (Hiden Analytical, Werrington, UK). Notably, N_2 was continuously flushed into the purge inlet (~150 ml min^{-1}) to avoid the accumulation of hazardous H_2 in the QMS.

All gases were measured with a Faraday detector and with a voltage, electron emission and electron energy of 830 V, 100 μA and 70 eV respectively. Seven gases were considered for the analysis, namely H_2, CO, CO_2, N_2, H_2O, O_2 and Ar. The measurement parameters for each gas are given in Table S2 (see Supplementary data). The background values for each gas except Ar were determined after flushing the QMS with pure Ar. Thereafter, calibration was performed with ambient air [from which the H_2O concentration was calculated from the room temperature, pressure and relative humidity using the Buck equation (Buck, 1981)], as well as with a

$CO_2/O_2/N_2$ gas cylinder and/or with syngas (CO, CO_2, H_2, N_2) artificially prepared with the highly accurate MFCs. All gases were purchased from PanGas (Dagmersellen, Switzerland). The calibration values were normalized with respect to N_2, which was present in the three calibration gases. The accuracy of the measurement was verified with gases of known composition before analysis, as shown in Supplementary data Fig. S1. The analysis time was in the range of 1 min.

The CO consumption rate n_{CO} in mol min^{-1} was calculated using N_2 as inert gas with the following equation (1)

$$n_{CO} = \frac{\frac{F_G}{V_m}\left(X_{CO_{out}}\frac{X_{N_2 in}}{X_{N_2 out}} - X_{CO_{in}}\right)}{100},\qquad(1)$$

where F_G is the total gas volume flow in ml min^{-1}, V_m the molar volume at 25°C in ml mol^{-1}, $X_{CO_{out}}$ the CO content in mol % in the off-gas measured by MS, $X_{N_{2in}}$ the N_2 portion in the syngas feed in mol %, $X_{N_{2out}}$ the N_2 content in mol % in the off-gas measured by MS, and $X_{CO_{in}}$ is the CO portion in syngas feed in mol %. The H_2 and CO_2 production rates (n_{H_2} and n_{CO_2}) were calculated accordingly using the modified versions of Eq. 1 by replacing CO-related parameters by H_2 and CO_2 ones, respectively.

Acknowledgements

The research leading to these results has received funding from the European Union's Seventh Framework Programme for research, technological development and demonstration under grant agreement no. 311815.

We thank to Aldo Vaccari and Michael Sequeira Carvalho (HES-SO), Andreas Hediger and Dieter Cueni (Infors AG) for technical help with the bioreactor setup and software issues, Jim Melling (HIDEN Analytical) for assisting in case of QMS issues, Antoine Fornage (HES-SO) for support with GC analytics and Immanuel Herbst (Infors AG) for drawing Fig. 2.

Conflict of Interest

Stephanie Karmann, Stéphanie Follonier, Sven Panke and Manfred Zinn declare that they have no conflict of interest. Daniel Egger and Dirk Hebel disclose that they are employed by Infors AG, the manufacturer of the bioreactors described in this manuscript.

References

Abu-Absi, N.R., Zamamiri, A., Kacmar, J., Balogh, S.J., and Srienc, F. (2003) Automated flow cytometry for acquisition of time-dependent population data. *Cytometry A* **51**: 87–96.

Abubackar, H.N., Veiga, M.C., and Kennes, C. (2012) Biological conversion of carbon monoxide to ethanol: Effect of pH, gas pressure, reducing agent and yeast extract. *Bioresour Technol* **114**: 518–522.

ATSDR Agency for Toxic Substances and Disease Registry (2012) Toxicological profile for carbon monoxide. URL https://www.atsdr.cdc.gov/ToxProfiles/tp201.pdf

Bengelsdorf, F.R., Straub, M., and Dürre, P. (2013) Bacterial synthesis gas (syngas) fermentation. *Environ Technol* **34**: 1639–1651.

Brandl, H., Gross, R.A., Lenz, R.W., and Fuller, R.C. (1988) *Pseudomonas oleovorans* as a source of poly (β-hydroxyalkanoates) for potential applications as biodegradable polyesters. *Appl Environ Microbiol* **54**: 1977–1982.

Broger, T., Odermatt, R.P., Huber, P., and Sonnleitner, B. (2011) Real-time on-line flow cytometry for bioprocess monitoring. *J Biotechnol* **154**: 240–247.

Buck, A.L. (1981) New equations for computing vapor pressure and enhancement factor. *J Appl Meteorol* **20**: 1527–1532.

Carius, L., Hädicke, O., and Grammel, H. (2013) Stepwise reduction of the culture redox potential allows the analysis of microaerobic metabolism and photosynthetic membrane synthesis in *Rhodospirillum rubrum*. *Biotechnol Bioeng* **110**: 573–585.

Chipman, D.C. (2009) Hybrid thermochemical/biological processing of biomass for the production of polyhydroxyalkanoates and hydrogen gas from *Rhodospirillum rubrum* cultured on synthesis gas. Graduate Theses and Dissertations, Iowa State University, Ames, Iowa. Paper 10945.

Do, Y.S., Smeenk, J., Broer, K.M., Kisting, C.J., Brown, R., Heindel, T.J., *et al.* (2007) Growth of *Rhodospirillum rubrum* on synthesis gas: conversion of CO to H_2 and poly-β-hydroxyalkanoate. *Biotechnol Bioeng* **97**: 279–286.

Dolan, M.C. (1985) Carbon monoxide poisoning. *Can Med Assoc J* **133**: 392–399.

Eurostat (2014) Municipal waste statistics. URL http://ec.europa.eu/eurostat/statistics-explained/index.php/Municipal_waste_statistics

Extrand, C.W., and Monson, L. (2006) Gas permeation resistance of a perfluoroalkoxy-tetrafluoroethylene copolymer. *J Appl Polym Sci* **100**: 2122–2125.

FDA (2004) Guidance for industry PAT — a framework for innovative pharmaceutical development, manufacturing, and quality assurance. URL http://www.fda.gov/downloads/drugs/guidances/ucm070305.pdf

Furrer, P., Hany, R., Rentsch, D., Grubelnik, A., Ruth, K., Panke, S., and Zinn, M. (2007) Quantitative analysis of bacterial medium-chain-length poly([R]-3-hydroxyalkanoates) by gas chromatography. *J Chromatogr A* **1143**: 199–206.

Hammes, F., Broger, T., Weilenmann, H.-U., Vital, M., Helbing, J., Bosshart, U., *et al.* (2012) Development and laboratory-scale testing of a fully automated online flow cytometer for drinking water analysis. *Cytometry A* **81**: 508–516.

Heo, J., Halbleib, C.M., and Ludden, P.W. (2001) Redox-dependent activation of CO dehydrogenase from *Rhodospirillum rubrum*. *Proc Natl Acad Sci USA* **98**: 7690–7693.

Joels, N., and Pugh, L.G.C.E. (1958) The complete dissociation curve of carboxyhaemoglobin. *J Physiol* **142**: 63–77.

Jones, S.T. (2007) Gas liquid mass transfer in an external airlift loop reactor for syngas fermentation. Retrospective Theses and Dissertations, Iowa State University, Ames, Iowa. Paper 15547

Kacmar, J., Carlson, R., Balogh, S.J., and Srienc, F. (2006) Staining and quantification of poly(3-hydroxybutyrate) in *Saccharomyces cerevisiae* and *Cupriavidus necator* cell populations using automated flow cytometry. *Cytometry A* **69**: 27–35.

Karmann, S., Follonier, S., Bassas-Galia, M., Panke, S., and Zinn, M. (2016) Robust at-line quantification of poly (3-hydroxyalkanoate) biosynthesis by flow cytometry using a BODIPY 493/503-SYTO 62 double-staining. *J Microbiol Methods* **131**: 166–171.

Kerby, R.L., Hong, S.S., Ensign, S.A., Coppoc, L.J., Ludden, P.W. and Roberts, G.P. (1992) Genetic and physiological characterization of the *Rhodospirillum rubrum* carbon monoxide dehydrogenase system. *J Bacteriol* **174**, 5284–5294.

Kerby, R.L., Ludden, P.W., and Roberts, G.P. (1995) Carbon monoxide-dependent growth of *Rhodospirillum rubrum*. *J Bacteriol* **177**: 2241–2244.

Kiviharju, K., Salonen, K., Moilanen, U., Meskanen, E., Leisola, M., and Eerikäinen, T. (2007) On-line biomass measurements in bioreactor cultivations: comparison study of two on-line probes. *J Ind Microbiol Biotechnol* **34**: 561–566.

Le, C.D., Kolaczkowski, S.T. and Mcclymont, D.W.J. (2015) Using quadrupole mass spectrometry for on-line gas analysis – gasification of biomass and refuse derived fuel. *Fuel* **139**, 337–345.

Lee, J.H., Lee, S.H., Yim, S.S., Kang, K.-H., Lee, S.Y., Park, S.J., and Jeong, K.J. (2013) Quantified high-throughput screening of *Escherichia coli* producing poly(3-hydroxybutyrate) based on FACS. *Appl Biochem Biotechnol* **170**: 1767–1779.

van de Loosdrecht, J., and Niemantsverdriet, J.W.H. (2012) Synthesis gas to hydrogen, methanol, and synthetic fuels. In *Chemical Energy Storage*. Schlögl, R. (ed.). Walter de Gruyter GmbH: Berlin, Germany, pp. 443–457.

Louthan, M.R., Caskey, G.R., Donovan, J.A., and Rawl, D.E. (1972) Hydrogen embrittlement of metals. *Phys B Condens Matter* **10**: 357–368.

Mesters, C. (2016) A selection of recent advances in C1 chemistry. *Annu Rev Chem Biomol Eng* **7**: 223–238.

Mohammadi, M., Najafpour, G.D., Younesi, H., Lahijani, P., Uzir, M.H., and Mohamed, A.R. (2011) Bioconversion of synthesis gas to second generation biofuels: a review. *Renew Sustain Energy Rev* **15**: 4255–4273.

Molitor, B., Richter, H., Martin, M.E., Jensen, R.O., Juminaga, A., Mihalcea, C., and Angenent, L.T. (2016) Carbon recovery by fermentation of CO-rich off gases – turning steel mills into biorefineries. *Bioresour Technol* **215**: 386–396.

Munasinghe, P.C., and Khanal, S.K. (2010) Biomass-derived syngas fermentation into biofuels: opportunities and challenges. *Bioresour Technol* **101**: 5013–5022.

Najafpour, G.D., and Younesi, H. (2007) Bioconversion of synthesis gas to hydrogen using a light-dependent photo-synthetic bacterium, *Rhodospirillum rubrum*. *World J Microbiol Biotechnol* **23**: 275–284.

NIOSH National Institute for Occupational Safety and Health (2016) International Chemical Safety Cards (ICSC) – carbon monoxide. URL https://www.cdc.gov/niosh/ipcsne ng/neng0023.html

Revelles, O., Tarazona, N., García, J.L., and Prieto, M.A. (2016) Carbon roadmap from syngas to polyhydroxyalkanoates in *Rhodospirillum rubrum*. *Environ Microbiol* **18**: 708–720.

Schultz, J.E., and Weaver, P.F. (1982) Fermentation and anaerobic respiration by *Rhodospirillum rubrum* and *Rhodopseudomonas capsulata*. *J Bacteriol* **149**: 181–190.

Sudesh, K., Abe, H., and Doi, Y. (2000) Synthesis, structure and properties of polyhydroxyalkanoates: biological polyesters. *Prog Polym Sci* **25**: 1503–1555.

Velghe, I., Carleer, R., Yperman, J., and Schreurs, S. (2011) Study of the pyrolysis of municipal solid waste for the production of valuable products. *J Anal Appl Pyrolysis* **92**: 366–375.

Wanner, U., and Egli, T. (1990) Dynamics of microbial growth and cell composition in batch culture. *FEMS Microbiol Rev* **6**: 19–43.

White, C.M., Steeper, R.R., and Lutz, A.E. (2006) The hydrogen-fueled internal combustion engine: a technical review. *Int J Hydrogen Energy* **31**: 1292–1305.

Younesi, H., Najafpour, G., Ku Ismail, K.S., Mohamed, A.R., and Kamaruddin, A.H. (2008) Biohydrogen production in a continuous stirred tank bioreactor from synthesis gas by anaerobic photosynthetic bacterium: *Rhodospirillum rubrum*. *Bioresour Technol* **99**: 2612–2619.

Zabetakis, M.G. (1965) Flammability characteristics of combustible gases and vapors, Bulletin 627, Bureau of Mines, U.S.A.

Zinn, M., Witholt, B., and Egli, T. (2001) Occurrence, synthesis and medical application of bacterial polyhydroxyalkanoate. *Adv Drug Deliv Rev* **53**: 5–21.

Comparative studies of the composition of bacterial microbiota associated with the ruminal content, ruminal epithelium and in the faeces of lactating dairy cows

Jun-hua Liu, Meng-ling Zhang, Rui-yang Zhang, Wei-yun Zhu and Sheng-yong Mao*

Laboratory of Gastrointestinal Microbiology, College of Animal Science and Technology, Nanjing Agricultural University, Nanjing, Jiangsu Province, China.

Summary

The objective of this research was to compare the composition of bacterial microbiota associated with the ruminal content (RC), ruminal epithelium (RE) and faeces of Holstein dairy cows. The RC, RE and faecal samples were collected from six Holstein dairy cows when the animals were slaughtered. Community compositions of bacterial 16S rRNA genes from RC, RE and faeces were determined using a MiSeq sequencing platform with bacterial-targeting universal primers 338F and 806R. UniFrac analysis revealed that the bacterial communities of RC, RE and faeces were clearly separated from each other. Statistically significant dissimilarities were observed between RC and faeces (*P* = 0.002), between RC and RE (*P* = 0.003), and between RE and faeces (*P* = 0.001). A assignment of sequences to taxa showed that the abundance of the predominant phyla Bacteroidetes was lower in RE than in RC, while a significant higher (*P* < 0.01) abundance of Proteobacteria was present in RE than in RC. When compared with the RC, the abundance of Firmicutes and Verrucomicrobia was higher in faeces, and RC contained a greater abundance of Bacteroidetes and Tenericutes. A higher proportions of *Butyrivibrio* and *Campylobacter* dominated RE as compared to RC. The faecal microbiota was less diverse than RC and dominated by genera *Turicibacter* and *Clostridium*. In general, these findings clearly demonstrated the striking compositional differences among RC, RE and faeces, indicating that bacterial communities are specific and adapted to the harbouring environment.

*For correspondence. E-mail maoshengyong@163.com

Introduction

In cattle, feed digestion occurs mainly in the rumen. In the rumen, microorganisms work to convert fermented feed into volatile fatty acids and microbial mass, thus providing nutrients for the animal. Additional fermentation also occurs in the hindgut of cattle, where the fermentable substrates are limited to slower digesting polymers such as crystalline starches that have escaped foregut digestion and absorption, as well as some secreted mucins (Vanhatalo and Ketoja, 1995). Hindgut symbiotic bacteria continue the fermentation process and also provide important vitamins for the host, such as vitamin K, thiamine and riboflavin (Godoy-Vitorino et al., 2012). However, feed ingredients in the hindgut are different than those in the rumen, which may result in a difference in the composition of microbial communities in the forestomach and the hindgut. Indeed, Aiple et al. (1992) reported that the ruminal inocula gave shorter lag times and produced more gas versus faecal inocula in *in vitro* fermentation, indicating that there are some differences in microbial activities and microbial composition between the rumen and the hindgut. In this study, we hypothesize that the hindgut may contain a different microbial community compared with the foregut; therefore, we have to compare the structure and composition of rumen and hindgut bacterial communities.

In addition to regional effects, the host's physiology may also affect the composition of gastrointestinal tract microbiota. The rumen bacterial microbiota are distinguished by three different subpopulations based on their localization: (i) the community in ruminal fluid; (ii) the community attached to solids; and (iii) the community attached to the ruminal epithelium (RE) (Cho et al., 2006). Among these, the ruminal epithelial adherent bacterial community performs a variety of functions necessary for host health, including the hydrolysis of urea, scavenging oxygen and recycling epithelial tissue (Cheng and Wallace, 1979). Early studies using microscopy and cultural techniques showed that bacterial community adhered to the RE were distinct from those associated with ruminal content (RC) (Cheng et al., 1979; McCowan et al., 1980). However, these culture-based techniques may underestimate the biodiversity of the epimural biofilm, because it can be difficult to distinguish between species that are closely related, and many members of this community are likely unculturable. As a result, numerous members of the epimural community remain unidentified. The advent of genetic techniques has revealed an extensive microbial diversity that was previously undetected with culture-dependent methods. For example, more recent work using cloning, sequencing and fingerprint profile analyses based on 16S rRNA sequences have shown both the diversity of ruminal epithelial community and its differences as compared to the community present in RC (Sadet *et al.*, 2007; Sadet-Bourgeteau *et al.*, 2010; Chen *et al.*, 2011). However, the information on RE is still limited because of the low throughput of the traditional 16S rRNA clone library method and fingerprint profiles. Determining 16S rRNA short variable tags using high-throughput sequencing technologies such as 454 pyrosequencing and MiSeq sequencing provided an unprecedented sequencing depth with tens of thousands of reads per sample. These methods regenerated people's interest in measuring and comparing the composition and richness of microbial taxa in RE samples (Licht *et al.*, 2007; Sayers *et al.*, 2009; Mao *et al.*, 2013a). Based on the 454 pyrosequencing techniques, Petri *et al.* (2013) reported that the genera *Atopobium*, *Desulfocurvus*, *Fervidicola*, *Lactobacillus* and *Olsenella* dominated the RE in beef cattle during subacute ruminal acidosis. However, the work carried out by Petri *et al.* (2013) was mainly focused on characterizing the composition of the adherent epithelial bacterial community during dietary adaptation from a forage-based diet to a grain-based diet, and did not explore the difference in the composition of bacterial microbiota between the RC and RE. In addition, the species composition of epithelial biofilms may be affected by other variables such as sex, diet, ruminal pH, aerotolerance, nutrient absorption, epithelial cell turnover, the passage of digesta and host communication (McCann *et al.*, 2014). Moreover, the typical diet of beef cattle is different from that of dairy cattle. Thus, the composition of RE-associated microbiota of lactating dairy cattle still needs to be explored. In this study, we characterized the composition of the microbial communities in the RE of lactating dairy cows.

Materials and methods

Animals and sample collection

Six Holstein dairy cattle aged 5 years (body weight: 612.9 ± 63.4 kg, milk yield: 19.4 ± 1.5 kg day^{-1}, 206–287 days in lactation) were used in this study. All animal care procedures were approved by the Institutional Animal Care and Use Committee of Zhejiang University prior to initiation of the experiment. The cows' diets (at 23 kg day^{-1} dry matter intake) were formulated to meet or exceed the energy requirements of Holstein cattle yielding 25 kg of milk per day with 3.50% milk fat and 3.10% true protein (Table S1). Diets were fed *ad libitum* as a total mixed ration to reduce the selection of dietary components. The cattle were fed at 7:00 a.m. and 6:00 p.m. (one-half of the total daily ration at each feeding). The experimental period was 86 days; the first 83 days were used for diet adaptation and the last 3 days were used for measurements. Throughout the experimental period, cattle were housed in tie stalls and fed *ad libitum*, and they were given free access to freshwater during the trial.

On day 84, 85 and 86, the cattle were slaughtered at 4–6 h after last feeding in a local slaughterhouse, with two cows slaughtered each day. The RE (50–100 g each), RC (250–300 g each) and faeces (250–300 g each) were collected within 30 min after slaughter. RE samples were rinsed three times with a sterile phosphate-buffered saline (pH 7.0) to remove the digesta, cut into 4–5 mm^2 samples, and scraped from the underlying tissue using a germ-free glass slide, immediately transferred into liquid nitrogen until DNA extraction. The collected RC (250–300 g each) and faeces (250–300 g each) were homogenized in a homogenizer instrument (Media, Foshan, China) with 18000r/s for 30 s. The homogenized RC and faeces were then sampled and immediately frozen in liquid nitrogen. The remaining samples were centrifuged at 2000 g and the supernatants were stored at −20°C for volatile fatty acid analysis. The amount of volatile fatty acid was measured using capillary column gas chromatography (GC-14B; Shimadzu, Kyoto, Japan; capillary column: 30 m × 0.32 mm × 0.25 mm film thickness; column temperature = 110°C; injector temperature = 180°C; detector temperature = 180°C) (Mao *et al.*, 2012).

DNA extraction, PCR amplification, illumina MiSeq sequencing and sequencing data processing

Three grams (wet weight) of homogenized samples of the RC (the ratio of rumen liquid and solid was about 1:2), RE and faeces from each cattle were used for the DNA extraction. The DNA was extracted by a bead-beating method using a mini-bead beater (Biospec Products, Bartlesville, OK, USA), followed by phenol–chloroform extraction (Mao *et al.*, 2012). The DNA was quantified using a Nanodrop spectrophotometer (Nyxor Biotech, Paris, France) following staining using a Quant-it Pico Green dsDNA kit (Invitrogen, Paisley, UK). The DNA samples were stored at −80°C until further processing.

The V3–V4 region of the bacteria 16S ribosomal RNA gene was amplified by PCR (95°C for 2 min, followed by 25 cycles at 95°C for 30 s, 55°C for 30 s and 72°C for

30 s with a final extension at 72°C for 5 min). The amplification used primers 338F (5′-barcode-ACTCCTRCGG-GAGGCAGCAG-3′) and 806R (5′-GGACTACCVGG GTATCTAAT-3′), where the barcode is an eight-base sequence unique to each sample. PCR was performed in triplicate 20 μl mixtures containing 4 μl of 5× FastPfu Buffer, 2 μl of 2.5 mM dNTPs, 0.8 μl of each primer (5 μM), 0.4 μl of FastPfu Polymerase and 10 ng of template DNA. Amplicons were extracted from 2% agarose gels and purified using the AxyPrep DNA Gel Extraction Kit (Axygen Biosciences, Union City, CA, USA) according to the manufacturer's instructions and quantified using QuantiFluor™ -ST (Promega, Madison, WI, USA). Purified amplicons were pooled in equimolar concentrations and paired-end sequenced (2 × 250) on an Illumina MiSeq platform according to standard protocols (Caporaso et al., 2012).

Raw FASTQ files were demultiplexed and quality-filtered using QIIME (version 1.70) (Campbell et al., 2010) with the following criteria: (i) The 250 bp reads were truncated at any site receiving an average quality score < 20 over a 10 bp sliding window, and truncated reads that were shorter than 50 bp were discarded. (ii) Exact barcode matching, two nucleotide mismatch in primer matching, and reads containing ambiguous characters were removed. (iii) Only sequences that overlap longer than 10 bp were assembled according to their overlap sequence. Reads which could not be assembled were discarded. Operational taxonomic units (OTUs) were clustered with 97% similarity cut-off using UPARSE (version 7.1 http://drive5.com/uparse/) and chimeric sequences were identified and removed using UCHIME (Edgar, 2010). The most abundant sequences within each OTU were designated as 'representative sequences', and were then aligned against the core set of Greengenes 13.5 (DeSantis et al., 2006) using PYNAST (Caporaso et al., 2010) with the default parameters set by QIIME. A PH Lane mask supplied by QIIME was used to remove the hypervariable regions from the aligned sequences. FASTTREE (Price et al., 2009) was used to create a phylogenetic tree of the representative sequences. Sequences were classified using the Ribosomal Database Project classifier with a standard minimum support threshold of 80% (Wang et al., 2007). Sequences identified as chloroplasts or mitochondria were removed from the analysis. Community diversity was estimated using the ACE, Chao1 and Shannon indices. The unweighted UniFrac distance method was used to perform a principal coordinate analysis (PCoA) (Lozupone and Knight, 2005), and an unweighted distance-based analysis of molecular variance (AMOVA) was conducted to assess significant differences between the samples using mothur v.1.29.0 (Schloss et al., 2009).

Use of quantitative real-time PCR to enumerate microbial community

The 16S rRNA gene copy number of the phyla Firmicutes, and Bacteroidetes were enumerated by quantitative PCR on an Applied Biosystems 7300 Real-Time PCR System (ABI, Foster City, CA, USA) using SYBR Green as the fluorescent dye. The reaction mixture (25 μl) consisted of 12.5 μl of IQ SYBR Green Supermix (Bio-Rad, Richmond, CA, USA), 0.2 μM of each primer set and 5 μl of the template DNA. The amount of DNA in each sample was determined in triplicate, and the mean values were calculated. The primers were selected on the basis of a careful review of published literature. Primers for universal bacteria were forward: 5′-CCTACGGGAGGCAGCAG-3′ and reverse: 5′-ATTACC GCGGCTGCTGG-3′ (Shinkai et al., 2007); primers for Bacteroidetes were forward: 5′-GGARCATGTGGTT-TAATTCGATGAT-3′ and reverse: 5′-AGCTGACGACAA CCATGCAG-3′ (Guo et al., 2008); primers for Firmicutes were forward: 5′-GGAGYATGTGGTTTAATTCGAAGCA-3′ and reverse: 5′-AGCTGACGACAACCATGCAC-3′ (Zhao et al., 2013). External standards were prepared by making 10-fold serial dilutions of purified plasmid DNA containing the 16S rRNA gene sequence of Streptococcus bovis. A standard curve was set up in every 96-well plate and all standard curves met the required standards of efficiency ($R^2 > 0.99$, 90% > E > 110%). Results were expressed as log10 numbers of marker loci gene copies per gram of RC, RE or faeces (wet weight).

Statistical analysis

The effects of sampling site on the bacterial prevalence and the VFA levels were analysed using a one-way ANOVA procedure of SPSS (SPSS v.16; SPSS, Chicago, IL, USA) according to the following equation: $Y_{ij} = \mu + S_i + e_{ij}$, where Y_{ij} was the observation (VFA data, bacterial density, and the relative abundance of a given bacterial phyla, genera, or species (in %), μ was the overall mean, S_i was the sampling site effect (i = 3), and e_{ij} was the residual error. All P-values obtained by one-way ANOVA analyses of the bacteria community were corrected for a false discovery rate (FDR) of 0.05 with the Benjamini–Hochberg method. FDR-corrected P-values below 0.05 ($q < 0.05$) were considered significant. When $q < 0.05$, Tukey's test was employed to determine significant differences among the sampling sites.

Results

Diversity of the bacterial community

In this study, 16S rRNA gene sequence analysis of the RC, RE and faeces samples generated a total of

637 405 quality sequences with an average of
48 480 ± 8324 sequences per sample. The overall
number of OTUs detected by the analysis was 4039
based on 97% nucleotide sequence identity between
reads. To assess whether our sampling effort provided
sufficient OTU coverage to accurately describe the bacterial composition of each region, sample-based and
individual-based rarefaction curves were generated for
each region (Fig. S1). The results showed that the
Good's coverage was greater than 0.97, implying that
our sampling effort was sufficient for the samples from
all animals (Table S2). The Shannon diversity index
and number of OTUs in RC were higher ($P < 0.05$)
than that in RE (Table 1). The Chao value and Shannon index were also higher ($P < 0.05$) in the RC than
that in the faeces. The Chao value and Shannon index
were higher ($P < 0.05$) in RE than that in faeces; while
the number of OTUs in RE was lower ($P < 0.05$) than
that in faeces.

When the bacterial composition of microbiota among
the RC, RE and faeces was compared using unweighted
Unifrac distance, the bacterial communities of RC, RE
and faeces clearly separated from each other (Fig. 1).

Table 1. The diversity and richness of bacterial community in the
RC, RE and faeces at 3% dissimilarity level ($n = 6$).

	RC	RE	Faeces	SEM[1]	P-value
Chao value	2477[a]	2439[a]	1732[b]	91	< 0.001
Shannon index	6.07[a]	5.50[b]	4.30[c]	0.18	< 0.001
Number of OTUs	1999[a]	1288[b]	1945[a]	83	< 0.001

1. Standard error of means.
Means without common letter differ, $P < 0.05$.

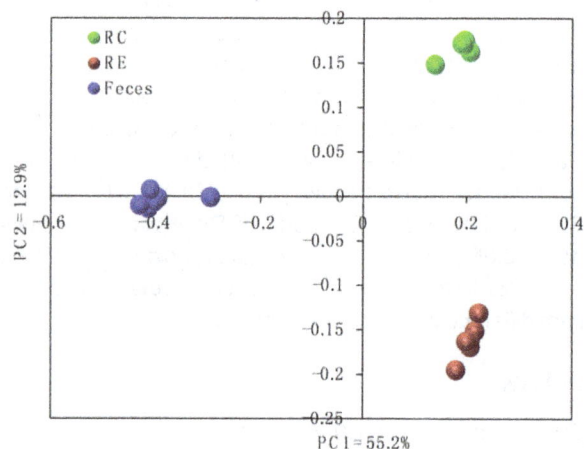

Fig. 1. PCoA of bacterial communities from the RC, RE and faecal
samples.

AMOVA was used to assess the statistical significance
of the spatial separation that was observed among the
different regions in PCoA plots. Statistically significant
dissimilarities were observed between RC and faeces
($P = 0.002$), between RC and RE ($P = 0.003$), and
between RE and faeces ($P = 0.001$).

The composition of the bacterial phyla, genera and OTUs

A total of 18 phyla were detected in all samples. Among
them, Firmicutes and Bacteroidetes were detected as
the dominant phyla regardless of sampling type
(Fig. 2A), but their ratio among the groups varied considerably.

At the phylum level, 17 different bacterial phyla were taxonomically identified in RC samples. The majority of the
sequences that were obtained belonged to Firmicutes
(53.25% ± 3.55%), Bacteroidetes (39.06% ± 3.29%) and
Proteobacteria (13.29% ± 0.26%) (Fig. 2A). The phyla
Firmicutes, Bacteroidetes, Proteobacteria, Actinobacteria,
Lentisphaerae, Tenericutes, Spirochaetae, Chloroflexi,
Cyanobacteria, Elusimicrobia, Synergistetes, Fibrobacteres and Fusobacteria were found in all samples. In RE
samples, 18 different bacterial phyla were identified, and
the majority of the sequences that were obtained belonged
to Firmicutes (48.03% ± 3.40%), Bacteroidetes (27.71%
± 3.18%) and Proteobacteria (11.43% ± 3.38%)
(Fig. 2A). In faecal samples, Firmicutes (87.62% ± 3.82%)
dominated all bacterial communities (Fig. 2A).

At the genus level, a total of 372 taxa were observed
across all samples; however, 61.6% of all sequences
were not identified at the genus level. For clarity and
visualization purposes, the top 50 bacterial taxa are presented in Fig. S2. The results showed that the predominant taxa in the collected samples included *Prevotella*,
Clostridium, *Turicibacter*, *Butyrivibrio*, *Succiniclasticum*,
Ruminococcus, *Mogibacterium*, *Campylobacter*, *Desulfobulbus*, *Syntrophococcus*, *Acetitomaculum*, *Treponema*, unclassified Mollicutes, unclassified Ruminococcaceae, unclassified Peptostreptococcaceae, unclassified Rikenellaceae, unclassified Christensenellaceae,
unclassified Prevotellaceae, unclassified Bacteroidales,
unclassified Lachnospiraceae and unclassified Clostridiales.

At the OTUs level, a total of 4859 OTUs were calculated at a 0.03 dissimilarity cut-off in combination across
all of these samples. The number of OTUs shared
among RC, RE and faeces was 1020 (Fig. S3). The
number of OTUs shared between the RC and RE was
2444, the number of OTUs shared between the RC and
faeces was 1260, whereas the number of OTUs shared
between the RE and faeces was 1165. The number of
OTUs specific to RC, RE and faeces was 416, 703 and
911 respectively.

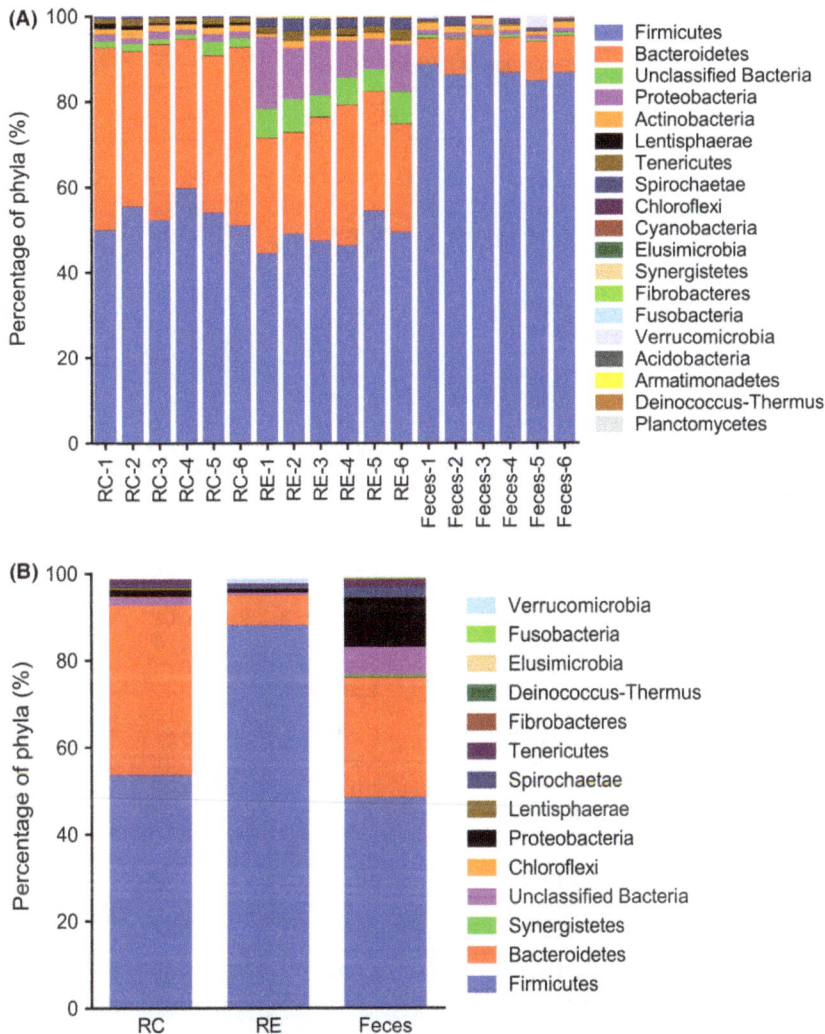

Fig. 2. Distributions of phyla. A. The distribution of phyla for each sample. B. Distribution of the phyla averaged across the RC and RE groups (only the phyla which was significantly affected by the sampling site were presented, $q < 0.05$). C. Distribution of the phyla averaged across the RC and faeces (only the phyla which were significantly affected by the sampling site were presented, $P < 0.05$). The numbers 1 to 6 refer to the six cattle.

The difference in the composition of bacterial microbiota between RC and RE samples

At the phylum level, a lower ($P < 0.05$) abundance of Bacteroidetes, Lentisphaerae and Chloroflexi was observed in RE compared with RC (Fig. 2B), and the RE showed a higher ($P < 0.05$) proportion of Proteobacteria, Spirochaetae, Synergistetes, Tenericutes, Fibrobacteres, Fusobacteria, Elusimicrobia, Deinococcus-Thermus and unclassified bacteria.

Table 2 shows the dominant taxa (the taxon with a relative abundance of ≥1% in at least one sampling site) significantly affected ($q < 0.05$) by the sampling sites. When compared with RE, the RC had a greater ($P < 0.05$) percentage of *Prevotella, Saccharofermentans, Succiniclasticum, Ruminococcus*, and some unclassified bacteria including unclassified Ruminococcaceae, unclassified Christensenellaceae and unclassified Bacteroidales, while

RE presented a higher ($P < 0.05$) abundance of *Butyrivibrio, Mogibacterium, Treponema, Syntrophococcus, Howardella, Campylobacter, Desulfovibrio, Desulfobulbus*, unclassified Clostridiales, unclassified Prevotellaceae, unclassified bacteria and unclassified Erysipelotrichaceae.

Of the 4859 OTUs detected in this study, 35 OTUs represented ≥ 1% of all sequences in at least one sampling site (Table 3). When compared with the RE, the relative abundance of 10 OTUs was higher in the RC ($P < 0.05$), while the RE presented a higher proportion of 16 OTUs ($P < 0.05$).

The difference in the composition of bacterial microbiota between RC and faecal samples

At the phylum level, when compared with RC (Fig. 2B), the abundance of Firmicutes and Verrucomicrobia was higher ($P < 0.05$) in faeces, while the RC contained a

Table 2. Dominant taxa calculated from collected samples of RC, RE and faeces ($n = 6$).

Classification	Percentage of total sequences			SEM[1]	q-value
	RC	RE	Faeces		
Firmicutes					
Butyrivibrio	3.34[b]	11.32[a]	2.00[b]	1.02	< 0.001
Clostridium	0.01[b]	0.02[b]	12.64[a]	1.47	< 0.001
Turicibacter	0.02[b]	0.01[b]	7.99[a]	0.94	< 0.001
Ruminococcus	2.43[a]	0.35[c]	1.51[b]	0.22	< 0.001
Cellulosilyticum	< 0.01[b]	< 0.01[b]	1.45[a]	0.21	< 0.001
Saccharofermentans	1.92[a]	0.27[b]	0.07[c]	0.21	< 0.001
Syntrophococcus	0.18[b]	2.90[a]	0.04[b]	0.34	< 0.001
Howardella	0.03[b]	1.41[a]	0.02[b]	0.19	< 0.001
Succiniclasticum	3.65[a]	2.17[b]	< 0.01[c]	0.38	< 0.001
Acetitomaculum	1.92[a]	1.34[ab]	0.88[b]	0.14	0.001
Mogibacterium	1.22[b]	2.39[a]	0.76[c]	0.19	< 0.001
Unclassified Christensenellaceae	11.66[a]	2.41[b]	3.50[b]	1.03	< 0.001
Unclassified Ruminococcaceae	15.26[a]	7.26[b]	18.04[a]	1.37	< 0.001
Unclassified Lachnospiraceae	5.48[a]	5.81[a]	1.95[b]	0.47	< 0.001
Unclassified Clostridiales	1.99[b]	6.19[a]	1.91[b]	0.51	< 0.001
Unclassified Peptostreptococcaceae	0.13[b]	0.04[b]	29.61[a]	3.46	< 0.001
Unclassified Erysipelotrichaceae	0.32[b]	1.35[a]	0.24[b]	0.13	< 0.001
Bacteroidetes					
Prevotella	16.94[a]	2.29[b]	0.09[c]	1.87	< 0.001
Unclassified Rikenellaceae	10.66[a]	9.43[a]	2.26[b]	0.94	< 0.001
Unclassified Prevotellaceae	4.02[b]	10.70[a]	0.76[c]	1.05	< 0.001
Unclassified Bacteroidales	6.81[a]	4.39[b]	2.40[c]	0.51	< 0.001
Proteobacteria					
Campylobacter	< 0.01[b]	3.72[a]	< 0.01[b]	0.56	0.001
Desulfovibrio	0.04[b]	1.38[a]	< 0.01[b]	0.16	< 0.001
Desulfobulbus	0.01[b]	3.13[a]	< 0.01[b]	0.39	< 0.001
Spirochaetae					
Treponema	0.4[c]	1.95[a]	0.72[b]	0.19	< 0.001
Verrucomicrobia					
Akkermansia	ND	< 0.01[b]	1.16[a]	0.19	0.007
Tenericutes					
Unclassified Mollicutes	1.22[a]	1.57[a]	0.37[b]	0.14	< 0.001
Unclassified					
Unclassified Bacteria	1.87[b]	6.44[a]	0.58[c]	0.64	< 0.001

1. Standard error of means.
Means without common letter differ, $q < 0.05$.

greater abundance ($P < 0.05$) of Bacteroidetes, Chloroflexi, Lentisphaerae, Tenericutes and unclassified bacteria.

At the genus level, when compared with RC, the abundance of Clostridium, Turicibacter, Cellulosilyticum, Akkermansia, Treponema and unclassified Peptostreptococcaceae was greater ($P < 0.05$) in faeces (Table 2), while the percentages of Ruminococcus, Acetitomaculum, Mogibacterium, Prevotella, Saccharofermentans, Succiniclasticum, unclassified Christensenellaceae, unclassified Bacteroidales, unclassified Rikenellaceae, unclassified Lachnospiraceae, unclassified Prevotellaceae, unclassified bacteria and unclassified Mollicutes were lower ($P < 0.05$) in faeces.

At the OTUs level, when compared with faeces (Table 3), the relative abundance of 10 OTUs was higher in RC ($P < 0.05$), while faeces showed a greater abundance of nine OTUs ($P < 0.05$).

The difference in the composition of bacterial microbiota between RE and faecal samples

At the phylum level, when compared with RE (Fig. 2B), the abundance of Firmicutes and Verrucomicrobia was higher ($P < 0.05$) in faeces, while the RE contained a greater abundance ($P < 0.05$) of Bacteroidetes, Synergistetes, unclassified bacteria, Proteobacteria, Spirochaetae, Tenericutes, Fibrobacteres, Deinococcus-Thermus, Elusimicrobia and Fusobacteria.

At the genus level, the abundance of Butyrivibrio, Campylobacter, Desulfobulbus, Desulfovibrio, Howardella, Mogibacterium, Prevotella, Succiniclasticum, Syntrophococcus, Treponema, unclassified bacteria, unclassified Bacteroidales, unclassified Clostridiales, unclassified Erysipelotrichaceae, unclassified Lachnospiraceae, unclassified Mollicutes, unclassified Prevotellaceae and unclassified Rikenellaceae in RE

Table 3. Dominant OTU calculated from collected samples of RC, RE and faeces ($n = 6$).

No. OTU ID	Classification	Percentage of total sequences			SEM[1]	q-value
		RC	RE	Faeces		
Bacteroidetes						
OTU1916	*Prevotella*	1.14[a]	0.18[b]	< 0.01[b]	0.13	< 0.001
OTU2537	*Prevotella*	1.54[a]	0.09[b]	ND	0.44	0.003
OTU6271	*Prevotella*	1.19[a]	0.06[b]	ND	0.82	< 0.001
OTU6764	Unclassified Prevotellaceae	0.06[b]	7.91[a]	< 0.01[c]	0.17	< 0.001
OTU3160	Unclassified Rikenellaceae	3.59[a]	0.60[b]	< 0.01[b]	0.30	< 0.001
OTU3169	Unclassified Rikenellaceae	0.02[b]	1.96[a]	ND	0.11	< 0.001
OTU5919	Unclassified Rikenellaceae	0.02[b]	1.28[a]	ND	0.16	< 0.001
Firmicutes						
OTU1531	*Butyrivibrio*	< 0.01[b]	2.47[a]	< 0.01[b]	0.56	< 0.001
OTU3217	*Butyrivibrio*	< 0.01[b]	2.48[a]	ND	0.13	< 0.001
OTU369	*Butyrivibrio*	0.85[b]	0.09[c]	1.21[a]	0.28	< 0.001
OTU6481	*Butyrivibrio*	ND	1.10	ND	0.16	< 0.001
OTU2081	*Cellulosilyticum*	< 0.01[b]	< 0.01[b]	1.11[a]	0.54	< 0.001
OTU604	*Clostridium*	0.01[b]	0.01[b]	11.60[a]	0.18	< 0.001
OTU2453	*Howardella*	0.03[b]	1.38[a]	0.02[b]	0.18	< 0.001
OTU4649	*Mogibacterium*	0.08[b]	1.71[a]	0.05[b]	0.24	< 0.001
OTU1687	*Saccharofermentans*	1.06[a]	0.15[b]	0.06[b]	0.15	< 0.001
OTU1719	*Succiniclasticum*	1.54[a]	0.43[b]	< 0.01[b]	0.40	< 0.001
OTU5108	*Succiniclasticum*	1.81[a]	1.27[b]	< 0.01[b]	0.28	< 0.001
OTU1315	*Turicibacter*	0.02[b]	0.01[b]	6.99[a]	0.29	< 0.001
OTU3497	Unclassified Christensenellaceae	1.00[a]	0.12[b]	0.05[b]	0.11	< 0.001
OTU583	Unclassified Christensenellaceae	4.70[a]	0.31[b]	0.12[b]	0.13	< 0.001
OTU1072	Unclassified Clostridiales	< 0.01[b]	1.03[a]	ND	0.12	< 0.001
OTU5630	Unclassified Clostridiales	< 0.01[b]	1.03[a]	ND	0.32	< 0.001
OTU1956	Unclassified Lachnospiraceae	0.12[b]	2.28[a]	< 0.01[b]	0.20	< 0.001
OTU2405	Unclassified Peptostreptococcaceae	0.01[b]	< 0.01[b]	1.52[a]	1.87	< 0.001
OTU4730	Unclassified Peptostreptococcaceae	0.06[b]	0.02[b]	16.00[a]	1.11	< 0.001
OTU4916	Unclassified Peptostreptococcaceae	0.06[b]	0.01[b]	9.54[a]	0.20	< 0.001
OTU120	Unclassified Ruminococcaceae	4.23[a]	1.18[b]	0.05[c]	0.12	< 0.001
OTU1888	Unclassified Ruminococcaceae	0.01[b]	< 0.01[b]	4.55[a]	0.52	< 0.001
OTU3720	Unclassified Ruminococcaceae	< 0.01[b]	< 0.01[b]	2.59[a]	0.15	< 0.001
Proteobacteria						
OTU2345	*Campylobacter*	< 0.01[b]	3.66[a]	< 0.01[b]	1.35	0.001
OTU3579	*Desulfobulbus*	< 0.01[b]	1.10[a]	ND	0.13	< 0.001
Spirochaetae						
OTU713	*Treponema*	< 0.01[b]	1.28[a]	ND	0.13	< 0.001

1. Standard error of means.
Means without common letter differ, $q < 0.05$.

was higher ($P < 0.05$) than those in faeces (Table 2), and the proportions of *Ruminococcus*, *Turicibacter*, *Clostridium*, *Cellulosilyticum*, *Akkermansia*, unclassified Peptostreptococcaceae and unclassified Ruminococcaceae were greater ($P < 0.05$) in faeces than in RE.

At the OTUs level, when compared with faeces (Table 3), the relative abundance of 17 OTUs was higher in RE ($P < 0.05$), while faeces showed a greater abundance of nine OTUs ($P < 0.05$).

The density of the microbiota in RC, RE and faeces

The real-time PCR results showed that the 16S rRNA gene copy numbers of total bacteria, phyla Firmicutes, and Bacteroidetes were higher ($P < 0.01$) in the RC and faeces than in RE (Table 4). The number of Firmicutes was greater ($P < 0.01$) in faeces than in RC, and no sig-

nificant differences ($P > 0.05$) were observed in the numbers of total bacteria and Bacteroidetes between the RC and faeces.

Discussion

Comparison of the composition of bacterial microbiota associated with the RC and RE

The RC-associated microbial composition of dairy cows has been extensively surveyed using next-generation sequencing technologies (Licht *et al.*, 2007; Rettedal *et al.*, 2009; Sayers *et al.*, 2009; Callaway *et al.*, 2010; Vasiljevic *et al.*, 2011; Evans *et al.*, 2012; Godoy-Vitorino *et al.*, 2012; Mao *et al.*, 2013a; Mao *et al.*, 2013b; Liu *et al.*, 2014). In the present study, cows were slaughtered to collect enough of the epimural bacteria for a high-quality community analysis. However, previous

Table 4. Population of total bacteria and phyla Bacteroidetes and Fimicutes in the RC, RE and faeces (n = 6) (log10 copy number of 16S RNA gene per gram of sample).

	RC	RE	Faeces	SEM[1]	P-value
Total bacteria	12.05[a]	10.58[b]	12.33[a]	0.192	< 0.01
Firmicutes	11.08[a]	9.42[c]	11.60[b]	0.231	< 0.01
Bacteroides	11.84[a]	9.35[b]	11.46[a]	0.273	< 0.01

1. Standard error of means.
Means without common letter differ, P < 0.05.

reports of bacterial community composition in slaughtered animals suffered from the fact that slaughter was performed long after the animals' last meal, which means that the RC bacterial community composition from slaughtered cows may be different from that from rumen fistulated cows. Current knowledge regarding the RE-associated bacteria in dairy cattle is limited compared with the bacterial ecology and diversity of the RC microbial community. In this study, we hypothesized that the chosen sampling site can affect the diversity, composition and structure of bacteria microbiota in dairy cattle. In line with the reports based on PCR-DGGE approach (Sadet *et al.*, 2007), our results based on PCoA and AMOVA clearly demonstrated that the bacterial communities attached to RE were different from those attached to the RC (Fig. 1). Furthermore, quantification by qPCR gave higher values in terms of 16S rRNA genes in RC than RE for total bacteria. These distinctions between the bacterial community attached to the RE and those in RC may be explained based on sampling sites. The RE situated at the interface between the host tissues, while the RC is in contact with a variety of substrates and other microscale conditions, and this heterogeneous environment could affect bacterial diversity, composition and density as shown in other ecosystems (Horner-Devine *et al.*, 2004).

The differences in bacterial composition between RE and RC communities were studied at the phylum, genus and OTUs levels. At the phylum level, the abundance of Bacteroidetes was higher in the RC than in the RE of dairy cattle, and similar result was also observed in the 16S rRNA gene copy number between the RC and RE; this may be explained by the diversity of substrate utilization by Bacteroidetes spp. It is known that some Bacteroidetes species are able to hydrolyse soluble polysaccharides found in plant cell walls (Power *et al.*, 2014). Thus, a higher abundance and number of Bacteroidetes spp. in RC will help to degrade and ferment organic matter (Salyers *et al.*, 1995). Consistent with the reports by Chen *et al.* (2011), our study showed that RE contained a greater abundance of Proteobacteria, and this was believed to be caused by the trace amounts of oxygen diffused through the tissue, which may favour a high density of Proteobacteria as many members of this phylum are microaerophiles or facultative anaerobes and hence not sensitive to oxygen toxicity. Indeed, oxygen consumption by the epimural community is thought to be beneficial to the oxygen sensitive anaerobes of the rumen microbial ecosystem (Cheng *et al.*, 1979).

At the genus level, *Prevotella* was predominant in the RC. Previous studies have revealed that different *Prevotella* spp. can utilize individual sugars, amino acids and small peptides for their growth (Fondevila and Dehority, 1996; Takahashi *et al.*, 2000). Thus, they can selectively utilize carbohydrates and proteins from diet, which further resulted in predominance in rumen microbial community. In line with the report by Li *et al.* (2012), our results also revealed that the other abundant genera in the RC included *Succiniclasticum*, *Ruminococcus* and *Saccharofermentans*. These three genera are known to play an important role in fibre degradation (*Ruminococcus* and *Saccharofermentans*) and propionate formation (*Succiniclasticum*); thus, the fact that a higher percentage of these genera was observed in the RC is reasonable for dairy cattle. Conversely, the genera *Butyrivibrio*, *Campylobacter*, *Desulfobulbus*, *Syntrophococcus* and *Mogibacterium* were prevalent in RE. Of these, the abundance of *Butyrivibrio* (belonging to Firmicutes) was much higher in RE samples compared with the RC. These epithelial butyrate producers may release butyrate close to the epithelium and so they may enhance butyrate bioavailability for the host, which may be particularly useful in proliferating rumen and reticulum epithelium (Siavoshian *et al.*, 2000). In addition, our results also revealed that the genus *Mogibacterium* dominated in RE, these results were consistent with the report by Li *et al.* (2012), who reported that the genus *Mogibacterium* was predominate in the RE of steers. This genus is associated with ammonia assimilation through the ruminal epithelial wall for phenylacetate biosynthesis (Nakazawa *et al.*, 2000), and phenylacetate may bind to glutamine to form phenylacetyl glutamine by rumen bacteria, which is very important in rumen ammonia metabolism and absorption (Wallace, 1979). Our data also revealed that a higher proportion of *Campylobacter*, a microaerophilic genus, dominated of the RE compared to the RC in dairy cows, indicating that its presence is consistent with oxygen availability at the epithelial surface, as mentioned earlier. Consistent with a previous study by Mao *et al.* (2013b), the current study showed that some unclassified bacteria including unclassified Christensenellaceae, unclassified Rikenellaceae, unclassified Prevotellaceae, unclassified Ruminococcaceae and unclassified Bacteroidales were the predominant bacterial taxa in the RC and RE samples. In the present study, the reason that a higher abundance of unclassified Christensenellaceae, unclassified Ruminococcaceae and unclassified Bacteroidales was observed in RC samples remains unknown, but it is apparently because of their important role in the ruminal microbial ecosystem. As the meta-analysis revealed that these unclassified bacterial groups were likely competitive in the rumen and that some of their species might have an important role in ruminal feed digestion (Kim *et al.*, 2011). In summary, these data revealed that the composition and the abundance of dominate genera varied considerably between the RC and RE, and the unique distribution of epithelium-attached bacterial composition is very likely due to host–bacterium interactions at the RE.

Comparison of the composition of bacterial microbiota between the RC and faeces

The gastrointestinal tract serves as a habitat for a diverse and dynamic community of bacterial species that can affect growth, health and well-being of the host. The faecal microbiota of cattle affects not only animal health but also food safety (Shanks *et al.*, 2011). Recently, some research using pyrosequencing methods revealed that diet greatly influences the faecal microbiota of cattle (Callaway *et al.*, 2010; Shanks *et al.*, 2011). Although these studies provided some insights into the community structure of faecal microbiota, the number of sequences (< 10 000 reads) analysed in these studies was limited. Lack of power in deep sequencing prohibits an understanding of the whole profile of the faecal bacterial community. In the present study, a total of 265 156 sequences were obtained from the six samples (44 192 sequences per sample), and the coverage was higher than 0.97 (Table S2), which indicates that our study presented a relatively whole profile of the faecal microbial community.

In line with previous studies, Firmicutes in this study constituted a major fraction of the total sequencing reads in cattle faeces (Fig. 2A). At the genus level, our observations revealed a significantly higher prevalence of *Turicibacter* (belonging to Firmicutes), *Treponema* and *Clostridium* in the faecal microbial community compared with the RC. Of these three, *Turicibacter* is a relatively unstudied genus. Recent reports of 16S rRNA gene and ribosomal intergenic spacer analysis data indicate the presence of *Turicibacter* bacteria in the rumen and faeces of cattle (Callaway *et al.*, 2010; Mao *et al.*, 2013a). *Turicibacter* has also been reported in pig, rat and goat hindguts (Licht *et al.*, 2007; Rettedal *et al.*, 2009; Liu *et al.*, 2014). Previous studies revealed that many *Treponema* spp. are associated with ulcerative mammary dermatitis and bovine digital dermatitis in cattle, and contagious ovine digital dermatitis in sheep (Sadet *et al.*, 2007; Evans *et al.*, 2012), which could have deleterious effects on the hindgut health. Conversely, some *Tre-*

ponema spp. can positively influence the host animal by improving the digestion of complex organic matter such as cellulose. Thus, the physiological function of this genus depends on which *Treponema* spp. are present. *Clostridium* spp. is a broad genus, and is ubiquitous in the gastrointestinal tract. Previous studies showed that Clostridia can both positively and negatively influence the host animal, and these effects are typically specifically associated with the individual *Clostridium* species that is involved (Kanauchi *et al.*, 2005). Many have negative influences on animal health, including species such as *C. perfringens*, *C. tetani*, *C. botulinum* and *C. difficile* (Songer, 1998; Attwood *et al.*, 2006), and they can also cause significant productivity problems including reducing the protein availability in fresh forage diets (Reilly and Attwood, 1998). Conversely, some *Clostridium* spp. may be beneficial and improve the digestion of complex organic matter such as cellulose, or even act as beneficial probiotics (Widyastuti *et al.*, 1992; Ozutsumi *et al.*, 2005).

Our results also revealed that the dominant genera *Prevotella*, *Succiniclasticum*, *Ruminococcus*, *Acetitomaculum*, *Saccharofermentans*, *Mogibacterium* and some unclassified bacteria including unclassified Christensenellaceae, unclassified Bacteroidales, unclassified Rikenellaceae, unclassified Lachnospiraceae, unclassified Prevotellaceae and unclassified Mollicutes were much higher in RC than in faeces, while the *Turicibacter*, *Clostridium* and unclassified Peptostreptococcaceae-dominated faeces, suggesting that the predominant community of microbial flora were different between the RC and faeces. In addition, a PCoA and AMOVA revealed that the structure of the ruminal bacterial community was significantly different from the faecal microbiota. However, despite these differences in the bacterial community of the microhabitats, no significant difference was found in the level of total volatile fatty acids between the RC and faeces (Fig. S4), indicating that faecal microbial communities tend to exhibit a similar function during fermentation of some substrates.

In the present study, our data showed that 1260 OTUs were shared between the rumen microbiota and the faecal microbiota (Fig. S4), indicating that these OTUs detected in the rumen might survive in the hindgut. Nevertheless, it is worthy of note that this speculation is based on a PCR assay, and an important drawback to PCR is the potential amplification of DNA from dead bacterial cells as well as from viable bacterial cells (Josephson *et al.*, 1993). Thus, part of the core OTUs detected in the hindgut might be from dead cells. In addition, as most of the shared OTUs cannot be identified at the genus level, the function of these species is not very clear. Therefore, new methods such as proteomics and transcriptomics methods should be applied to elucidate their function and activity in the cattle gastrointestinal microbial ecosystem.

In conclusion, our findings clearly demonstrated the striking compositional differences between RC, RE and faeces, indicating that bacterial communities are specific and adapted to the harbouring environment.

Acknowledgements

Financial support for this study was through the National Basic Research Program of the China Ministry of Science and Technology (2011CB100801).

Funding information

No funding information provided.

Conflicts of interest

None of the authors of this paper has a financial or personal relationship with other people or organizations that could inappropriately influence or bias the content of the paper.

References

Aiple, K.P., Steingass, H., and Menke, K.H. (1992) Suitability of a buffered fecal suspension as the inoculum in the Hohenheim Gas Test. 1. Modification of the method and its ability in the prediction of organic-matter digestibility and metabolizable energy content of ruminant feeds compared with rumen fluid as inoculum. *J Anim Physiol Anim Nutr*, **67:** 57–66.

Attwood, G., Li, D., Pacheco, D., and Tavendale, M. (2006) Production of indolic compounds by rumen bacteria isolated from grazing ruminants. *J Appl Microbiol*, **100:** 1261–1271.

Callaway, T.R., Dowd, S.E., Edrington, T.S., Anderson, R.C., Krueger, N., Bauer, N., *et al.* (2010) Evaluation of bacterial diversity in the rumen and feces of cattle fed different levels of dried distillers grains plus solubles using bacterial tag-encoded FLX amplicon pyrosequencing. *J Anim Sci*, **88:** 3977–3983.

Campbell, B.J., Polson, S.W., Hanson, T.E., Mack, M.C., and Schuur, E.A. (2010) The effect of nutrient deposition on bacterial communities in Arctic tundra soil. *Environ Microbiol*, **12:** 1842–1854.

Caporaso, J.G., Bittinger, K., Bushman, F.D., DeSantis, T.Z., Andersen, G.L., and Knight, R. (2010) PyNAST: a flexible tool for aligning sequences to a template alignment. *Bioinformatics*, **26:** 266–267.

Caporaso, J.G., Lauber, C.L., Walters, W.A., Berg-Lyons, D., Huntley, J., Fierer, N., *et al.* (2012) Ultra-high-throughput microbial community analysis on the Illumina HiSeq and MiSeq platforms. *ISME J*, **6:** 1621–1624.

Chen, Y., Penner, G.B., Li, M., Oba, M., and Guan, L.L. (2011) Changes in bacterial diversity associated with

epithelial tissue in the beef cow rumen during the transition to a high-grain diet. *Appl Environ Microbiol*, **77:** 5770–5781.

Cheng, K.J., and Wallace, R.J. (1979) The mechanism of passage of endogenous urea through the rumen wall and the role of ureolytic epithelial bacteria in the urea flux. *Br J Nutr*, **42:** 553–557.

Cheng, K., McCowan, R., and Costerton, J. (1979) Adherent epithelial bacteria in ruminants and their roles in digestive tract function. *Am J Clin Nutr*, **32:** 139–148.

Cho, S.J., Cho, K.M., Shin, E.C., Lim, W.J., Hong, S.Y., Choi, B.R., *et al.* (2006) 16S rDNA analysis of bacterial diversity in three fractions of cow rumen. *J Microbiol Biotechnol*, **16:** 92–101.

DeSantis, T.Z., Hugenholtz, P., Larsen, N., Rojas, M., Brodie, E.L., Keller, K., *et al.* (2006) Greengenes, a chimera-checked 16S rRNA gene database and workbench compatible with ARB. *Appl Environ Microbiol*, **72:** 5069–5072.

Edgar, R.C. (2010) Search and clustering orders of magnitude faster than BLAST. *Bioinformatics*, **26:** 2460–2461.

Evans, N.J., Brown, J.M., Hartley, C., Smith, R.F., and Carter, S.D. (2012) Antimicrobial susceptibility testing of bovine digital dermatitis treponemes identifies macrolides for in vivo efficacy testing. *Vet Microbiol*, **160:** 496–500.

Godoy-Vitorino, F., Goldfarb, K.C., Karaoz, U., Leal, S., Garcia-Amado, M.A., Hugenholtz, P., *et al.* (2012) Comparative analyses of foregut and hindgut bacterial communities in hoatzins and cows. *ISME J*, **6:** 531–541.

Fondevila, M., and Dehority, B.A. (1996) Interactions between *Fibrobacter succinogenes, Prevotella ruminicola*, and *Ruminococcus flavefaciens* in the digestion of cellulose from forages. *J Anim Science*, **74:** 678–684.

Guo, X., Xia, X., Tang, R., Zhou, J., Zhao, H., and Wang, K. (2008) Development of a real-time PCR method for Firmicutes and Bacteroidetes in faeces and its application to quantify intestinal population of obese and lean pigs. *Lett Appl Microbiol*, **47:** 367–373.

Horner-Devine, M.C., Lage, M., Hughes, J.B., and Bohannan, B.J. (2004) A taxa–area relationship for bacteria. *Nature*, **432:** 750–753.

Josephson, K.L., Gerba, C.P., and Pepper, I.L. (1993) Polymerase chain reaction detection of nonviable bacterial pathogens. *Appl Environ Microbiol*, **59:** 3513–3515.

Kanauchi, O., Matsumoto, Y., Matsumura, M., Fukuoka, M., and Bamba, T. (2005) The beneficial effects of microflora, especially obligate anaerobes, and their products on the colonic environment in inflammatory bowel disease. *Curr Pharm Design*, **11:** 1047–1053.

Kim, M., Morrison, M., and Yu, Z. (2011) Status of the phylogenetic diversity census of ruminal microbiomes. *FEMS Microbiol Ecol*, **76:** 49–63.

Li, M., Zhou, M., Adamowicz, E., Basarab, J.A., and Guan, L.L. (2012) Characterization of bovine ruminal epithelial bacterial communities using 16S rRNA sequencing, PCR-DGGE, and qRT-PCR analysis. *Vet Microbiol*, **155:** 72–80.

Licht, T.R., Madsen, B., and Wilcks, A. (2007) Selection of bacteria originating from a human intestinal microbiota in the gut of previously germ-free rats. *FEMS Microbiol Lett*, **277:** 205–209.

Liu, J., Xu, T., Zhu, W., and Mao, S. (2014) High-grain feeding alters caecal bacterial microbiota composition and fermentation and results in caecal mucosal injury in goats. *Br J Nutr*, **112:** 416–427.

Lozupone, C., and Knight, R. (2005) UniFrac: a new phylogenetic method for comparing microbial communities. *Appl Environ Microbiol*, **71:** 8228–8235.

Mao, S., Zhang, R., Wang, D., and Zhu, W. (2012) The diversity of the fecal bacterial community and its relationship with the concentration of volatile fatty acids in the feces during subacute rumen acidosis in dairy cows. *BMC Vet Res*, **8:** 237.

Mao, S., Huo, W., and Zhu, W. (2013a) Use of pyrosequencing to characterize the microbiota in the ileum of goats fed with increasing proportion of dietary grain. *Curr Microbiol*, **67:** 341–350.

Mao, S., Zhang, R., Wang, D., and Zhu, W. (2013b) Impact of subacute ruminal acidosis (SARA) adaptation on rumen microbiota in dairy cattle using pyrosequencing. *Anaerobe*, **24:** 12–19.

McCann, J.C., Wickersham, T.A., and Loor, J.J. (2014) High-throughput methods redefine the rumen microbiome and its relationship with nutrition and metabolism. *Bioinfor Biol Insights*, **8:** 109.

McCowan, R., Cheng, K., and Costerton, J. (1980) Adherent bacterial populations on the bovine rumen wall: distribution patterns of adherent bacteria. *Appl Environ Microbiol*, **39:** 233–241.

Nakazawa, F., Sato, M., Poco, S.E., Hashimura, T., Ikeda, T., Kalfas, S., et al. (2000) Description of Mogibacterium pumilum gen. nov., sp nov and Mogibacterium vescum gen. nov., sp nov., and reclassification of Eubacterium timidum (Holdeman et al. 1980) as Mogibacterium timidum gen. nov., comb. nov. *Int J Syst Evol Micr*, **50:** 679–688.

Ozutsumi, Y., Hayashi, H., Sakamoto, M., Itabashi, H., and Benno, Y. (2005) Culture-independent analysis of fecal microbiota in cattle. *Biosci Biotechnol Biochem*, **69:** 1793–1797.

Petri, R.M., Schwaiger, T., Penner, G.B., Beauchemin, K.A., Forster, R.J., McKinnon, J.J., and McAllister, T.A. (2013) Changes in the rumen epimural bacterial diversity of beef cattle as affected by diet and induced ruminal acidosis. *Appl Environ Microbiol*, **79:** 3744–3755.

Power, S.E., O'Toole, P.W., Stanton, C., Ross, R.P., and Fitzgerald, G.F. (2014) Intestinal microbiota, diet and health. *Br J Nutr*, **111:** 387–402.

Price, M.N., Dehal, P.S., and Arkin, A.P. (2009) FastTree: computing large minimum evolution trees with profiles instead of a distance matrix. *Mol Biol Evol*, **26:** 1641–1650.

Reilly, K., and Attwood, G.T. (1998) Detection of Clostridium proteoclasticum and closely related strains in the rumen by competitive PCR. *Appl Environ Microbiol*, **64:** 907–913.

Rettedal, E., Vilain, S., Lindblom, S., Lehnert, K., Scofield, C., George, S., et al. (2009) Alteration of the ileal microbiota of weanling piglets by the growth-promoting antibiotic chlortetracycline. *Appl Environ Microbiol*, **75:** 5489–5495.

Sadet, S., Martin, C., Meunier, B., and Morgavi, D.P. (2007) PCR-DGGE analysis reveals a distinct diversity in the bacterial population attached to the rumen epithelium. *Animal*, **1:** 939–944.

Sadet-Bourgeteau, S., Martin, C., and Morgavi, D. (2010) Bacterial diversity dynamics in rumen epithelium of wethers fed forage and mixed concentrate forage diets. *Vet Microbiol*, **146:** 98–104.

Salyers, A.A., Shoemaker, N.B., and Li, L.-Y. (1995) In the driver's seat: the Bacteroides conjugative transposons and the elements they mobilize. *J Bacteriol*, **177:** 5727.

Sayers, G., Marques, P.X., Evans, N.J., O'Grady, L., Doherty, M.L., Carter, S.D., and Nally, J.E. (2009) Identification of spirochetes associated with contagious ovine digital dermatitis. *J Clin Microbiol*, **47:** 1199–1201.

Schloss, P.D., Westcott, S.L., Ryabin, T., Hall, J.R., Hartmann, M., Hollister, E.B., et al. (2009) Introducing mothur: open-source, platform-independent, community-supported software for describing and comparing microbial communities. *Appl Environ Microbiol*, **75:** 7537–7541.

Shanks, O.C., Kelty, C.A., Archibeque, S., Jenkins, M., Newton, R.J., McLellan, S.L., et al. (2011) Community structures of fecal bacteria in cattle from different animal feeding operations. *Appl Environ Microbiol*, **77:** 2992–3001.

Shinkai, T., Matsumoto, N., and Kobayashi, Y. (2007) Ecological characterization of three different phylogenetic groups belonging to the cellulolytic bacterial species Fibrobacter-succinogenes in the rumen. *Anim Sci J*, **78:** 503–511.

Siavoshian, S., Segain, J., Kornprobst, M., Bonnet, C., Cherbut, C., Galmiche, J., and Blottiere, H. (2000) Butyrate and trichostatin A effects on the proliferation/differentiation of human intestinal epithelial cells: induction of cyclin D3 and p21 expression. *Gut*, **46:** 507–514.

Songer, J.G. (1998) Clostridial diseases of small ruminants. *Vet Res*, **29:** 219–232.

Takahashi, N., Sato, T. and Yamada, T. (2000) Metabolic Pathways for Cytotoxic End Product Formation from Glutamate-and Aspartate-Containing Peptides by *Porphyromonas gingivalis*. *J Bacteriol*, **182:** 4704–4710.

Vanhatalo, A., and Ketoja, E. (1995) The role of the large-intestine in post-ruminal digestion of feeds as measured by the mobile-bag method in cattle. *Br J Nutr*, **73:** 491–505.

Vasiljevic, N., Wu, K., Brentnall, A.R., Kim, D.C., Thorat, M.A., Kudahetti, S.C., et al. (2011) Absolute quantitation of DNA methylation of 28 candidate genes in prostate cancer using pyrosequencing. *Dis Markers*, **30:** 151–161.

Wallace, R. (1979) Effect of ammonia concentration on the composition, hydrolytic activity and nitrogen metabolism of the microbial flora of the rumen. *J Appl Bacteriol*, **47:** 443–455.

Wang, Q., Garrity, G.M., Tiedje, J.M., and Cole, J.R. (2007) Naive Bayesian classifier for rapid assignment of rRNA sequences into the new bacterial taxonomy. *Appl Environ Microbiol*, **73:** 5261–5267.

Widyastuti, Y., Lee, S.K., Suzuki, K., and Mitsuoka, T. (1992) Isolation and characterization of rice-straw degrading clostridia from cattle rumen. *J Vet Med Sci*, **54:** 185–188.

Zhao, X., Guo, Y., Guo, S., and Tan, J. (2013) Effects of *Clostridium butyricum* and *Enterococcus faecium* on growth performance, lipid metabolism, and cecalmicrobiota of broiler chickens. *Appl Microbiol Biot*, **97:** 6477–6488.

8

Microbial protein: future sustainable food supply route with low environmental footprint

Silvio Matassa,[1,2] Nico Boon,[1] Ilje Pikaar[3] and Willy Verstraete[1,2,4,*]

[1] Center of Microbial Ecology and Technology (CMET), Ghent University, Coupure Links 653, 9000 Gent, Belgium.
[2] Avecom NV, Industrieweg 122P, 9032 Wondelgem, Belgium.
[3] The School of Civil Engineering, The University of Queensland, St. Lucia, QLD 4072, Australia.
[4] KWR Watercycle Research Institute, Post Box 1072, 3430, BB Nieuwegein, The Netherlands.

Summary

Microbial biotechnology has a long history of producing feeds and foods. The key feature of today's market economy is that protein production by conventional agriculture based food supply chains is becoming a major issue in terms of global environmental pollution such as diffuse nutrient and greenhouse gas emissions, land use and water footprint. Time has come to re-assess the current potentials of producing protein-rich feed or food additives in the form of algae, yeasts, fungi and plain bacterial cellular biomass, producible with a lower environmental footprint compared with other plant or animal-based alternatives. A major driver is the need to no longer disintegrate but rather upgrade a variety of low-value organic and inorganic side streams in our current non-cyclic economy. In this context, microbial bioconversions of such valuable matters to nutritive microbial cells and cell components are a powerful asset. The worldwide market of animal protein is of the order of several hundred million tons per year, that of plant protein several billion tons of protein per year; hence, the expansion of the production of microbial protein does not pose disruptive challenges towards the process of the latter. Besides protein as nutritive compounds, also other cellular components such as lipids (single cell oil), polyhydroxybuthyrate, exopolymeric saccharides, carotenoids, ectorines, (pro)vitamins and essential amino acids can be of value for the growing domain of novel nutrition. In order for microbial protein as feed or food to become a major and sustainable alternative, addressing the challenges of creating awareness and achieving public and broader regulatory acceptance are real and need to be addressed with care and expedience.

Introduction

From the times when our ancestors decided to settle, growing crops and domesticating animals became consolidated practices allowing constant feed and food production. As human civilization proceeded, new strategies of securing food supply have continuously been discovered, consolidated and improved. The major driver of such process was the need to provide resilience towards the changing elements of nature, continuously threatening food supply (Berglund, 2003).

The current anthropogenic pressure on earth's finite resources and the concomitant dynamics of climate change, generate serious concerns about the resilience of the contemporary agricultural feed/food chains (Godfray et al., 2010). In view of the still growing world population towards 10 billion in 2050 (Ezeh et al., 2012), it has been calculated that the world will need to produce about 70% more food calories than in 2006 (Ranganathan, 2013). Therefore, there is a need to find reliable alternative solutions, able to strengthen future food security while minimizing the impact on the global sustainability.

Microorganisms have always been central in basic food processing techniques, for instance converting fibres into edible food when fermenting dough to produce bread, or milk into cheese, allowing its long-term preservation (Caplice and Fitzgerald, 1999). They have been often used as direct food source, as it is the case for yeast or algae. The latter, together with bacteria, constitute the microbial actors involved in processing food. They can also be used directly as feed or food source (Anupama and Ravindra, 2000). The term 'microbe' is used here in the broad connotation of bacteria, fungi, yeast and algae.

In the early 1960s, when public awareness grew in respect to the impeding global demographic boom, the

*For correspondence. E-mail willy.verstraete@UGent.be

need to search for alternatives to sustainably feed a growing population corresponded in major efforts to develop alternative feed and food sources (Goldberg, 1985). Several attempts were made to develop and bring to practice the production of high-quality protein additives from microorganisms, known as microbial protein (MP), or single cell protein (SCP), mainly by using abundant and low cost hydrocarbon substrates such as methanol and methane (Goldberg, 1985). The Imperial Chemical Industries (ICI) were the first to bring to full scale production and commercialization a MP product called Pruteen®, produced from methanol oxidation by means of *Methylophilus methylotrphus* (Westlake, 1986). Besides industrially developed hydrocarbon-based MP, researchers investigated a whole range of other possibilities to produce MP, including the use of natural or artificial light, molecular hydrogen and many different organic substrates such as by-products from the sugar industry as well as other food processing residues or even food wastes (Anupama and Ravindra, 2000). Despite being well accepted and successful in many feed trials with livestock, the actual and definitive breakthrough of MP in the animal feed market was hampered by the low prices achieved by more conventional protein sources such as soybean and fishmeal in the late 1970s as well as the fairly underdeveloped state of fermentation technology. Concomitantly, the rising oil prices in the subsequent decades led to the end of the ICI enterprise because of the relatively high costs of MP production and the consequent competitive disadvantage towards other cheaper more 'natural' alternatives (Øverland et al., 2010).

In recent years, however, research and development around MP is regaining momentum both in the scientific and industrial domains. The steep increase in the prices of fishmeal (from about $500 per ton in the 1990s to $1500 to $2500 in recent years), together with the environmental pressure of soybean production on land and water use in the tropical areas of the globe justify the re-examination of the microbial alternative (Kupferschmidt, 2015).

In the present article, we align the possibilities offered as well as the challenges to be faced by the use of MP production as a biotechnological tool to help securing nutritive protein supply in the years to come. Threatened by forthcoming population growth, climate change and agricultural unsustainability, mankind must seek, once more, new forms of adaptation to safeguard itself.

Microbial protein: feed, food and further

MP as feed

The main driver leading to the renaissance of MP as a source of feed is indubitably the aquaculture sector. Fish farming currently provides about 50% of world's fish food supply, and it is projected to grow further, becoming a key sector in the supply of high-quality protein for the global population. In this context, scientific research and the industrial applications have found in MP a powerful ally. Aquaculture accounts nowadays for more than 73% of the global fishmeal consumption, with wild fish capture clearly unable to provide enough high-quality feed for such a fast growing sector (The World Bank, 2013). Production of MP from natural gas has recently received a great deal of attention, with innovative fermentation processes allowing high volumetric productivities (3–4 kg MP dry matter (DM) per m^3 reactor volume per hour) by continuous cultures of *Methylococcus capsulatus*, marketed under the name of FeedKind® (Unibio, 2016). The latter level of productivity has a physical footprint which is a factor 1000, or more, smaller than any conventional vegetable protein production system (Matassa et al., 2015a). Besides achieving feasible industrial scale production and costs competitiveness with fishmeal, the final MP product is comparable to fishmeal in terms of essential amino acid profile and overall nutritive value (Øverland et al., 2010). Being tested in numerous feed trials with different fish species, resulting in promising perspectives, full-scale production is currently ongoing, with a production of up to 80 000 ton DM/year foreseen in the near future (see Table 2).

In addition to aquaculture, the MP product has also been successfully tested in feed trials with terrestrial animals including major livestock like ruminants, pigs and chickens, broadening its potential market applications (Øverland et al., 2010). In this case though, the relatively low price of soybean meal and the abundant and well-established use as main protein additive in livestock production of the latter, still counteract the application of natural gas-based MP as replacement of substantial percentages of feed composed by fishmeal.

An alternative route to produce MP consists of recovering valuable nutrients from various side streams of the food industry, for instance feed and food processing water (Lee et al., 2015). In this case, the use of heterotrophic microorganisms such as yeast and bacteria allows to convert the organic carbon and the nutrients (N, P) in the waste or processing waters into MP (Anupama and Ravindra, 2000). Microbial protein produced along this line might constitute a valuable and competitive route to produce a substitute for soy protein for animal feed. Indeed it should be possible to generate such MP at costs which take into account the revenue from the avoidance of the treatment of the mineral nutrients (N, P) present in side (waste) streams. As a matter of fact, dissipation of reactive nitrogen back to atmosphere as dinitrogen gas by means of the conventional nitrification–denitrification pathway comes to a cost of about 2–

3 Euro per kg nitrogen-N, while capture of phosphorous in the wastewater line costs of the order of 7 Euro per kg P (Levlin and Hultman, 2003). Note that the market value of proteinaceous nitrogen from vegetable sources is at the current price of some 1.1–1.6 Euro per kg dry weight protein (see Table 1) corresponding to some 6 Euro per kg proteinaceous N. When this microbial proteinaceous N is converted to high-value animal protein, one can attain an end-value of the same order and even up to 14 Euro per kg dry weight protein, in case of fish.

An example of the implementation of such approach to food processing water is the study published by Lee et al. (2015), where the effluent of a brewery is used as feedstock for the production of SCP-MP. The latter study relates to a technology implemented by Nutrinsic, dealing with a production volume of 5000 ton DM/year from a brewery effluent (see Table 2). Also, at present in Belgium, a first full-scale MP production installation is under construction dealing with the upgrading of potato process waters. It should be in production by 2016 at the level of 5000 ton MP per year (Valpromic NV, pers. comm.). Yet, for the latter bacterial-based MP products, there are so far no clear cut data in terms of their putative market size or market values. Nevertheless, the sector is attracting growing interests from investors dealing with novel aspects of the cyclic economy (Nutrinsic, 2014).

MP as food

Microbial protein is an alternative source of high-quality protein able to replace animal protein like fishmeal in livestock nutrition and aquaculture. Going one step higher in the food chain, MP is meeting the FAO/WHO requirements in terms of essential amino acid scoring pattern for human nutrition (Fig. 1) and therefore, also humans could benefit greatly from the use of MP directly as food.

Algae are reported to have supported the life of ancient populations living close to the sea for millennia, providing a constant source of protein and vitamins. Algae and microalgae are currently used as food and food supplements in food industry (Anupama and Ravindra, 2000; Becker, 2007), with a global production achieving 9000 ton DM/year (see Table 2) with a market value estimated about 2.4 billion Euro with a projected yearly growth of 10%.

Table 1. Production volumes and price of various animal and vegetable protein sources.

Protein source	Production volume (Mton DM/y)	Farm gate price ($/kg DM)	Average protein content (% DW)	Price per unit protein ($/kg protein DM)	Ref
Animal					
Fish	66.7	2.07	15–20	10–14	Waite et al. (2014)
Pork	108.5	1.54	20	7.7	Waite et al. (2014)
Chicken	92.7	1.43	31	4.6	Waite et al. (2014)
Beef	62.7	2.70	25	10.8	Waite et al. (2014)
Vegetable					
Soybean	320.2	0.37	35	1.1	Indexmundi (2016a); USDA (2015)
Wheat	712.7	0.19	12	1.6	FAO, 2015; Indexmundi (2016b)

Table 2. Overview of current production volumes and market sizes for different microbial protein. Hyphens indicate that values were not available.

Organisms	Production volume (ton DM/y)	Production costs (Euro/kg DM)	Global market value (Billion Euro)	Yearly growth (% per year)	Remarks	Ref
Yeast	3 000 000	–	9.2	7.9	Mostly commercialized as baker's yeast and for ethanol fermentation. Global market value projected to 2019	Kellershohn and Russell (2015)
Algae (microalgae)	9000	4–25	2.4	10	Besides feed and food, derivatives are also used	Enzing et al. (2014)
Mycoprotein (Quorn®)	25 000	–	0.214	20	Investments for a plant of 22000 tons per year were done in 2015	Beer (2015)
Bacteria (Profloc®)	5000	1–1.1	–	–		Nutrinsic (2015)
Bacteria (FeedKind®)	80 000	–	–	–	Commercial production foreseen on 2016	Byrne (2016)
Valpromic	5000	–	–	–		Personal communication

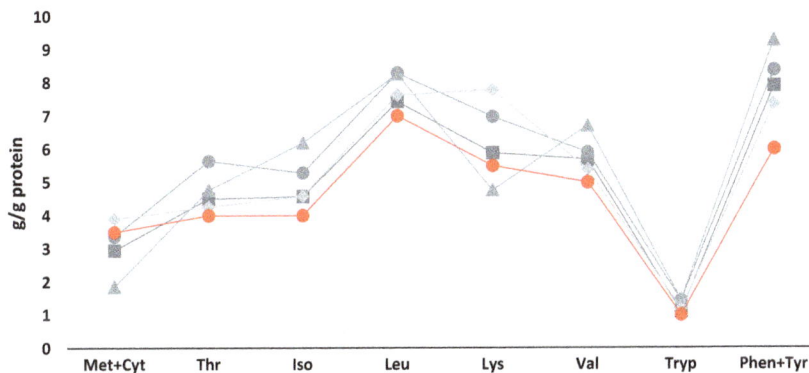

Fig. 1. Essential amino acid scoring pattern of microbial protein from bacteria (Pseudomonas/Methylophilus spp.) (—■—), yeast (Candida spp.) (—●—), algae (Spirulina maxima (—▲—), compared with the high-quality animal protein from fishmeal (◇) as well as to the FAO/WHO standard (—●—) for amino acid scoring pattern for human nutrition. Source: Harper (1981); Tacon (1987)

Yeast, being at the base of food processing since the first bread was baked, or grapes fermented, can also be used as direct food source, as it was the case e.g. with the massive campaign of yeast production and supply, first to the army, then to the whole population during World War II (Khachatourians and Arora, 2002). Currently, yeast is a major player in the microbial derived production of products for food as well as for other applications. Baker's yeast and alcohols fermentation are the two main processes employing yeast, with a projected global market value for 2019 of up to 9.2 billion Euro and an annual growth forecast of 7.9% (see Table 2).

Fungi are also a suitable alternative and have also made their way as human food. Quorn™ is the most successful example of the so called mycoprotein, which is commercialized and sold in some 15 countries worldwide (Wiebe, 2004). Mycoproteins are particularly suited to reproduce the taste and consistency of meat; this explains their success as alternative to conventional animal-based products. Currently, mycoprotein production supporting Quorn™ products manufacturing amounts to 25 000 ton DM/year, with a global market value of about 214 million Euro, prospected to grow with 20% annually in the coming years.

Added value applications

Besides being rich in nutritive protein, microorganisms offer the possibility of producing a broad variety of added-value products, suitable for both animal and human nutrition (Vandamme and Revuelta, 2016). Table 3 summarizes the average amount of protein producible by algae, fungi and bacteria, as well as other possible added-value products already investigated or produced from microorganisms.

Certain microalgae and cyanobacteria are primary producers of microbial oil, suitable as substitutes for

vegetable oil in food supplements. Particularly, the high concentration of fatty acids can replace fatty acids otherwise derived from rape seed, soy, sunflower oil and palm oil. The purification of omega-3 fatty acid can offer even higher value applications, e.g. for clinical purposes, eicosapentanoic acid and decosahexaenoic acid, normally obtained from fish oil, can be also concentrated and purified from naturally omega-3-accumulating microalgae. Vitamins such as vitamin B12 and provitamin A are also important high-value products obtainable from edible algae, conferring additional nutritional benefits in livestock production. Carbohydrates, which can be accumulated up to 70% of the cell dry weight by many algal species are also of nutritional value, but the major research effort so far was directed towards the use of algae for biofuel, biogas or biohydrogen generation (Jones and Mayfield, 2012; Draaisma et al., 2013).

Fungi, mainly yeast-like fungi are the main agents involved in the saccharification of fibres from corn, as well as fermentation of other organic substrates. While processing corn fibre with yeast like *Aureobasidium*, xylose, arabinose and glucose can be produced at different relative concentrations depending on the pre-treatment of the fibres feedstock. The sugars can then be further fermented in bioethanol, xylitol and pullulan. Xylitol and pullulan find particularly application as food additive for their specific property of flavour-enhancing and binding agents. Besides sugars and sugar-derivatives, yeasts like *Phaffa rhodozyma* (now Xanthophyllomyces rodochrous) can be used to produce valuable carotenoid pigments like astaxanthin, mainly used in aquaculture as feed supplement for salmon (Leathers, 2003).

Bacteria are a versatile group of microorganisms able to produce a large array of added-value bio-products. Biopolymers such as polyhydroxyalkanes (PHA) are named to be biological alternatives to petroleum-based chemicals to produce plastics. Yet, so far large-scale

Table 3. Overview of different microorganism for MP and added-value product formation.

Microorganism	Average crude protein content (% CDW)	Nutritional value	Added value by-products (% CDW)	Remarks	Ref.
Algae	40–60	Compares favourably to egg, soy and what protein. Cell wall digestibility is an issue	• Microbial oil (50–70%) • Carbohydrates (up to 70%) • Vitamins • …	Triacylglycerides (TAG) can replace partly vegetable oils in food products. Poly unsaturated fatty acids (PUFA) are of interest for health applications	Draaisma *et al.* (2013); Harun *et al.* (2010)
Fungi (Filamentous and Yeast)	30–70	Amino acids and digestibility of mycoprotein is similar to egg and milk	• Carbohydrates • Pullulan • Xylitol • Astaxanthin • …	High unsaturated/saturated fatty acids and low fat content makes them highly suitable for human nutrition	Thrane (2007)
Bacteria	50–83	Amino acids and digestibility is similar to those of fishmeal	• Internal storage polymers (PHB) • Ectoine • Lipids • Extracellular polysaccharides • Growth media and vitamins • …		Strong *et al.* (2015)

applications are to the best of our knowledge, not industrially established. Recently, other applications for PHA/polyhydroxybuthyrate (PHB) such as in the medical field are emerging. Of interest for aquaculture is the ongoing research on the prebiotic effects of PHB when used as feed supplement, offering an interesting alternative to antibiotics (Defoirdt *et al.*, 2007; De Schryver *et al.*, 2010). Another interesting niche product which can be derived from bacteria is osmo-protectants such as glutamate and ectoine (Lentzen and Schwarz, 2006). The latter is a high-value cyclic imino acid used in cosmetic formulation, but which has found application also in aquaculture as highly active protectant against oxidative stress. Bacteria can also produce relevant amounts of lipids, commonly employed in biofuel production. High-quality membrane-derived lipids can also be employed as human health supplement, being already tested as effective in reducing plasma cholesterol during animal tests (Strong *et al.*, 2015).

Forthcoming challenges

The extensive use of MP products as partial replacement of conventional protein feed additives such as soybean and fishmeal can offer the opportunity of decreasing part of the environmental pressure that these products exert on land and water use. A recent report of the British Carbon Trust evaluated the environmental impact of FeedKind® protein, a bacterial MP feed additive produced from natural gas (see *MP as feed*). The report evaluated two FeedKind® commercial products in terms of greenhouse gas, land and water use, comparing them with soybean and fishmeal (Cumberlege *et al.*,

2016). In terms of freshwater consumption, the report shows an average value of about 29 m^3 per ton MP produced. A more detailed analysis shows that this 29 m^3 is for about 80% determined by the vegetable oil used as binding agent to produce a MP-pelletized product. If the latter major contribution is excluded by producing a simple straightforward protein powder, the actual freshwater requirement comes down to the order of 1 m^3 per ton MP. From Fig. 2, it can be derived that this value about water foot print is about 20 and 140 times lower than fishmeal and soybean meal respectively.

The same trend is observable for the required land. The value of 52 m^2 per ton MP is in fact due to vegetable oil for the production of the pelletized form, whereas no arable land is required in case of the powdered MP. Compared with the 6655 m^2 land per ton protein required for the production of soybean meal concentrate, the value of quasi zero land foot print reveals how the land footprint of MP production is a major benefit in respect to conventional agricultural-based protein production. Fishmeal of course requires minimal amount of land for its processing, yet the dramatic impact of wild fish capture on ocean ecosystems is well known and documented (Pauly *et al.*, 2005).

Finally, the above-mentioned report also analyses the carbon footprint of FeedKind®. The value of 5.8 ton CO_{2eq}/ton MP is mainly due to the natural gas necessary for the metabolism of the bacteria involved in the biological fermentation process. This value would be as low as 1.7 ton CO_{2eq}/ton MP in case biogas and renewable energy is used in place of fossil fuels to power the reactor-based production and downstream processing of the final MP product. For fishmeal and soybean meal concentrate, the report indicates values of 2.6 and 0.8 ton CO_{2eq}/ton protein respectively. Nevertheless, if the spared agricultural land in case of MP production would be accounted for its recovered carbon capture potential, the overall benefit in avoided carbon emissions from MP production could possibly be much higher.

In this context, an interesting alternative to natural gas-based MP is represented by autotrophic microorganisms such as algae or hydrogen-oxidizing bacteria. If algae offer the great advantage of being able to use sun light to fix carbon dioxide into biomass, the main drawbacks of such process are the high land footprint required together with the technical challenges of downstream processing of poorly concentrated algal biomass (Majid et al., 2014). On the other hand, the hydrogen gas needed to fix carbon dioxide into bacterial biomass by means of hydrogen-oxidizing bacteria is more expensive resource, but the land footprint and the biomass concentrations achievable with modern fermentation technologies outscore those of the algal platform. In case the production of the latter is connected to hydrogen generated by means of renewable energies (solar, wind, etc.), this allows an elegant platform for MP production and concomitant carbon dioxide capture (Matassa et al., 2015b).

To the scientific community, it is evident how a dedicated industrial production of MP can represent a key biotechnological tool to curb down the environmental impact of the current feed and food chain assuring the necessary amounts of nutritive protein for mankind. Clearly, significant efforts are warranted to bring this to practice at relevant scales. A key feature is to deal with the aspects of public awareness. At present, the mere economic market rules justify the application of MP in feed for livestock only in some niche applications such as aquaculture. Yet, if the externalized environmental costs of the current feed/food production system would be taken into account and made clear to the broader public (including decision makers in political institutions), the MP route would result in a more rational alternative, able to offer immediate advantages in terms of water and land use, with direct consequences on increased carbon capture potential of ecosystems restored by better agricultural land use (Galloway and Leach, 2016). An important aspect, in this sense, relates to nutrients flows, and principally the excessive input in our biosphere of reactive nitrogen species (NH_4^+, NO_2^-, NO_3^-) produced by fixing atmospheric N_2 gas by means of the Haber–Bosch process. Compared with the proposed sustainable boundary of 35 Mton N_2 fixed per year, the current 121 Mton actually converted into reactive nitrogen surpass the sustainability boundary of almost 3.5 times (Rockstrom et al., 2009). Moreover, if the economic benefit in agricultural production ranges between 20 and 80 billion Euro per year, the annual costs (including damages to both environment and human health) of N pollution by agriculture have been estimated in the range of 35–230 billion Euro per year (Van Grinsven et al., 2013). It has been recently demonstrated how the high nitrogen inefficiency of the soil–plant system could be mitigated by MP production (Matassa et al., 2015a), decreasing significantly the impact of eutrophication, nitrous oxide emissions and ecosystems disturbance due to unbalanced anthropogenic nitrogen inputs.

Besides the awareness of the overall environmental benefit of MP production for feed and food, the development of higher value by-products will allow boosting and bolstering the MP biotech platform. Thus, a more powerful penetration into the market of microbial-based product as replacement of chemically derived ones, as discussed above, will be possible. This will play in favour of establishing a public mindset more open and prone to acceptance towards microbial derived products.

Obviously, barriers must be overcome in order to allow a widespread adoption of the MP biotechnology. Besides the official legal recognition of some MP products as feed and food (Øverland et al., 2010), further openings are warranted in terms of used carbon and nutrient sources recovery and their up-cycling into edible MP products as part of the cyclic economy. This will impact drastically on how efficiently our current society makes use of its precious primary resources.

Conclusions

- Microbial protein qualifies as an excellent source of nutritive proteins, but other cellular components can also be of increasing importance, driving new developments for microbial-based by-products. In the context

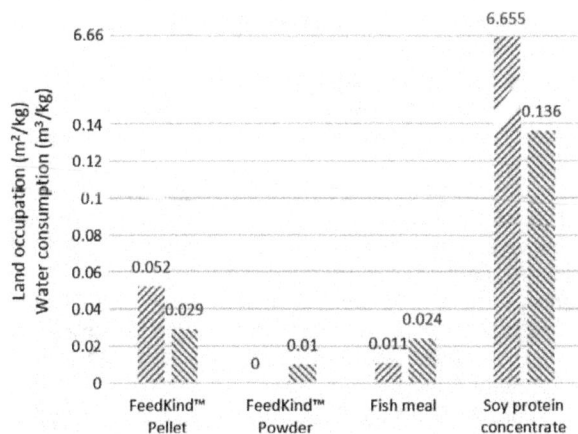

Fig. 2. Land (▨) and freshwater (▨) requirements of MP compared with fishmeal and soy protein concentrate. The values are normalized to the protein content of each product. Source: Cumberlege et al. (2016)

of the need to generate new biotechnological processes and products, this line of research and development, particularly in countries where new opportunities are warranted, should be explored with great care (Timmis *et al.*, 2014).

- The current conventional agricultural based supply route for nutritive animal proteins have a high environmental impact; they should be reconsidered and their externalized environmental costs should be assessed and benchmarked to those of MP.
- Upgrading various valuable nutrients (N, P) and nutritive resources (organic carbon), the production of heterotrophic microbes as food and feed certainly has gaining renewed industrial interest, particularly in the context of the cyclic economy.
- The overall transition of the public mindset towards widespread acceptance and appreciation of MP as main supply route for feed and food needs to be prepared in the near future with care and foresight particularly in terms of quality and regulatory issues.

References

Anupama, and Ravindra, P. (2000) Value-added food: single cell protein. *Biotechnol Adv* 18: 459–479.

Becker, E.W. (2007) Micro-algae as a source of protein. *Biotechnol Adv* 25: 207–210.

Beer, E. (2015) "No success model" for new protein ingredients [WWW Document]. URL http://www.foodnavigator.com/Market-Trends/No-success-model-for-new-protein-ingredients

Berglund, E. (2003) Human impact and climate changes—synchronous events and a causal link? *Quat. Int.* 105: 7–12.

Byrne, J. (2016) Calysta says gas to fishmeal replacement protein on path to commercialization [WWW Document]. URL http://www.feednavigator.com/R-D/Calysta-says-gas-to-fishmeal-replacement-protein-on-path-to-commercialization

Caplice, E., and Fitzgerald, G.F. (1999) Food fermentations: role of microorganisms in food production and preservation. *Int J Food Microbiol* 50: 131–149. doi:10.1016/S0168-1605(99)00082-3.

Cumberlege, T., Blenkinsopp, T. and Clark, J. (2016) Assessment of environmental impact of FeedKind protein Carbon Trust, URL https://www.carbontrust.com/media/672719/calysta-feedkind.pdf.

De Schryver, P., Sinha, A.K., Kunwar, P.S., Baruah, K., Verstraete, W., Boon, N., *et al.* (2010) Poly-beta-hydroxybutyrate (PHB) increases growth performance and intestinal bacterial range-weighted richness in juvenile European sea bass, Dicentrarchus labrax. *Appl Microbiol Biotechnol* 86: 1535–1541. doi:10.1007/s00253-009-2414-9.

Defoirdt, T., Boon, N., Sorgeloos, P., Verstraete, W., and Bossier, P. (2007) Alternatives to antibiotics to control bacterial infections: luminescent vibriosis in aquaculture as an example. *Trends Biotechnol* 25: 472–479. doi:10.1016/j.tibtech.2007.08.001.

Draaisma, R.B., Wijffels, R.H., Slegers, P.M.E., Brentner, L.B., Roy, A., and Barbosa, M.J. (2013) Food commodities from microalgae. *Curr Opin Biotechnol* 24: 169–177. doi:10.1016/j.copbio.2012.09.012.

Enzing, C., Ploeg, M., Barbosa, M. and Sijtsma, L. (2014) Microalgae-based products for the food and feed sector: an outlook for Europe. doi:10.2791/3339.

Ezeh, A.C., Bongaarts, J., and Mberu, B. (2012) Global population trends and policy options. *Lancet* 380: 142–148. doi:10.1016/S0140-6736(12)60696-5.

FAO (2015) World food situation [WWW Document]. URL http://www.fao.org/worldfoodsituation/csdb/en/

Galloway, J.N., and Leach, A.M. (2016) Sustainability: your feet's too big. *Nat Geosci* 9: 97–98.

Godfray, H.C.J., Beddington, J.R., Crute, I.R., Haddad, L., Lawrence, D., Muir, J.F., *et al.* (2010) Food security: the challenge of feeding 9 billion people. *Science* 327: 812–818.

Goldberg, I. (1985) *Single Cell Protein, Biotechnology Monographs.* Springer-Verlag, Berlin.

Harper, A. (1981) Amino Acid Scoring Patterns [WWW Document]. URL http://www.fao.org/3/contents/aa7e1ca5-4634-51bf-a465-5bf12d5cec2d/M3013E00.HTM

Harun, R., Singh, M., Forde, G.M., and Danquah, M.K. (2010) Bioprocess engineering of microalgae to produce a variety of consumer products. *Renew Sustain Energy Rev* 14: 1037–1047. doi:10.1016/j.rser.2009.11.004.

Indexmundi (2016a) Soybeans [WWW Document]. URL http://www.indexmundi.com/commodities/?commodity=soybeans.

Indexmundi (2016b) Wheat [WWW Document]. URL http://www.indexmundi.com/commodities/?commodity=wheat.

Jones, C.S., and Mayfield, S.P. (2012) Algae biofuels: versatility for the future of bioenergy. *Curr Opin Biotechnol* 23: 346–351. doi:10.1016/j.copbio.2011.10.013.

Kellershohn, J. and Russell, I. (2015) Yeast biotechnology. In *Advances in Food Biotechnology.* John Wiley & Sons, pp. 303–310. doi:10.1002/9781118864463.ch18.

Khachatourians, G.G. and Arora, D.K. (2002) *Applied Mycology and Biotechnology*, Volume 2. Amsterdam: Agriculture and Food Production.

Kupferschmidt, K. (2015) Why insects could be the ideal animal feed [WWW Document]. *Science.* doi:10.1126/science.aad4709.

Leathers, T.D. (2003) Bioconversions of maize residues to value-added coproducts using yeast-like fungi. *FEMS Yeast Res* 3: 133–140. doi:10.1016/S1567-1356(03)00003-5.

Lee, J.Z., Logan, A., Terry, S., and Spear, J.R. (2015) Microbial response to single-cell protein production and brewery wastewater treatment. *Microb Biotechnol* 8: 65–76. doi:10.1111/1751-7915.12128.

Lentzen, G., and Schwarz, T. (2006) Extremolytes: natural compounds from extremophiles for versatile applications. *Appl Microbiol Biotechnol* 72: 623–634. doi:10.1007/s00253-006-0553-9.

Levlin, E. and Hultman, B. (2003) Phosphorus recovery from phosphate rich side-streams in wastewater treatment plants. In Polish Swedish Seminar, Gdansk March. pp. 47–56.

Majid, M., Shafqat, S., Inam, H., Hashmi, U. and Kazi, A.G. (2014) *Biomass and Bioenergy: Processing and*

Properties. Hakeem, R.K., Jawaid, M. and Rashid, U. (eds). Cham, Switzerland: Springer International Publishing, pp. 207–224. doi:10.1007/978-3-319-07641-6_13.

Matassa, S., Batstone, D.J., Huelsen, T., Schnoor, J.L., and Verstraete, W. (2015a) Can direct conversion of used nitrogen to new feed and protein help feed the world? *Environ Sci Technol* **49:** 5247–5254. doi:10.1021/es505432w.

Matassa, S., Boon, N., and Verstraete, W. (2015b) Resource recovery from used water: the manufacturing abilities of hydrogen-oxidizing bacteria. *Water Res* **68:** 467–478. doi:10.1016/j.watres.2014.10.028.

Nutrinsic (2014) Nutrinsic Corporation Raises $12 [WWW Document]. URL http://www.marketwired.com/press-rele ase/nutrinsic-corporation-raises-127-million-to-fuel-growth-1938044.htm

Nutrinsic (2015) Nutrinsic Announces Grand Opening of First US ProFloc(TM) Facility. [WWW Document]. URL http://www.marketwired.com/press-release/-2020797.htm

Øverland, M., Tauson, A.-H., Shearer, K., and Skrede, A. (2010) Evaluation of methane-utilising bacteria products as feed ingredients for monogastric animals. *Arch. Anim. Nutr.* **64:** 171–189. doi:10.1080/17450391003691534.

Pauly, D., Watson, R., and Alder, J. (2005) Global trends in world fisheries: impacts on marine ecosystems and food security. *Philos Trans R Soc Lond B Biol Sci* **360:** 5–12.

Ranganathan, J. (2013) The Global Food Challenge Explained in 18 Graphics [WWW Document]. World Resour. Inst. URL http://www.wri.org/blog/2013/12/global-food-challenge-explained-18-graphics

Rockstrom, J., Steffen, W., Noone, K., Persson, A., Chapin, F.S., Lambin, E.F., *et al.* (2009) A safe operating space for humanity. *Nature* **461:** 472–475.

Strong, P.J., Xie, S., and Clarke, W.P. (2015) Methane as a resource: can the methanotrophs add value? *Environ Sci Technol* **49:** 4001–4018. doi:10.1021/es504242n.

Tacon, G.J.A. (1987) The nutrition and feeding of farmed fish and shrimp - a training manual 2. Nutrient sources and composition. FAO. URL http://www.fao.org/3/contents/d66b3e1f-c059-50fa-9ba2-717e9940b7f1/AB470E00.htm

The World Bank (2013) FISH TO 2030 Prospects for Fisheries and Aquaculture. doi:83177-GLB.

Thrane, U. (2007) Fungal protein for food. In *Food Mycology, Mycology*. CRC Press, pp. 353–360. doi:10.1201/9781420020984.ch18.

Timmis, K., de Lorenzo, V., Verstraete, W., Garcia, J.L., Ramos, J.L., Santos, H., *et al.* (2014) Pipelines for New Chemicals: a strategy to create new value chains and stimulate innovation-based economic revival in Southern European countries. *Environ Microbiol* **16:** 9–18. doi:10.1111/1462-2920.12337.

Unibio (2016) Introduction _ Unibio [WWW Document]. URL http://www.unibio.dk/technology/introduction.

USDA (2015) World agricultural supply and demand estimates. *United States Dep. Agric.* doi:WASDE-525.

Van Grinsven, H.J.M., Holland, M., Jacobsen, B.H., Klimont, Z., Sutton, M., and Willems, W.J. (2013) Costs and benefits of nitrogen for europe and implications for mitigation. *Environ Sci Technol* **47:** 3571–3579.

Vandamme, E.J. and Revuelta, J.L. (eds.) (2016) *Industrial Biotechnology of Vitamins, Biopigments, and Antioxidants*. Weinheim, Germany: Wiley-VCH, pp. 560.

Waite, R., Beveridge, M., Brummett, R.E., Castine, S., Chaiyawannakarn, N., Kaushik, S., *et al.* (2014) Improving productivity and environmental performance of aquaculture. pp. 1–60. doi:10.5657/FAS.2014.0001.

Westlake, R. (1986) Large-scale continuous production of single cell protein. *Chemie Ing. Tech.* **58:** 934–937. doi:10.1002/cite.330581203.

Wiebe, M.G. (2004) QuornTM Myco-protein - Overview of a successful fungal product. *Mycologist* **18:** 17–20. doi:10.1017/S0269915X04001089.

9

Towards the development of multifunctional molecular indicators combining soil biogeochemical and microbiological variables to predict the ecological integrity of silvicultural practices

Vincent Peck,[1] Liliana Quiza,[1] Jean-Philippe Buffet,[1] Mondher Khdhiri,[1] Audrey-Anne Durand,[1] Alain Paquette,[2] Nelson Thiffault,[2,3] Christian Messier,[2,4] Nadyre Beaulieu,[5] Claude Guertin[1,*] and Philippe Constant[1,**]

[1]INRS-Institut Armand-Frappier, 531 boulevard des Prairies, Laval, Québec Canada H7V 1B7.
[2]Centre d'étude de la forêt, Université du Québec à Montréal, Case postale 8888, succursale Centre-ville, Montréal, Québec Canada H3C 3P8.
[3]Direction de la recherche forestière, Ministère des Forêts, de la Faune et des Parcs, 2700 Einstein, Québec, Québec Canada G1P 3W8.
[4]Institut des Sciences de la Forêt Tempérée (ISFORT), Université du Québec en Outaouais (UQO), 58 rue Principale, Ripon, Québec Canada J0V 1V0.
[5]Produits Forestiers Résolu, 2419 Route 155 sud, La Tuque, Québec Canada G9X 3N8.

For correspondence. *E-mail Claude.Guertin@iaf.inrs.ca

**E-mail   Philippe.Constant@iaf.inrs.ca

Funding Information This research was supported by a grant from the Natural Sciences and Engineering Research Council of Canada (NSERC) Engage Program to PC (grant EGP 463612-14).

Summary

The impact of mechanical site preparation (MSP) on soil biogeochemical structure in young larch plantations was investigated. Soil samples were collected in replicated plots comprising simple trenching, double trenching, mounding and inverting site preparation. Unlogged natural mixed forest areas were used as a reference. Analysis of soil nutrients, abundance of bacteria and gas exchanges unveiled no significant difference among the plots. However, inverting site preparation resulted in higher variations of gas exchanges when compared with trenching, mounding and unlogged natural forest. A combination of the biological and physicochemical variables was used to define a multifunctional classi-

fication of the soil samples into four distinct groups categorized as a function of their deviation from baseline ecological conditions. According to this classification model, simple trenching was the approach that represented the lowest ecological risk potential at the microsite level. No relationship was observed between MSP method and soil bacterial community structure as assessed by high-throughput sequencing of bacterial 16S rRNA gene; however, indicator genotypes were identified for each multifunctional soil class. This is the first identification of multifunctional molecular indicators for baseline and disturbed ecological conditions in soil, demonstrating the potential of applied microbial ecology to guide silvicultural practices and ecological risk assessment.

Introduction

Tree plantations are gaining increased interest to protect natural forests, restore ecosystem services and meet various social needs. For instance, more than 15% of current wood production is supported by tree plantations and this contribution is expected to rise in the future (Paquette and Messier, 2009). Mechanical site preparation (MSP) is a common practice to improve seedling performance in tree plantations dedicated to intensive wood production and forest regeneration. This approach, applied following clear-cut or variable retention logging, involves the utilization of heavy machineries to break soil structure in order to improve soil physical conditions limiting tree growth and control competing vegetation (Löf et al., 2012). Soil scarification and mounding are the most usual MSP methods. Scarification consists in mixing organic and mineral upper layers of soil by trenching. This technique increases soil aeration and temperature, while favouring nutrient availability by accelerating nitrogen mineralization and limiting the invasion of competing vegetation (Prévost, 1992; Thiffault and Jobidon, 2006). Mounding consists in creating elevated planting spots where soil is excavated and deposited on the ground, next to the ditch, while inverting the soil horizons to get mineral soil on the top and an organic layer at the bottom (Sutton, 1993). As an alternative to mounding to increase worker safety for following operations such as cleaning and thinning, excavated soil can also be placed back into the original ditch, resulting in inverting site preparation. Mounding treatments provide the beneficial effects on soil aeration, temperature, competing vegetation control and nutrient availability observed with scarification and are particularly suitable to get seedling spots free from water logging conditions in wet areas (Kabrick et al., 2005; Simon et al., 2013). Mounding and inversion using an excavator are also expected to be less damaging to the environment than regular trenching, disturbing a lesser proportion of the treated area (especially inversion) and avoiding the creation of linear trenches that may cause soil erosion (Buitrago et al., 2015).

Increased soil temperature, followed by the stimulation of organic carbon mineralization and nutrient release are the prevailing soil disturbances caused by MSP (Jandl et al., 2007). In contrast to plant diversity, which is expected to be resilient to this management practice (Haeussler et al., 2004), the resilience, resistance or vulnerability of soil microbiome to these marked soil disturbances remains unknown. This question is of critical importance in ecosystem-based forest management where establishment of tree plantations must exert minimal alteration or even restore ecosystem services sustained in natural forests (Martin et al., 2014). Because microorganisms play a crucial role in global biogeochemical cycles and closely interact with vegetation through nutrient transfers as well as water retention, the composition of soil microbial communities can be seen as an indicator of soil ecosystem functioning. For instance, high microbial biodiversity was shown to promote soil multiple ecological functions including plant diversity and nutrient cycling (Wagg et al., 2014), resistance to the invasion and survival of allochthonous pathogen species (van Elsas et al., 2012) as well as resistance and resilience of microbial community structure to certain environmental stress (Tardy et al., 2014).

The overarching objective of this study was to compare the metabolic activity and composition of soil bacterial community between recently clear-cut and site prepared plots within an hybrid larch (Larix × marschlinsii Coaz) plantation and unlogged natural mixed boreal forest conservation areas to assess the impact of different MSP on soil ecosystem functioning a few years following treatments. The rationale for this approach is the identification of soil biogeochemical processes and microbes affected by MSP to select silvicultural practices offering the best early environmental performance at the microsite level. We tested the hypothesis that both the conversion of natural forest to a hybrid larch monoculture and the intensification of MSP treatments reduce the

activity and the taxonomic diversity of soil bacterial community in the early years after this conversion. We finally combined soil biogeochemical and microbiological data sets to explore the relevance of microbial molecular indicators to predict the environmental impact and guide silviculture practices. In contrast to previous investigations comparing single parameter such as soil enzyme activities between reference sites and managed forests (e.g. Staddon et al., 1999), we have combined multiple biogeochemical variables as a metric to classify soil under an original multifunctional system and searched for indicator 16S rRNA gene sequences restricted to specific soil multifunctional classes. Biogeochemical variables were selected to get a broad classification system, including gaseous exchange involving taxonomically diverse and specialized microbial guilds. Under this framework, we expected to identify indicators for baseline or disturbed ecological conditions in soil and to explore the potential of soil bacterial community monitoring as a promising approach to predict the ecological integrity of soil in the early stage of intensively managed tree plantations.

Results

Soil physicochemical properties, gaseous exchanges and abundance of bacteria

Conversion of unlogged natural forest to a hybrid larch plantation caused no significant change in measured soil physicochemical properties at the microsite level (Table 1). Soil carbon and nitrogen concentrations were significantly and positively correlated (Pearson, $P < 0.0001$) and C:N stoichiometry ranged from 18 to 32 among samples. All tested soil represented a net sink for H_2 and CO and net source for CO_2 (Table 1). Even though gaseous exchanges were not affected by MSP procedures, some treatments induced high variations between replicates. In general, the coefficient of variation (CV) of gas exchanges increased as a function of the intensification of the MSP techniques (Fig. 1). For instance, H_2, CO and CO_2 exchanges measured in soil samples collected in excavated mounds displayed CV ranging from 58% to 62% while the same variables measured in unlogged natural forest samples were characterized by CV of 7–32%. Trace gas turnover showed no significant relationship with soil carbon and nitrogen concentrations (Pearson, $P > 0.05$) but a negative correlation was observed between soil respiration and pH (Pearson, $P = 0.03$). H_2 uptake rates increased as function of CO uptake (Pearson, $P < 0.0001$) and CO_2 production (Pearson, $P = 0.02$) activities. CO_2 production rates were not related to measured CO uptake activities (Pearson, $P > 0.05$). The abundance of bacteria was proportional to CO_2 production rate in soil (Pearson, $P = 0.01$).

Microbial Bioremediation: Processes, Techniques and Applications

Table 1. Physicochemical properties, trace gas exchanges, bacterial 16S rRNA gene abundance and bacterial diversity in soil. Averages are represented with standard deviations in parentheses[a].

Sites	Carbon (%)	Sand (%)	Silt (%)	Clay (%)	Nitrogen (%)	pH	H_2 oxidation (nmol $g_{(dw)}^{-1}$ h^{-1})	CO oxidation (nmol $g_{(dw)}^{-1}$ h^{-1})	CO_2 production (µmol $g_{(dw)}^{-1}$ h^{-1})	16S rRNA gene (copies $g_{(dw)}^{-1}$)	Alpha diversity (Shannon)	Alpha diversity (ACE)
Unlogged natural	4.9 (1.1)	83 (3)	13 (3)	4 (1)	0.22 (0.04)	4.3 (0.1)	6.3 (0.4)	4.5 (0.6)	4.1 (1.3)	7.9 (1.3) $\times 10^9$	6.2 (0.08)	2950 (242)
Simple	5.4 (1.3)	86 (3)	13 (3)	1 (1)	0.27 (0.06)	4.6 (0.2)	5.8 (1.6)	4.5 (2.0)	3.1 (1.4)	4.9 (3.6) $\times 10^9$	6.3 (0.2)	2895 (565)
Double intensive	4.1 (1.2)	84 (2)	14 (2)	2 (0)	0.21 (0.06)	4.7 (0.2)	5.2 (1.4)	3.8 (1.7)	1.5 (0.5)	2.0 (2.6) $\times 10^9$	6.2 (0.3)	2617 (732)
Inversion	3.0 (1.4)	76 (2)	22 (2)	2 (0)	0.13 (0.04)	4.7 (0.2)	3.0 (2.3)	2.1 (1.4)	1.4 (0.2)	1.5 (1.4) $\times 10^9$	6.1 (0.5)	2268 (985)
Mound	4.3 (0.8)	85 (8)	14 (8)	1 (1)	0.17 (0.03)	4.6 (0.2)	6.6 (4.1)	4.8 (1.8)	1.1 (2.3)	1.8 (2.1) $\times 10^9$	6.0 (0.2)	2247 (877)

a. No variable showed significant difference between the treatments (ANOVA, $P > 0.05$).

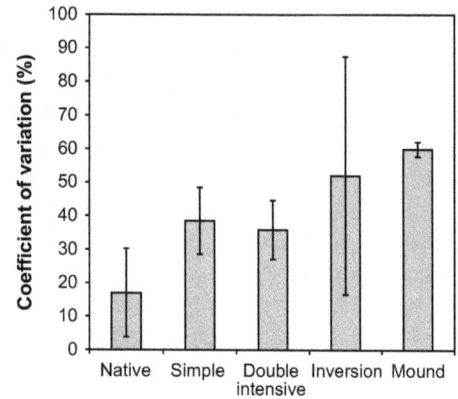

Fig. 1. Coefficient of variation (CV) for gaseous exchanges. The bars represent the average and standard deviation of CV measured for H_2, CO and CO_2.

Multifunctional soil classification

Soil physicochemical properties, gaseous exchanges and bacteria abundance variables were utilized to define a multifunctional soil classification. Under this classification approach, the distribution of each variable was considered to compute a distance matrix measuring the association between soil samples (Fig. 2A). Four different multifunctional classes were identified in the clustering analysis. The level of disturbance characterizing each class was defined on the basis of their Euclidean distance from the unlogged natural forest plots. Firstly, the soil sample constituting class I (M-B) represented the MSP treatment that resulted in the most intense disturbance of baseline ecological functions. Secondly, soil samples included in class II (S-A, S-B) belong to the category of MSP treatments that caused slight deviations from baseline soil ecological functions. Soil samples belonging to class III (N-B, N-A, N-C) correspond to baseline of soil ecological functions. Finally, class IV (S-C, I-A, I-C, I-B, M-A, M-C, D-A, D-B and D-C) encompass soils for which the MSP treatments caused important alteration of baseline ecological functions at the microsite level. The unlogged natural forest was the sole condition for which replicated composite samples exhibited treatment-specific, conserved multifunctional profile. Taken together, this classification model indicates that soil multifunctional profile observed in inversion and mound excavations plots were those showing the strongest deviation from unlogged natural forest at the microsite level. With the exception of one replicate (S-C), simple trenching (S-A and S-B) was the treatment exerting the lowest incidence on soil multifunctional profile (Fig. 2A).

A principal component analysis was computed to represent the position of sampling sites in a reduced space defined by the measured variables. The ordination space

(A)

(B)

(C)

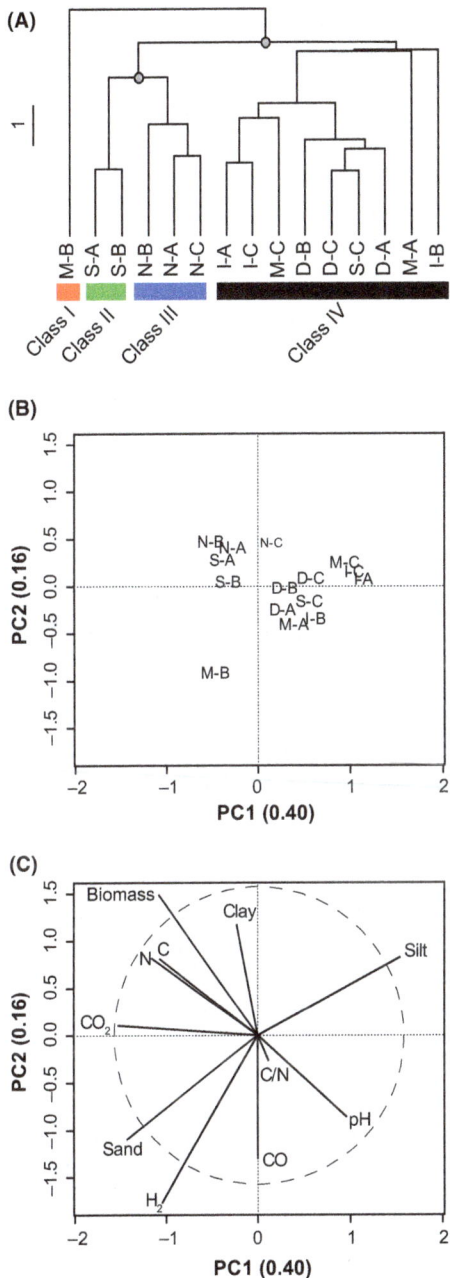

Fig. 2. Multifunctional comparison of soil samples according to their physicochemical (C, N, C:N ratio, pH, texture), trace gas exchanges (H_2, CO, CO_2) and abundance of bacteria (16S rRNA gene abundance; labelled as 'Biomass' in the plot) profiles. (A) UPGMA agglomerative clustering of soil samples according to a Euclidean distance matrix calculated with standardized variables. The grey circles denote the nodes delineating the four multifunctional classes significantly discriminated by SIMPROF permutation procedure ($P < 0.05$). Colour labels show the assignation of the soil samples to their multifunctional class (red; class I, green; class II, blue; class II and black; class IV). The scale bar represents the Euclidean distance in the dendrogram. (B) Principal component analysis showing the distribution of sampling sites in a reduced space defined by soil physicochemical properties, gaseous exchanges and abundance of bacteria. The colours used to present soil samples correspond to the clusters identified in the UPGMA (A). (C) Variables defining the distribution of soil samples along the first and the second axis are represented along with the equilibrium circle of descriptors showing the contribution of variables to the formation of the reduced space. The detection limit of the qPCR assay was utilized to estimate the abundance bacteria in sample M-A for which the low yield of the DNA extraction procedure precluded qPCR and bacterial 16S rRNA gene profiling (see the Material and Methods section for more details).

Indeed, CO_2 efflux was 3.9 µmol $g_{(dw)}^{-1}$ h^{-1} in S-A and S-B, while a value of 1.6 µmol $g_{(dw)}^{-1}$ h^{-1} was measured in S-C. The sample M-B displayed the highest H_2 and CO uptake rates, explaining its position in the ordination space (Fig. 2B).

Soil bacterial community taxonomic structure

Quality control, classification and equalization of the 16S rRNA gene sequence libraries yielded 5451 bacterial OTU (97% identity threshold). Overall, 50% of the sequences belonged to *Proteobacteria* mostly represented by *alpha-* (74%) and *delta-Proteobacteria* (11%). The *Acidobacteria* and *Actinobacteria* were the two other phyla dominating the bacterial communities with 19% and 9% relative abundance respectively (Fig. S1). Neither the conversion of unlogged natural mixed forest to a hybrid larch monoculture nor MSP treatments caused significant alteration at the microsite level in bacterial OTU richness as evaluated by the Shannon diversity index and ACE estimator (Table 1). In general, beta diversity defined as the variability in OTU composition among replicated plots measured by multivariate dispersion showed more variations in the larch plantation than in unlogged natural mixed forest conservation areas (Fig. S2). Agglomerative clustering of the samples according to their microbial community profile showed that soil samples collected in different treatments could not be discriminated on the basis of OTU composition (Fig. 3A). Unlogged natural forest clustered together with M-B, while all other clusters were composed either of unique plots or plots originating from different MSP procedures. A redundancy analysis (RDA) was performed to infer the relationship between 16S rRNA gene profiles

defined by the first two components explained 56% of the variation observed (Fig. 2B). Five variables defined the reduced space represented by both axes (Fig. 2C). The gradient associated to CO_2 production rate explained the distribution of plots along the first component, while abundance of bacteria and H_2 uptake rate defined the distribution along the second axis. Soil CO_2 respiration was the most preponderant variable responsible for the clustering of S-A and S-B trenching plots with unlogged natural forest samples and the clustering of S-C with mounding, inverting and intensive trenching plots.

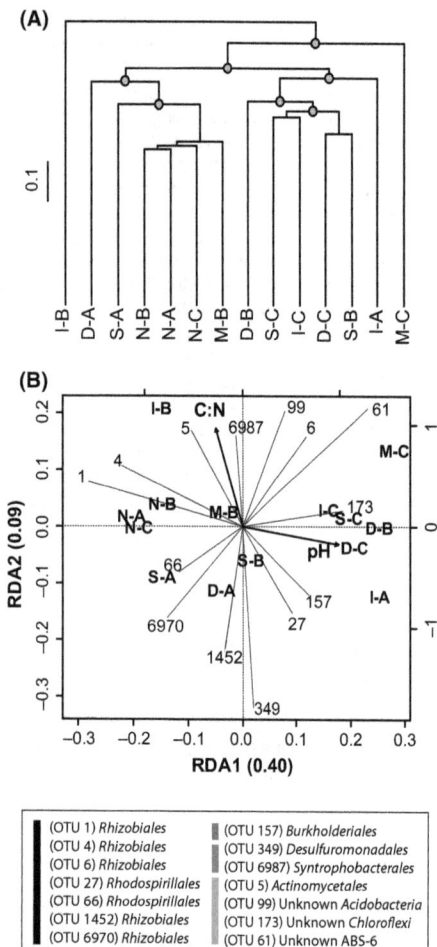

Fig. 3. Comparison of soil samples according to their 16S rRNA gene profile. (A) UPGMA agglomerative clustering of soil samples derived from a matrix of Euclidean distance calculated after Hellinger transformation of OTU (97% identity threshold) absolute abundance. The grey circles denote the nodes delineating the four groups of samples significantly discriminated by SIMPROF permutation procedure ($P < 0.05$). The scale bar represents the Euclidean distance in the dendrogram. Colour labels show the assignation of the soil samples to their multifunctional class (red; class I, green; class II, blue; class II and black; class IV). (B) Parsimonious RDA triplot of Hellinger-transformed OTU absolute frequency matrix explained by soil pH and C:N ratio. Only the 14 OTUs displaying extreme distribution in the reduced space are depicted for clarity. These OTUs are identified in the legend with colour bars discriminating α-*Proteobacteria* (black), β-*Proteobacteria* (blue), δ-*Proteobacteria* (red) and other phyla (orange), as determined using the Greengene reference database V13_8_99 (McDonald *et al.*, 2012). The colour labels used to present soil samples in the RDA triplot correspond to the clusters identified in the UPGMA (Fig. 4A). The sample M-A is absent due the low yield of the DNA extraction procedure for this soil (see the Material and Methods section for more details).

and environmental variables (Fig. 3B). The most parsimonious model to explain variation in the distribution of 16S rRNA gene sequences included soil C:N stoichiometry and pH. The other variables being redundant to soil C:N and pH, their addition to the analysis increased the

variance inflation factor unduly, and they were therefore ignored in the analysis. The first two canonical axes explained 49% of the total variance of bacterial OTU distribution. Significance of the RDA was confirmed with 1000 permutations of OTU data matrix ($P = 0.001$). Soil pH played an important role in the dispersion of the samples along the first axis, while C:N discriminated the samples along the second.

Microbial molecular indicator for soil multifunctional classes

The bacterial community profiles derived from the soil samples were classified within the four classes defined by the clustering of soil according to the multifunctional classification model (Fig. 2A). In total, 693 OTU were ubiquitously detected in all soil samples (Fig. S3). However, the search for indicator OTU unveiled coherence between the distribution of several members of the rare biosphere comprising less than 0.1% of the bacterial communities and soil multifunctional classes (Fig. 4). Indicator OTU encompassed a broad taxonomic diversity and *Proteobacteria* was the only phylum represented in the four multifunctional classes. In the case of multifunctional classes represented by more than one indicator, the OTU displaying the highest abundance was identified as representative indicator. Two indicators are of particular interest for this study. Firstly, the OTU 3283 (classified as a member of the order *Myxococcales*) was considered as a disturbance indicator because it was detected in soil samples for which the multifunctional classification (class IV) diverged from baseline ecological functions observed in unlogged natural mixed forest (Fig. 4). Furthermore, distribution of this candidate disturbance indicator was consistent with the environmental variables defining the multifunctional soil classification model. According to the unconstrained principal component analysis (PCA) ordination analysis, soil CO_2 respiration was a preponderant variable discriminating soil samples classified in class IV from those belonging to class II and class III (Fig. 2). As expected, the abundance of 16S rRNA gene sequences belonging to OTU 3283 showed a negative correlation with CO_2 soil respiration (Table 2). Secondly, the OTU 398 (classified as a member of the order *Rhodospirillales*) was considered as an indicator for baseline ecological conditions in soil because it was only detected in unlogged natural mixed forest conservation areas (Fig. 4). Distribution of this OTU was related to soil pH, bacteria abundance and soil clay content (Table 2). This result is in agreement with the contribution of these variables to the distribution of unlogged natural mixed forest soil samples in the PCA utilized to identify the factors defining the structure of the multifunctional classification model (Fig. 2B and C). In

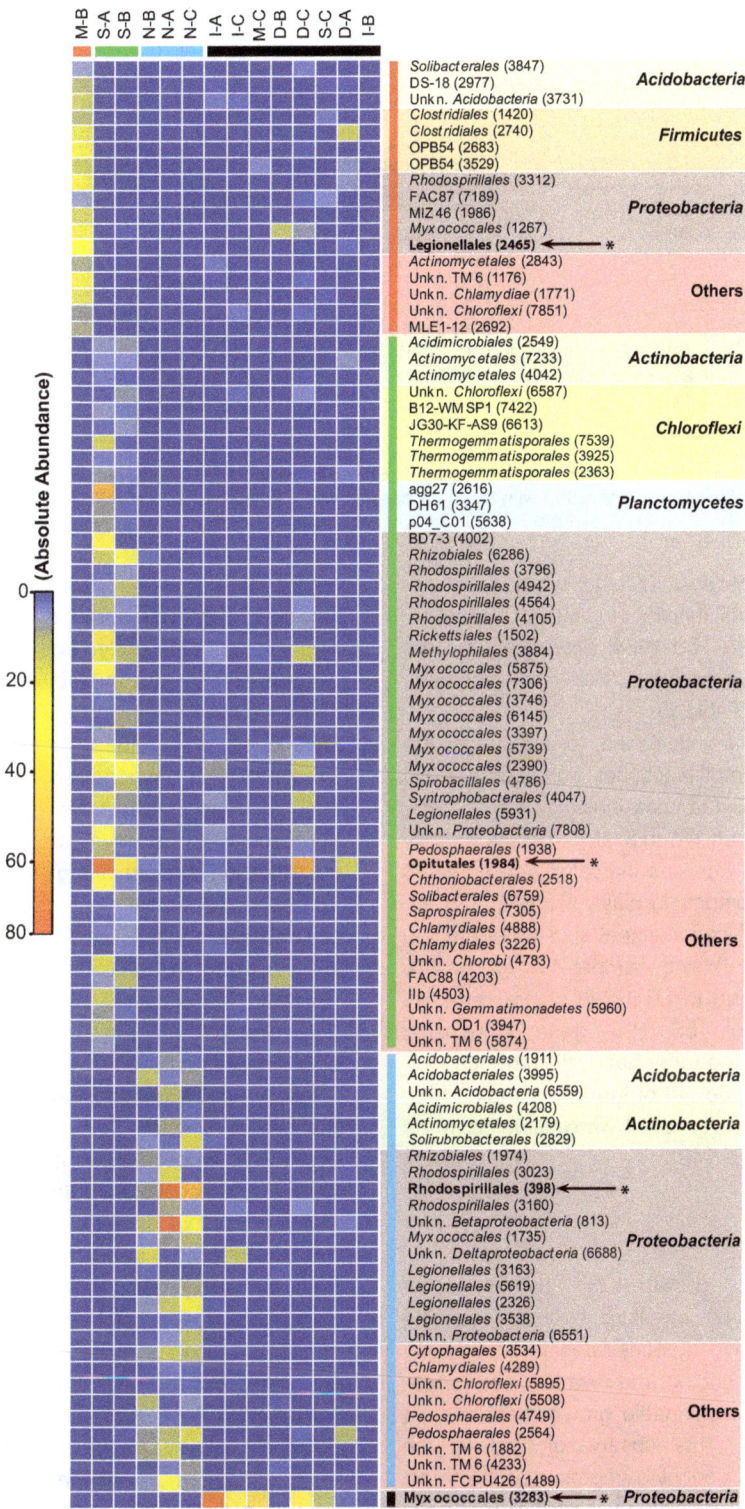

Fig. 4. Identification of soil multifunctional molecular indicators. The heatmap shows the absolute abundance of OTUs detected in the soil samples categorized into the four multifunctional classes previously defined (Fig. 2A). Colour bars show the assignation of the soil samples to their multifunctional class (red; class I, green; class II, blue; class II and black; class IV). Taxonomic assignation of the OTU was done using the Greengene reference database V13_8_99 (McDonald *et al.*, 2012). Representative indicators are highlighted with bold characters and are identified with an arrow and asterisk (← *).

Table 2. Spearman correlation between the abundance of the representative indicator ribotype selected for the multifunctional classes identified in this study and soil biological, physical and chemical variables (see Table 1).

Variables	Selected multifunctional indicators			
	Class I OTU 2465 (*Legionellales*)	Class II OTU 1984 (*Opitutales*)	Class II OTU 398 (*Rhodospirillales*)	Class IV OTU 3283 (*Myxococcales*)
H_2	0.34 (0.24)	0.03 (0.92)	0.25 (0.38)	**−0.67 (0.009)**
CO	0.05 (0.88)	−0.41 (0.15)	0.01 (0.98)	−0.53 (0.053)
CO_2	0.14 (0.64)	0.16 (0.60)	0.49 (0.08)	**−0.77 (0.001)**
Bacteria abundance	−0.11 (0.70)	0.30 (0.29)	**0.68 (0.007)**	−0.46 (0.101)
C	−0.46 (0.10)	0.22 (0.44)	0.21 (0.47)	**−0.59 (0.026)**
N	−0.36 (0.21)	0.30 (0.29)	0.21 (0.47)	**−0.62 (0.017)**
C/N	−0.46 (0.10)	−0.45 (0.11)	0.40 (0.16)	−0.10 (0.74)
pH	0.13 (0.65)	−0.02 (0.95)	**−0.73 (0.003)**	**0.61 (0.020)**
Sand	0.43 (0.12)	0.44 (0.11)	0.06 (0.84)	**−0.72 (0.003)**
Clay	−0.07 (0.80)	−0.44 (0.12)	**0.81 (0.0005)**	−0.13 (0.65)
Silt	−0.39 (0.16)	−0.38 (0.19)	−0.21 (0.47)	**0.77 (0.001)**

Spearman rho correlation coefficients are presented with the significance levels (α-value) in brackets. Bold characters represent significant correlation.

contrast to these two indicators, OTU for which the distribution was specific to multifunctional class I (OTU 2465; *Legionellales*) or class II (OTU 1984; *Opitutales*) did not show any significant correlation with the variables measured in this study (Table 2).

Two PCR assays were designed to challenge the indicator for baseline multifunctional conditions (OTU 398; class III) and the disturbance indicator (OTU 3283; class IV) identified through the indicator species statistical analysis (Table S1). Baseline indicator was detected in the three samples comprising class III (N-A, N-B, N-C), but weak PCR amplification signal also was detected for sample M-B (class I) and sample I-A (class IV), where no reads belonging to OTU 398 where retrieved from the high-throughput 16S rRNA gene sequencing analysis (Fig. S4A). On the other hand, disturbance indicator was detected in three out of the five samples from which reads assigned to OTU 3283 were detected in the sequencing analysis (Fig. S4B).

Discussion

Monitoring of early growth of hybrid larch (*Larix* × *marschlinsii* Coaz) seedlings over two growing seasons following planting demonstrated that trenching (simple and double), mounding and inverting site preparation resulted in undistinguishable growth performance (Buitrago *et al.*, 2015). This observation led to the conclusion that simple trenching represents the most economically attractive silvicultural practice for early growth of hybrid larch. Because soil microorganisms are at the core of key ecological functions such as nutrients transfer and biogeochemical cycles, we investigated the impact of MSP on the metabolic activity and composition of soil microbial communities. Our hypothesis that both the conversion of natural forest to a hybrid larch

monoculture and the intensification of MSP treatments reduce the activity and the taxonomic diversity of soil microbiome in the early years of this conversion was not verified. Neither the microbial activities nor the diversity indices showed significant difference between the MSP techniques and natural unlogged forest. Beside the examination of environmental variables and the distribution of microorganisms in soil, the originality of our approach was the development of a multifunctional soil classification system along with molecular indicators to assess the potential ecological risk of MSP techniques on ecosystem functions. The integration of biogeochemical and microbiological variables – instead of considering variables individually – was shown to be essential to determine the response of soil bacterial community to silvicultural practices.

An important question in conservation biology is the number of species that must be protected to ensure ecosystem functioning. A milestone in the field of biodiversity and ecosystem functioning was the observation that the minimal number of species required to support multiple functions is much higher than estimates derived from the analysis of individual processes (Hector and Bagchi, 2007; Gamfeldt *et al.*, 2008). As ecosystems are conserved for their multiple natural services, these studies highlighted the need for integrated approaches to set reasonable conservation objectives (Balvanera *et al.*, 2014). In this study, we selected multiple gas turnover as a subset of ecosystem functions. H_2 and CO soil uptake activities are catalysed by different guilds of bacterial species that contribute to mitigate the global emissions of these climate-relevant trace gases in the atmosphere. High-affinity, H_2-oxidizing bacteria are mostly represented by specialized *Actinobacteria* and some representatives of the *Proteobacteria*, *Chloroflexi* and *Acidobacteria* (Constant *et al.*, 2011; Meredith *et al.*,

2013). These bacteria are responsible for 80% of the global sink of atmospheric H_2 (Constant et al., 2009; Ehhalt and Rohrer, 2009). Soil survey showed the importance of carbon and nitrogen contents to explain the activity of this functional group in soil (Gödde et al., 2000; Khdhiri et al., 2015). Carboxydovore bacteria, encompassing the Actinobacteria, Proteobacteria and Chloroflexi, are responsible for the soil uptake of atmospheric CO (King and Weber, 2007; King and King, 2014; Quiza et al., 2014). Even though no systematic investigation about the impact of soil nutrients on CO uptake rate has been reported, organic matter-rich soils displayed higher potential CO uptake activity than desert soils along vegetation transects (Weber and King, 2010). The activity of carboxydovore bacteria contributes to 15% of the global sink of atmospheric CO, while scavenging CO produced by biological and abiotic reactions in soil. In contrast to H_2 and CO, soil CO_2 respiration is supported by a broad diversity of bacteria and fungi thriving in soil using a heterotrophic growth metabolism. In addition to the availability of soil organic matter, factors influencing soil respiration rate are complex, including soil nutrient stoichiometry (Drake et al., 2013). Including H_2, CO and CO_2 exchanges in multifunctional classification model was, therefore, a relevant approach to consider important ecosystem services that are provided by taxonomically diverse microorganisms. Addition of further environmental variables such as pH, C, N, abundance of bacteria and soil texture defining soil biogeochemical structure provided a supplementary dimension to the multifunctional classification system. Consideration of other environmental variables such as organic matter and nutrient turnover in the future is not precluded as this could refine the soil multifunctional classification system to satisfy other ecosystem functions prioritized in sustainable forest management.

In contrast to the comparative analysis of individual environmental parameters that unveiled no significant impact of MSP due to the variance of the observations, the multifunctional classification approach led to the identification of MSP treatment plots for which ecological functions differed from baseline conditions as measured in unlogged natural forest nearby. According to this classification, double trenching, mounding and inverting site preparation were the less sustainable MSP techniques at the microsite level. The low environmental performance of mounding and inversion plots at this scale was also supported by the higher coefficients of variation in the gaseous exchanges measured in these soils than those observed in trenching and unlogged natural mixed forest (Fig. 1). This finding, indicating that three distinct ecological services were sensitive to mounding and inverting, is probably due to the heterogeneity caused by the heavy machinery used for site preparation. The

procedure involved the transfer of excavated soil and was prone to large variations due to the presence of rocks, resulting in variable volume and depth of the excavations in addition to potential variation in the horizontal and lateral distribution of displaced organic and mineral soil horizons and ground microtopography. Our results are relevant at the microsite level, i.e. local environmental conditions directly influencing planted seedling physiology and growth. Although mounding and inverting intensively disturb the soil profile, thus having significant local impacts on soil functions, trenching furrows affect a higher proportion of the planted site. Determining which treatment is best in terms of overall ecological impacts will require further work, including scaling up and interpreting these impacts at the stand level, taking into account the proportion of disturbed soil in each treatment.

Soil microbial diversity was neither related to MSP treatment nor multifunctional classes identified in this study. Soil pH, rather than MSP treatments, was the most important parameter to explain the composition of soil microbial communities (Fig. 3). Indeed, variation in pH was shown as a preponderant variable explaining the composition of soil microbiome and any silvicultural treatment resulting in marked pH alteration are expected to influence the distribution of dominant lineages (Lauber et al., 2009). The most important aspect of soil microbial community analysis was the clustering of the 16S rRNA gene profiles according to the multifunctional soil classification model, resulting in the identification of indicator OTU. Those indicators represented members of the rare biosphere in soil, for which the incidence of environmental variables on their distribution could not be observed using ordination techniques parameterized with the whole microbial community profiles (Fig. 3). Two relevant indicators were identified. The OTU 3283 (Myxococcales) detected in soil samples that were divergent from the baseline conditions observed in unlogged natural mixed forests could be used as a diagnostic tool to assess the impact of MSP on ecosystem services. On the other hand, the OTU 398 (Rhodospirillales) was associated to baseline soil multifunctional attributes. Bioindicators of soil quality involving exoenzyme activity measurements (Staddon et al., 1999) and other approaches involving the monitoring of arthropods (Pearce and Venier, 2006) and small mammals (Pearce and Venier, 2005) as bioindicators of sustainable forest management have already been developed. In contrast to these metrics requiring relatively laborious monitoring efforts, molecular diagnostic tools, specific to selected indicator OTU, are fast and easily integrated with abiotic and biological factors in the environment. In the context of the hybrid larch plantation investigated here, monitoring of the identified indicators, especially the indicator for baseline multifunctional soil attributes, could be used for environmental certification and ecological risk assessment after adequate scaling up addressing key requirements of ecological indicators.

The molecular indicators identified in this study are relevant at the stand level, with the abiotic and biotic conditions that prevailed in soil at the time of sampling. The experimental plan was not designed to address the consistency of molecular indicators over time, space and sites. Further investigations are needed to challenge the indicators in other hybrid larch as well as other fast-growing tree plantations and assess their specificity (Bartell, 2006). Longer term studies are also needed to assess whether spatial and temporal variations of soil biogeochemical processes in intensive silviculture can be related to alterations of molecular indicator distribution profiles. Indeed, it is expected that indicator OTU identified in this study are not universal due to the impact of abiotic and biological gradients shaping soil microbial communities in soil. Nevertheless, our data show that the distribution of soil microbial communities is not random, with the distribution of some members restricted to specific multifunctional soil classes, at the stand level. Combination of soil bacterial, fungal and archaeal community taxonomic and functional profiles through metagenomic surveys should be considered in future attempts to develop and validate molecular indicators for sustainable forest management. Finally, PCR diagnostic assays targeting the baseline and stress indicators clearly demonstrated that a composite of indicator OTU rather than a single indicator would be needed for predicting soil multifunctional classes (Fig. S4). This could be achieved through the adaptation of the indicator species modelling approach used to select the minimal subset of species necessary to predict species richness of invertebrate and vertebrate assemblages to microbial ecology (Thomson et al., 2007; Azeria et al., 2009).

In conclusion, we showed that MSP treatments influence the overall signature of soil biogeochemical structure at the microsite level, suggesting that mounding and inverting site preparation could pose higher potential ecological risk for ecological functions in soil than trenching at the local scale. Further work is needed to scale up and interpret these results at the landscape level and over longer period of time, taking into account the proportion of disturbed soil in each treatment and the overall productivity of each treatment. In effect, the lower ecological risk potential of a less productive treatment could be cancelled by the need of converting a larger area into fast-growing plantations to make up for the lesser productivity. Chronosequences, spatial variability, and different tree plantation types will need further attention in future investigations to elucidate the short term impact of MSP on soil biogeochemical structure and

challenge the molecular indicators identified in this study. Because of functional redundancy in microorganisms, these studies would be essential to challenge and define limitations of the indicators. Nevertheless, this work demonstrates the relevance of applied ecology to evaluate the sustainability of silvicultural practices.

Experimental Procedures

Sampling site

The study site was located near La Tuque (Québec, Canada; 47° 37′ 19″ N, 72° 49′ 55″ W), about 250 km north of Montréal. The experimental area, dominated by balsam fir (Abies balsamea (L.) Mill.), paper birch (Betula papyrifera Marsh.), yellow birch (B. alleghaniensis Britt.), red maple (Acer rubrum L.) and black spruce (Picea mariana (Mill.) BSP) was harvested in 2009 (clear-cut with 5% retention) prior to the installation of an experiment to investigate the impact of different MSP techniques on the growth performance of hybrid larch planted in April 2010 (Buitrago et al., 2015). The experimental design consisted of a complete block design comprising three replicated blocks (Fig. 5). Briefly, each block was separated into four plots randomly assigned to different MSP treatments encompassing trenching and mounding. Trenching was performed with rotary discs mixing mineral and organic soil horizons. This treatment was applied either as a simple (simple plots; composite samples S-A, S-B, S-C) or double trenching (double intensive plots; composite samples D-A, D-B, D-C) corresponding to an increasing gradient of soil mixing (Buitrago et al., 2015). An excavator was used for the mounding treatments consisting in inverting the soil horizons to place mineral soil on the top and the organic layer at the bottom. Excavated soil was either placed back to the original ditch (inversion plots; composite samples I-A, I-B, I-C) or on soil surface next to the ditch generated by the excavation (mound plots; composite samples M-A, M-B, M-C). Seedlings were planted in the hinge position of the trenching furrows or near the centre of excavated mounds. Finally, plots consisting of non-harvested, non-planted natural mixed forest areas (retention areas located within the three experimental blocks, and consisting in approximately 500–600 m^2 plots) were used as reference in this study (unlogged natural plots; composite samples N-A, N-B, N-C). These unlogged forest soils were selected as reference plots based on third-party forest certification criteria (e.g. Forest Stewardship Council), where reference forests in the vicinity of forest management units are used to measure the impact of management plans on ecosystem integrity.

The experimental area was visited in July 2014 for soil sample collection. In total, 10 replicated soil samples were collected in each plot (3 blocks × 5 plots × 10

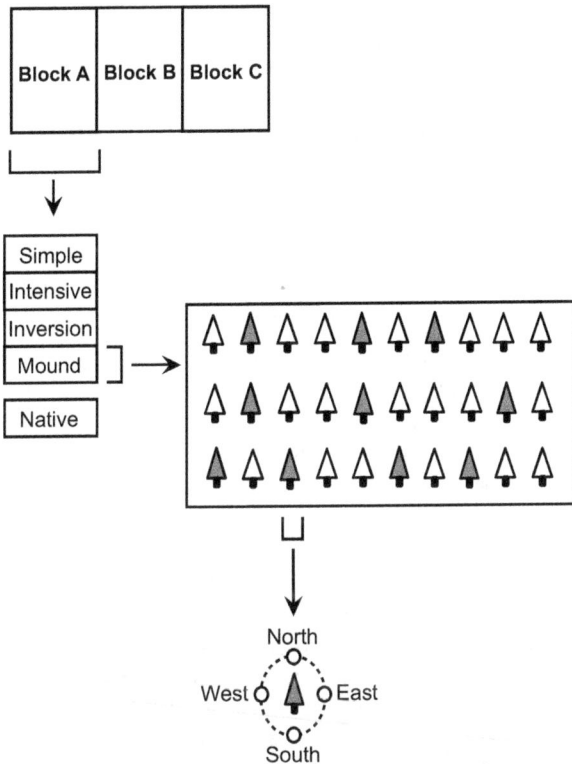

Fig. 5. Schematic representation of the sampling design. Five treatment plots were replicated in three blocks. Soil samples were collected at 10 locations from each plots (illustrated with grey trees). Soil was collected in proximity of the stem, at one of the four cardinal points. Replicated soil samples were pooled to obtain one composite sample per replicated plot. The (native) treatment consisted of non-harvested, non-planted natural mixed forest areas located within the experimental blocks.

replicates = 150 samples in total). Individual trees for which information regarding early growth parameters was available (Buitrago et al., 2015) were first randomly selected and the A-horizon (0–10 cm) was collected within a 15 cm radius of the stem at one of the four cardinal points also determined randomly (Fig. 5). Each replicate sample consisted of ~200 g soil stored on ice in Whirl-Pak® bags. Composite samples (3 blocks × 5 plots = 15 composites) were prepared on the site less than 6 h after collection. Approximately, 100 g soil from each replicate were thoroughly mixed in a plastic bucket and transferred to plastic bags. All the samples were then immediately stored at 4°C. The 15 composite samples were processed within 3 weeks for DNA extraction, nutrient analyses and gas exchanges measurements. Soil texture was determined using the hydrometer method, and particle size distribution assigned soil samples to the loamy sand textural class (Elghamry and Elashkar, 1962). Because soil samples were collected in a 15-cm radius of the stem, the results reported in this study are representative of a microscale of the whole experimental larch plantation landscape area.

Physicochemical analyses

Composite soil samples were dried at 20°C for 48 h, homogenized (2 mm sieve) and pulverized with mortar and pestle. Soil pH was analysed in suspensions using a 1:2.5 (w/v) soil-to-water ratio (MP220 pH-meter; Mettler Toledo, Mississauga, ON, Canada). Total soil carbon and nitrogen contents were determined using an elemental combustion system (ECS 4010; Costech Analytical Technologies Inc., Valencia, CA, USA). Analyses were performed with 18 to 123 mg pulverized soil samples and certified atropine standard ($C_{17}H_{23}NO_3$; Costech Analytical Technologies Inc.) containing 70.86% total carbon and 4.84% total nitrogen was used for calibration. Relative errors were lower than 2.8% for total carbon and 1.1% for total nitrogen analyses as observed with repeated analyses of atropine standard.

Gaseous exchanges

Composite samples were dried at 20°C for 48 h and homogenized (2 mm sieve) and soil water content was adjusted to 20% water holding capacity. A defined amount of composite soil sample (20 g) was transferred in a 500 ml (nominal volume) Gibco® glass bottle (Wheaton Industries Inc., Millville, NJ, USA) fitted with foam plugs to allow gaseous exchanges between soil and atmosphere, while avoiding microcosm contamination with airborne particles. Soil microcosms were then transferred to an environmental chamber (MLR-350®; Sanyo, Osaka, Japan) and incubated 3 days in the dark, at 25°C and 50% relative air moisture. H_2, CO and CO_2 soil-to-air exchanges were measured using gas chromatography assays. Briefly, soil microcosm foam plugs were replaced with gastight caps equipped with butyl septa. A defined volume of two air mixtures containing either 469 ± 9 ppm H_2 (GTS-Welco, Minersville, PA, USA) or 508 ± 10 ppm CO (GTS-Welco) was injected to the static headspace of the microcosms, resulting in H_2 and CO levels of 2.5–3 ppmv. Headspace samples (10 ml) were collected with a Pressure Lok® gastight glass syringe (VICI® Precision Sampling Inc., Baton Rouge, LA, USA) and injected through the injection port of a gas chromatograph equipped with a reduction gas detector (ta3000R; Ametek Process Instruments®, Newark, DE, USA). The first-order oxidation rates were calculated by integrating the H_2 and CO mole fraction time series measured over a 1-h period, using at least five concentration points for data integration. Soil CO_2 respiration was measured in the soil microcosms fitted with gastight caps and flushed with ambient laboratory air. Linear CO_2 mole fraction time series was measured over a 72-h period, with four concentration points for data integration using a gas chromatograph equipped with a methanizer and a flame ionization detector (7890B GC System; Agilent Technologies, Mississauga, ON, Canada). Sealed microcosms were incubated in the environmental chamber during the measurements. The reproducibility of gas analyses was assessed before each set of experiments by repeated analysis of three certified standard gas mixtures: (i) 2.13 ± 0.11 ppmv H_2 balance air (GTS-Welco), (ii) 2.05 ± 0.10 ppmv CO balance air (GTS-Welco) and (iii) 610 ± 12 ppmv CO_2, 5 ± 0.1 ppmv CH_4, 1 ± 0.02 ppmv N_2O balance nitrogen (Agilent Technologies), and standard deviations were $< 5\%$. No significant gaseous exchanges were observed for blank measurements of empty microcosms. Because of the occurrence of simultaneous production and consumption of trace gases in nature and their dependence on temperature and moisture, gaseous exchanges presented in this study must be considered as potential activities. Gaseous exchanges were expressed in mol of gas h^{-1} g^{-1} of soil on a dry-weight basis; with soil water content measured using standard gravimetric method.

Abundance of bacteria

The abundance of bacteria was estimated by quantification of 16S rRNA gene of bacteria in soil by a universal bacterial qPCR assay (Fierer et al., 2005). Genomic DNA was extracted from 0.5 to 3.0 $g_{(dw)}$ composite soil samples using a combination of chemical and mechanical cell lysis procedure (Constant et al., 2008). DNA was precipitated with 2 volumes 96% ethanol and polyvinylpolypyrrolidone spin column was used for final purification (Berthelet et al., 1996). Purified DNA extracts from two technical replicates were pooled (200 µl in total) and kept frozen at $-20°C$ before qPCR. The reactions were performed using 5 µl of diluted genomic DNA (1:100 and 1:500) and no quantification bias due to Taq polymerase inhibitors was observed. The qPCR assay was based on a standard curve prepared by using triplicate 10-fold dilutions of PCR-amplified standard DNA. Genomic DNA of Burkholderia xenovorans LB400 served as template for 16S rRNA gene standard DNA. PCR products were purified (E.Z.N.A. Cycle Pure Kit, Omega Bio-Tek®, Norcross, GA, USA) and quantified using the Quantifluor™ dsDNA System (Promega, Madison, WI, USA) according to the instructions of the manufacturers. Standard curves encompassing 10^2 to 10^8 copies $µl^{-1}$ of standard DNA were prepared and displayed linear relationship between the signal and the logarithm copy number with reaction efficiencies of 0.99 ($r^2 = 0.99$). The Perfecta SYBR Green Fast Mix (Quanta Biosciences®, Gaithersburg, MD, USA) was used for the qPCR performed in a Rotor-Gene 6000 qPCR cycler (Corbett Life Science®, Concorde, NSW, Australia). For

unknown reasons, the extraction procedure was not successful for the sample M-A. Neither the utilization of FastDNA SPIN kit for Soil® (MP Biomedicals, Solon, OH, USA), modifications of the extraction buffer, nor increased amount of soil in the lysis procedure improved the yield of the extraction for sample M-A. This sample thus was not considered for qPCR and bacterial 16S rRNA gene ribotyping analyses.

Bacterial 16S rRNA gene sequencing

PCR amplification of the V6-V8 regions of 16S rRNA, libraries preparation and Illumina MiSeq 250 bp paired-ends sequencing reactions were performed by the technical staff of Centre d'Innovation Génome Québec et Université McGill, resulting in 6 274 978 raw sequences. Paired-end reads were merged using the software Flash (Magoč and Salzberg, 2011) with minimum and maximum overlap length between the two reads of 20 and 250 bases respectively. The maximum proportion of mismatched base pairs tolerated in the overlap region was 30%. The 6 073 727 merged reads were processed using the software UPARSE (Edgar, 2013). Briefly, sequences were truncated to a uniform length of 420, representing the length of more than 97.5% of the sequences in the database. Reads with a low-quality score were removed using 2.0 as the maximum expected error value. The remaining 4 884 667 high-quality reads were de-replicated, sorted by size and singletons were removed. The remaining unique reads were clustered into 8248 operational taxonomic units (OTU) with the UPARSE-OTU greedy clustering method using a 97% identity threshold. The UPARSE-REF algorithm detected and removed 98 006 chimeric sequences during clustering procedure. Furthermore, 191 chimeric OTU were removed with the software UCHIME run against ChimeraSlayer 'gold' reference database (Edgar et al., 2011). The final database contained 4 228 736 sequences. Libraries were normalized to the sequencing effort of the smallest 16S rRNA gene library (166 040 sequences) to avoid biases in comparative analyses introduced by sampling depth. The software QIIME version 1.8.0 (Caporaso et al., 2010) was used to perform equalization of the libraries and to eliminate OTU comprising less than eight representative sequences, corresponding to a threshold of 0.005% of the total number of reads per library. QIIME was also used to pick and align one representative sequences for each OTU to assign a taxonomic classification using the Greengene reference database V13_8_99 (McDonald et al., 2012). The resulting OTU table comprising the abundance and taxonomic affiliation (phylum, class, order and family levels) of the OTUs in the samples was utilized to compute alpha diversity (i.e. species richness with Ace and

Shannon indices) and the multivariate dispersion of the OTU as a measure of beta diversity (Anderson et al., 2006) with the packages 'fossil' and 'vegan' implemented in the software R (Oksanen et al., 2012; Vavrek, 2012). Raw sequences were deposited to the Sequence Read Archive of the National Center for Biotechnology Information under the Bioproject PRJNA280109.

Statistical analyses

Statistical analyses were performed using the software R (R Core Development Team, 2008). The impact of MSP treatments on soil biogeochemical properties was tested using one-way analysis of variance and *post hoc* Bonferroni-corrected *t*-test. Shapiro–Wilk normality test was applied to assess normal distribution of data before the ANOVA. Distribution of soil pH, carbon content and nitrogen content followed a normal distribution, while the distribution of other variables was normalized by logarithmic (H_2, CO, CO_2 exchanges) or square root (abundance of bacteria) transformations. Normalized variables also were used for Pearson correlation analysis. Cluster analysis was computed using the function 'hclust' in the package 'stats' (R Core Development Team, 2014) to explore sampling site similarities defined by variations in gas exchange (H_2, CO and CO_2), pH, abundance of bacteria as well as total carbon and total nitrogen contents measured in soil. The environmental variables were standardized before the analysis by subtracting individual values by the average and dividing them by the standard deviation. A Euclidean distance matrix was used to generate a UPGMA agglomerative clustering of the samples according to their biogeochemical profile. The identification of statistically different soil multifunctional classes was done by performing 999 permutations of the environmental variables data set separately across the samples and comparing the observed similarity score of each cluster against the expected values under the null hypothesis using the similarity profile tool (SIMPROF) implemented in the package 'clustsig' (Clarke et al., 2008). PCA was used to explore sampling sites partitioning in a reduced space defined by environmental gradients. Meaningful ordination axes whose eigenvalues were larger than the average of all eigenvalues were selected for biological interpretation. Equilibrium circle of descriptor with the radius $\sqrt{d/p}$ (where d is the number of dimensions of the reduced space: 2 and p is the total space: 10) was plotted to identify variables significantly contributing to the axes defining the position of sampling sites. Discrimination of the samples according to their ribotyping profile was performed after Hellinger transformation of the OTU table to avoid unduly relationships between explanatory variables and 16S rRNA gene composition supported by the high weight of rare species (Legendre and Gallagher,

2001). UPGMA clustering analysis was conducted to compare the samples according to their ribotyping profile. RDA was computed using standardized environmental variables with the package 'vegan' (Oksanen et al., 2012), according to the comprehensive procedure described by Borcard and colleagues (Borcard et al., 2011). The most parsimonious constrained model to explain bacterial ribotyping profile was obtained by forward selection of the environmental variables (Blanchet et al., 2008) and permutation tests ($n = 1000$) were performed to assess the significance of the RDA. Indicator OTU characterizing UPGMA agglomerative clustering of the samples according to their multifunctional classification were identified using the indicator species analysis procedure implemented in the package 'indicspecies' (Dufrêne and Legendre, 1997). Minimal significance level (alpha) of the indicator OTU was 0.05, tested against 999 random permutations of samples among the biogeochemical clusters.

Acknowledgements

This research was supported by a grant from the Natural Sciences and Engineering Research Council of Canada (NSERC) Engage Program to PC (grant EGP 463612-14). The work of VP was supported by a NSERC – Undergraduate Student Research Awards. The authors are grateful to the personnel staff of *Centre d'Innovation Génome Québec et Université McGill* for preparation of 16S rRNA gene libraries and sequencing services.

Conflict of Interest

All the authors have no conflict of interest to declare.

References

Anderson, M.J., Ellingsen, K.E., and McArdle, B.H. (2006) Multivariate dispersion as a measure of beta diversity. *Ecol Lett*, **9:** 683–693.

Azeria, E.T., Fortin, D., Hébert, C., Peres-Neto, P., Pothier, D., and Ruel, J.-C. (2009) Using null model analysis of species co-occurrences to deconstruct biodiversity patterns and select indicator species. *Diversity Distrib*, **15:** 958–971.

Balvanera, P., Siddique, I., Dee, L., Paquette, A., Isbell, F., Gonzalez, A., et al. (2014) Linking biodiversity and ecosystem services: current uncertainties and the necessary next steps. *Bioscience*, **64:** 49–57.

Bartell, S.M. (2006) Biomarkers, bioindicators, and ecological risk assessment- A brief review and evaluation. *Environ Bioindicators*, **1:** 60–73.

Berthelet, M., Whyte, L.G., and Greer, C.W. (1996) Rapid, direct extraction of DNA from soils for PCR analysis using polyvinylpolypyrrolidone spin columns. *FEMS Microbiol Lett*, **138:** 17–22.

Blanchet, F.G., Legendre, P., and Borcard, D. (2008) Forward selection of explanatory variables. *Ecology*, **89**: 2623–2632.

Borcard, D., Gillet, F. and Legendre, P. (2011) *Numerical Ecology with R*. New York: Springer.

Buitrago, M., Paquette, A., Thiffault, N., Bélanger, N., and Messier, C. (2015) Early performance of planted hybrid larch: effects of mechanical site preparation and planting depth. *New Forest*, **46**: 319–337.

Caporaso, J.G., Kuczynski, J., Stombaugh, J., Bittinger, K., Bushman, F.D., Costello, E.K., *et al.* (2010) QIIME allows analysis of high-throughput community sequencing data. *Nat Meth*, **7**: 335–336.

Clarke, K.R., Somerfield, P.J., and Gorley, R.N. (2008) Testing of null hypotheses in exploratory community analyses: similarity profiles and biota-environment linkage. *J Exp Mar Biol Ecol*, **366**: 56–69.

Constant, P., Poissant, L., and Villemur, R. (2008) Isolation of *Streptomyces* sp. PCB7, the first microorganism demonstrating high-affinity uptake of tropospheric H$_2$. *ISME J*, **2**: 1066–1076.

Constant, P., Poissant, L., and Villemur, R. (2009) Tropospheric H$_2$ budget and the response of its soil uptake under the changing environment. *Sci Total Environ*, **407**: 1809–1823.

Constant, P., Chowdhury, S.P., Hesse, L., Pratscher, J., and Conrad, R. (2011) Genome data mining and soil survey for the novel group 5 [NiFe]-hydrogenase to explore the diversity and ecological importance of presumptive high affinity H$_2$-oxidizing bacteria. *Appl Environ Microbiol*, **77**: 6027–6035.

Drake, J.E., Darby, B.A., Giasson, M.-A., Kramer, M.A., Phillips, R.P., and Finzi, A.C. (2013) Stoichiometry constrains microbial response to root exudation- insights from a model and a field experiment in a temperate forest. *Biogeosciences*, **10**: 821–838.

Dufrêne, M., and Legendre, P. (1997) Species assemblages and indicator species: the need for a flexible asymetrical approach. *Ecol Monogr*, **67**: 345–366.

Edgar, R.C. (2013) UPARSE: highly accurate OTU sequences from microbial amplicon reads. *Nat Meth*, **10**: 996–998.

Edgar, R.C., Haas, B.J., Clemente, J.C., Quince, C., and Knight, R. (2011) UCHIME improves sensitivity and speed of chimera detection. *Bioinformatics*, **27**: 2194–2200.

Ehhalt, D.H., and Rohrer, F. (2009) The tropospheric cycle of H$_2$: a critical review. *Tellus B*, **61**: 500–535.

Elghamry, W., and Elashkar, M. (1962) Simplified textural classification triangles. *Soil Sci Soc Am J*, **26**: 612–613.

van Elsas, J.D., Chiurazzi, M., Mallon, C.A., Elhottová, D., Krištůfek, V., and Salles, J.F. (2012) Microbial diversity determines the invasion of soil by a bacterial pathogen. *Proc Natl Acad Sci USA*, **109**: 1159–1164.

Fierer, N., Jackson, J.A., Vilgalys, R., and Jackson, R.B. (2005) Assessment of soil microbial community structure by use of taxon-specific quantitative PCR assays. *Appl Environ Microbiol*, **71**: 4117–4120.

Gamfeldt, L., Hillebrand, H., and Jonsson, P.R. (2008) Multiple functions increase the importance of biodiversity for overall ecosystem functioning. *Ecology*, **89**: 1223–1231.

Gödde, M., Meuser, K., and Conrad, R. (2000) Hydrogen consumption and carbon monoxide production in soils with different properties. *Biol Fertil Soils*, **32**: 129–134.

Haeussler, S., Bartemucci, P., and Bedford, L. (2004) Succession and resilience in boreal mixedwood plant communities 15–16 years after silvicultural site preparation. *For Ecol Manage*, **199**: 349–370.

Hector, A., and Bagchi, R. (2007) Biodiversity and ecosystem multifunctionality. *Nature*, **448**: 188–190.

Jandl, R., Lindner, M., Vesterdal, L., Bauwens, B., Baritz, R., Hagedorn, F., *et al.* (2007) How strongly can forest management influence soil carbon sequestration? *Geoderma*, **137**: 253–268.

Kabrick, J.M., Dey, D.C., Sambeek, J.W.V., Wallendorf, M., and Gold, M.A. (2005) Soil properties and growth of swamp white oak and pin oak on bedded soils in the lower Missouri River floodplain. *For Ecol Manage*, **204**: 315–327.

Khdhiri, M., Hesse, L., Popa, M.E., Quiza, L., Lalonde, I., Meredith, L.K., *et al.* (2015) Soil carbon content and relative abundance of high affinity H$_2$-oxidizing bacteria predict atmospheric H$_2$ soil uptake activity better than soil microbial community composition. *Soil Biol Biochem*, **85**: 1–9.

King, C.E., and King, G.M. (2014) Description of *Thermogemmatispora carboxidivorans* sp. nov., a novel carbon monoxide-oxidizing member of the *Ktedonobacteria* isolated from a geothermally-heated biofilm, and analysis of carbon monoxide oxidation by members of the *Ktedonobacteria*. *Int J Syst Evol Microbiol*, **64**: 1244–1251.

King, G.M., and Weber, C.F. (2007) Distribution, diversity and ecology of aerobic CO-oxidizing bacteria. *Nat Rev Micro*, **5**: 107–118.

Lauber, C.L., Hamady, M., Knight, R., and Fierer, N. (2009) Pyrosequencing-based assessment of soil pH as a predictor of soil bacterial community structure at the continental scale. *Appl Environ Microbiol*, **75**: 5111–5120.

Legendre, P., and Gallagher, E. (2001) Ecologically meaningful transformations for ordination of species data. *Oecologia*, **129**: 271–280.

Löf, M., Dey, D., Navarro, R., and Jacobs, D. (2012) Mechanical site preparation for forest restoration. *New Forest*, **43**: 825–848.

Magoč, T., and Salzberg, S.L. (2011) FLASH: fast length adjustment of short reads to improve genome assemblies. *Bioinformatics*, **27**: 2957–2963.

Martin, B., Marc, L., Nelson, T., Alain, P., Luc, L., Louis, B., *et al.* (2014) Issues and solutions for intensive plantation silviculture in a context of ecosystem management. *For Chronicle*, **90**: 748–762.

McDonald, D., Price, M.N., Goodrich, J., Nawrocki, E.P., DeSantis, T.Z., Probst, A., *et al.* (2012) An improved Greengenes taxonomy with explicit ranks for ecological and evolutionary analyses of bacteria and archaea. *ISME J*, **6**: 610–618.

Meredith, L.K., Rao, D., Bosak, T., Klepac-Ceraj, V., Tada, K.R., Hansel, C.M., *et al.* (2013) Consumption of atmospheric hydrogen during the life cycle of soil-dwelling actinobacteria. *Environ Microbiol Rep*, **6**: 226–238.

Oksanen, J., Blanchet, F., Kindt, R., Legendre, P., Minchin, P., O'Hara, R., *et al.* (2012) Vegan: community ecology package. R package version 2.0-4. URL http://cran.r-project.org/package=vegan.

Paquette, A., and Messier, C. (2009) The role of plantations in managing the world's forests in the Anthropocene. *Front Ecol Environ*, **8:** 27–34.

Pearce, J., and Venier, L. (2005) Small mammals as bioindicators of sustainable boreal forest management. *For Ecol Manage*, **208:** 153–175.

Pearce, J.L., and Venier, L.A. (2006) The use of ground beetles (Coleoptera: Carabidae) and spiders (Araneae) as bioindicators of sustainable forest management: a review. *Ecol Indic*, **6:** 780–793.

Prévost, M. (1992) Effets du scarifiage sur les propriétés du sol, la croissance des semis et la compétition: revue des connaissances actuelles et perspectives de recherches au Québec. *Ann For Sci*, **49:** 277–296.

Quiza, L., Lalonde, I., Guertin, C., and Constant, P. (2014) Land-use influences the distribution and activity of high affinity CO-oxidizing bacteria associated to type I-*coxL* genotype in soil. *Front Microbiol*, **5:** 1–15. doi:10.3389/fmicb.2014.00271.

R Core Development Team (2008) R: A language and environment for statistical computing. Vienna, Austria: Computing RFfS.

R Core Development Team (2014) The R Stats Package - version 3.1.3. URL https://stat.ethz.ch/R-manual/R-devel/library/stats/html/hclust.html.

Simon, B.-G., David, P., Christian, M., and Nicolas, B. (2013) Root production of hybrid poplars and nitrogen mineralization improve following mounding of boreal Podzols. *Can J For Res*, **43:** 1092–1103.

Staddon, W., Duchesne, L., and Trevors, J. (1999) The role of microbial indicators of soil quality in ecological forest management. *Forest Chron*, **75:** 81–86.

Sutton, R.F. (1993) Mounding site preparation: a review of European and North American experience. *New Forest*, **7:** 151–192.

Tardy, V., Mathieu, O., Lévêque, J., Terrat, S., Chabbi, A., Lemanceau, P., et al. (2014) Stability of soil microbial structure and activity depends on microbial diversity. *Environ Microbiol Rep*, **6:** 173–183.

Thiffault, N., and Jobidon, R. (2006) How to shift unproductive *Kalmia angustifolia – Rhododendron groenlandicum* heath to productive conifer plantation. *Can J For Res*, **36:** 2364–2376.

Thomson, J.R., Fleishman, E., Nally, R.M., and Dobkin, D.S. (2007) Comparison of predictor sets for species richness and the number of rare species of butterflies and birds. *J Biogeogr*, **34:** 90–101.

Vavrek, M.J. (2012) Fossil: Palaeoecological and palaeogeographical analysis tools - R Package version 0.3.7. URL http://cran.r-project.org/web/packages/fossil/index.html.

Wagg, C., Bender, S.F., Widmer, F., and van der Heijden, M.G.A. (2014) Soil biodiversity and soil community composition determine ecosystem multifunctionality. *Proc Ntl Acad Sci*, **111:** 5266–5270.

Weber, C.F., and King, G.M. (2010) Distribution and diversity of carbon monoxide-oxidizing bacteria and bulk bacterial communities across a succession gradient on a Hawaiian volcanic deposit. *Environ Microbiol*, **12:** 1855–1867.

The sequence capture by hybridization: a new approach for revealing the potential of mono-aromatic hydrocarbons bioattenuation in a deep oligotrophic aquifer

Magali Ranchou-Peyruse,[1] Cyrielle Gasc,[2]
Marion Guignard,[1] Thomas Aüllo,[3] David Dequidt,[4]
Pierre Peyret[2] and Anthony Ranchou-Peyruse[1],*

[1]Université de Pau et des Pays de l'Adour, Equipe
Environnement et Microbiologie, IPREM-CNRS 5254,
F-64013 Pau, France.
[2]Université d'Auvergne, EA 4678 CIDAM, 63001
Clermont-Ferrand, France.
[3]TIGF – Transport et Infrastructures Gaz France, 40
Avenue de l'Europe, CS20522, 64000 Pau, France.
[4]STORENGY – Geosciences Department,
Bois-Colombes, France.

Summary

The formation water of a deep aquifer (853 m of depth) used for geological storage of natural gas was sampled to assess the mono-aromatic hydrocarbons attenuation potential of the indigenous microbiota. The study of bacterial diversity suggests that Firmicutes and, in particular, sulphate-reducing bacteria (*Peptococcaceae*) predominate in this microbial community. The capacity of the microbial community to biodegrade toluene and *m*- and *p*-xylenes was demonstrated using a culture-based approach after several hundred days of incubation. In order to reveal the potential for biodegradation of these compounds within a shorter time frame, an innovative approach named the solution hybrid selection method, which combines sequence capture by hybridization and next-generation sequencing, was applied to the same original water sample. The *bssA* and *bssA*-like genes were investigated as they are considered good biomarkers for the potential of toluene and xylene biodegradation. Unlike a PCR approach which failed to detect these genes directly from formation water, this innovative strategy demonstrated the presence of the *bssA* and *bssA*-like genes in this oligotrophic ecosystem, probably harboured by *Peptococcaceae*. The sequence capture by hybridization shows significant potential to reveal the presence of genes of functional interest which have low-level representation in the biosphere.

*For correspondence. E-mail anthony.ranchou-peyruse@univ-pau.fr

Funding Information
Storengy and TIGF are acknowledged for funding for this research project.

Introduction

The degradation of toluene and *m*, *p* and *o*-xylenes (TX) under anoxic conditions has been demonstrated in numerous marine and continental environments. It can be associated with the reduction of nitrate, sulphate, iron and CO_2 (Dolfing *et al.*, 1990; Beller and Spormann, 1997; Harms *et al.*, 1999; Kane *et al.*, 2002; Kube *et al.*, 2004; Morasch *et al.*, 2004; Morasch and Meckenstock, 2005; Washer and Edwards, 2007; Aüllo *et al.*, 2016). In all cases, the initial addition of fumarate to the hydrocarbon molecule is catalysed by either benzylsuccinate synthase (toluene) or benzylsuccinate synthase-like enzymes (xylenes). For about 15 years, the *bssA* gene, which encodes for the alpha subunit of this protein, has been used as a biomarker for the biodegradation of TX under anoxic conditions. Beller *et al.* (2002) were the first to design primers targeting this gene, using available sequences from isolated strains. These primers then preferentially targeted nitrate-reducing Betaproteobacteria. Subsequently, other primer sets enabled sulphate-reducing, iron-reducing and syntrophic bacteria to be targeted (Winderl *et al.*, 2007; Beller *et al.*, 2008; Staats *et al.*, 2011; Fowler *et al.*, 2014). There was a difficulty in amplifying the *bssA* gene of some sulphate-reducing bacteria belonging to the *Clostridia* which was partly solved by the subsequent design of specific primers for this class (von Netzer *et al.*, 2013; for the latest review on *bssA*-like gene diversity, see von Netzer *et al.*, 2016).

In oligotrophic and stable environments, like deep continental aquifers (−500 to −1200 m), the input of exogenous TX represents a potential source of carbon which is likely to modify the growth and survival strategies of microorganisms. Over the last several years, the potential for biodegradation of monoaromatic hydrocarbons in such environments has been demonstrated through culture-based approaches (Morasch et al., 2004; Berlendis et al., 2010; Aüllo et al., 2016). The low biomass present in these ecosystems makes the demonstration of biodegradation capacities difficult (i.e. detection of bssA genes) in formation water (FW). It is currently difficult, indeed impossible, to assess the degradation potential of FW sampled from a deep aquifer without having recourse to laboratory-based degradation assays. In our previous studies, amplifications of the bssA gene using the primers cited previously have unfortunately often proved to be unsuccessful, as a result of the few targets available and problems of non-specific amplification (Aüllo et al., 2016). Next-generation sequencing (NGS) techniques allow a deeper analysis of genetic diversity and thus offer the possibility of dispensing with culture-based techniques. However, diversity analyses are very often unable to identify microbial populations with low-level representation. The sequence capture by hybridization (Gasc et al., 2016), therefore, constitutes an alternative for the efficient detection of rare or unknown sequences in metagenomic samples (Denonfoux et al., 2013; Bragalini et al., 2014; Biderre-Petit et al., 2016).

In the context of this study, we sought to demonstrate the presence of bssA genes in a deep aquifer used to store natural gas, which is associated with trace amounts of other hydrocarbons, in order to reveal a potential for natural bioattenuation of TX. We hypothesized that the Clostridia, in particular members of the Peptococcaceae family, play a key role in the degradation of TX in deep aquifers (Basso et al., 2009; Berlendis et al., 2010; Aüllo et al., 2016). It is known that some members of Peptococcaceae are capable of degrading mono-aromatic hydrocarbons directly or via syntrophic relationships (Morasch et al., 2004; Taubert et al., 2012). A conventional approach to amplify the bssA gene using the sets of primers available in the literature is compared with the sequence capture by hybridization approach. This study represents the first instance of its use in this type of environment and in an industrial context.

Materials and methods

Sampling

In 2011, water samples were obtained from a deep aquifer (853 m of depth) used for geological storage of natural gas (aquifer 1, Paris Basin, France). Several physicochemical parameters of the FW are indicated in Table 1 (IPL Santé, Environnement Durables, Ile De France). After the tubing was cleaned as described previously (Basso et al., 2005), the biomass from the FW was collected at the wellhead by filtration through 70 Sterivex® filters (EMD; Millipore, Molsheim, France),

Table 1. Physico-chemical parameters and constituents of the formation water sampled from a deep aquifer (aquifer 1 in this study) at 853 m of depth below groundwater and analysed at atmospheric pressure.

Physico-chemical parameters	
Temperature (°C)	36
pH	8.25
Conductivity at 25°C ($\mu S\ cm^{-1}$)	6000
Redox potential (mV)	−363
Pressure (bars)	93
Total suspended solids (mg l^{-1})	16.0
Constituents[a]	
Carbonates (mg l^{-1})	< 20
Sulphates (mg l^{-1})	2186.5
Ammonium (mg l^{-1})	1.75
Calcium (mg l^{-1})	25.1
Magnesium (mg l^{-1})	15.9
Sodium (mg l^{-1})	1400
Potassium (mg l^{-1})	34.0
Chloride (mg l^{-1})	260
Silicates (mg $SiO_2\ l^{-1}$)	15.8
Phosphorus (mg l^{-1})	< 0.05
Nitrates (mg l^{-1})	< 2
Fluoride (mg l^{-1})	1.51
Barium (mg l^{-1})	0.013
Total iron ($\mu g\ l^{-1}$)	3200
Ferrous iron ($\mu g\ l^{-1}$)	< 100
Manganese ($\mu g\ l^{-1}$)	44
Organic carbon (mg l^{-1})	1.0

a. Arsenic, cadmium, chrome, copper, tin, mercury, lead, vanadium and zinc were also measured, but were below the limits of detection.

while maintaining anoxic conditions. Filters were used to collect the microbial biomass over a period of 6.5 h from 500 l of FW. At the end of sampling, the filters were placed in bags maintaining anaerobiosis (GasPak™ EZ; BD, Franklin Lakes, NJ, USA). Several litres of FW were also sampled in sterile glass bottles and degassed with nitrogen in the laboratory prior to subsequently preparing and running microbial cultures. The set of samples (filters and water) were immediately placed at 4°C and transported to the laboratory. The samples to be used for molecular analysis were then frozen at −20°C and those for the cultures used the following day.

Biodegradation assays

Two biodegradation assays of a mixture of toluene, m-, p- and o- xylenes, here referred to as TX (with a final concentration of 10 ppm for each hydrocarbon; Sigma-Aldrich, Saint Quentin Fallavier, France) were carried out at the same time with FW alone (FW condition) or enriched with the concentrated biomass (FWCB condition). The concentrated biomass was obtained for the latter condition by resuspending 67 Sterivex® filters (EMD; Millipore) in 250 ml of anoxic FW under agitation. Ten per cent (v/v) of this concentrated biomass was used to inoculate the FWCB. For each condition, that is, without addition of concentrated biomass (FW+TX) and with addition of concentrated biomass (FWCB+TX), an abiotic control was performed by the addition of 5% v/v of 1 M HCl. The four microcosms, with a final volume of 40 ml, were prepared in 100 ml Wheaton serum bottles sealed with butyl rubber stoppers (Bellco Glass, Vineland, NJ). All manipulations were performed in a glovebox (Getinge La Calhene, France) in an atmosphere of 95% N_2 and 5% H_2. The hydrogen is necessary for the palladium catalyst can react to remove oxygen traces. All cultures were incubated at 37°C, in the dark without agitation. Periodically, TX degradation was monitored in the four microcosms using SPME/GC/FID as described in Aüllo et al. (2016).

Bacterial enumeration

Eighteen millilitres of FW were fixed on site with 2 ml 10% borax-buffered formaldehyde (37%; Sigma-Aldrich) and stored at 4°C until quantification. Furthermore, after resuspension of the Sterivex™ filters in FW for the biodegradation assays, 18 ml of the concentrated biomass was sampled and fixed the same way. In all cases, 500 µl of 4′,6′-diamidino-2-phenylindole (DAPI; Sigma-Aldrich) stock solution (200 µg ml^{-1}) was added to 10 ml of fixed sample, and then vacuum filtration was performed using 0.2 µm pore-size black polycarbonate filters (Millipore). Ten fields were selected at random for each filter, and the cells were counted on an Axio Observer.Z1 inverted microscope (Zeiss, Oberkochen, Germany) equipped with a 63× oil immersion objective (Plan APO, N.A. 1.4, M27). Images were obtained with a Zeiss Axiocam 506 mono CCD camera via the Zeiss ZEN 2012 interface.

Bacterial community analysis

Microbial diversity was investigated on a Sterivex™ filter previously stored at −20°C. The filter was crushed in liquid nitrogen and the DNA extracted with the Powersoil DNA Isolation kit (MoBio Laboratories, Inc., Carlsbad, CA, USA) following the manufacturers' specifications. Genomic DNA was sent to a commercial company (MR DNA, Shallowater, TX, USA). The hypervariable V4 region of the 16S rRNA gene was amplified with PCR primers 515/806 and sequenced by the MiSeq 2 × 300 bp run (Illumina, San Diego, CA, USA). The construction of DNA libraries with the amplicons and the sequencing were performed following the manufacturers' specifications. The sequence data were subsequently processed through the analysis pipeline designed by MR DNA. At first, the sequences were demultiplexed and the barcodes and primers were removed before eliminating all sequences smaller than 150 bp. Sequences displaying ambiguous bases were also eliminated. Operational taxonomic units (OTUs) were defined by clustering sequences displaying 97% similarity. Thus, singleton sequences and chimeras

were eliminated. The final OTUs were subsequently classified taxonomically using BLASTn against a curated database derived from RDPII and NCBI.

Cloning and sequencing of benzylsuccinate synthase alpha-subunit (bssA) genes

PCR amplifications of the bssA gene were performed for different sets of primers found in the literature. The primer sets 7772F/8546R (Winderl et al., 2007), 7768F/8543R (von Netzer et al., 2013), bss3F/bssAR (Staats et al., 2011), and 997F/1230R (Brow et al., 2013) were used to target, respectively, the bssA gene of Deltaproteobacteria, Clostridia, Geobacter and nitrate-reducing Betaproteobacteria (Table S1). For each positive amplification, the PCR fragment corresponding to the expected size was purified on an agarose gel and cloned using the TOPO TA Cloning kit (Fisher Scientific, Hampton, NH, USA) before being sequenced by Qiagen Genomic Services (Hilden, Germany).

Capture probe design and synthesis

A set of five 31- to 38-mers degenerate probes covering the bssA gene was designed from nine bssA nucleic sequences from strains belonging to Deltaproteobacteria, Firmicutes (EF123665, EF123667, EU780921, FO203503, EF123663, EF123662) and environmental sequences obtained with 7772F/8546R primers set. These three latter amplicons were obtained at the end of TX degradation from enrichment cultures with autochthonous microbiota of different deep aquifers (aquifer 1: KX576576, aquifer 2: KX576577, aquifer 3: KX576575). The deep aquifer 2 was studied by Berlendis et al. (2010) (Paris Basin). The third aquifer (aquifer 3) is located in the Aquitaine Basin (southwest of France). These sequences were processed using the KASpOD software (Parisot et al., 2012) (Table S1). Adaptor sequences were added at each extremity of the probe to enable their PCR amplification, resulting in "ATCGCAC CAGCGTGT-N$_{(31-38)}$-CACTGCGGCTCCTCA" sequences, with N$_{(31-38)}$ representing the bssA-specific capture probe. Biotinylated RNA capture probes were then synthesized as described by Ribière et al. (2016).

Preparation of biological samples and libraries

Next-generation sequencing libraries were constructed on genomic DNA extracted directly from aquifer 1 FW after concentration by means of ethanol precipitation. Libraries were prepared using the Nextera XT Kit (Illumina) using the manufacturer's instructions (Genoscreen, Lille, France) and were PCR amplified using the GC-RICH PCR system kit (RocheDiagnostics GmbH, Mannheim,

Germany) with primers complementary to the library adapters to obtain sufficient amounts of DNA to run the sequence capture by hybridization.

Hybridization capture and elution

Solution hybrid selection (SHS) was conducted on FW sample according to the protocol described by Ribière et al. (2016). Briefly, 500 ng of heat denatured libraries was hybridized to the set of biotinylated RNA probes for 24 h at 65°C. Probe/target heterodimers were trapped by streptavidin-coated paramagnetic beads (Dynabeads M-280 Steptavidin, Invitrogen, Carlsbag, CA, USA). After several washing steps, the captured targets were eluted from the beads using NaOH and purified using AMPure beads (Beckman Coulter Genomics, Takeley, Essex, United Kingdom). Enriched products were PCR amplified using primers complementary to the library adapters and purified again using AMPure beads (Beckman Coulter Genomics). To increase the enrichment, a second round of hybridization and amplification was performed using the obtained captured products.

Illumina MiSeq sequencing and data analysis

Captured DNA products were sequenced using a single MiSeq 2 × 300 bp run (Illumina) according to the manufacturer's specifications (Genoscreen). All raw reads were scanned for library adaptors and quality filtered using PRINSEQ-lite PERL script (Schmieder and Edwards, 2011) prior to assembly and analysis. The clean reads were assembled de novo using IDBA-UD (v1.1.1) (Peng et al., 2012). Contigs generated were combined for a second round of assembly using CAP3 to generate longer contigs (Huang and Madan, 1999). The amino acid (AA) sequences were deduced from the final assembled nucleotide contigs and then aligned with reference open-reading frames sourced from public databases using MEGA version 6 (Tamura et al., 2013). The phylogenetic tree was constructed with the same software using the neighbour-joining method. The bootstrap analysis was performed for 1000 replicates.

Nucleotide sequence accession numbers

Sixty-five sequences (16S rRNA and bssA genes) have been submitted to GenBank under accession numbers KX576572 to KX576636.

Results and discussion

Characterization of the site

In the autumn of 2011, the FW of a deep aquifer (aquifer 1) used for geological storage of natural gas was sampled after cleaning of the sampling well, to remove the microorganisms growing as a biofilm on the tubing and the wellhead (see Basso et al., 2009 for Scheme of principle of a natural gas underground storage in an aquifer). During this study, the methane bubble was located close to the well and the presence of methane, originally dissolved at reservoir depth, was observed in the sampled water. At the time of sampling, the water was at a temperature of 36°C, with a pH of 8.25, and no oxygen was detected and a redox potential of −363 mV (Table 1). This aquifer displayed a total organic carbon concentration (TOC; 1 mg l^{-1}) typical of this type of environment which has been little impacted by human activity (Pedersen, 1997; Sahl et al., 2008). Meckenstock et al. (2015) report in their review dedicated to microbial clean-up in contaminated aquifers that only 0.5–5% of this TOC is likely to be directly used as carbon source by the microbial community which makes this deep aquifer an oligotrophic environment. The analysis of the chemical composition of this water reveals a sulphate content of 2186.5 mg l^{-1}, a total iron content of 3200 µg l^{-1} but no nitrate. These data suggest that the indigenous microbial community is probably dominated by sulphate-reducing bacteria. Sulphate reduction is indeed an important metabolic process in deep subsurface environments and particularly in aquifers (Detmers et al., 2004; Amend and Teske, 2005; Bombach et al., 2010; Itävaara et al., 2011) where it has been demonstrated that sulphate-reducing bacteria also play a key role in the degradation of mono-aromatic hydrocarbons (Basso et al., 2009; Berlendis et al., 2010; Aüllo et al., 2016). The concentrated biomass derived from the filtration of FW was subjected to a diversity analysis in order to determine the predominant bacterial groups. For this purpose, the V4 region of the sequence of the 16S rRNA gene was targeted by NGS (no archaea were detected). The taxa detected in this study are regularly found in studies of deep subsurface ecosystems. Figure 1 shows that Firmicutes were clearly predominant (61.1%) among the 62 OTUs found (similarity ≥ 97%), with 29.3% of Peptococcaceae (Carboxydothermus, candidatus Desulforudis, Pelotomaculum and Desulfotomaculum) and 31.2% of sequences belonging to Thermoanaerobacteraceae (Thermoanaerobacter, Moorella, Sporotomaculum and Ammonifex). The detection thermophilic microorganisms signatures (e.g. Moorella or Thermoanaerobacter) in a mesothermic environment may be surprising. However, similar results have already been reported in the study of another mesothermic aquifer (Berlendis et al., 2010; aquifer 2 in this study). Taking into account of the diversity observed, most sulphate reduction must be carried out by Firmicutes since the Deltaproteobacteria, which include a large number of sulphate-reducing bacteria, represent less than 1% of total sequences. The

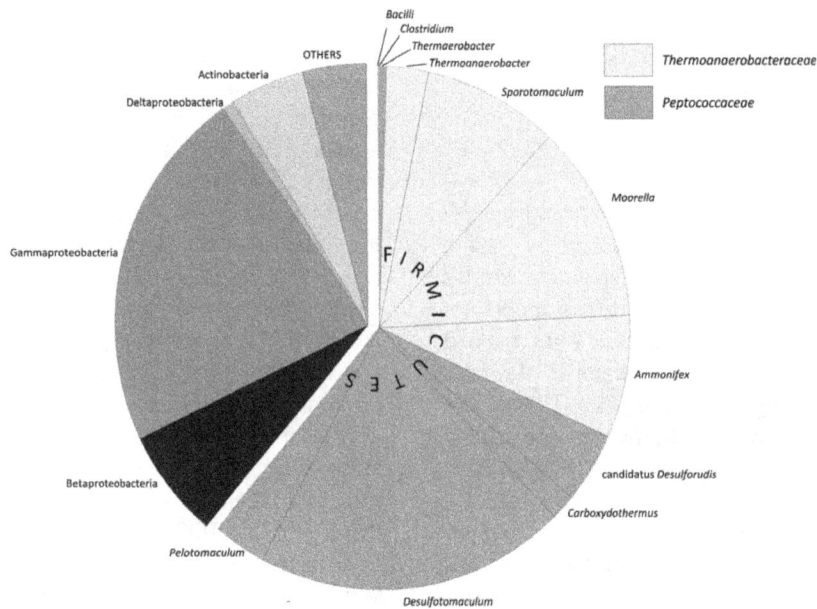

Fig. 1. Composition of the bacterial community in the formation water collected from a deep aquifer (−853 m). The pie chart represents the percentage of each taxon within the whole microbial species analysed.

dominance of the sulphate-reducing community by Firmicutes has been shown in a 120 m deep aquifer displaying low sulphate concentrations (≤ 17.1 mg l^{-1}; Detmers et al., 2004). Salinity seems to play a role in the distribution of Deltaproteobacteria and sulphate-reducing Firmicutes (Leloup et al., 2005). Here, the dominance of Firmicutes can probably be further explained by their capacity to sporulate, their capacity to survive in oligotrophic environments and their metabolic versatility (Sass et al., 1997; Spring and Rosenzweig, 2006; Orsi et al., 2016). Moreover, some studies have suggested or demonstrated the key role of some *Peptococcaceae*, and in particular representatives of the genera *Desulfotomaculum*, *Desulfosporosinus* and *Pelotomaculum*, in the degradation of BTEX (Robertson et al., 2000; Liu et al., 2004; Morasch et al., 2004; Cupples, 2011; Abu Laban et al., 2015).

Biodegradation assays

The day after sampling, assays of biodegradation of toluene and the 3 xylene isomers were initiated either directly on the water obtained from the site (FW+TX) or in water supplemented with concentrated indigenous bacterial biomass (FWCB+TX). A cell-count carried out in FW using epifluorescence (DAPI) showed a cell concentration of $4.5 \times 10^5 \pm 2.3 \times 10^5$ cells ml^{-1}, while the concentration of microorganisms in FWCB was four times greater; 1.8×10^6 cells ml^{-1}. While it is difficult to quantify microorganisms attached to the mineral matrix in deep aquifers, it is commonly accepted that the indigenous biomass growing in a biofilm is very largely

dominant compared with pelagic microorganisms (Whitman et al., 1998; Griebler and Lueders, 2009). Even if the diversity between the pelagic and the attached microorganisms can differ (Röling and van Verseveld, 2002), the biomass supplemented condition tended to simulate the influence of the bacterial concentration in our biodegradation assays. In the two microcosms of this study, toluene and m- and p- xylenes (Tm, pX) were degraded simultaneously (Fig. 1). Toluene is assumed to be the most easily biodegraded aromatic hydrocarbon in anoxic conditions, which induces that it is degraded first and therefore delay the degradation of other hydrocarbons (Haag et al., 1991; Edwards et al., 1992; Phelps and Young, 1999; Meckenstock et al., 2004; Morasch et al., 2004). However, this interpretation must be tempered since it is evident that the biodegradation potential depends on the pool of key genes present. Indeed simultaneous biodegradation of toluene and xylene has already been observed (Herrmann et al., 2009; Shah, 2014). It is interesting to observe that while the m- and p-xylenes were degraded, o-xylene resisted degradation throughout the incubation period. In the FW degradation assay, the degradation of Tm, pX began after a lag phase of about 800 days. Degradation was almost four times faster (237 days) when the biomass was concentrated fourfold, which implies that the number of microorganisms influences Tm, pX degradation. This result suggests that the in situ biodegradation process, with a supposed higher concentration of microorganisms in biofilm structures, could be faster than revealed by biodegradation studies in the laboratory (Botton and Parson, 2006; Berlendis et al., 2010; Higashioka et al.,

2012; Aüllo *et al.*, 2016). Indeed, only the pelagic fraction of the aquifers can be harvested, which underestimates the potential for in situ biodegradation.

BssA *detection assay by PCR*

Numerous studies have shown that the *bssA* and *bssA*-like genes can serve as biomarkers for the in situ anaerobic degradation of toluene and xylenes (Krieger *et al.*, 1999; Andreoni and Gianfreda, 2007; Kazy *et al.*, 2010; Cury *et al.*, 2015). These studies are most frequently conducted in surface environments or in shallow aquifers. Even though the diversity study based on 16S rRNA gene indicates a predominance of Firmicutes and the water composition suggests that sulphate reduction could play a major role in this aquifer, the choice was made to be as exhaustive as possible regarding the metabolisms involved in bioattenuation. For this purpose, four primer pairs designed to target nitrate-reducing bacteria (Brow *et al.*, 2013), iron-reducing bacteria (Staats *et al.*, 2011), sulphate-reducing bacteria belonging to the Proteobacteria or Clostridiales (Winderl *et al.*, 2007; von Netzer *et al.*, 2013) were tested as part of this study (Table S1). The *bssA* gene was not amplified by PCR in the DNA extracted from FW samples whether supplemented or not supplemented with concentrated biomass, although the degradation results obtained subsequently in the microcosms demonstrated the capacity to degrade the T*m*, *p*X. Our team has been trying to demonstrate the presence of this gene directly in FW for some years without success. Three times only this gene was amplified by the primers of Winderl *et al.* (2007) in microcosms from three different aquifers (aquifers 1, 2 and 3), after incubation periods of more than 100 days (data not shown). It should be noted that the amplicons corresponding to the expected size had to be systematically purified on an agarose gel in order to eliminate numerous non-specific amplifications before sequencing. Although this amplification approach has proven its value in surface environments and shallow aquifers, it did not seem appropriate for the constraints of our study sites (trace mono-aromatic hydrocarbons, low biomass and impossibility of recovering biofilms in this industrial context). Therefore, we had to develop a new molecular biological approach enabling detection of the *bssA* gene directly in FW while conducting time-consuming biodegradation assays.

BssA *detection assay using sequence capture by hybridization approach*

In 2013, Denonfoux and co-workers successfully combined a SHS method with next-generation sequencing for the first time, in order to capture a biomarker gene in a complex metagenome. The methodology was used to explore the methanogenic communities present in a lacustrine environment by targeting the methyl coenzyme M reductase subunit A (*mcrA*) gene with a set of non-overlapping probes, which targeted both known sequences and potential undescribed variants of the *mcrA* gene. The *mcrA* sequences represented more than 40% of the obtained sequences after two cycles of capture, revealing enrichment compared with shotgun sequencing, in which only 0.003% of the sequences corresponded to the target gene. In addition, because *mcrA* and 16S rRNA gene phylogenies are congruent, this approach allowed the methanogen community to be described and revealed higher diversity than previously observed with other methods. Indeed, hybridization capture recovered sequences from the Methanobacteriales order, belonging to the rare biosphere, which were not detected through direct sample sequencing due to the sequencing depth, or through PCR amplification, due to possible primer bias. This method appeared to be appropriate for our constraints given its sensitivity, and the fact that it does not need amplification step, which is often unfruitful in this type of study. The SHS probes (Table S1) were designed from *bssA* sequences of sulphate-reducing bacteria belonging to the Deltaproteobacteria and the Firmicutes, as well as the three sequences previously amplified with the primers of Winderl *et al.* (2007) in enrichments with mono-aromatic hydrocarbons (KX576575, KX576576, KX576577). After analysis of reads (Table S2), 498 500 reads proved close to *bssA* and *bssA*-like genes and were grouped into three *bssA* homologous contigs. The contig sequence_10944 alone includes 98% of the *bssA* reads indicating that such *bssA* gene dominates in the ecosystem. Sequence capture by hybridization gives quantitative results close to that obtained by qPCR as demonstrates by Denonfoux *et al.* (2013). The three contigs obtained by sequence capture and the three *bssA* amplicons obtained in this study were compared with sequences from pure strains and environmental samples deposited in international databases after having been translated into AA sequences. The results are presented as a phylogenetic tree constructed from the comparison of a 90AA region. The main contig, contig sequence_10944 (aquifer 1) and the amplicons obtained from the sulphate-reducing enrichments from FWs (aquifers 2 and 3) are closed to *bssA* sequences defined as sensu stricto by Acosta-González *et al.* (2013). The dominant *bssA* gene obtained by SHS is close to the BF clone obtained from an enrichment described as degrading benzene and dominated by *Peptococcaceae*-related Gram-positive microorganisms (87% identity, 276AA). In our current state of knowledge, the reason for the presence of this gene in this enrichment described by Abu Laban *et al.* (2010) cannot be explained since the initial benzene biodegradation step does not seem to involve the addition of a fumarate molecule (enabled by the benzylsuccinate synthase) but a carboxylation as described by the authors. In our case, the procurement of a *bssA* gene associated with the *Peptococcaceae* family is consistent with the diversity data obtained in this study (Fig. 1) and tends to confirm the supposed role of *Peptococcaceae* in the degradation of T*m*, *p*X in this deep aquifer. In referring to the broad dominance of Firmicutes over Deltaproteobacteria, we can hypothesize that these are the principal sulphate-reducing bacteria in this deep aquifer. Several studies have provided evidence for a positive correlation between the dominance of sulphate-reducing bacteria affiliated to *Peptococcaceae* in subsurface environments and depth (Moser *et al.*, 2005; Chivian *et al.*, 2008; Itävaara *et al.*, 2011; Guan *et al.*, 2013). These microorganisms play a major role in the carbon cycle in deep environments via the recycling of organic material, which is, in our context, the degradation of mono-aromatic hydrocarbons. To date, only two strains of *Peptococcaceae* (*Desulfotomaculum* sp. Ox39 and *Desulfosporosinus meridiei*) have been described as being able to degrade toluene and/or xylenes (Liu and Garcia-Dominguez, 2004; Morasch *et al.*, 2004). However, several studies using non culture-based approaches tend to demonstrate that these microorganisms are often playing a key role in the degradation of mono-aromatic hydrocarbons. In the case of a gas condensate-contaminated aquifer, *Desulfosporosinus* sp. was shown to initiate toluene degradation (Fowler *et al.*, 2012, 2014). The key role of the *Peptococcaceae* has also been demonstrated in other environments and in sulphate-reducing conditions and/or methanogenesis (Abu Laban *et al.*, 2009; Winderl *et al.*, 2010; Pilloni *et al.*, 2011; Sun and Cupples, 2012; Sun *et al.*, 2014; Abu Laban *et al.*, 2015; Tan *et al.*, 2015). It is interesting to note that the *bssA* amplicon obtained for the

enrichment culture with FW from the aquifer 1 at the end of degradation (Fig. 2) is phylogenetically located in the OX39-homologues cluster as described by von Netzer *et al.* (2013). The genes present in this cluster could be involved in the degradation of xylenes (Herrmann *et al.*, 2009; Bragalini *et al.*, 2014). This sequence is close to the *Desulfotomaculum* sp. Ox39 *bssA*-like gene (74% similarity, 227AA) and to environmental sequences obtained from enrichments derived from contaminated aquifers (Herrmann *et al.*, 2009; von Netzer *et al.*, 2013). Finally, the two last contigs that represent only 2% of the bssA homologous sequences obtained by SHS (Table S2), contig sequence_48572 and contig sequence_31410, form a separate cluster located between the *assA* and the *bssA* genes. The existence of deeply branching bacteria was also found in the DNA-SIP study performed on a sample from the Testfeld Süd aquifer contaminated by hydrocarbons (Winderl *et al.*, 2010), suggesting that a large part of the diversity of *bssA* sequences sensu stricto and sensu lato, and *bssA* homologues (*assA*, *nmsA*, *hbsA*) is still to be discovered. The contig sequence_48572 is close to the *Desulfobacula toluolica bssA* sequence (42% similarity, 115AA; ABM92935). As regard the contig sequence_31410 is close to a *Desulfotomaculum* sp. 46_20 alkylsuccinate synthase obtained from an oil reservoir in Alaska (46%/ 138AA; KUK63464), but also to a pyruvate-formate lyase derived from a strain affiliated to *Peptococcaceae* from a Opalinus Clay rock porewater BRC-3 borehole (39%/ 142AA; KJS47223) and to a glycyl radical enzyme of the strain *Desulfosporosinus* sp. BRH_c37, also obtained from the BRC-3 site (38%/143AA; KUO70645).

Conclusions

The study of the potential for TX degradation in oligotrophic environments such as deep aquifers is very

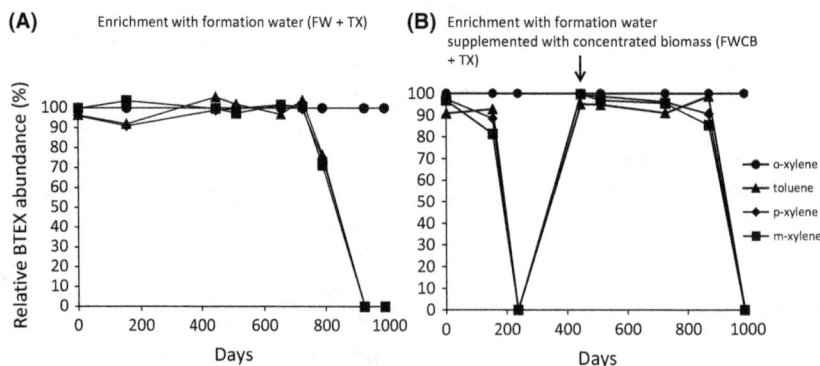

Fig. 2. Degradation of mono-aromatic hydrocarbons (toluene, *m*- and *p*-xylenes) during incubation of formation water collected anoxically to protect autochthonous microbiota, FW (A) or with the formation water supplemented with concentrated biomass, FWCB (B). Filled circles: *o*-xylene, filled triangles: toluene, filled diamonds: *p*-xylene, filled squares: *m*-xylene. Arrow indicates at day 447 the addition of toluene, *m*- and *p*-xylenes (10 ppm). Start levels of mono-aromatic hydrocarbons were 10 ppm.

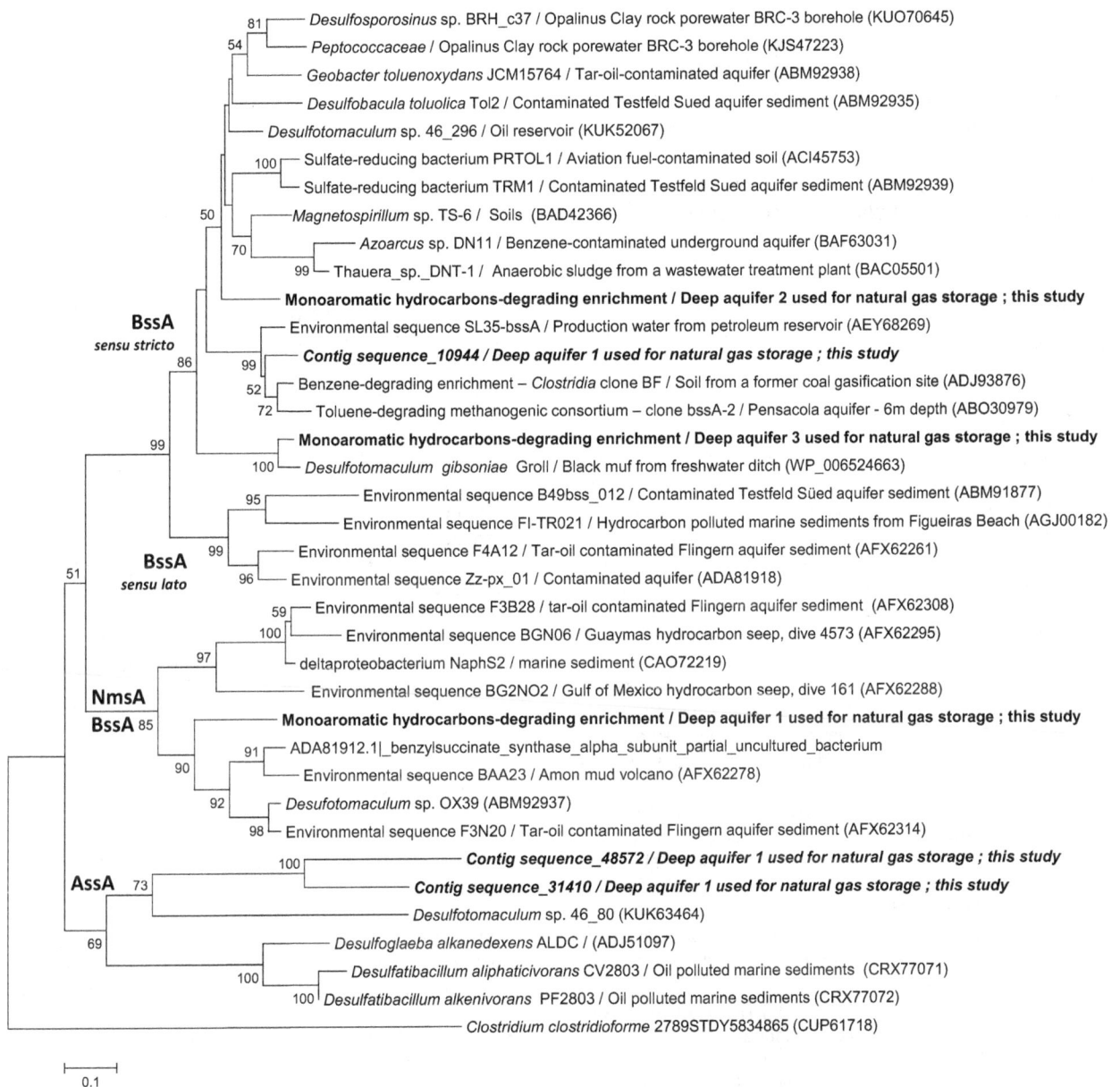

Fig. 3. Phylogenetic tree based on partial *bssA*-like amino acid sequences from deep aquifers used for geological natural gas storage (in bold; this study) compared with sequences from pure strains, enrichment cultures or environments retrieved in the databases. Sequences in bold were obtained by PCR from genomic DNA from mono-aromatic hydrocarbon-degrading enrichment cultures (aquifers 1, 2 and 3). Sequences in bold and italic were obtained by SHS method directly from formation water (aquifer 1). The evolutionary distances were computed using the Poisson correction method. Evolutionary analyses were conducted in MEGA6 with a bootstrap test of 1000 replicates.

difficult using conventional cultural approaches, as they require long incubation periods (several months to several years) and/or biomass. Biomarker detection allowing the evaluation of the biodegradation potential of an ecosystem or the monitoring of bioremediation operations is necessary tools for environmental engineering. Currently, the *bssA* and *bssA*-like genes represent excellent biomarkers for the degradation of some mono-aromatic hydrocarbons. However, no primer sets tested in the study enabled demonstration of the presence of

bssA genes directly in FW, while enrichments subsequently showed that the metabolic potential was present.

Direct sequencing of metagenomic samples in recent years has allowed for increased precision in microbial diversity analyses but only dominant taxa could be revealed. The sequence capture by hybridization approach used in this study proved its efficiency for the specific capture of targeted *bssA* sequences. Indeed, this was the only method that enabled *bssA* and *bssA*-like sequences to be obtained directly from FW. Therefore, this method constitutes a major asset in developing a clearer understanding of ecosystems and in monitoring bioattenuation phenomena in the context of mature environmental engineering. The *bssA* sequences and the diversity analyses based on the 16S rDNA sequences once again revealed the key role of *Peptococcaceae* in the degradation of mono-aromatic hydrocarbons in deep continental aquifers.

Conflict of interest

None declared.

References

Abu Laban, N.A., Selesi, D., Jobelius, C., and Meckenstock, R.U. (2009) Anaerobic benzene degradation by Gram-positive sulfate-reducing bacteria. *FEMS Microbiol Ecol* **68:** 300–311.

Abu Laban, N.A., Selesi, D., Rattel, T., Tischler, P., and Meckenstock, R.U. (2010) Identification of enzymes involved in anaerobic benzene degradation by a strictly anaerobic iron-reducing enrichment culture. *Environ Microbiol* **12:** 2783–2796.

Abu Laban, N.A., Dao, A., and Foght, J. (2015) DNA stable-isotope probing of oil sands tailing pond enrichment cultures reveals different key players for toluene degradation under methanogenic and sulfidogenic conditions. *FEMS Microbiol Ecol* **91:** fiv039.

Acosta-González, A., Rosselló-Móra, R., and Marqués, S. (2013) Diversity of benzylsuccinate synthase-like (*bssA*) genes in hydrocarbon-polluted marine sediments suggests substrate-dependent clustering. *Appl Environ Microb* **79:** 3667–3676.

Amend, J.P., and Teske, A. (2005) Expanding frontiers in deep subsurface microbiology. *Palaeogeogr Palaeoclimatol Palaeoecol* **219:** 131–155.

Andreoni, V., and Gianfreda, L. (2007) Bioremediation and monitoring of aromatic-polluted habitats. *Appl Microbiol Biotechnol* **76:** 287–308.

Aüllo, T., Berlendis, S., Lascourrèges, J.-F., Dessort, D., Duclerc, D., Saint-Laurent, S., et al. (2016) New bio-indicators for long term natural attenuation of monoaromatic compounds in deep terrestrial aquifers. *Front Microbiol* **7:** 122.

Basso, O., Lascourrèges, J.-F., Jarry, M., and Magot, M. (2005) The effect of cleaning and disinfecting the sampling well on the microbial communities of deep subsurface water samples. *Environ Microbiol* **7:** 13–21.

Basso, O., Lascourrèges, J.-F., Le Borgne, F., Le Goff, C., and Magot, M. (2009) Characterization by culture and molecular analysis of the microbial diversity of a deep subsurface gas storage aquifer. *Res Microbiol* **160:** 107–116.

Beller, H.R., and Spormann, A.M. (1997) Benzylsuccinate formation as a means of anaerobic toluene activation by sulfate-reducing strain PRTOL1. *Appl Environ Microb* **63:** 3729–3731.

Beller, H.R., Kane, S.R., Legler, T.C., and Alvarez, P.J.J. (2002) A real-time polymerase chain reaction method for monitoring anaerobic, hydrocarbon-degrading bacteria based on a catabolic gene. *Environ Sci Technol* **36:** 3977–3984.

Beller, H.R., Kane, S.R., Legler, T.C., McKelvie, J.R., Lollar, B.S., Pearson, F., et al. (2008) Comparative assessments of benzene, toluene, and xylene natural attenuation by quantitative polymerase chain reaction analysis of a catabolic gene, signature metabolites, and compound-specific isotope analysis. *Environ Sci Technol* **42:** 6065–6072.

Berlendis, S., Lascourrèges, J.-F., Schraauwers, B., Sivadon, P., and Magot, M. (2010) Anaerobic biodegradation of BTEX by original bacterial communities from an underground gas storage aquifer. *Environ Sci Technol* **44:** 3621–3628.

Biderre-Petit, C., Dugat-Bony, E., Mege, M., Parisot, N., Adrian, L., Moné, A., et al. (2016) Distribution of *Dehalococcoidia* in the anaerobic deep water of a remote meromictic crater lake and detection of *Dehalococcoidia*-derived reductive dehalogenase homologous genes. *PLoS ONE* **11:** e0145558. doi:10.1371/journal.pone.0145558.

Bombach, P., Chatzinotas, A., Neu, T.R., Kästner, M., Lueders, T., and Vogt, C. (2010) Enrichment and characterization of a sulfate-reducing toluene-degrading microbial consortium by combining in situ microcosms and stable isotope probing techniques. *FEMS Microbiol Ecol* **71:** 237–246.

Botton, S., and Parson, J.R. (2006) Degradation of BTEX compounds under iron-reducing conditions in contaminated aquifer microcosms. *Environ Chem* **25:** 2630–2638.

Bozinovski, D., Taubert, M., Kleinsteuber, S., Richnow, H.-H., von Bergen, M., Vogt, C., and Seifert, J. (2014) Metaproteogenomic analysis of a sulfate-reducing enrichment culture reveals genomic organization of key enzymes in the *m*-xylene degradation pathway and metabolic activity of proteobacteria. *Syst Appl Microbiol* **37:** 488–501.

Bragalini, C., Ribière, C., Parisot, N., Vallon, L., Prudent, E., Peyretaillade, E., et al. (2014) Solution hybrid selection capture for the recovery of functional full-length eukaryotic cDNAs from complex environmental samples. *DNA Res* **21:** 685–694.

Brow, C.N., O'Brien, J.R., Johnson, R.L., and Simon, H.M. (2013) Assessment of anaerobic toluene biodegradation activity by *bssA* transcript/gene ratios. *Appl Environ Microb* **79:** 5338–5344.

Chivian, D., Brodie, E.L., Alm, E., Culley, D.E., Dehal, P.S., DeSantis, T.Z., et al. (2008) Environmental genomics reveals a single-species ecosystem deep within earth. *Science* **322:** 275–278.

Cupples, A.M. (2011) The use of nucleic acid based stable isotope probing to identify the microorganisms responsible for anaerobic benzene and toluene biodegradation. *J Microbiol Meth* **85:** 83–91.

Cury, J.C., Jurelevicius, D.A., Villela, H.D.M., Jesus, H.E., Peixoto, R.S., Schaefer, C.E.G.R., et al. (2015) Microbial diversity and hydrocarbon depletion in low and high diesel-polluted soil samples from Keller Peninsula, South Shetland Islands. *Antarc Sci* 27: 263–273.

Denonfoux, J., Parisot, N., Dugat-Bony, E., Biderre-Petit, C., Boucher, D., Morgavi, D.P., *et al.* (2013) Gene capture coupled to high-throughput sequencing as a strategy for targeted metagenome exploration. *DNA Res* 20: 185–196.

Detmers, J., Strauss, H., Schulte, U., Bergmann, A., Knittel, K., and Kuever, J. (2004) FISH shows that *Desulfotomaculum* spp. are the dominating sulfate-reducing bacteria in a pristine aquifer. *Microbiol Ecol* 47: 236–242.

Dolfing, J., Zeyer, J., Binder-Eicher, P., and Schwarzenbach, R. (1990) Isolation and characterization of a bacterium that mineralizes toluene in the absence of molecular oxygen. *Arch Microbiol* 154: 336–341.

Edwards, E.A., Wills, L.E., Reinhard, M., and Grbić-Galić, D. (1992) Anaerobic degradation of toluene and xylene by aquifer microorganisms under sulfate-reducing conditions. *Appl Environ Microb* 58: 794–800.

Fowler, S.J., Dong, X., Sensen, C.W., Suflita, J.M., and Gieg, L.M. (2012) Methanogenic toluene metabolism: community structure and intermediates. *Environ Microbiol* 14: 754–764.

Fowler, S.J., Gutierrez-Zamora, M.L., Manefield, M., and Gieg, L.M. (2014) Identification of toluene degraders in a methanogenic enrichment culture. *FEMS Microbiol Ecol* 89: 625–636.

Gasc, C., Peyretaille, E., and Peyret, P. (2016) Sequence capture by hybridization to explore modern and ancien genomic diversity in model and nonmodel organisms. *Nucleic Acids Res* 44: 4504–4518.

Griebler, C., and Lueders, T. (2009) Microbial biodiversity in groundwater ecosystems. *Freshwat Biol* 54: 649–677.

Guan, J., Xia, L.P., Wang, L.Y., Liu, J.F., Gu, J.D., and Mu, B.Z. (2013) Diversity and distribution of sulfate-reducing bacteria in four petroleum reservoirs detected by using 16s rRNA and *dsrAB* genes. *Int Biodeterior Biodegr* 76: 58–66.

Haag, F., Reinhard, M., and McCarty, P.L. (1991) Degradation of toluene and *p*-xylene in anaerobic microcosms: evidence for sulfate as terminal electron acceptor. *Environ Toxicol Chem* 10: 1379–1389.

Harms, G., Zengler, K., Rabus, R., Aeckersberg, F., Minz, D., Rosselló-Mora, R., and Widdel, F. (1999) Anaerobic oxidation of *o*-xylene, *m*-xylene, and homologous alkylbenzenes by new types of sulfate-reducing bacteria. *Appl Environ Microbiol* 65: 999–1004.

Herrmann, S., Vogt, C., Fischer, A., Kuppardt, A., and Richnow, H.H. (2009) Characterization of anaerobic xylene biodegradation by two-dimensional isotope fractionation analysis. *Environ Microbiol Rep* 1: 535–544.

Higashioka, Y., Kojima, H., and Fukui, M. (2012) Isolation and characterization of novel sulfate-reducing bacterium capable of anaerobic degradation of *p*-xylene. *Microbes environ* 27: 273–277.

Huang, X., and Madan, A. (1999) CAP3: a DNA sequence assembly program. *Genome Res* 9: 868–877.

Itävaara, M., Nyyssönen, M., Kapanen, A., Nousiainen, A., Ahonen, L., and Kukkonen, I. (2011) Characterization of bacterial diversity to a depth of 1500 m in the Outokumpu deep borehole, Fennoscandian Shield. *FEMS Microbiol Ecol* 77: 295–309.

Kane, S.R., Beller, H.R., Legler, T.C., and Anderson, R.T. (2002) Biochemical and genetic evidence of benzylsuccinate synthase in toluene-degrading, ferric iron-reducing *Geobacter metallireducens*. *Biodegradation* 13: 149–154.

Kazy, S.K., Monier, A.L., and Alvarez, P.J.J. (2010) Assessing the correlation between anaerobic toluene degradation activity and *bssA* concentrations in hydrocarbon-contaminated aquifer material. *Biodegradation* 21: 793–800.

Krieger, C.J., Beller, H.R., Reinhard, M., and Spormann, A.M. (1999) Initial reactions in anaerobic oxidation of *m*-xylene by the denitrifying bacterium *Azoarcus* sp strain T. *J Bacteriol* 181: 6403–6410.

Kube, M., Heider, J., Amann, J., Hufnagel, P., Kuhner, S., Beck, A., *et al.* (2004) Genes involved in the anaerobic degradation of toluene in a denitrifying bacterium, strain Ebn1. *Arch Microbiol* 181: 182–194.

Leloup, J., Petit, F., Boust, D., Deloffre, J., Bally, G., Clarisse, O., and Quillet, L. (2005) Dynamics of sulfate-reducing microorganisms (*dsrAB* genes) in two contrasting mudflats of the Seine Estuary (France). *Microb Ecol* 50: 307–314.

Liu, A., Garcia-Dominguez, E., Rhine, E.D. and Young, L.Y. (2004) A novel arsenate respiring isolate that can utilize aromatic substrates. *FEMS Microbiol Ecol* 48: 323–332.

Meckenstock, R.U., Safinowski, M., and Griebler, C. (2004) Anaerobic degradation of polycyclic aromatic hydrocarbons. *FEMS Microbiol Ecol* 47: 381–386.

Meckenstock, R.U., Elsner, M., Griebler, C., Lueders, T., Stumpp, C., Aamand, J., *et al.* (2015) Biodegradation: updating the concepts of control for microbial cleanup in contaminated aquifers. *Environ Sci Technol Lett* 49: 7073–7081.

Morasch, B., and Meckenstock, R.U. (2005) Anaerobic degradation of *p*-xylene by a sulfate-reducing enrichment culture. *Curr Microbiol* 51: 127–130.

Morasch, B., Schink, B., Tebbe, C.C., and Meckenstock, R.U. (2004) Degradation of *o*-xylene and *m*-xylene by a novel sulfate-reducer belonging to the genus *Desulfotomaculum*. *Arch Microbiol* 181: 407–417.

Moser, D.P., Gihring, T.M., Brockman, F.J., Fredrickson, J.K., Balkwill, D.L., Dollhopf, M.E., *et al.* (2005) *Desulfotomaculum* and *Methanobacterium* spp. dominate a 4- to 5- kilometer-deep fault. *Appl Environ Microbiol* 71: 8773–8783.

von Netzer, F., Pilloni, G., Kleindienst, S., Kruger, M., Knittel, K., Grundger, F., and Lueders, T. (2013) Enhanced gene detection assays for fumarate-adding enzymes allow uncovering of anaerobic hydrocarbon degraders in terrestrial and marine systems. *Appl Environ Microbiol* 79: 543–552.

von Netzer, F., Kuntze, K., Vogt, C., Richnow, H.H., Boll, M., and Lueders, T. (2016) Functional gene markers for fumarate-adding and dearomatizing key enzymes in anaerobic aromatic hydrocarbon degradation in terrestrial environments. *J Mol Microbiol Biotechnol* 26: 180–194.

Orsi, W.D., Jørgensen, B.B., and Biddle, J.F. (2016) Transcriptional analysis of sulfate reducing and chemolithoautotrophic sulfur oxidizing bacteria in the deep subseafloor. *Environ Microbiol Rep* 8: 452–460.

Parisot, N., Denonfoux, J., Dugat-Bony, E., Peyret, P., and Peyretaillade, E. (2012) KASpOD-a web service for highly specific and explorative oligonucleotide design. *Bioinfor-matics* **28:** 3161–3162.

Pedersen, K. (1997) Microbial life in deep granitic rock. *FEMS Microbiol Rev* **20:** 399–414.

Peng, Y., Leung, H.C.M., Yiu, S.M., and Chin, F.Y.L. (2012) IDBA-UD: a de novo assembler for single-cell and metagenomic sequencing data with highly uneven depth. *Bioinformatics* **28:** 1420–1428.

Phelps, C.D., and Young, L. (1999) Anaerobic biodegradation of BTEX and gasoline in various aquatic sediments. *Biodegradation* **10:** 15–25.

Pilloni, G., von Netzer, F., Engel, M., and Lueders, T. (2011) Electron acceptor-dependent identification of key anaerobic toluene degraders at a tar-oil-contaminated aquifer by pyro-SIP. *FEMS Microbiol Ecol* **78:** 165–175.

Ribière, C., Beugnot, R., Parisot, N., Gasc, C., Defois, C., Denonfoux, J., *et al.* (2016) Targeted gene capture by hybridization to Illuminate ecosystem functioning. *Microb Environ Gen* **1399:** 167–182.

Röling, W.F., and van Verseveld, H.W. (2002) Natural attenuation: what does the subsurface have in store? *Biodegradation* **13:** 53–64.

Robertson, W.J., Franzmann, P.D., and Mee, B.J. (2000) Spore-forming, *Desulfosporosinus*-like sulphate-reducing bacteria from a shallow aquifer contaminated with gazolene. *J Appl Microbiol* **88:** 248–259.

Sahl, J.W., Schmidt, R., Swanner, E.D., Mandernack, K.W., Templeton, A.S., Kieft, T.L., *et al.* (2008) Subsurface microbial diversity in deep-granitic-fracture water in Colorado. *Appl Environ Microbiol* **74:** 143–152.

Sass, H., Cypionka, H., and Babenzien, H.-D. (1997) Vertical distribution of sulfate-reducing bacteria at the oxic-anoxic interface in sediments of the oligotrophic Lake Stechlin. *FEMS Microbial Ecol* **22:** 245–255.

Schmieder, R., and Edwards, R. (2011) Quality control and preprocessing of metagenomic datasets. *Bioinformatics* **27:** 863–864.

Shah, M. (2014) An application of sequencing batch reactors in microbial degradation of benzene, toluene, & xylene under anoxic and micro aerobic condition. *J Appl Environ Microbiol* **2:** 231–236.

Spring, S., and Rosenzweig, F. (2006) The genera *Desulfitobacterium* and *Desulfosporosinus*: taxonomy. *Prokaryotes.* **4:** 771–786.

Staats, M., Braster, M., and Roling, W.F.M. (2011) Molecular diversity and distribution of aromatic hydrocarbon-degrading anaerobes across a landfill leachate plume. *Environ Microbiol* **13:** 1216–1227.

Sun, W., and Cupples, A.M. (2012) Diversity of five anaerobic toluene degrading microbial communities investigated using stable isotope probing (SIP). *Appl Environ Microbiol* **72:** 972–980.

Sun, W., Sun, X., and Cupples, A.M. (2014) Identification of *Desulfosporosinus* as toluene-assimilating microorganisms from a methanogenic consortium. *Int Biodeter Biodegr* **88:** 13–19.

Tan, B., Jane Fowler, S., Abu Laben, N., Dong, X., Sensen, C.W., Foght, J. *et al.* (2015) Comparative analysis of metagenomes from three methagenomic hydrocarbon-degrading enrichment cultures with 41 environmental samples. *ISME J* **9:** 2028–2045.

Tamura, K., Stecher, G., Peterson, D., Filipski, A., and Kumar, S. (2013) MEGA6: molecular evolutionary genetics analysis version 6.0. *Mol Biol Evol* **30:** 2725–2729.

Taubert, M., Vogt, C., Wubet, T., Kleinsteuber, S., Tarkka, M.T., Harms, H., *et al.* (2012) Protein-SIP enables time-resolved analysis of the carbon flux in a sulfate-reducing, benzene-degrading microbial consortium. *ISME J* **6:** 2291–2301.

Washer, C.E., and Edwards, E.A. (2007) Identification and expression of benzylsuccinate synthase genes in a toluene-degrading methanogenic consortium. *Appl Environ Microbiol* **73:** 1367–1369.

Whitman, W.B., Coleman, D.C., and Wiebe, W.J. (1998) Prokaryotes: the unseen majority. *Proc Natl Acad Sci* **95:** 6578–6583.

Winderl, C., Schaefer, S., and Lueders, T. (2007) Detection of anaerobic toluene and hydrocarbon degraders in contaminated aquifers using benzylsuccinate synthase (*bssA*) genes as a functional marker. *Environ Microbiol* **9:** 1035–1046.

Winderl, C., Penning, H., von Netzer, F., Meckenstock, R.U., and Lueders, T. (2010) DNA-SIP identifies sulfate-reducing *Clostridia* as important toluene degraders in tar-oil-contaminated aquifer sediment. *ISME J* **4:** 1314–1325.

11

Microbial synthesis of a novel terpolyester P(LA-*co*-3HB-*co*-3HP) from low-cost substrates

Yilin Ren,[1] Dechuan Meng,[1] Linping Wu,[2]
Jinchun Chen,[1] Qiong Wu[1] and Guo-Qiang Chen[1,3,4,]*

[1]*Center for Synthetic and Systems Biology, School of Life Science, Tsinghua-Peking Center for Life Sciences, Tsinghua University, Beijing 100084, China.*
[2]*Guangzhou Institutes of Biomedicine and Health, Chinese Academy of Sciences, Guangzhou 510530, People's Republic of China.*
[3]*Center for Nano and Micro Mechanics, Tsinghua University, Beijing 100084, China.*
[4]*MOE Key Lab of Industrial Biocatalysis, Dept Chemical Engineering, Tsinghua University, Beijing 100084, China.*

Summary

Polylactide (PLA) is a bio-based plastic commonly synthesized by chemical catalytic reaction using lactic acid (LA) as a substrate. Here, novel LA-containing terpolyesters, namely, P[LA-*co*-3-hydroxybutyrate (3HB)-*co*-3-hydroxypropionate (3HP)], short as PLBP, were successfully synthesized for the first time by a recombinant *Escherichia coli* harbouring polyhydroxyalkanoate (PHA) synthase from *Pseudomonas stutzeri* (PhaC1$_{Ps}$) with 4-point mutations at E130D, S325T, S477G and Q481K, and 3-hydroxypropionyl-CoA (3HP-CoA) synthesis pathway from glycerol, 3-hydroxybutyryl-CoA (3HB-CoA) as well as lactyl-CoA (LA-CoA) pathways from glucose. Combining these pathways with the PHA synthase mutant phaC1$_{Ps}$ (E130D S325T S477G Q481K), the random terpolyester P(LA-*co*-3HB-*co*-3HP), or PLBP, was structurally confirmed by nuclear magnetic resonance to consist of 2 mol% LA, 90 mol% 3HB, and 8 mol% 3HP respectively.

*For correspondence. E-mail: chengq@mail.tsinghua.edu.cn

Funding information
Plasmids pSEVA351 and pBHR68 were kindly donated by Professor Victor de Lorenzo of CSIC/Spain and Professor Alexander Steinbüchel of Műnster Univ/Germany respectively. This research was financially supported by the independent research program of Tsinghua University (grant no. 2015THZ10) and National Natural Science Foundation of China (grant no. 31430003). The Dutch Polymer Institute also provides partial funding to support this study.

Remarkably, the PLBP terpolyester was produced from low-cost sustainable glycerol and glucose. Monomer ratios of PLBP could be regulated by ratios of glycerol to glucose. Other terpolyester thermal and mechanical properties can be manipulated by adjusting the monomer ratios. More PLBP applications are to be expected.

Introduction

As increasing global warming and plastic pollution threaten human sustainability, materials from renewable biomass are attracting attention due to their biodegradability and environmentally friendliness. Polylactide (PLA) is a representative of bio-based biodegradable polyester synthesized via combination of microbial lactic acid (LA) fermentation and chemical polymerization of lactide (Sudesh and Iwata, 2008; Nampoothiri *et al.*, 2010; Park *et al.*, 2012a,b). PLA has been used in areas of biomedical implants, food packaging and drug delivery. However, the complicated synthetic process including fermentation for LA production, LA purification, lactide esterification and lactide ring-opening polymerization increases the cost, meanwhile the heavy metal residues in the final polymer products could limit its food or medical applications. In addition, PLA poor thermal and mechanical properties are also adverse to its large-scale applications.

Polyhydroxyalkanoates (PHA) are a family of diverse polyesters synthesized by a variety of bacteria as intracellular carbon and energy storage compounds (Li *et al.*, 2007; Chen and Hajnal, 2015; Koller and Rodríguez-Contreras, 2015). As biorenewable and biodegradable materials, the diversity of PHA provides different physical properties to suit various applications (Chen, 2009; Brigham and Sinskey, 2012; Koller, 2014). Copolymerization of LA with other hydroxyalkanoate (HA) monomers via microbial synthesis is one of the effective methods to improve the physical properties of PLA or PHA (Li *et al.*, 2010).

Recently, several engineered PHA synthases able to utilize lactyl-CoA (LA-CoA) and 3-hydroxybutyric-CoA (3HB-CoA) as substrates were reported (Taguchi *et al.*, 2008; Yang *et al.*, 2011; Ochi *et al.*, 2013). To deliver LA-CoA in recombinant *Escherichia coli*, propionyl-CoA transferases (Pct) in alanine fermentation pathway of several organisms including *Clostridium propionicum* and *Megasphaera elsdenii* were expressed in host strains respectively. A 6 mol% LA fraction in the copolyester was achieved.[15] Subsequently, efforts were made on increasing the LA ratio in the copolyester by regulating the metabolic flux or evolving the PHA synthase (Jung *et al.*, 2010; Yamada *et al.*, 2010; Shozui *et al.*, 2011), and also by the uses of different organisms in addition to commonly used *E. coli*, including *Corynebacterium glutamicum*, *Ralstonia eutropha* and *Sinorhizobium meliloti* (Song *et al.*, 2012; Park *et al.*, 2013; Tran and Charles, 2015). In addition, low-cost carbon substrates, such as glucose and xylose, were applied to synthesize LA-based copolyesters (Park *et al.*, 2012a,b; Nduko *et al.*, 2014; Salamanca-Cardona *et al.*, 2014), which could reduce production cost and facilitate its industrialization.

As mentioned, PLA has the major deficiencies of poor flexibility, ductility and thermal resistance, and copolymerization is possibly effective to improve the PLA properties. Therefore, LA copolyesters were investigated with P(LA-co-3HB) as a representative (Taguchi *et al.*, 2008; Jung *et al.*, 2010; Yamada *et al.*, 2010; Shozui *et al.*, 2011; Yang *et al.*, 2011; Ochi *et al.*, 2013). However, mechanical properties of PHB are similar to PLA, therefore, other monomers such as 3-hydroxyvalerates (3HV), 3-hydroxyhexanoate (3HHx) and glycolate (GA) were introduced into the LA copolyesters for property improvements (Shozui *et al.*, 2010a,b; Choi *et al.*, 2016; Li *et al.*, 2016). Remarkably, poly(3-hydroxypropionate) (P3HP), a relatively new PHA family member, has become very interesting due to its strong mechanical properties including an elongation at break of more than 600% and Young's modulus of 3 GPa (Andreeßen and Steinbüchel, 2010; Zhou *et al.*, 2011a,b; Meng *et al.*, 2012). Thus, 3HP monomers in LA copolyesters could very likely compensate for the shortcomings of PLA. This study attempted to biosynthesize a LA-containing terpolyester consisting of LA, 3HP and 3HB with improved properties over PLA homopolyester.

Results and discussion

Engineering a pathway for biosynthesis of P(LA-co-3HB-co-3HP) from unrelated carbon source

For the above-mentioned purpose, the substrate specificity of PHA synthase is the most important factor determining the monomer constituents incorporated into PHA. A mutant PHA synthase from *Pseudomonas stutzeri* strain 1317 (PhaC1$_{Ps}$) with 4-point mutations at E130D, S325T, S477G and Q481K was prepared as stated below, which could incorporate LA and other 3-HAs into PLBP terpolyester. In addition, a 3HP synthetic pathway

combined with LA and 3HB pathways were constructed to supply the monomers for the terpolyester synthesis (Fig. 1).

The engineering pathway allows adjustments of 3HB, LA and 3HP monomer ratios in the terpolyesters by feeding various ratios of glucose to glycerol. When a higher LA ratio is needed, LA can be added to the culture to supply additional LA precursor.

Engineering a PHA synthase able to polymerize LA

The substrate specificity into PHA synthase is the most important factor determining the monomer constituents incorporated into PHA. Common PHA synthases are capable of polymerizing the 3-hydroxyalkanoates (3HAs) CoA with variable carbon chain lengths (Bernd, 2003). LA, which is the monomer of PLA, is a 2HA, which is not accepted by natural PHA synthases. To synthesize PLBP terpolyesters, a mutant of PHA synthase, able to incorporate LA and other 3-HAs at the same time was obtained via site-directed mutagenesis on type II PHA synthase PhaC1$_{Ps}$ of *P. stutzeri* 1317 which has already a versatile substrate specificity (Park *et al.*, 2015). Five mutation sites on phaC1$_{ps}$ (phaC2$_{ps}$), including E130D,

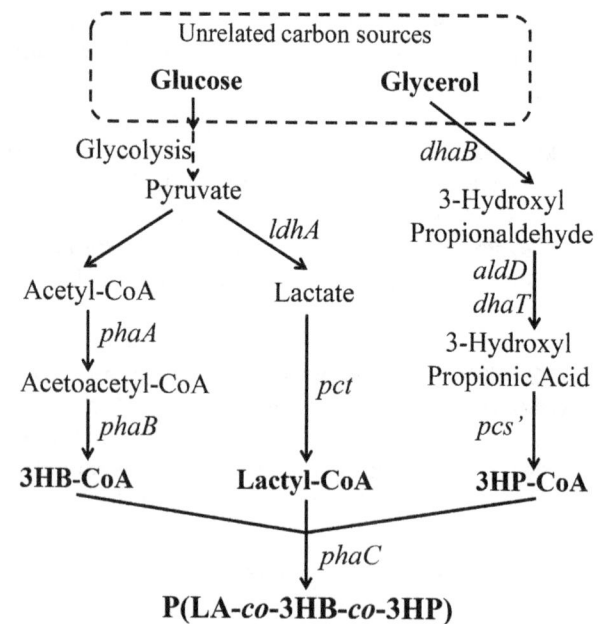

Fig. 1. Metabolically engineered pathways for production of terpolyester P(LA-*co*-3HB-*co*-3HP) or PLBP from unrelated carbon source. Enzymes encoded by each gene are described below: *phaA*, β-ketothiolase; *phaB*, NADPH-dependent acetoacetyl-CoA reductase; *ldhA*, lactate dehydrogenase; *pct*, propionyl-CoA transferase; *dhaB*, glycerol dehydratase; *dhaT*, 1,3-propanediol dehydrogenase; *aldD*, aldehyde dehydrogenase; *pcs'*, propanoyl-CoA synthetase; *phaC*, PHA synthase from *Pseudomonas stutzeri* strain 1317 (PhaC1$_{Ps}$) with 4-point mutations at E130D, S325T, S477G and Q481K.

S325(326)T, F392(393)S, S477(478)G and Q481(482)K, were selected based on the alignment result with the previous LA-polymerizing enzymes (Fig. 2) (Taguchi *et al.*, 2008; Jung *et al.*, 2010; Yamada *et al.*, 2010; Yang *et al.*, 2011; Chuah *et al.*, 2013; Ochi *et al.*, 2013). With various combinations of these site mutations, up to 20 *phaC1_Ps* and *phaC2_Ps* variants were constructed (Table S1). To compare their activities towards LA polymerization, the 3HB-CoA synthetic pathway including genes of β-ketothiolase (*phaA*), NADPH-dependent acetoacetyl-CoA reductase (*phaB*) of *R. eutropha*, LA-CoA synthetic pathway including lactate dehydrogenase (*ldhA*) of *E. coli* and propionyl-CoA transferase (*pct*) of *Clostridium propionicum* were constructed into a series of plasmids termed pBLPCAB-X (X represents the specific *phaC* variant in the plasmid). pBLPCAB-Xs were derived from pBHR68 (kindly donated by Prof Steinbüchel of Münster Univ/Germany), with the three genes of *ldhA*, *pct* and *phaC* variant inserted downstream the P_re promoter. The LA polymerization activities of typical *phaC* variants were determined by expressing the corresponding plasmids in *E. coli* S17-1 (Table 1).

When analysing the PHA accumulation capacity, single-point mutation on *phaC_Ps* was found insufficient for LA polymerizations (data not shown). By combining Q481K and S325T mutations in *phaC1_Ps*, the recombinant started to synthesize P(2.81% LA-*co*-3HB) copolyester, confirming two-point mutant *phaC1_Ps* (Q481K S325T) capable of polymerizing LA into PHA. Additional mutations on E130D and S477G to the above two points mutant increased the LA specificity as the LA ratio in the PHA copolyester produced by the recombinant expressing *phaC1_Ps* (Q481K S325T E130D S477G) increased to over 5% (Table 1). However, mutation F392S had a negative effect on LA incorporation, as F392S added to *phaC1_Ps* (Q481K S325T E130D S477G) reduced LA ratio in the copolyester to 3% (Table 1). Interestingly, *phaC2_Ps* showed no activity towards LA-CoA no matter what point mutations were introduced. As a result, the LA polymerizing mutant enzyme PhaC1_Ps (Q481K

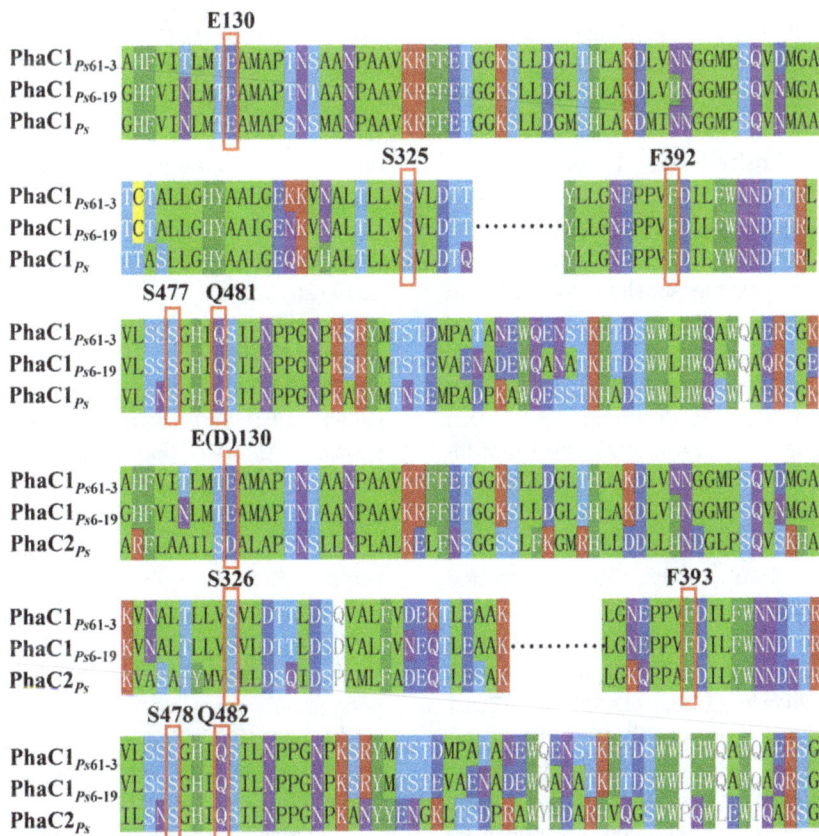

Fig. 2. Alignment of *phaC1_Ps* and *phaC2_Ps* with previous LA polymerizing enzymes reported (Taguchi *et al.*, 2008; Yamada *et al.*, 2010; Yang *et al.*, 2011). *phaC1_Ps* and *phaC2_Ps* from *Pseudomonas stutzeri* were aligned with reported LA polymerizing enzymes which were *phaC1_Ps61-3* from *Pseudomonas sp.* 61-3 and *phaC1_Ps6-19* from *Pseudomonas sp.* MEBL6-19. Several constitutive sites were found dominating the LA polymerizing capacity (in the red frames). Abbreviations: phaC1_Ps, *Pseudomonas stutzeri* phaC1; phaC2_Ps, *Pseudomonas stutzeri* phaC2; phaC1_Ps61-3, *Pseudomonas sp.* 61-3 phaC1; phaC1_Ps6-19, *Pseudomonas sp.* MEBL6-19 phaC1; E, glutamic acid; D, aspartic acid; S, serine; F, phenylalanine; Q, glutamine. Five mutation sites on *phaC1_ps* (*phaC2_ps*), including E130D, S325(326)T, F392(393)S, S477(478)G and Q481(482)K, were selected based on the alignment result with the LA polymerizing enzymes.

Table 1. LA polymerization activities of typical *phaC* variants.

E. coli	phaC$_{Ps}$ variants	CDM (g L^{-1})	PHA/CDM (wt%)	LA (mol%)
S-pBLPCAB1	phaC1$_{Ps}$	4.03 ± 0.27	39.22 ± 0.58	0
S-pBLPCAB1-2	phaC1$_{Ps}$ (Q481K S325T)	4.18 ± 0.13	40.12 ± 0.23	2.81 ± 0.30
S-pBLPCAB1-4	phaC1$_{Ps}$ (Q481K S325T E130D S477G)	3.13 ± 0.31	36.94 ± 0.38	5.01 ± 1.24
S-pBLPCAB1-5	phaC1$_{Ps}$ (Q481K S325T E130D S477G F392S)	3.01 ± 0.22	37.51 ± 0.23	3.01 ± 0.72
S-pBLPCAB2	phaC2$_{Ps}$	3.24 ± 0.09	41.50 ± 0.61	0
S-pBLPCAB2-2	phaC2$_{Ps}$ (Q482K S326T)	4.15 ± 0.43	37.71 ± 0.49	0
S-pBLPCAB2-3	phaC2$_{Ps}$ (Q482K S326T S478G)	3.76 ± 0.52	36.13 ± 0.77	0

Recombinant strains were cultivated for 48 h in shake flasks. The data are the averages of three parallel experiments. LA, lactate; CDM, cell dry mass.

S325T E130D S477G) with the highest efficiency was obtained compared with other *phaC* wild type or mutant enzymes (Table 1 and Table S2). We therefore named the plasmid pBLPCAB1-4-containing *phaC1$_{Ps}$* (Q481K S325T E130D S477G) plasmid pLA in further studies.

Construction of an effective PLBP synthetic system

Aimed to produce a novel terpolyester PLBP, three synthetic pathways for each constituent were constructed (Fig. 1), including PHB and PLA synthetic routes that were reported. The P3HP synthetic pathway was focused on as it would improve the mechanical properties of the terpolyesters.

P3HP could be synthesized from 1, 3-propanediol, glycerol and glucose (Andreeßen and Steinbüchel, 2010; Zhou et al., 2011a,b; Meng et al., 2015). Glycerol was chosen in this study to regulate 3HP ratio in the copolyesters, whereas glucose was the substrate for PLA and 3HB synthesis. The 3HP synthetic pathway from glycerol consisted of genes encoding glycerol dehydratase (*dhaB*) of *Klebsiella pneumoniae*, 1,3-propanediol dehydrogenase (*dhaT*) and aldehyde dehydrogenase (*aldD*) of *Pseudomonas putida* KT2442 and ACS domain of trifunctional propionyl-CoA synthetase (*pcs'*) functioning as a CoA ligase of *Chloroflexus aurantiacus* (Andreeßen and Steinbüchel, 2010; Zhou et al., 2011a,b; Meng et al., 2012). The p3HP1p plasmid was constructed based on the pSEVA351 for P3HP production and contained genes of *dhaT, aldD, dhaB* and *pcs'* (Silva-Rocha et al., 2013). Gene fragments *dhaT-aldD, pcs* and *dhaB* were amplified from plasmid pZQ03, pDC02 and pZQ01 respectively. Subsequently, the pSEVA351 backbone was ligated with these three fragments by Gibson assembly. When strengthen the expression of dominant gene *dhaB* controlling glycerol utilization, the optimized plasmid p3HP2p exhibited enhanced efficiency in P3HP synthesis when it was co-expressed with pBHR68 in recombinant *E. coli* S17-1 (Table S3).

Finally, recombinant *E. coli* S-LA-3HP harbouring plasmids pLA and p3HP2p was obtained. When cultivated in LB medium supplemented with 20 g L^{-1} glucose and 10 g L^{-1} glycerol, a new terpolyester P(90.41 mol% 3HB-co-7.78 mol% 3HP-co-1.81 mol% LA) was successfully synthesized for the first time from unrelated carbon sources glucose and glycerol.

Nuclear magnetic resonance analyses of the PLBP terpolyester

The composition and monomer sequence distribution of P(3HB-co-3HP-co-LA) was confirmed by nuclear magnetic resonance (NMR) (Fig. 3). From the ^{1}H NMR spectra (Fig. 3A), there were not only well-characterized proton resonances of B(2), B(3) and B(4) in 3HB units (3HB abbreviated as B, 3HP monomer abbreviated as P and LA abbreviated as A) but also additional four proton resonances assigned to the 3HP units such as P(2), P(3) and LA units including A(2) and A(3) with identical intensities based on the previous studies (Park et al., 2013, 2015; Meng et al., 2015). The molar ratio of 3HB, 3HP and LA in the PHA copolyester was 90.41%, 7.78% and 1.81%, respectively, as calculated using the intensity of B(3), P(3) and A(2). The individual carbon species in 3HB, 3HP and LA monomer were also identified by specific ^{13}C NMR (Fig. 3B). The expanded spectra of individual splitting resonance of carboxyl carbon B(1), P(1) and A(1) in the copolyester were split into multiple peaks (Fig. 3C), which were assigned to the 3HB-centred (B(1)*B), 3HP-centred (P(1)*P), LA-centred (A(1)*A) and the three units comonomer sequences [B(1)*P(1)*A(1)] (N*M represents the interaction of monomer N and M) (Meng et al., 2015; Park et al., 2015). This phenomenon was due to different sequence arrangements of 3HB, 3HP and LA monomers in the polymer chains, which is common in random PHA copolyesters (Hu et al., 2011; Tripathi et al., 2012).

The tacticity distribution of tri-block copolyester was studied via ^{13}C NMR based on various stereosequences. The good resolution of methylene regions B(2) was chosen as an example for analysis. Two sharp peaks corresponding to B(2)*B and B(2)*P*A were observed (Fig. 3D). No split in each peak was visible, indicating that the stereo-sequence was isotactic. If the polymer is syndiotactic or atactic, some diad or quadruple peaks should be observed in ^{13}C NMR (Kemnitzer et al., 1993; Hocking and Marchessault, 1995). The above detailed analysis of ^{1}H NMR and ^{13}C NMR spectra clearly indicated that PHA was a random copolyester P(3HB-co-3HP-co-LA) with an isotactic microstructure.

Regulation of the monomer ratios in PLBP terpolyesters

Recombinant E. coli S17-1 harbouring genes described in Fig. 1 could produce terpolyesters PLBP consisting of

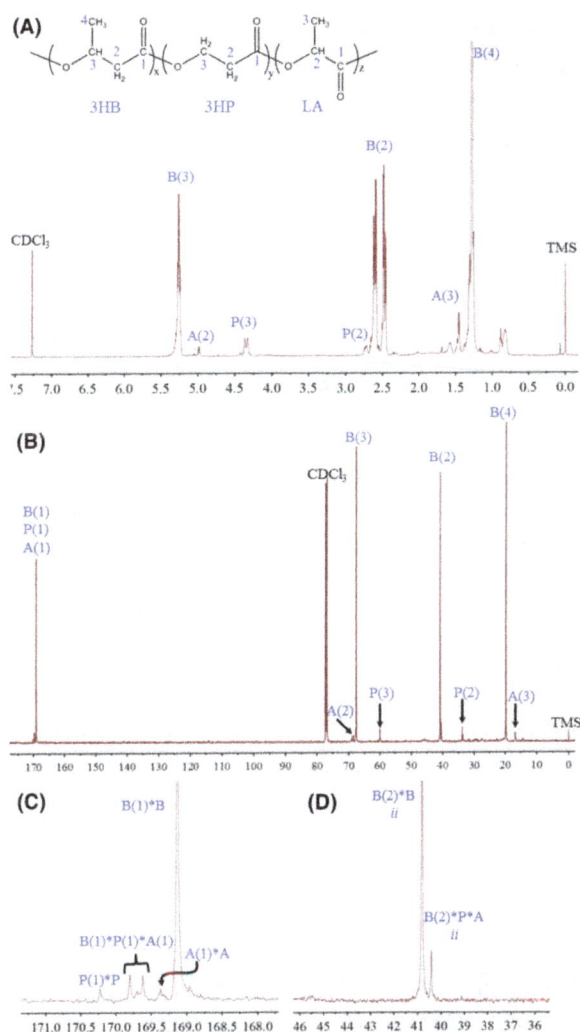

Fig. 3. NMR analysis of PLBP terpolyester. ^{1}H NMR spectra (A) and ^{13}C NMR spectra (B) of random copolyester P(3HB-co-3HP-co-LA) containing 90.41 mol% 3HB, 7.78 mol% 3HP and 1.81 mol% LA, respectively, and its expanded ^{13}C NMR spectra of carboxyl carbon [B(1), P(1), A(1)] area (C) and methylene regions (D) in the terpolyester. B, P and A refer to 3HB, 3HP and LA; numbering schemes were the same as molecular formulations of polyester indicated in the inset in (A). N*M represents the interaction of monomer N and M. "i" indicates "isotactic." Chemical shifts are in ppm and tetramethylsilane (TMS) was employed as an internal chemical shift standard.

LA, 3HB and 3HP when grown in a mixture of glucose and glycerol as substrates (Table 2). The terpolyesters PLBP were fully synthesized from unrelated carbon sources, which was more economic for PHA synthesis (Hermann-Krauss et al., 2013; Povolo et al., 2013). Compositions of monomers LA, 3HB and 3HP in PLBP could be adjusted by changing the glucose-to-glycerol ratios.

Lactic acid ratio was as low as 1.81% in the terpolyester when PLBP was produced by the wild-type E. coli S17-1 harbouring genes described in Fig. 1. To increase the LA ratio in PLBP, the synthetic pathway was optimized and the competitive pathways of LA were deleted. Gene ldhA was inserted in pLA plasmid and was overexpressed to produce more LA as ldhA functions in converting pyruvate into LA (Jung et al., 2010). However, it was found that overexpression of ldhA alone could not enhance the LA content in the terpolyester (Table S4) as a gene dld could convert LA back into pyruvate (Dym et al., 2000; Choi et al., 2016). Unexpectedly, the recombinant strain-containing plasmid pLA' without ldhA, replacing pLA plasmid of the PLBP synthetic system, synthesized a PLBP terpolyester with an increased LA ratio under the same culture conditions (Fig. S1.2). The deletion of ldhA in pLA plasmid resulted in a transcriptional change in those genes downstream the same promoter for the ones that were closer to the promoter (Fig. S1.1). It was proven that the transcriptional level of pct was correlated with the LA ratio in PLBP terpolyesters from the RT-PCR results (Fig. S1.2). It suggested that PCT-mediated CoA transfer reaction that transformed LA into LA-CoA was the time-limiting step of LA polymerization. Meanwhile, to improve LA production, a competitive pathway, pyruvate–formate pathway, was weakened. Gene pflA, the activator of pyruvate–formate lyase, was knocked out in E. coli S17-1 (Zhu and Shimizu, 2004; Shozui et al., 2010b; Zhou et al., 2011a,b). The gene pflA knockout lead to an obvious decrease in formate production of the strain, whereas the lactate concentration in the culture medium increased oppositely (Fig. S2).

Escherichia coli S17-1 ΔpflA harbouring two plasmids of pLA' and p3HP2p was constructed and studied in shake flasks using different substrates combinations (Table 2 and Fig. 4). Obviously, the LA ratio in PLBP terpolyester produced by this recombinant was increased compared with the wild-type E. coli S17-1. The substrates added to the culture were utilized simultaneously indicating the synthesis of random copolyesters (Fig. 4). In general, several tendencies could be summarized from the data: there was a positive correlation between LA ratio and glucose concentration. Especially when glucose concentration was decreased to 10 g L^{-1}, the LA ratio was down to 0.79% because most of the glucose

Table 2. PLBP production using various substrate concentrations by *E. coli* S17-1 *ΔpflA* harbouring two plasmids of pLA' and p3HP2p.

Glu (g L⁻¹)	Gly (g L⁻¹)	LA (g L⁻¹)	CDM (g L⁻¹)	PHA (wt%)	3HP (mol%)	LA (mol%)
20	10	0	3.70 ± 0.12	40.94 ± 0.44	9.85 ± 1.43	10.74 ± 2.22
20	5	0	3.88 ± 0.3	45.17 ± 1.08	8.04 ± 0.77	13.10 ± 1.23
20	2	0	4.79 ± 0.08	53.31 ± 2.12	5.42 ± 0.8	9.18 ± 0.31
30	10	0	5.21 ± 0.72	59.04 ± 1.07	8.93 ± 0.87	11.53 ± 2.15
10	10	0	3.05 ± 0.41	38.67 ± 1.7	12.03 ± 0.39	0.79 ± 0.55
20	10	2	1.43 ± 0.19	45.06 ± 3.50	10.27 ± 0.69	23.54 ± 1.95
20	10	5	0.98 ± 0.23	42.67 ± 0.56	8.95 ± 2.3	27.78 ± 0.72

The recombinant was cultivated in LB medium supplemented with different concentrations of glucose (glu), glycerol (gly) and LA for 48 h in shake flasks. The data are the averages of three parallel experiments.
CDM, cell dry mass; PHA, the terpolyester; 3HP, 3-hydroxypropionate; LA, lactate.

was utilized for cell growth; when extra D, L-LA was added into the medium, LA ratio increased sharply to more than 27%, whereas the cell dry mass (CDM) was deceased oppositely due to the toxicity of LA; as the substrate of P3HP, glycerol affected 3HP ratio in PLBP terpolyester in a similar relationship as that between glucose and LA ratio. Interestingly, 3HP ratio reached its peak value of 12.03% when glucose was reduced to 10 g L⁻¹. Synthesis of PLBP terpolyester with variable compositions was achieved.

At a high glucose concentration of 30 g L⁻¹, cells grew to over 5.2 g L⁻¹ containing close to 60% PHA, this was both the highest cell growth and the highest PHA accumulation compared with other low glucose concentration (Table 2). The terpolyester consisted of 9% 3HP, 12% LA and 79% 3HB. The result indicated that high concentration of glucose favoured formation of cell mass and PHA, especially PHB. At the lowest glycerol concentration of 2 g L⁻¹, 3HP had the lowest content of 5% in the terpolyester. LA addition improved LA ratios in the terpolyester, but it had negative effect on cell growth (Table 2). Hence, several strategies exist to

change the composition of the monomer level by adjusting the ratio of fed carbon sources.

Physical characterization of PLBP with different monomer compositions

Two types of PLBPs were extracted and the compositions of them were determined by GC. The physical properties of PLBPs including molecular mass, thermal properties and mechanical properties were studied (Table 3). The weight average molecular mass of PLBP was less than 2×10^5, which was lower compared with its homopolyester and P(LA-*co*-3HB) copolyester, yet approximated the same as weight average molecular mass of PLBV terpolyesters (Shozui *et al.*, 2010b). In addition, PLBPs inherited the ductility of P3HP as it had an improved elongation at break of over 100% when the 3HP ratio reached 15%. The thermal parameters including T_m, T_g and ΔH_m of PLBP were in between of those of individual homopolyesters (PLA, P3HB and P3HP) as blocks. Interestingly, PLBPs had two T_m and the lower one was close to the T_m of PHB. Possibly, the presence of 3HB-rich segments in the terpolyester led to this phenomenon. The incorporation of 3HP into P(LA-*co*-3HB) significantly improved the tensile strength and elongation at break compared with its copolyester P(LA-*co*-3HB).

Conclusions

Recombinant *E. coli* S17-1 *ΔpflA* consisting of three synthetic pathways of lactyl-CoA, 3-hydroxypropionyl-CoA and 3-hydroxybutyryl-CoA was able to polymerize the three monomers to form random terpolyester P(LA-*co*-3HB-*co*-3HP) or PLBP under the catalysis of phaC1ₚₛ (Q481K S325T E130D S477G) cloned from *P. stutzeri* strain 1317 with four specific point mutations. The terpolyester compositions can be controlled by changing the ratios of three substrates including glucose, glycerol and LA. The terpolyesters have changing properties depending on monomer ratios,

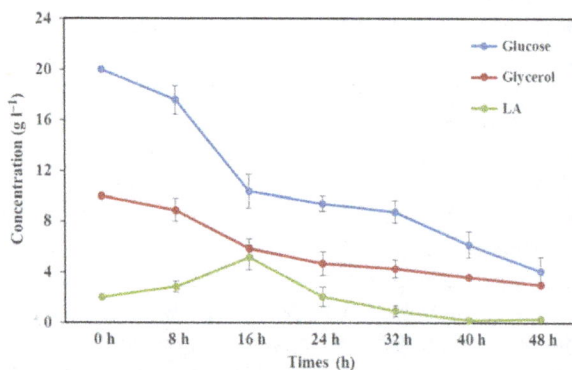

Fig. 4. The concentration of glucose, glycerol and lactate in shake-flask studies. *E. coli* S17-1 *ΔpflA* harbouring two plasmids of pLA' and p3HP2p was cultured in LB medium supplemented with 20 g L⁻¹ glucose, 10 g L⁻¹ glycerol and 2 g L⁻¹ lactate. Blue line, glucose; red line, glycerol; green line, lactate. Error bars represent the standard deviation of experiments conducted in triplicates.

especially an enhanced elongation at break compared with the homopolyesters of 3HB and LA as well as their copolyesters.

Experimental procedures

Microorganism, plasmid and shake-flask culture conditions

All the microorganisms and plasmids used in this study are listed in Table 4. *E. coli* Trans 1T1 and JM109 were used as the host strains for genetic manipulation. *E. coli* S17-1 was used for polymer production (Simon *et al.*, 1983). All the *E. coli* strains were cultured in LB medium. LB medium contains (g L^{-1}): 10 tryptone, 5 yeast extract and 10 NaCl. Glycerol and glucose were added into LB medium as carbon sources. The seed cultures stored at $-80°C$ were inoculated into LB medium and cultivated at 37°C for 12 h at 200 rpm/min on a rotary shaker for reactivation (BBT-14-BJQH042, INFORS HT, Hong Kong, China). Subsequently, the seed cultures were inoculated into 500 mL conical flasks containing 50 mL LB medium with an inoculation volume ratio of 5%.

Table 3. Physical characterization of various PLBP terpolyesters.

LA	3HB	3HP	M_w (10^4)	M_n (10^4)	M_w/M_n	Tensile strength (MPa)	Young's modulus (MPa)	Elongation at break (%)	T_g (°C)	T_m (°C)	ΔH_m (J g^{-1})
100[e]	0	0	20	–	–	52 ± 2	1020	2	60	153	9.2
0	100[f]	0	–	–	–	18 ± 0.7	1470 ± 78	3 ± 0.4	7.1	131.8	–
0	0	100[g]	–	30	–	28.3	333.3	683.5	−21.5	78	54
15[h]	85	0	82	34.2	2.4	10 ± 0	194 ± 5	75 ± 2	−9, 19	149, 167	0.6, 3.2
1.8	90.4	7.8[i]	15.3	9.1	1.68	18 ± 2	332.8 ± 8.5	15.3 ± 4.5	−5.9	129, 150	3.3, 32
7.2	79.8	13[j]	11.7	7.6	1.54	12.5 ± 1.3	231.4 ± 9.7	100.9 ± 12	−2	132, 154	2.7, 39

Monomer ratio[a] (mol %); Molecular mass[b]; Mechanical properties[c]; Thermal properties[d]

[a]Determined by gas chromatography.
[b]M_w, weight-averaged Molecular mass; M_n, number-averaged Molecular mass; M_w/M_n; polydispersity; the unit of M_w and M_n is Da.
[c]The values are the averages of at least three independent measurements.
[d]T_g, glass-transition temperature; T_m, melting temperature; ΔH_m, enthalpy of fusion.
[e]PLA was chemically synthesized (Zaman *et al.*, 2011).
[f]P3HB was synthesized by bacteria (Li *et al.*, 2011).
[g]P3HP was synthesized by bacteria (Zhou *et al.*, 2011a,b).
[h]P(LA-*co*-3HB) was produced by recombinant *E. coli* (Yamada *et al.*, 2011).
[i]Sample weight of P(90.4 mol% 3HB-co-7.8 mol% 3HP-co-1.8 mol% LA) was 18.2 mg.
[j]Sample weight of P(79.8 mol% 3HB-co-13 mol% 3HP-co-7.2 mol% LA) was 17.5 mg.

Table 4. Strains and plasmids used in this study.

Strains/plasmids	Description	Reference/source
E. coli Trans1-T1	Expression host	TransGen Biotech
E. coli S17-1	*recA*, harbours the *tra* genes of plasmid RP4 in the chromosome; *proA*, *thi-1*	Simon *et al.* (1983)
S-NC	*E. coli* S17-1 harbours pBluescript SK$^-$ plasmid	This study
S-BL	*E. coli* S17-1 harbours pBL plasmid	This study
S-LA	*E. coli* S17-1 harbours pLA plasmid	This study
S-LA'	*E. coli* S17-1 harbours pLA' plasmid	This study
pBHR68	Derivative of pBluescript SK$^-$ containing the 5.2-kb *SmaI/EcoRI* fragment comprising the PHA operon from *Ralstonia eutropha*	Spiekermann *et al.* (1999)
pSEVA351	Cloning vector, RSF1010 replicon, Cm^R	Silva-Rocha *et al.* (2013)
pBluescript SK$^-$	The commonly used commercial plasmid	TransGenBiotech
pZQ 03	Derivative of pBHR68, *phaC* and *pcs'* under the control of *lac* promoter, Amp^R	Zhou *et al.* (2011a,b)
pDC02	Derivative of pBHR68, *phaC*, *dhaB*, *gpp*, *gpd*, *gdrAB* and *pduP* under the control of P$_{Re}$ promoter derived from *Ralstonia eutropha pha* operon, Amp^R	Meng *et al.* (2015)
pLA	Derivative of pBHR68, *phaC1$_{Ps}$* (S325T Q481K E130D S477G), *pct* and *ldhA* was inserted into backbone	This study
pLA'	Derivative of pBHR68, *phaC1$_{Ps}$* ((S325T Q481K E130D S477G) and *pct* was inserted into backbone	This study
p3HP1p	Derivative of pSEVA351, *dhaB*, *dhaT*, *aldD* and *pcs'* was inserted into backbone downstream 1 P$_{re}$ promoters	This study
p3HP2p	Derivative of pSEVA351, *dhaB*, *dhaT*, *aldD* and *pcs'* was inserted into backbone with 2 P$_{re}$ promoters	This study
pBL	Derivative of pBluescript SK$^-$, *ldhA* was inserted into backbone	This study

When an antibiotic selection pressure was required, the medium was supplemented with ampicillin (100 μg mL^{-1}), kanamycin (50 μg mL^{-1}) or chloramphenicol (34 μg mL^{-1}). To increase the LA ratio in the terpolyester, various amounts of D, L-LA (1, 2 or 5 g L^{-1}) were added into the LB medium along with NaOH-modulating pH to neutral. When P3HP was the constituent of the polymer, 5 μM vitaminB$_{12}$ (VB$_{12}$) was added into the medium to maintain the activity of glycerol dehydratase (dhaB).

PHA analysis, extraction and purification

Cells were harvested by centrifugation (CR21 GIII; Hitachi, Tokyo, Japan) at 10 000 rpm/min for 8 min, then washed with distilled water and centrifuged again. CDM were measured after the concentrated cells were lyophilized at −65°C with five times air pressure (LGJ-10C; SiHuanKeXue, Beijing, China). Thirty to fourty milligrams of lyophilized cells was used for the transesterification reaction in which 2 mL of transesterification mixture and 2 mL of chloroform were added in each transesterification test tube (Kato et al., 1996). After a 4 h transesterification at 100°C, the PHA content and monomer compositions of the cells were assayed by gas chromatograph (GC-2014; Shimadzu, Suzhou, China) (Ouyang et al., 2007). The intracellular polymers were extracted using a Soxhlet extractor (Soxtec 2050; Foss, Hilleroed, Denmark). The extracted PHA was purified via precipitation when mixed with the 10-folds volume of ice-cold ethanol and dissolved in chloroform for film casting.

Metabolic flux analysis

Concentrations of LA, glycerol and glucose were determined using a high-performance liquid chromatography (HPLC) (LC-20A; Shimadzu) equipped with an ion exchange column (Aminexs HPX-87H; Bio-Rad, 7.8 × 300 mm^2, Hercules, California, USA) and a refractive index detector (RID-10A; Shimadzu). Gene pct transcriptional level was assayed using RT-PCR. Total RNA was extracted from recombinant E. coli strains using RNA prep pure Cell/Bacteria Kit (Tiangen, Beijing, China). cDNA was synthesized using Fastquant RT Kit (Tiangen) and then real-time PCR (RT-PCR) was carried out for mRNA analysis with SuperReal PreMix (SYBR Green; Tiangen). 16S rRNA was used as the inner standard. The experimental procedures were described (Lv et al., 2015).

NMR analysis on PHA

The ^{1}H and ^{13}C spectra were obtained using a JEOL JNM-ECA 600 NMR spectrometer to measure the polymer compositions, the chemical microstructures and the monomer sequences. Tetramethylsilane was used as the internal standard.

Molecular mass and other properties of PHA

Molecular mass was determined via gel permeation chromatography (GPC Spectra System P2000; Shimadzu) equipped with a Shimadzu HSG60 column at 40°C. The melting temperature (T_m), enthalpy of fusion (ΔH_m) and glass-transition temperature (T_g) were measured via differential scanning calorimetry (DSC-60; Shimadzu) in a temperature ranging from −80°C to 200°C under a nitrogen atmosphere of 50 mL min^{-1}. Thermal stabilities of the materials were studied by a thermogravimetric analyser (TA-Q50; TA Instrument, New Castle, Delaware, USA). Three to five milligrams of each sample was loaded at temperature ranging from 10 to 400°C in a nitrogen atmosphere of 50 mL min^{-1} (Shen et al., 2009). Mechanical properties were studied using a materials testing machine (INSTRON 3365; Instron, Grove City, Ohio, USA) at room temperature at a speed of 5 mm min^{-1}.

References

Andreeßen, B., and Steinbüchel, A. (2010) Biosynthesis and biodegradation of 3-hydroxypropionate-containing polyesters. Appl Environ Microbiol 76: 4919–4925.

Bernd, H. (2003) Polyester synthases: natural catalysts for plastics. Biochem J 376: 15–33.

Brigham, C.J., and Sinskey, A.J. (2012) Applications of polyhydroxyalkanoates in the medical industry. Int J Biotechnol Wellness Ind 1: 52.

Chen, G.Q. (2009) A microbial polyhydroxyalkanoates (PHA) based bio- and materials industry. Chem Soc Rev 38: 2434–2446.

Chen, G.-Q., and Hajnal, I. (2015) The 'PHAome'. Trend Biotechnol 33: 559–564.

Choi, S.Y., Park, S.J., Kim, W.J., Yang, J.E., Lee, H., Shin, J., and Lee, S.Y. (2016) One-step fermentative production of poly (lactate-co-glycolate) from carbohydrates in Escherichia coli. Nat Biotechnol 34: 435–440.

Chuah, J.-A., Tomizawa, S., Yamada, M., Tsuge, T., Doi, Y., Sudesh, K., and Numata, K. (2013) Characterization of site-specific mutations in a short-chain-length/medium-chain-length polyhydroxyalkanoate synthase: in vivo and in vitro studies of enzymatic activity and substrate specificity. Appl Environ Microb 79: 3813–3821.

Dym, O., Pratt, E.A., Ho, C., and Eisenberg, D. (2000) The crystal structure of D-lactate dehydrogenase, a peripheral membrane respiratory enzyme. Proc Natl Acad Sci USA 97: 9413–9418.

Hermann-Krauss, C., Koller, M., Muhr, A., Fasl, H., Stelzer, F., and Braunegg, G. (2013) Archaeal production of polyhydroxyalkanoate (PHA) co- and terpolyesters from biodiesel industry-derived by-products. Archaea 2013: 465–466.

Hocking, P.J., and Marchessault, R.H. (1995) Microstructure of Poly [(R, S)-. beta.-hydroxybutyrate] by 13C NMR. Macromolecules 28: 6401–6409.

Hu, D., Chung, A.-L., Wu, L.-P., Zhang, X., Wu, Q., Chen, J.-C., and Chen, G.-Q. (2011) Biosynthesis and characterization of polyhydroxyalkanoate block copolymer P3HB-b-P4HB. *Biomacromolecules* **12**: 3166–3173.

Jung, Y.K., Kim, T.Y., Park, S.J., and Lee, S.Y. (2010) Metabolic engineering of *Escherichia coli* for the production of polylactic acid and its copolymers. *Biotechnol Bioeng* **105**: 161–171.

Kato, M., Bao, H., Kang, C.-K., Fukui, T., and Doi, Y. (1996) Production of a novel copolyester of 3-hydroxybutyric acid and medium-chain-length 3-hydroxyalkanoic acids by *Pseudomonas* sp. 61-3 from sugars. *Appl Microbiol Biotechnol* **45**: 363–370.

Kemnitzer, J.E., McCarthy, S.P., and Gross, R.A. (1993) Preparation of predominantly syndiotactic poly (. beta.-hydroxybutyrate) by the tributyltin methoxide catalyzed ring-opening polymerization of racemic. beta.-butyrolactone. *Macromolecules* **26**: 1221–1229.

Koller, M. (2014) Poly (hydroxyalkanoates) for food packaging: application and attempts towards implementation. *Appl Food Biotechnol* **1**: 3–15.

Koller, M., and Rodríguez-Contreras, A. (2015) Techniques for tracing PHA-producing organisms and for qualitative and quantitative analysis of intra-and extracellular PHA. *Eng Life Sci* **15**: 558–581.

Li, R., Zhang, H., and Qi, Q. (2007) The production of polyhydroxyalkanoates in recombinant *Escherichia coli*. *Bioresour Technol* **98**: 2313–2320.

Li, Z.-J., Shi, Z.-Y., Jian, J., Guo, Y.-Y., Wu, Q., and Chen, G.-Q. (2010) Production of poly (3-hydroxybutyrate-co-4-hydroxybutyrate) from unrelated carbon sources by metabolically engineered *Escherichia coli*. *Metab Eng* **12**: 352–359.

Li, S.Y., Dong, C.L., Wang, S.Y., Ye, H.M., and Chen, G.-Q. (2011) Microbial production of polyhydroxyalkanoate block copolymer by recombinant *Pseudomonas putida*. *Appl Microbiol Biotechnol* **90**: 659–669.

Li, Z.-J., Qiao, K., Shi, W., Pereira, B., Zhang, H., Olsen, B.D., and Stephanopoulos, G. (2016) Biosynthesis of poly (glycolate-co-lactate-co-3-hydroxybutyrate) from glucose by metabolically engineered *Escherichia coli*. *Metab Eng* **35**: 1–8.

Lv, L., Ren, Y.-L., Chen, J.-C., Wu, Q., and Chen, G.-Q. (2015) Application of CRISPRi for prokaryotic metabolic engineering involving multiple genes, a case study: controllable P (3HB-co-4HB) biosynthesis. *Metab Eng* **29**: 160–168.

Meng, D.-C., Shi, Z.-Y., Wu, L.-P., Zhou, Q., Wu, Q., Chen, J.-C., and Chen, G.-Q. (2012) Production and characterization of poly (3-hydroxypropionate-co-4-hydroxybutyrate) with fully controllable structures by recombinant *Escherichia coli* containing an engineered pathway. *Metab Eng* **14**: 317–324.

Meng, D.-C., Wang, Y., Wu, L.-P., Shen, R., Chen, J.-C., Wu, Q., and Chen, G.-Q. (2015) Production of poly (3-hydroxypropionate) and poly (3-hydroxybutyrate-co-3-hydroxypropionate) from glucose by engineering *Escherichia coli*. *Metab Eng* **29**: 189–195.

Nampoothiri, K.M., Nair, N.R., and John, R.P. (2010) An overview of the recent developments in polylactide (PLA) research. *Bioresour Technol* **101**: 8493–8501.

Nduko, J.M., Matsumoto, K.I., Ooi, T., and Taguchi, S. (2014) Enhanced production of poly (lactate-co-3-hydroxybutyrate) from xylose in engineered *Escherichia coli* overexpressing a galactitol transporter. *Appl Microbiol Biotechnol* **98**: 2453–2460.

Ochi, A., Matsumoto, K.I., Ooba, T., Sakai, K., Tsuge, T., and Taguchi, S. (2013) Engineering of class I lactate-polymerizing polyhydroxyalkanoate synthases from *Ralstonia eutropha* that synthesize lactate-based polyester with a block nature. *Appl Microbiol Biotechnol* **97**: 3441–3447.

Ouyang, S.-P., Luo, R.C., Chen, S.-S., Liu, Q., Chung, A., Wu, Q., and Chen, G.-Q. (2007) Production of polyhydroxyalkanoates with high 3-hydroxydodecanoate monomer content by fadB and fadA knockout mutant of *Pseudomonas putida* KT2442. *Biomacromolecules* **8**: 2504–2511.

Park, S.J., Lee, S.Y., Kim, T.W., Jung, Y.K., and Yang, T.H. (2012a) Biosynthesis of lactate-containing polyesters by metabolically engineered bacteria. *Biotechnol J* **7**: 199–212.

Park, S.J., Lee, T.W., Lim, S.-C., Kim, T.W., Lee, H., Kim, M.K., *et al.* (2012b) Biosynthesis of polyhydroxyalkanoates containing 2-hydroxybutyrate from unrelated carbon source by metabolically engineered *Escherichia coli*. *Appl Microbiol Biotechnol* **93**: 273–283.

Park, S.J., Jang, Y.-A., Lee, H., Park, A.-R., Yang, J.E., Shin, J., *et al.* (2013) Metabolic engineering of *Ralstonia eutropha* for the biosynthesis of 2-hydroxyacid-containing polyhydroxyalkanoates. *Metab Eng* **20**: 20–28.

Park, S.J., Jang, Y.A., Noh, W., Oh, Y.H., Lee, H., David, Y., *et al.* (2015) Metabolic engineering of *Ralstonia eutropha* for the production of polyhydroxyalkanoates from sucrose. *Biotechnol Bioeng* **112**: 638–643.

Povolo, S., Romanelli, M.G., Basaglia, M., Ilieva, V.I., Corti, A., Morelli, A., *et al.* (2013) Polyhydroxyalkanoate biosynthesis by *Hydrogenophaga pseudoflava* DSM1034 from structurally unrelated carbon sources. *New Biotechnol* **30**: 629–634.

Salamanca-Cardona, L., Ashe, C.S., Stipanovic, A.J., and Nomura, C.T. (2014) Enhanced production of polyhydroxyalkanoates (PHAs) from beechwood xylan by recombinant *Escherichia coli*. *Appl Microbiol Biotechnol* **98**: 831–842.

Shen, X.-W., Yang, Y., Jian, J., Wu, Q., and Chen, G.-Q. (2009) Production and characterization of homopolymer poly (3-hydroxyvalerate)(PHV) accumulated by wild type and recombinant *Aeromonas hydrophila* strain 4AK4. *Bioresour Technol* **100**: 4296–4299.

Shozui, F., Matsumoto, K.I., Motohashi, R., Yamada, M., and Taguchi, S. (2010a) Establishment of a metabolic pathway to introduce the 3-hydroxyhexanoate unit into LA-based polyesters via a reverse reaction of β-oxidation in *Escherichia coli* LS5218. *Polym Degrad Stab* **95**: 1340–1344.

Shozui, F., Matsumoto, K.I., Nakai, T., Yamada, M., and Taguchi, S. (2010b) Biosynthesis of novel terpolymers poly (lactate-co-3-hydroxybutyrate-co-3-hydroxyvalerate) s in lactate-overproducing mutant *Escherichia coli* JW0885 by feeding propionate as a precursor of 3-hydroxyvalerate. *Appl Microbiol Biotechnol* **85**: 949–954.

Shozui, F., Matsumoto, K.I., Motohashi, R., Sun, J., Satoh, T., Kakuchi, T., and Taguchi, S. (2011) Biosynthesis of a lactate (LA)-based polyester with a 96 mol% LA fraction and its application to stereocomplex formation. *Polym Degrad Stab* **96**: 499–504.

Silva-Rocha, R., Martínez-García, E., Calles, B., Chavarría, M., Arce-Rodríguez, A., de las Heras, A., *et al.* (2013) The Standard European Vector Architecture (SEVA): a coherent platform for the analysis and deployment of complex prokaryotic phenotypes. *Nucleic Acids Res* **41**: D666–D675.

Simon, R., Priefer, U., and Pühler, A. (1983) A broad host range mobilization system for in vivo genetic engineering: transposon mutagenesis in gram negative bacteria. *Nat Biotechnol* **1**: 784–791.

Song, Y., Matsumoto, K.I., Yamada, M., Gohda, A., Brigham, C.J., Sinskey, A.J., and Taguchi, S. (2012) Engineered *Corynebacterium glutamicum* as an endotoxin-free platform strain for lactate-based polyester production. *Appl Microbiol Biotechnol* **93**: 1917–1925.

Spiekermann, P., Rehm, B.H., Kalscheuer, R., Baumeister, D., and Steinbüchel, A. (1999) A sensitive, viable-colony staining method using Nile red for direct screening of bacteria that accumulate polyhydroxyalkanoic acids and other lipid storage compounds. *Arch Microbiol* **171**: 73–80.

Sudesh, K., and Iwata, T. (2008) Sustainability of biobased and biodegradable plastics. *CLEAN Soil Air Water* **36**: 433–442.

Taguchi, S., Yamada, M., Matsumoto, K.I., Tajima, K., Satoh, Y., Munekata, M., *et al.* (2008) A microbial factory for lactate-based polyesters using a lactate-polymerizing enzyme. *Proc Natl Acad Sci USA* **105**: 17323–17327.

Tran, T.T., and Charles, T.C. (2015) Genome-engineered *Sinorhizobium meliloti* for the production of poly (lactic-co-3-hydroxybutyric) acid copolymer. *Can J Microbiol* **62**: 1–9.

Tripathi, L., Wu, L.-P., Chen, J., and Chen, G.-Q. (2012) Synthesis of Diblock copolymer poly-3-hydroxybutyrate-block-poly-3-hydroxyhexanoate [PHB-b-PHHx] by a beta-oxidation weakened *Pseudomonas putida* KT2442. *Microb Cell Fact* **11**: 44.

Yamada, M., Matsumoto, K.I., Shimizu, K., Uramoto, S., Nakai, T., Shozui, F., and Taguchi, S. (2010) Adjustable mutations in lactate (LA)-polymerizing enzyme for the microbial production of LA-based polyesters with tailor-made monomer composition. *Biomacromolecules* **11**: 815–819.

Yamada, M., Matsumoto, K.I., Uramoto, S., Motohashi, R., Abe, H., and Taguchi, S. (2011) Lactate fraction dependent mechanical properties of semitransparent poly (lactate-co-3-hydroxybutyrate) s produced by control of lactyl-

CoA monomer fluxes in recombinant *Escherichia coli*. *J Biotechnol* **154**: 255–260.

Yang, T.H., Jung, Y.K., Kang, H.O., Kim, T.W., Park, S.J., and Lee, S.Y. (2011) Tailor-made type II Pseudomonas PHA synthases and their use for the biosynthesis of polylactic acid and its copolymer in recombinant *Escherichia coli*. *Appl Microbiol Biotechnol* **90**: 603–614.

Zaman, H.U., Song, J.C., Park, L.-S., Kang, I.-K., Park, S.-Y., Kwak, G., *et al.* (2011) Poly (lactic acid) blends with desired end-use properties by addition of thermoplastic polyester elastomer and MDI. *Polym Bull* **67**: 187–198.

Zhou, L., Zuo, Z.-R., Chen, X.-Z., Niu, D.-D., Tian, K.-M., Prior, B.A., *et al.* (2011a) Evaluation of genetic manipulation strategies on D-lactate production by *Escherichia coli*. *Curr Microbiol* **62**: 981–989.

Zhou, Q., Shi, Z.Y., Meng, D.C., Wu, Q., Chen, J.C., and Chen, G.Q. (2011b) Production of 3-hydroxypropionate homopolymer and poly(3-hydroxypropionate-co-4-hydroxybutyrate) copolymer by recombinant *Escherichia coli*. *Metab Eng* **13**: 777–785.

Zhu, J., and Shimizu, K. (2004) The effect of pfl gene knockout on the metabolism for optically pure D-lactate production by *Escherichia coli*. *Appl Microbiol Biotechnol* **64**: 367–375.

12

Syngas obtained by microwave pyrolysis of household wastes as feedstock for polyhydroxyalkanoate production in *Rhodospirillum rubrum*

Olga Revelles,[1] Daniel Beneroso,[2,†]
J. Angel Menéndez,[2] Ana Arenillas,[2] J. Luis García[1]
and M. Auxiliadora Prieto[1,*]

[1]*Centro de Investigaciones Biológicas, CSIC, C/ Ramiro de Maeztu, 9, 28040 Madrid, Spain.*
[2]*Instituto Nacional del Carbón, CSIC, Apartado 73, 33080 Oviedo, Spain.*

Summary

The massive production of urban and agricultural wastes has promoted a clear need for alternative processes of disposal and waste management. The potential use of municipal solid wastes (MSW) as feedstock for the production of polyhydroxyalkanoates (PHA) by a process known as syngas fermentation is considered herein as an attractive bio-economic strategy to reduce these wastes. In this work, we have evaluated the potential of *Rhodospirillum rubrum* as microbial cell factory for the synthesis of PHA from syngas produced by microwave pyrolysis of the MSW organic fraction from a European city (Seville). Growth rate, uptake rate, biomass yield and PHA production from syngas in *R. rubrum* have been analysed. The results revealed the strong robustness of this syngas fermentation where the purity of the syngas is not a critical

*For correspondence. E-mail auxi@cib.csic.es

[†]Present address: Microwave Process Engineering Research Group, Faculty of Engineering, The University of Nottingham, Nottingham, NG7 2RD, UK.

Funding Information
Research leading to these results has received funding from the European Union's Seventh Framework Programme for Research, Technological Development and Demonstration under grant agreement no. 311815 (SYNPOL), and from the Comunidad de Madrid (P2013/MIT2807). D. B. also acknowledges the financial support received from PCTI and FICYT of the Government of the Principado de Asturias. The technical support of Inmaculada Calvillo and Ana Valencia is very much appreciated.

constraint for PHA production. Microwave-induced pyrolysis is a tangible alternative to standard pyrolysis, because it can reduce cost in terms of energy and time as well as increase syngas production, providing a satisfactory PHA yield.

Introduction

The development of a bio-based industry requires the production of chemicals and biomaterials to be settled on the exploitation of non-fossil carbon resources. Municipal solid wastes (MSW) and other organic waste resources, such as agricultural residues and sewage sludge contain significant reusable carbon fractions suitable for eco-efficient valorization processes. These resources take advantage over fossil sources as they are abundantly available, do not require additional production costs and are substrates that do not compete with human nutrition chain.

In this context, thermochemical conversion technologies of wastes, other than combustion, such as pyrolysis and gasification are becoming widely accepted as suitable valorization alternatives for non-fossil resources (Arena, 2012; Tanigaki *et al.*, 2013). For instance, the gasification of organic wastes produces syngas ($CO+H_2$) that can be used as chemical platform for the production of bulk chemicals such as ethanol, butanol, acetic acid or butyric acid (Munasinghe and Khanal, 2010). In this regard, the microbial bioconversion of thermochemically processed wastes presents a highly attractive potential nowadays (Koutinas *et al.*, 2014; Drzyzga *et al.*, 2015) (Fig. 1). Of special interest is the development of the chemical industry devoted to produce bio-based and biodegradable plastics, such as polyhydroxyalkanoates (PHA), with many applications including biomedical uses (Reddy *et al.*, 2003; Chen and Wu, 2005; Madbouly *et al.*, 2014; Drzyzga *et al.*, 2015). These bioplastics represent an alternative to oil-derived plastics and are degraded by many microorganisms, having gained importance in the last years (Keshavarz and Roy, 2010; Nikodinovic-Runic *et al.*, 2013).

The production of syngas from biomass and organic wastes is mainly accomplished by means of gasification

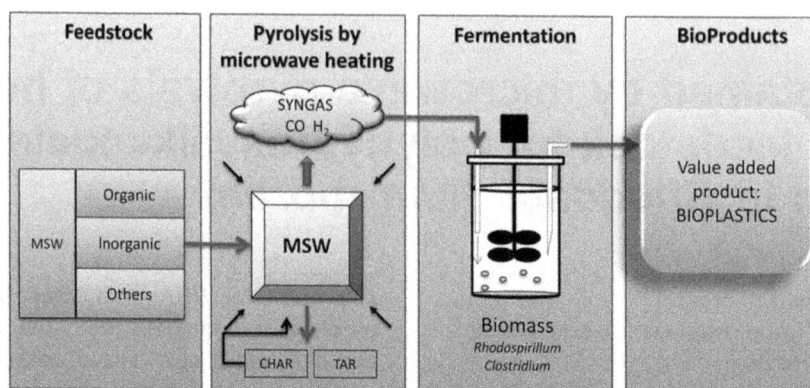

Fig. 1. Sustainable bioconversion of MSW into PHB process. MSW from Seville landfill were collected and subjected to MIP-induced pyrolysis in a microwave oven. The obtained syngas is further fermented by *R. rubrum*-producing PHB.

processes (Mahinpey and Gomez, 2016), although new emerging thermal conversion technologies, such as the microwave-induced pyrolysis (MIP) has been reported as an effective and greener way to increase the waste conversion to H_2 and CO (Budarin *et al.*, 2011; Beneroso *et al.*, 2013, 2014, 2015a,b,c). The attractiveness behind MIP is the fact that microwaves are able to induce an instantaneous and volumetric heating within the samples and thus, overcoming heat transfer constraints as a result of the low thermal conductivity of biomass and biowastes (Beneroso *et al.*, 2015b; Beneroso and Fidalgo, 2016a). In addition to syngas and tars, a carbonaceous solid fraction (known as char) is produced during MIP. This fraction has a low value owing to its high ash content, although this can be used as a solid fuel or for soil amendment purposes. Beneroso *et al.* (2016b) has recently reported the potential use of char as an additive to soils, and particularly, the char obtained from MSW can be added to soil for carbon sequestration without negatively affecting the soil fertility. Therefore, MIP is an attractive alternative within thermochemical technologies for syngas production (Beneroso *et al.*, 2015b).

Despite the toxicity of CO for most organisms, several microorganisms can use CO as source of carbon and/or energy for growth. *Rhodospirillum rubrum*, which is the type strain for the *Rhodospirillaceae* family, is able to grow under a broad variety of conditions, including syngas (Do *et al.*, 2007). This bacterium can utilize CO under anaerobic conditions as a sole carbon and energy source in the presence or absence of light (Kerby *et al.*, 1995; Revelles *et al.*, 2016a,b). It is important to emphasize that *R. rubrum* has a photosynthetic apparatus involved in the photosynthesis on anaerobic conditions. When exposed to CO, both a CODH and a CO-insensitive hydrogenase are induced catalysing the oxidation of CO into CO_2 and H_2. Approximately, 40% of the carbon fraction of syngas (CO_2 and CO) is assimilated into biomass. Although *R. rubrum* is a highly versatile bacterium

that possess the Calvin–Benson–Bassham cycle, it has been recently shown that other carboxylases than Rubisco are actively incorporating CO_2 from syngas into biomass (Revelles *et al.*, 2016a,b). Furthermore, *R. rubrum* is an attractive syngas utilizer for its ability to produce PHA as an energy storage molecule during the fermentation process. We recently reported around 20% of poly(3-hydroxybutyrate) (PHB) production on this microorganism during syngas fermentation both in darkness and/or light (Revelles *et al.*, 2016a,b). Other studies reported the heterologous overexpression of different genes encoding pyridine nucleotide transhydrogenases (pntAB, udhA) and acetoacetyl-CoA reductases (PhaB) in *R. rubrum* S1, during the synthesis of the copolymer poly(3-hydroxybutyrate-*co*-3-hydroxyvalerate) P(HB-HV) from syngas (Heinrich *et al.*, 2015). This engineered bacterium was pointed out as a promising production strain for syngas-derived, second-generation biopolymers, increasing the potential of this strain for industry application.

The utilization of MSW-derived syngas for the sustainable production of PHB is shown in this work. Of particular interest is the innovative use of MIP to achieve this goal, as this technology provides a syngas able to induce an efficient growth of *R. rubrum*, reducing cost in terms of energy and time at a much higher syngas productivity compared with syngas from conventional pyrolysis technologies.

Results and discussion

Growth of R. rubrum *in microwaving versus synthetic syngas*

An organic fraction from a MSW provided by ABENGOA BIOENERGÍA from a landfill in Seville (Spain) was used in this study. The MSW fraction was subjected to removal of moisture and inert solids, such as glass or metals. After this pre-treatment, the fraction size was reduced to 1–3 mm. The characterization of the sample can be found somewhere else (Beneroso et al., 2015a). To prepare syngas, the waste was subjected to MIP in a microwave oven (Beneroso et al., 2015a). The char from previous pyrolysis experiments was used as microwave receptor to induce the pyrolysis due to the low capacity of organic wastes to absorb microwaves. When the microwaves pass through the sample, they are absorbed by the receptor and the temperature increases, allowing the heat to be conducted to the waste to reach a temperature high enough to start the pyrolysis. As the pyrolysis proceeds, the waste is carbonized and is then able to absorb microwaves, so that from that point on it can be directly heated by microwave radiation. The detailed methodology has been described previously (Beneroso et al., 2015a). Table 1 shows the syngas composition of the MIP compared with the standard synthetic syngas used for bacterial fermentation (Revelles et al., 2016a,b).

Starter cultures of R. rubrum (ATCC 11170) were grown under anaerobic conditions on SYN medium (Revelles et al., 2016a) supplemented with 15 mM fructose at 30°C during 48 h until stationary phase (OD$_{600}$ 1.5). This culture was used as pre-inoculum for syngas fermentation.

Syngas experiments were done in bottles of 100 mL containing 20 mL of SYN medium supplemented with 10 mM acetate as we had previously demonstrated that the presence of acetate during syngas fermentation favours PHB accumulation (Revelles et al., 2016a,b). Before adding syngas, the closed degasified serum vials were subjected to 1 min vacuum-purge and the atmosphere was further saturated with syngas, either synthetic

or microwave-derived syngas, to 1 atm of pressure. This procedure for syngas feeding was repeated every day.

R. rubrum is a highly versatile bacterium that can ferment syngas either in dark or in light. On anaerobic conditions when the cells are exposed to light, the photosynthesis becomes active being this an extra energy source. Therefore, the potential impact of light on cell physiology during syngas fermentation was also evaluated. A syngas sample obtained from the MIP process and the synthetic syngas were used as substrates for syngas fermentation in parallel cultures of SYN medium supplemented with 10 mM of acetate in darkness and in light. A very detailed protocol for culturing R. rubrum in syngas both in darkness and in light can be found in Revelles et al. (2016a,b).

Table 2 shows the fermentation kinetic parameters, such as growth rate, μ; acetate uptake rate, Qs; and biomass dry weight production yield. None difference in the growth rate μ was observed among the two different syngas tested in light or darkness, i.e., 0.030 h^{-1} and 0.022 h^{-1} respectively. In both syngas, a higher growth rate was detected in light. It is worthy to stress that light has a positive impact in the growth rate, likely due to the contribution of the photosynthesis in the energetic status of the cell. The small presence of hydrocarbons (methane, ethylene and ethane) within the MIP-derived syngas (Table 1) does not have a harmful effect on bacterial growth. By high-performance liquid chromatography analysis, the presence of extracellular products such as malate, succinate, propionate, pyruvate and/or formate were measured, but none of them were detected in the supernatant. Interestingly, when MIP-derived syngas was used in the fermentation process, 1.5- to 2-fold higher acetate uptake rate was found compared with synthetic syngas in light and darkness respectively (Table 2). This effect was observed in the yield of biomass from acetate as well, being 1.5-fold higher in MIP-derived syngas than in synthetic syngas (Table 2).

Table 1. Gas composition of syngas.

Gases	Syngas composition (vol.%)[a]	
	Synthetic	MIP
CO	40	27
CO$_2$	10	6
H$_2$	40	37
N$_2$[b]	10	26
CH$_4$+C$_2$H$_4$+C$_2$H$_6$	0	4

a. Synthetic syngas was provided by Air Liquide (Air Liquide, www.airliquide.com). Composition of microwaving syngas was determined in a gas chromatograph (GC, Agilent 7890A) equipped with a TCD and two columns connected in series (80/100 Porapak Q and 70/80 Molesieve 13X). The initial oven temperature was 30°C, which was maintained with an isothermic step of 5 min. It was then programmed with a rate of 25°C min^{-1} until reached 180°C. The injector and detector temperatures were 150 and 250°C respectively. Helium (Air Liquide, www.airliquide.com) was used as carrier gas. Prior to the measurements, the gas analyser was calibrated by a standard gas and a calibration curve was established. The calculation for gas concentration was carried out using the GC data analysis software (ChemStation rev. B.04.03-SP1; Agilent Technologies, Santa Clara, CA 95051, United States).
b. The synthetic gas was balanced with N$_2$, while in the mixture of gases from MIP, the N$_2$ content is due to the N$_2$ used as carrier gas but not released in the pyrolysis process.

CO consumption and PHB production during the syngas fermentation

CO assimilation in R. rubrum is associated to the CODH enzyme that catalyses the oxidation of CO into CO$_2$. When the cells are growing on syngas plus acetate, the carbon fraction of syngas is incorporated into biomass and PHB by the activity of different carboxylases of the central carbon metabolism (Revelles et al., 2016a,b). Acetate assimilation involves two steps: the ferredoxin-dependent pyruvate synthase (PFOR) enzyme and the ethylmalonyl-CoA pathway (Revelles et al., 2016a,b). The use of acetate during syngas fermentation is critical to achieve PHB accumulation, where approximately 40% of PHB total carbon backbone comes from the carbon

Table 2. Kinetic growth parameters of *R. rubrum* with syngas in light and darkness.

Kinetic growth	Synthetic		MIP	
	Light	Darkness	Light	Darkness
μ (h^{-1})	0.029 ± 0.005	0.021 ± 0.005	0.031 ± 0.005	0.022 ± 0.005
Qs (mmol gDW^{-1} h^{-1})	1.51 ± 0.05	1.43 ± 0.05	2.05 ± 0.20	3.10 ± 0.15
Q_{CO} ($mmol_{CO}$ gDW^{-1} h^{-1})	0.01	ND	0.005	ND
Y (gDW g^{-1})	0.23 ± 0.05	0.17 ± 0.01	0.25 ± 0.05	0.13 ± 0.05
PHB (% CDW)	20 ± 5	28 ± 10	16 ± 1	10 ± 1

Parameters: μ (h^{-1}), specific growth rate; Q_s (mmol gDW^{-1} h^{-1}), acetate uptake rate; Y (gDW g^{-1}), biomass dry weight production yield; PHB (% cell dry weight). Values represent the mean $\pm$ standard deviation of three independent biological replicates. The growth rate (μ) was determined from log-linear regression of time-dependent changes in optical density at 600 nm (OD600), measured with a spectrophotometer (UV-VIS Spectrophotometer Shimatzu UV mini 1240) with appropriate dilutions when needed. To calculate specific biomass yields, correlation factors between cell dry weights and optical density (gCDW/OD600) were established for each condition. Acetate disappearance was quantified using an high-performance liquid chromatography system (GILSON), equipped with an Aminex HPX-87H column and a mobile phase of 2.5 mM H_2SO_4 solution at a 0.6 mL min^{-1} flow rate operated at 40°C. ND, non-determined.

fraction of syngas and the remaining fraction comes from acetate (Revelles *et al.*, 2016a,b). PHB is produced by a biosynthetic pathway consisting of three different enzymatic reactions from acetyl-CoA. The first step is the condensation of two acetyl-CoA molecules into acetoacetyl-CoA by 3-ketothiolase (catalyses by the enzyme PhaA). Then, acetoacetyl-CoA reductase (PhaB) allows the reduction of acetoacetyl-CoA by NADH to 3-hydroxybutyryl-CoA. Finally, the (*R*)-3-hydroxybutyryl-CoA monomers are polymerized into PHB by PHB synthase (PhaC).

The CO consumption at the end of the growth curve for each different syngas fermentation process was monitored. As no difference was registered between light and darkness, the CO consumption was analysed on light. A sample (1 mL) from the head-space of the culture bottle was taken at time zero and after 48 h of incubation in light. The differences in the final CO values were measured using a gas chromatograph equipped with a thermal conductivity detector (TCD). The results on CO consumption as percentage of gas converted and the uptake of CO are shown in Tables 2 and 3 respectively. The CO conversion registered was of about 40% in both synthetic and MIP-derived syngas. Although in both cases the % of CO conversion does not differ, the final amount of metabolized CO is half for MIP syngas than synthetic syngas (Table 3, 0.19 versus 0.44 mmoles of CO respectively). The lower CO uptake (see Table 2) determined for MIP syngas is compensated with a higher acetate uptake (Qs) observed in this syngas to keep a constant growth rate.

PHB production was quantified from cells harvested from cultures grown in syngas by centrifugation at 8000 *g* for 15 min at 4°C (Eppendorf Centrifuge 5810R). Cells were then washed twice in distilled water and lyophilized in Cryodos-50 (Telstar, Terrasa, Spain) at −56°C and 10^{-2} mbar. Furthermore, PHB monomer composition and cellular PHB content were determined by gas

Table 3. CO consumption during syngas fermentation.

Gas	% Conversion[a]		mmol of CO[b]	
	Synthetic	MIP	Synthetic	MIP
CO	39 ± 4	37 ± 8	0.44 ± 0.05	0.19 ± 0.1

a. The differences in the final gas composition are given as percentage of gas consumed.
b. mmol of CO consumed at the end of the growth.
A gas chromatograph (GC, Agilent 7890A) was equipped with a thermal conductivity detector (TCD) and two columns connected in series (80/100 Porapak Q and 70/80 Molesieve 13X) as described for 1.
Values represent the mean $\pm$ SD of three independent biological replicates.

chromatography (GC) of the methanolysed polyester. Methanolysis was carried out by suspending 5–10 mg of lyophilized cells in 0.5 mL of chloroform and 2 mL of methanol containing 15% sulfuric acid and 0.5 mg mL^{-1} of 3-methylbenzoic acid (internal standard), followed by an incubation at 80°C for 7 h. After cooling, 1 mL of demineralized water and 1 mL of chloroform were added and the organic phase containing the methyl esters was analysed as described previously (de Eugenio *et al.*, 2010). A standard curve from 0.5 to 2 mg of PHB (Cat: 36,350-2; Sigma, San Luis, Misuri, United States) was used to interpolate sample data.

Table 2 shows the PHB content of cells of *R. rubrum* from MSW-derived syngas versus synthetic syngas. When light was used as an extra source of energy, the percentage of PHB detected remained similar regardless of the source of syngas ($16 \pm 5\%$ and $20 \pm 1\%$ respectively). But, a small reduction was detected in cells grown on darkness ($28 \pm 10\%$ and $10 \pm 1\%$ respectively). Furthermore, granules of PHB can be clearly observed in *R. rubrum* grown in MIP-derived syngas by transmission electronic microscopy during syngas fermentation when using MIP-derived syngas (Fig. 2).

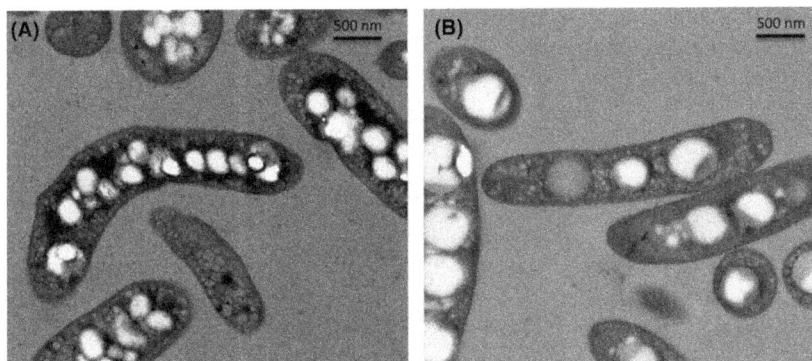

Fig. 2. Transmission electron micrograph (TEM) of *R. rubrum* growing in medium SYN with acetate fed microwave-induced pyrolysis syngas (A) and synthetic syngas (B) both containing PHB granules. Culture was harvested, washed twice in PBS and fixed in 5% (w/v) glutaraldehyde in the same solution. The cells were incubated with 2.5% (w/v) OsO$_4$ for 1 h, gradually dehydrated in ethanol solutions and propylene oxide and embedded in Epon 812 resin. Ultrathin sections (50–70 nm) were cut and observed using a Jeol-1230 electron microscope.

In this work, it has been shown that the use of wastes-derived syngas did not affect the growth rate of *R. rubrum* being of ~0.02 μ^{-1}, but an increase in the acetate uptake rate was found indicating a higher consumption of acetate that will compensate the lower CO consumption observed in MIP syngas, although CO conversion does not change. The lower CO consumption could be explained by the lower percentage of CO presence in MIP syngas. Interestingly, the accumulation of PHB observed in cells grown with MIP-derived syngas was similar to that found on synthetic syngas, especially when they were exposed to light. These results are a proof of concept of the potential of pyrolysed MSW as feedstock for the production of bioplastics. Furthermore, the robustness and versatility of *R. rubrum* to accumulate PHB regardless of the composition of the syngas increases the industrial interest of this bacterium to be used in syngas fermentation without introducing a previous costly step for syngas purification. The capability of this microorganism to synthesize other polymers different from PHA of industrial interest from different carbon sources has been recently shown in an engineering strain (Heinrich *et al.*, 2015) broadening its potential to be used in biopolymers production industry. In summary, this work represents the first experimental demonstration that MIP of MSW followed by syngas fermentation can become a valuable environmental-friendly strategy for waste treatment rendering high value-added products such as bioplastics.

Conclusion

The MSW generation has boosted with the increase in economic growth and population as well as with the living standards of developing countries. Finding a solution to manage MSW is a complex issue that needs to assess different social aspects including environmental impact and economic cost. One of the main bottle necks of using MSW as feedstocks is the complexity of the organic fraction composition, which hinders its application in conventional bioprocesses using microbial monocultures. In this study, the potential use of MSW as feedstock for the production of PHB by *R. rubrum* has been demonstrated by subjecting MSW to an emerging thermochemical process based on microwave heating, which enables the production of syngas useful as a derived feedstock. The effects of using syngas as feedstock on the physiology of *R. rubrum* was investigated demonstrating the robustness of this bacterium as biocatalyst, as the growth, syngas consumption and PHB production are not affected by the putative contaminants present in MSW-derived syngas. Our results expand the potential of this hybrid technology to be implemented in waste management programmes. Within this general strategy, the use of MIP provides an additional advantage versus conventional oven pyrolytic processes, because it reduces time and energy for syngas synthesis, hence reducing cost and permitting a much higher syngas volumetric productivity, without affecting syngas fermentation.

Conflict of interest

The authors declare no conflict of interest.

References

Arena, U. (2012) Process and technological aspects of municipal solid waste gasification. A review. *Waste Management* 32: 625–639.

Beneroso, D. and Fidalgo, B. (2016a) Microwave technology for syngas production from renewable sources. In *Syngas: Production, Energing Technologies and Ecological Impacts*. Myers, R. (ed.) Hauppauge, NY (USA): Nova Science Publishers, pp. 117–152.

Beneroso, D., Bermúdez, J.M., Arenillas, A., and Menéndez, J.A. (2013) Microwave pyrolysis of microalgae for high syngas production. *Bioresour Technol* **144**: 240–246.

Beneroso, D., Bermúdez, J.M., Arenillas, A., and Menéndez, J.A. (2014) Influence of the microwave absorbent and moisture content on the microwave pyrolysis of an organic municipal solid waste. *J Anal Appl Pyrol* **105**: 234–240.

Beneroso, D., Bermúdez, J.M., Arenillas, A., and Menéndez, J.A. (2015a) Influence of carrier gas on microwave-induced pyrolysis. *J Anal Appl Pyrol* **113**: 153–157.

Beneroso, D., Bermúdez, J.M., Arenillas, A., and Menéndez, J.A. (2015b) Comparing the composition of the synthesis-gas obtained from the pyrolysis of different organic residues for a potential use in the synthesis of bioplastics. *J Anal Appl Pyrol* **111**: 55–63.

Beneroso, D., Bermúdez, J.M., Arenillas, A. and Menéndez, J.A. (2015c) Microwave pyrolysis of organic wastes for syngas-derived biopolymers production. In *Production of Biofuels and Chemicals with Microwave*, Vol. 3. Fang, Z., Smith, J.R.L. and Qi, X. (eds). The Netherlands: Springer, pp. 99–127.

Beneroso, D., Arenillas, A., Montes-Morán, M.A., Bermúdez, J.M., Mortier, N., Verstichel, S., and Menéndez, J.A. (2016b) Ecotoxicity tests on solid residues from microwave induced pyrolysis of different organic residues: an addendum. *J Anal Appl Pyrol* (in press).

Budarin, V.L., Shuttleworth, P.S., Dodson, J.R., Hunt, A.J., Lanigan, B., Marriott, R., *et al.* (2011) Use of green chemical technologies in an integrated biorefinery. *Energy Environ Sci* **4**: 471–479.

Chen, G.-Q., and Wu, Q. (2005) The application of polyhydroxyalkanoates as tissue engineering materials. *Biomaterials* **26**: 6565–6578.

Do, Y.S., Smeenk, J., Broer, K.M., Kisting, C.J., Brown, R., Heindel, T.J., *et al.* (2007) Growth of *Rhodospirillum rubrum* on synthesis gas: conversion of CO to H_2 and poly-β-hydroxyalkanoate. *Biotechnol Bioeng* **97**: 279–286.

Drzyzga, O., Revelles, O., Durante, G., Diaz, E., Garcia, J.L., and Prieto, M.A. (2015) New challenges for syngas fermentation: bioplastics. *J Chem Technol Biotechnol* **90**: 1735–1751.

de Eugenio, L.I., Galan, B., Escapa, I.F., Maestro, B., Sanz, J.M., Garcia, J.L., and Prieto, M.A. (2010) The PhaD regulator controls the simultaneous expression of the pha genes involved in polyhydroxyalkanoate metabolism and turnover in *Pseudomonas putida* KT2442. *Environ Microbiol* **12**: 1591–1603.

Heinrich, D., Raberg, M., and Steinbüchel, A. (2015) Synthesis of poly(3-hydroxybutyrate-co-3-hydroxyvalerate) from unrelated carbon sources in engineered *Rhodospirillum rubrum*. *FEMS Microbiol Lett* **362**: fnv038.

Kerby, R.L., Ludden, P.W., and Roberts, G.P. (1995) Carbon monoxide-dependent growth of *Rhodospirillum rubrum*. *J Bacteriol* **177**: 2241–2244.

Keshavarz, T., and Roy, I. (2010) Polyhydroxyalkanoates: bioplastics with a green agenda. *Curr Opin Microbiol* **13**: 321–326.

Koutinas, A.A., Vlysidis, A., Pleissner, D., Kopsahelis, N., Lopez Garcia, I., Kookos, I.K., *et al.* (2014) Valorization of industrial waste and by-product streams via fermentation for the production of chemicals and biopolymers. *Chem Soc Rev* **43**: 2587–2627.

Madbouly, S.A., Schrader, J.A., Srinivasan, G., Liu, K., McCabe, K.G., Grewell, D., *et al.* (2014) Biodegradation behavior of bacterial-based polyhydroxyalkanoate (PHA) and DDGS composites. *Green Chem* **16**: 1911–1920.

Mahinpey, N., and Gomez, A. (2016) Review of gasification fundamentals and new findings: reactors, feedstock, and kinetic studies. *Chem Eng Sci* **148**: 14–31.

Munasinghe, P.C., and Khanal, S.K. (2010) Biomass-derived syngas fermentation into biofuels: opportunities and challenges. *Bioresour Technol* **101**: 5013–5022.

Nikodinovic-Runic, J., Guzik, M., Kenny, S.T., Babu, R., Werker, A., and Connor, K.E.O. (2013) Chapter four - carbon-rich wastes as feedstocks for biodegradable polymer (polyhydroxyalkanoate) production using bacteria. *Adv Appl Microbiol* **84**: 139–200.

Reddy, C.S.K., Ghai, R., Rashmi, and Kalia, V.C. (2003) Polyhydroxyalkanoates: an overview. *Bioresour Technol* **87**: 137–146.

Revelles, O., Calvillo, I., Prieto, A. and Prieto, M.A. (2016a) *Syngas Fermentation for Polyhydroxyalkanoate Production in Rhodospirillum rubrum*. Hydrocarbon and Lipid Microbiology Protocols. Springer-Verlag Berlin Heidelberg: Springer Protocols Handbooks. ISBN 978-3-662-49131-7.

Revelles, O., Tarazona, N., Garcia, J.L., and Prieto, M.A. (2016b) Carbon road map: from syngas to PHA. *Envirom Microbiol* **18**: 708–720.

Tanigaki, N., Fujinaga, Y., Kajiyama, H., and Ishida, Y. (2013) Operating and environmental performances of commercial-scale waste gasification and melting technology. *Waste Manage Res* **31**: 1118–1124.

Water-, pH- and temperature relations of germination for the extreme xerophiles *Xeromyces bisporus* (FRR 0025), *Aspergillus penicillioides* (JH06THJ) and *Eurotium halophilicum* (FRR 2471)

Andrew Stevenson,[1] Philip G. Hamill,[1]
Jan Dijksterhuis[2] and John E. Hallsworth[1,*]

[1]*Institute for Global Food Security, School of Biological Sciences, MBC, Queen's University Belfast, Belfast BT9 7BL, UK.*
[2]*CBS-KNAW Fungal Biodiversity Centre, Uppsalalaan 8, CT 3584, Utrecht, The Netherlands.*

Summary

Water activity, temperature and pH are determinants for biotic activity of cellular systems, biosphere function and, indeed, for all life processes. This study was carried out at high concentrations of glycerol, which concurrently reduces water activity and acts as a stress protectant, to characterize the biophysical capabilities of the most extremely xerophilic organisms known. These were the fungal xerophiles: *Xeromyces bisporus* (FRR 0025), *Aspergillus penicillioides* (JH06THJ) and *Eurotium halophilicum* (FRR 2471). High-glycerol spores were produced and germination was determined using 38 media in the 0.995–0.637 water activity range, 33 media in the 2.80–9.80 pH range and 10 incubation temperatures, from 2 to 50°C. Water activity was modified by supplementing media with glycerol+sucrose, glycerol+NaCl and glycerol+NaCl+sucrose which are known to be biologically permissive for *X. bisporus*, *A. penicillioides* and *E. halophilicum* respectively. The windows and rates for spore germination were quantified for water activity, pH and temperature; symmetry/asymmetry of the germination profiles were then determined in relation to *supra*- and sub-optimal conditions; and pH- and temperature optima for extreme xerophilicity were quantified. The windows for spore germination were ~1 to 0.637 water activity, pH 2.80–9.80 and > 10 and < 44°C, depending on strain. Germination profiles in relation to water activity and temperature were asymmetrical because conditions known to entropically disorder cellular macromolecules, i.e. *supra*-optimal water activity and high temperatures, were severely inhibitory. Implications of these processes were considered in relation to the *in-situ* ecology of extreme conditions and environments; the study also raises a number of unanswered questions which suggest the need for new lines of experimentation.

Introduction

Extremely xerophilic fungi act as pioneers under hostile conditions, and can catalyse ecosystem development at low water-availability. For instance, they can drive saprotrophic processes in arid soils, salt-saturated brines and other water-constrained environments such as stone surfaces, paper and other artefacts and high-solute foodstuffs. Understanding xerophile behaviour in relation to environmental constraints also facilitates our understanding of the biophysical limits for life and addresses important questions relating to the issue of habitability of hostile environments (e.g. saline and high-sugar habitats; Lievens *et al.*, 2015; Stevenson *et al.*, 2015a,b), including those which lie at the heart of the astrobiology field (Rummel *et al.*, 2014).

Over the past 100-year period, there have been occasional reports of germination and/or mycelial growth of extreme xerophiles at extremely low water-activity (≥ 0.710) (Stevenson *et al.*, 2015a,b). The majority of these provide single data-points at which researchers have observed important processes, such as spore germination (e.g. Pitt and Christian, 1968). Fungal germination is not only a seminal moment in the fungal life-cycle but it also facilitates dispersal and enables colonization of new substrates and habitats; it can enable development of new colonies in isolation from other microbes and/or bring fungi into contact with other living systems. Germination also represents a key biophysical event,

*For correspondence. E-mail j.hallsworth@qub.ac.uk

Funding Information
Funding was supplied by the Department of Agriculture, Environment and Rural Affairs (Northern Ireland) who supported A. Stevenson and P. G. Hamill, and Biotechnology and Biological Sciences Research Council (BBSRC, United Kingdom) project BBF0034711.

whereby life exits dormancy to recover from a period of hostile conditions, such as prolonged desiccation or exposure to extremes of chaotropicity or temperature (Hallsworth et al., 2003a; Wyatt et al., 2015a,b; Yakimov et al., 2015).

A series of studies has been carried out to systematically identify the most-xerophilic fungal genera and strains known to science (Williams and Hallsworth, 2009; Stevenson and Hallsworth, 2014; Stevenson et al., 2015a,b, in press). For the three most-xerophilic genera of fungi, the strains able to grow and/or germinate at the lowest water activities are Xeromyces bisporus (FRR 0025), Aspergillus penicillioides (JH06THJ) and Eurotium halophilicum (FRR 2471). In the current study, these strains were used as models for the biophysical fringe of Earth's biosphere. The specific aims were to: (i) determine windows of water activity, pH and temperature which are permissive for spore germination; (ii) quantify rates of germination and germ-tube extension in relation to these parameters; (iii) characterize the symmetry and/or asymmetry of the germination profiles in relation to supra- and sub-optimal conditions; (iv) determine the pH- and temperature optima for xerophilicity (i.e. those values which facilitate germination at the lowest water-activities); and (v) consider the implications of these germination processes for the in-situ ecology of extreme conditions and environments. We put forward key, unanswered questions which suggest the need for new lines of experimentation.

Results and discussion

Biotic windows for germination

Spores were produced in high-glycerol media (5.5 M glycerol, 0.821 water activity) and contained up to 15% w/v glycerol (Stevenson et al., in press). Strains produced D-shaped ascospores (X. bisporus), conidia (A. penicillioides) or a mixture of ascospores and conidia (E. halophilicum). For E. halophilicum, germination was assessed for conidia only. The most-permissive medium for germination of each xerophile strain (i.e. that which enables germination at low water-activity) are glycerol+sucrose, glycerol+NaCl and glycerol+NaCl+sucrose for X. bisporus (FRR 0025), A. penicillioides (JH06THJ) and E. halophilicum (FRR 2471) respectively (Stevenson et al., in press); these media were used as the basis for the current study (Tables 1–3).

To determine the windows of, and kinetics for, germination in relation to water activity, ranges of stressor concentrations were used for each strain (Table 1). The pH values of these media were close to neutral (i.e. in the range 6.3–7.2; Table 1) and plates were incubated at 30°C (see Experimental procedures). Collectively, the water activity windows for biotic activity of the xerophile strains (according to germination assays), as might be

Table 1. Culture media used for germination assays of Xeromyces bisporus FRR 0025, Aspergillus penicillioides JH06THJ and Eurotium halophilicum FRR 2471 to characterize germination performance over the entire water-activity window (see Fig. 1)[a].

Water activity[b]	Stressor type and concentration (M)[a]	pH
Xeromyces bisporus FRR 0025		
0.995	None	7.20
0.972	Glycerol (0.50)+sucrose (0.10)	7.20
0.920	Glycerol (1.70)+sucrose (0.20)	7.00
0.884	Glycerol (2.40)+sucrose (0.25)	6.90
0.849	Glycerol (3.00)+sucrose (0.35)	6.90
0.799	Glycerol (3.60)+sucrose (0.35)	6.70
0.762	Glycerol (4.30)+sucrose (0.40)	6.50
0.723	Glycerol (5.00)+sucrose (0.45)	6.50
0.699	Glycerol (5.50)+sucrose (0.50)	6.50
0.682	Glycerol (5.50)+sucrose (0.60)	6.50
0.674	Glycerol (5.50)+sucrose (0.65)	6.50
0.637	Glycerol (5.50)+sucrose (0.80)	6.30
Aspergillus penicillioides JH06THJ		
0.995	None	7.20
0.949	Glycerol (1.00)+NaCl (0.20)	7.10
0.917	Glycerol (1.50)+NaCl (0.30)	7.10
0.880	Glycerol (2.00)+NaCl (0.40)	6.90
0.868	Glycerol (2.50)+NaCl (0.40)	6.90
0.841	Glycerol (3.00)+NaCl (0.50)	6.80
0.824	Glycerol (3.50)+NaCl (0.50)	6.80
0.802	Glycerol (4.00)+NaCl (0.60)	6.80
0.787	Glycerol (4.50)+NaCl (0.70)	6.80
0.764	Glycerol (5.00)+NaCl (0.80)	6.80
0.741	Glycerol (5.50)+NaCl (1.00)	6.80
0.709	Glycerol (5.50)+NaCl (1.50)	6.70
0.692	Glycerol (5.50)+NaCl (1.60)	6.80
0.668	Glycerol (5.50)+NaCl (1.70)	6.80
0.640	Glycerol (5.50)+NaCl (1.80)	6.70
Eurotium halophilicum FRR 2471		
0.995	None	7.20
0.961	Glycerol (0.80)+NaCl (0.10)+sucrose (0.10)	7.20
0.939	Glycerol (1.00)+NaCl (0.20)+sucrose (0.20)	7.20
0.900	Glycerol (2.00)+NaCl (0.20)+sucrose (0.20)	7.00
0.875	Glycerol (2.50)+NaCl (0.30)+sucrose (0.30)	6.80
0.839	Glycerol (3.00)+NaCl (0.30)+sucrose (0.30)	6.90
0.823	Glycerol (3.50)+NaCl (0.40)+sucrose (0.40)	6.70
0.805	Glycerol (4.00)+NaCl (0.40)+sucrose (0.40)	6.70
0.771	Glycerol (4.50)+NaCl (0.50)+sucrose (0.50)	6.70
0.738	Glycerol (5.00)+NaCl (0.50)+sucrose (0.50)	6.80
0.701	Glycerol (5.50)+NaCl (0.50)+sucrose (0.30)	6.70
0.685	Glycerol (5.50)+NaCl (0.50)+sucrose (0.50)	6.70
0.651	Glycerol (5.50)+NaCl (0.80)+sucrose (0.50)	6.60

[a]All media were based on MYPiA: 1% malt extract, 1% yeast extract, 0.1% KH_2PO_4 and 1.5% (w/v) agar (Williams and Hallsworth, 2009).
[b]The water activity of each medium was measured at the temperature at which plates were incubated (30°C) and replicate values were within ± 0.002 units for the 1 to 0.900 water-activity range and ± 0.001 units for the 0.900 to 0.600 water-activity range (see Experimental procedures).

expected, spanned a range (0.995–0.637; Fig. 1) which was virtually equivalent to the water-activity window for life on Earth (Stevenson et al., 2015a,b). There were, however, some slight differences between strains. Within the 30-day time-period of the study, X. bisporus (FRR 0025) did not germinate at 0.995 or 0.972 water activity, and E. halophilicum (FRR 2471) did not germinate at 0.995 water activity; by contrast, A. penicillioides (JH06THJ) was able to germinate on all media in the

Table 2. Culture media used for germination assays of *Xeromyces bisporus* FRR 0025, *Aspergillus penicillioides* JH06THJ and *Eurotium halophilicum* FRR 2471 to characterize germination performance over the entire pH window (see Fig. 3)[a].

pH of culture medium[b]	Stressor type and concentration (M)[a]	Water activity[c]
Xeromyces bisporus FRR 0025		
2.90	Glycerol (5.5)+sucrose (0.3)	0.720
3.50	Glycerol (5.5)+sucrose (0.3)	0.719
4.60	Glycerol (5.5)+sucrose (0.3)	0.722
5.30	Glycerol (5.5)+sucrose (0.3)	0.721
5.80	Glycerol (5.5)+sucrose (0.3)	0.722
6.40	Glycerol (5.5)+sucrose (0.3)	0.722
7.10	Glycerol (5.5)+sucrose (0.3)	0.722
7.70	Glycerol (5.5)+sucrose (0.3)	0.719
8.10	Glycerol (5.5)+sucrose (0.3)	0.720
8.80	Glycerol (5.5)+sucrose (0.3)	0.718
9.60	Glycerol (5.5)+sucrose (0.3)	0.717
Aspergillus penicillioides JH06THJ		
3.00	Glycerol (5.5)+NaCl (1.2)	0.728
3.60	Glycerol (5.5)+NaCl (1.2)	0.725
4.70	Glycerol (5.5)+NaCl (1.2)	0.724
5.50	Glycerol (5.5)+NaCl (1.2)	0.727
6.00	Glycerol (5.5)+NaCl (1.2)	0.726
6.50	Glycerol (5.5)+NaCl (1.2)	0.729
7.00	Glycerol (5.5)+NaCl (1.2)	0.729
7.70	Glycerol (5.5)+NaCl (1.2)	0.724
8.40	Glycerol (5.5)+NaCl (1.2)	0.722
9.00	Glycerol (5.5)+NaCl (1.2)	0.726
9.80	Glycerol (5.5)+NaCl (1.2)	0.723
Eurotium halophilicum FRR 2471		
2.80	Glycerol (5.5)+NaCl (0.25)+sucrose (0.25)	0.723
3.70	Glycerol (5.5)+NaCl (0.25)+sucrose (0.25)	0.721
4.50	Glycerol (5.5)+NaCl (0.25)+sucrose (0.25)	0.724
5.50	Glycerol (5.5)+NaCl (0.25)+sucrose (0.25)	0.723
6.00	Glycerol (5.5)+NaCl (0.25)+sucrose (0.25)	0.725
6.40	Glycerol (5.5)+NaCl (0.25)+sucrose (0.25)	0.725
7.00	Glycerol (5.5)+NaCl (0.25)+sucrose (0.25)	0.726
7.50	Glycerol (5.5)+NaCl (0.25)+sucrose (0.25)	0.724
8.20	Glycerol (5.5)+NaCl (0.25)+sucrose (0.25)	0.721
8.90	Glycerol (5.5)+NaCl (0.25)+sucrose (0.25)	0.724
9.50	Glycerol (5.5)+NaCl (0.25)+sucrose (0.25)	0.719

[a]All media were based on MYPiA: 1% malt extract, 1% yeast extract, 0.1% KH_2PO_4 and 1.5% (w/v) agar (Williams and Hallsworth, 2009).
[b]Media were buffered by addition of citric acid/Na_2PO_4 (3.00, 3.75, 4.75 and 5.50), PIPES/NaOH (6.0, 6.5 and 7.2) or HEPES/NaOH (7.8, 8.5, 9.2, 10.1) pre-autoclave. The pH of liquid media was measured using a Mettler Toledo Seven Easy pH-probe (Mettler Toledo, Greifensee, Switzerland) and for solid media, post-autoclave using Fisherbrand colour-fixed pH indicator strips were used (Fisher Scientific, Leicestershire, UK).
[c]The water activity of each medium was measured at the same temperature at which plates were incubated (30°C) and replicate values were within ± 0.001 water activity (see *Experimental procedures*).

Table 3. Culture media used for germination assays of *Xeromyces bisporus* FRR 0025, *Aspergillus penicillioides* JH06THJ and *Eurotium halophilicum* FRR 2471 to characterize germination performance over the entire temperature window (see Fig. 4)[a].

Incubation temperature (°C)	Stressor type and concentration (M)[a]	Water activity[b]
Xeromyces bisporus FRR 0025		
2	Glycerol (5.5)+sucrose (0.3)	0.720
6	Glycerol (5.5)+sucrose (0.3)	0.719
10	Glycerol (5.5)+sucrose (0.3)	0.722
15	Glycerol (5.5)+sucrose (0.3)	0.721
20	Glycerol (5.5)+sucrose (0.3)	0.722
25	Glycerol (5.5)+sucrose (0.3)	0.722
30	Glycerol (5.5)+sucrose (0.3)	0.722
37	Glycerol (5.5)+sucrose (0.3)	0.719
44	Glycerol (5.5)+sucrose (0.3)	0.720
50	Glycerol (5.5)+sucrose (0.3)	0.718
Aspergillus penicillioides JH06THJ		
2	Glycerol (5.5)+NaCl (1.2)	0.728
6	Glycerol (5.5)+NaCl (1.2)	0.725
10	Glycerol (5.5)+NaCl (1.2)	0.724
15	Glycerol (5.5)+NaCl (1.2)	0.727
20	Glycerol (5.5)+NaCl (1.2)	0.726
25	Glycerol (5.5)+NaCl (1.2)	0.729
30	Glycerol (5.5)+NaCl (1.2)	0.729
37	Glycerol (5.5)+NaCl (1.2)	0.724
44	Glycerol (5.5)+NaCl (1.2)	0.722
50	Glycerol (5.5)+NaCl (1.2)	0.726
Eurotium halophilicum FRR 2471		
2	Glycerol (5.5)+NaCl (0.25)+sucrose (0.25)	0.723
6	Glycerol (5.5)+NaCl (0.25)+sucrose (0.25)	0.721
10	Glycerol (5.5)+NaCl (0.25)+sucrose (0.25)	0.724
15	Glycerol (5.5)+NaCl (0.25)+sucrose (0.25)	0.723
20	Glycerol (5.5)+NaCl (0.25)+sucrose (0.25)	0.725
25	Glycerol (5.5)+NaCl (0.25)+sucrose (0.25)	0.725
30	Glycerol (5.5)+NaCl (0.25)+sucrose (0.25)	0.726
37	Glycerol (5.5)+NaCl (0.25)+sucrose (0.25)	0.724
44	Glycerol (5.5)+NaCl (0.25)+sucrose (0.25)	0.721
50	Glycerol (5.5)+NaCl (0.25)+sucrose (0.25)	0.724

[a]All media were based on MYPiA: 1% malt extract, 1% yeast extract, 0.1% KH_2PO_4 and 1.5% (w/v) agar (Williams and Hallsworth, 2009).
[b]The water activity of each medium was measured at the same temperature at which plates were incubated and replicate values were within ± 0.001 water activity (see *Experimental procedures*).

water-activity range tested (0.995–0.640) (Fig. 1). An earlier study of xerophile germination reported limits of 0.820–0.740 for *X. bisporus* (FRR 2347), 0.780–0.740 for *A. penicillioides* (FRR 3772) and 0.700 for *Eurotium repens* (strain FRR 382), depending on pH, indicating that the strains used in the current study were considerably more xerophilic (Fig. 2; Gock *et al.*, 2003). Furthermore, *X. bisporus* (FRR 0025) was not only able to germinate at 0.637 water activity, but it also did so at a

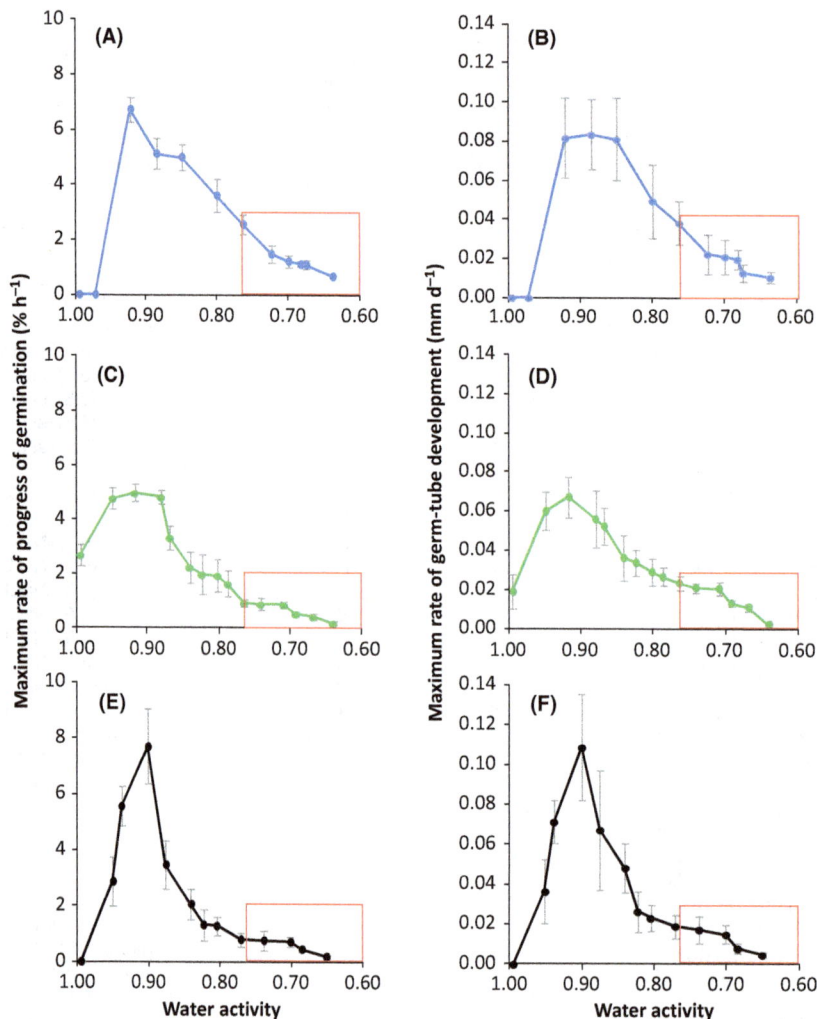

Fig. 1. Maximum rates of spore germination (% of total h^{-1}) and germ-tube development for *Xeromyces bisporus* FRR 0025 (A and B), *Aspergillus penicillioides* JH06THJ (C and D) and *Eurotium halophilicum* FRR 2471 (E and F) over a range of water activity, on malt-extract, yeast-extract phosphate agar (MYPiA) supplemented with diverse stressor(s) and incubated at 30°C. For *X. bisporus*, *A. penicillioides* and *E. halophilicum*, media were supplemented with glycerol+sucrose, glycerol+NaCl or glycerol+NaCl+sucrose (respectively) over a range of concentrations to give a range of water activities from 0.995 to 0.637 (see Table 1). The red box indicates the water-activity window selected for an additional study carried out to assess the potency of glycerol as a determinant for the water-activity limit for life (*Experimental procedures*; Stevenson *et al.*, in press). Maximum rates of germination and germ-tube development were determined from the curves (data not shown) and grey bars indicate standard errors.

reasonable rate implying that the actual water-activity window is more extensive (Fig. 1A and B).

Earlier studies have suggested a higher water-activity limit for mycelial extension than that observed for germination (Williams and Hallsworth, 2009). However, the current study indicates water-activity minima for germination which are more or less equivalent to those reported for mycelial extension (Williams and Hallsworth, 2009; Stevenson *et al.*, 2015a); possibly a consequence of the comparable types of (high-glycerol) media used in these three studies. Without biotechnological interventions, many fungi have a water-activity minimum for germination in the range 0.940–0.935 (Hallsworth and Magan, 1995; Hallsworth *et al.*, 2003a,b; Lahouar *et al.*, 2016). There are a few microbes known to have a water-activity

window for biotic activity that is comparable to that for the xerophiles in the current study (Stevenson *et al.*, 2015a).

The pH windows for germination were considerable, spanning a range which was ~6,7 and 7.5 pH-units wide for *X. bisporus* (FRR 0025), *A. penicillioides* (JH06THJ) and *E. halophilicum* (FRR 2471) respectively (Fig. 3). There was, however, a marked difference between strains. *X. bisporus* (FRR 0025) was unable to germinate at pH 8.8 but did so at a relatively high rate at pH 2.9, the lowest value tested; *A. penicillioides* (JH06THJ) did not germinate at pH 9.8 but did so at a reasonable rate at pH 3.0, the lowest value tested; and *E. halophilicum* (FRR 2471) germinated at a reasonable rates at pH 9.5 and 2.8, the most extreme values tested (Fig. 3).

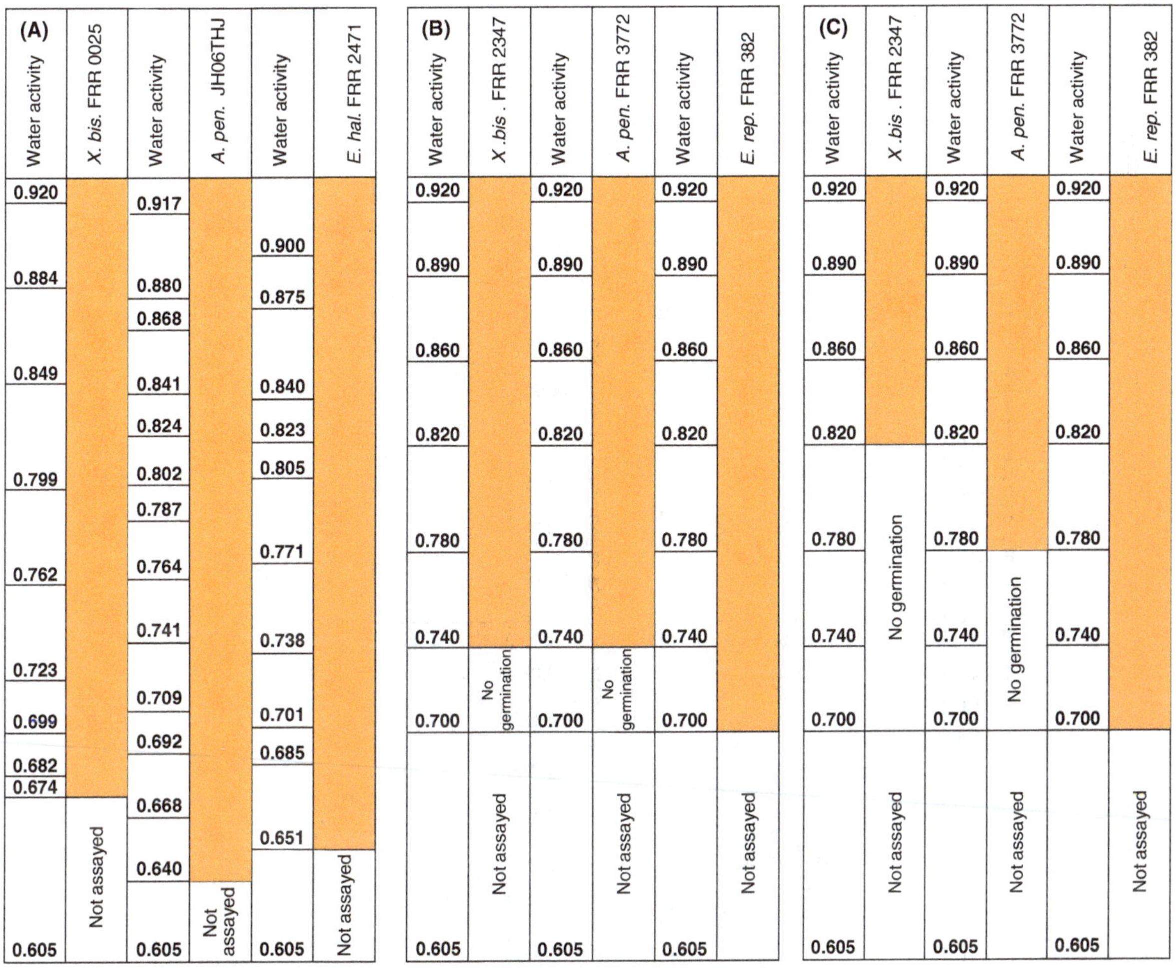

Fig. 2. Comparisons of ability to germinate for diverse xerophile strains at sub-optimal water-activity values at 30°C by 30 days: (A) *X. bisporus* (FRR 0025), *A. penicillioides* (JH06THJ) and *E. halophilicum* (FRR 2471) (current study) and (B and C) *X. bisporus* (FRR 2347), *A. penicillioides* (FRR 3722) and *Eurotium repens* (FRR 382) (Gock *et al.*, 2003); media were in the pH range 6.3–7.2 (A), and at pH 6.5 (B) and (C). Orange shading indicates germination within the 30-day period.

The three temperature windows for germination of the xerophile strains were relatively similar, and spanned a range of approximately 30°C (Fig. 4). However, *A. penicillioides* appeared slightly more capable at low temperature; it was able to germinate at 15°C (Fig. 4C and D). The psychrotolerance of closely related strains was characterized in an earlier study which found that, in the presence of chaotropic substances (Ball and Hallsworth, 2015), mycelial growth occurred close to 0°C (Chin *et al.*, 2010). The 30°C-wide temperature windows for germination of the strains in the current study were not exceptionally wide by comparison with those of some bacterial strains (Santos *et al.*, 2015).

Optimum conditions for xerophilicity

For the three strains, optimum rates of germination and germ-tube development were observed under similar conditions (Figs 1, 3 and 4). Generally, the optimum pH

for germination lay between 4.5 and 7.7 (Fig. 3) and the optimum germination temperature was 30°C, although *A. penicillioides* (JH06THJ) was equally capable at 37°C (Fig. 4). The pH- and temperature optima are consistent with those reported for germination and mycelia growth of xerophiles (Gock *et al.*, 2003; Williams and Hallsworth, 2009). However, rates of germination and germ-tube development were slow for *X. bisporus* (FRR 0025) and *E. halophilicum* (FRR 2471) in pH- and temperature assays (Figs 3 and 4) because the water activity of germination media was sub-optimal (i.e. in the range 0.729–0.717; Tables 2 and 3; Fig. 1). Germination and mycelial growth of xerophiles typically occur at the lowest water-activity at ~30°C and pH values in the range 5.5–7.5 (Pitt, 1975; Gock *et al.*, 2003; Williams and Hallsworth, 2009), and this is consistent with the high levels of xerophilicity observed in the current study at 30°C and in the pH range 6.30–7.20 (Table 1; Fig. 1).

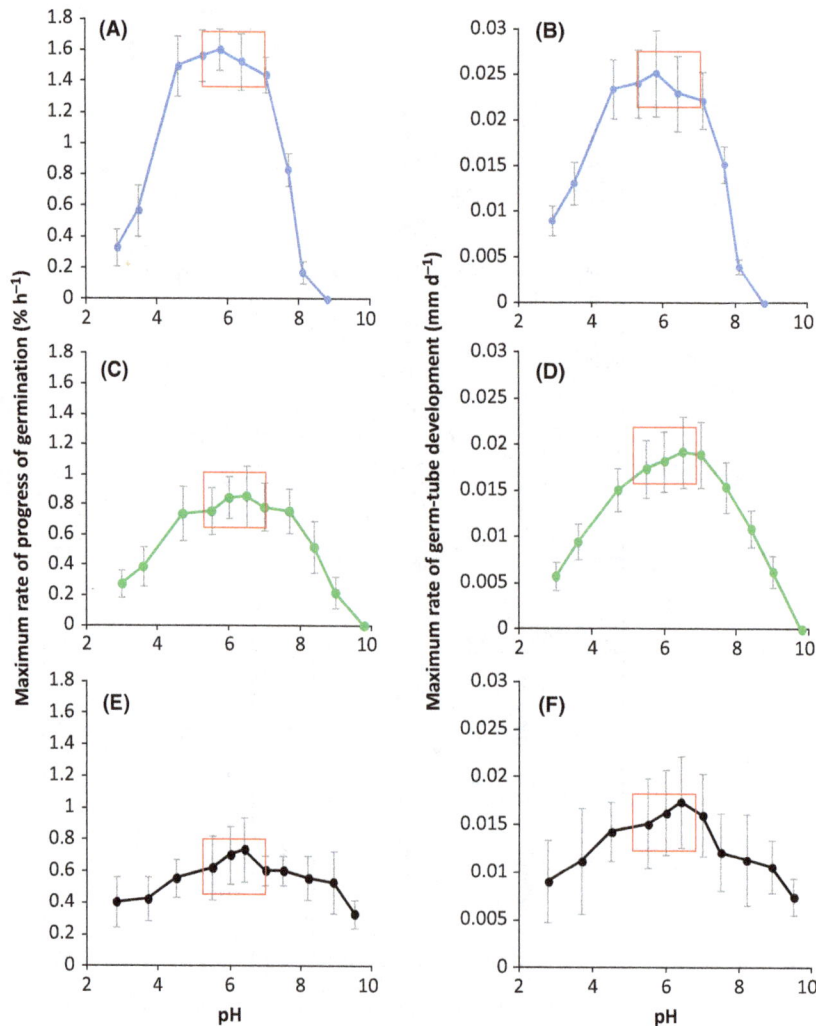

Fig. 3. Maximum rates of spore germination (% of total h^{-1}) and germ-tube development for *Xeromyces bisporus* FRR 0025 (A and B), *Aspergillus penicillioides* JH06THJ (C and D) and *Eurotium halophilicum* (E and F) over a range of pH values, on malt-extract, yeast-extract phosphate agar (MYPiA) supplemented with diverse stressor(s), buffered and incubated at 30°C. For *X. bisporus*, media were supplemented with glycerol (5.5 M)+sucrose (0.4 M); for *A. penicillioides* with glycerol (5.5 M)+ NaCl (1.2 M) and for *E. halophilicum* with glycerol (5.5 M)+NaCl (0.25 M)+sucrose (0.25 M), and buffered to give pH values from 2.80 to 9.80 (see Table 2). The red box indicates the pH window selected for an additional study carried out to assess the potency of glycerol as a determinant for the water-activity limit for life (*Experimental procedures*; Stevenson *et al.*, in press). Maximum rates of germination and germ-tube development were determined from the curves (data not shown) and grey bars indicate standard errors.

Asymmetrical response to supra- *and sub-optimal conditions*

For sub- and *supra*-optimal pH values, the decreases in germination and germ-tube development were more or less equivalent (Fig. 3). By contrast, the germination curves for water activity and temperature are asymmetrical (Figs 1 and 4). Asymmetry can be observed in biological systems at various levels; from the chirality of metabolites (Neville, 1976; Clark, 1977) to asymmetrical stress mechanisms, growth kinetics or dynamics of ecosystem development (current study; McCammick *et al.*, 2010; Cray *et al.*, 2013a). In terms of cellular stress

parameters, conditions which entropically disorder macro-molecular systems are the most severely inhibitory; e.g. *supra*-optimal water activities, temperatures or extreme chaotropicity (Figs 1 and 4; Hallsworth and Magan, 1994; Hallsworth *et al.*, 1998, 2003b; Hallsworth *et al.*, 2007; Bell *et al.*, 2013; Cray *et al.*, 2013a, 2015). By contrast, low temperature, low water-activity and kosmotropic substances induce a more gradual decrease in microbial growth/metabolism (Figs 1 and 4; Chin *et al.*, 2010).

There was considerable variation in germination and germ-tube development within each spore population assayed, as indicated by error bars (Figs 1, 3 and 4). For fungal xerophiles, both intraspecies and intrastrain

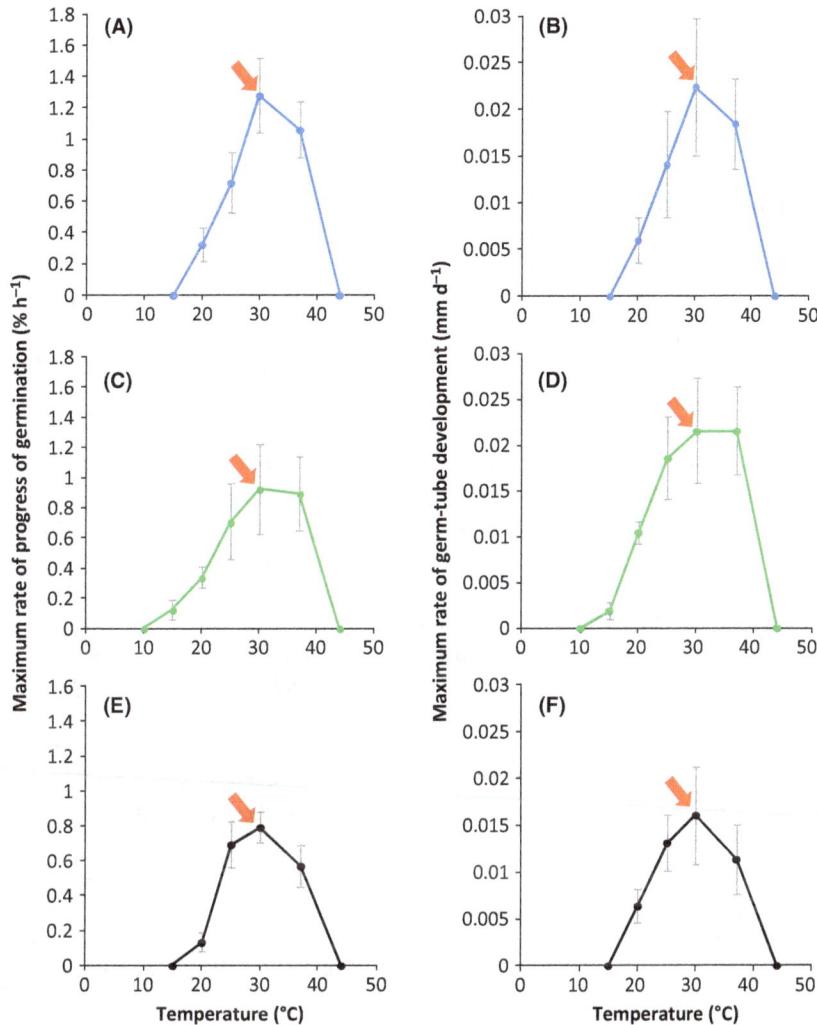

Fig. 4. Maximum rates of spore germination (% of total h^{-1}) and germ-tube development for *Xeromyces bisporus* FRR 0025 (A and B), *Aspergillus penicillioides* JH06THJ (C and D) and *Eurotium halophilicum* (E and F) over a range of temperatures on malt-extract, yeast-extract phosphate agar (MYPiA) supplemented with diverse stressor(s) and incubated between 2 and 50°C (see Table 3). For *X. bisporus*, media were supplemented with glycerol (5.5 M)+sucrose (0.4 M); for *A. penicillioides* with glycerol (5.5 M)+ NaCl (1.2 M) and for *E. halophilicum* with glycerol (5.5 M)+NaCl (0.25 M)+sucrose (0.25 M). The red arrow indicates the temperature at which an additional study was carried out to assess the potency of glycerol as a determinant for the water-activity limit for life (*Experimental procedures*; Stevenson *et al.*, in press). Maximum rates of germination and germ-tube development were determined from the curves (data not shown) and grey bars indicate standard errors.

variability of spore phenotype/behaviour (see also Stevenson *et al.*, in press) exemplify the natural variation inherent to biological systems (e.g. Hallsworth *et al.*, 2003a; Cray *et al.*, 2013a); a phenomenon which is even observed in populations of spores which have been harvested from an individual fungal colony. This is illustrated, for instance, by variation of survival which was demonstrated by the inactivation kinetics of fungal spores exposed to extreme stresses (Dijksterhuis *et al.*, 2013). In addition, a population of spores has a distribution of lag-phase times and rates of germination; both of these increase at low water-activity. Dagnas *et al.* (2015) described an increase in the spread of germination times for conidia of *Penicillium corylophilum* treated with

essential oils within red-cabbage seed extract. The increase of variability within a cell population can act in such a way that enables the population to overcome/circumvent stress. For instance, heterogeneity of the germination process even occurs within the multicellular macroconidia of *Fusarium culmorum*, where germ tubes are formed on the apical cells in the majority of the cases (Chitarra *et al.*, 2005). Interestingly, if apical cells were killed, the centrally located cells germinated giving the macroconidium a second chance (Chitarra *et al.*, 2005). Such variation, observed in populations of diverse types of propagule, is an inherent part of the ecology of both plant - and microbial species and may enhance the prevalence of a population within a specific habitat (Hill,

1977; Chitarra et al., 2005; Cray et al., 2013a; Oren and Hallsworth, 2014). Even the diversity of expressed genes can increase upon stress, as observed in *Aspergillus niger* conidia treated with the antifungal polyene natamycin (van Leeuwen et al., 2013). Phenotypic heterogeneity has also been reported for genetically homogeneous populations of bacteria (Touzain et al., 2010).

Concluding remarks

Under extreme-yet-permissive water-activity regimes, the germination of some of the most extremophilic microbes on Earth was characterized. Accordingly, the water activity windows for germination were extraordinarily wide (~1 to 0.651–0.637). The asymmetry of their germination kinetics reflected entropy-level stress mechanisms which operate on cellular macromolecules. The biotic windows of these strains, kinetics of germination and germ-tube development, symmetry/asymmetry of their stress phenotype and the inherent variation in behaviour within spore populations can act as determinants for ecological processes. Fundamental knowledge of xerophile behaviour based on this and other studies[1] suggests that these strains may have potential as model systems which can be used to address scientific questions in the fields of extremophile biology, biosphere function, food spoilage, astrobiology, biophysics and synthetic biology. The current study gave rise to a number of unanswered questions: might glycerol be able to catalyse germination at the water-activity limit for life (see *Experimental procedures*); under some conditions, can glycerol enable fungal germination at < 0.600 water activity; and can glycerol enhance the habitability of hostile environments. *A. penicillioides*, and other *Aspergillus* spp., are regularly found as contaminants of space craft, are highly tolerant to salt and low temperatures, and can function under microaerophilic or anaerobic conditions. It is pertinent, therefore to ask: what are the implications of abiotic glycerol, which has been identified in extraterrestrial locations, for potential contamination of other planetary bodies with terrestrial microbes during space-exploration missions?

Experimental procedures

Fungal strains, media and culture conditions

A. penicillioides strain JH06THJ was isolated by Williams and Hallsworth (2009) and is available from the corresponding author of the current article. *E. halophilicum*[2] strain FRR 2471 and *X. bisporus* strain FRR 0025 were

obtained from CSIRO Food and Nutritional Sciences Culture Collection (North Ryde, NSW, Australia). Cultures were maintained on malt-extract, yeast-extract phosphate agar (MYPiA; 10 g malt extract, 10 g yeast extract, 1 g anhydrous K_2HPO_4, agar 15 g l^{-1}) supplemented with 5.5 M glycerol (0.821 water activity) at 30°C.

Production of spores; germination assays

Spores were obtained from cultures incubated on MYPiA+glycerol (5.5 M) for 10–14 days for *A. penicillioides* and 21–28 days for *X. bisporus* and *E. halophilicum*. Spores were harvested from colonies growing on MYPiA+glycerol (5.5 M) media by covering Petri plates with sterile solutions of 5.5 M glycerol (15 ml); aerial spores were then dislodged by gently brushing with a sterile glass rod. The resulting suspension was passed through sterile glass-wool twice, to remove hyphal fragments as described in earlier studies (Hallsworth and Magan, 1995; Chin et al., 2010). Spore suspensions were then adjusted to a final concentration of 1×10^6 spores ml^{-1}. Inoculation of germination media was carried out by pipetting the spore suspension (150 µl) onto the medium; the suspension was then distributed across the agar surface using a sterilize glass spreader.

Germination was assessed by removing a 4-mm agar disc, and immediately quantifying percentage germination, spore diameter and germ-tube length using a light microscope. Plates were immediately resealed and placed back in the incubator after removal of the agar discs. Percentage germination was determined via counts of 200 spores, and 50 individual germinated spores were measured for germ-tube length; spores with germ-tubes longer than their diameter were considered to have germinated (Hallsworth and Magan, 1995). In each case, percentage germination and mean germ-tube length were determined for isolated spores and were not assessed for any spores located in clumps. Assessments were made at least daily over a 30-day period.

The germination process was characterized over the entire windows of water activity, temperature and pH for these three xerophile strains. Water relations for germination were assessed on MYPiA-based media supplemented, using a range of concentrations, with glycerol+sucrose for *X. bisporus* FRR 0025, glycerol+NaCl for *A. penicillioides* JH06THJ and glycerol+NaCl+sucrose *E. halophilicum* FRR 2471, as these media were highly permissive for germination of these strains in the low water-activity range (Stevenson et al., in press). For the determination of germination rates over the pH range, media were supplemented with: glycerol (5.5 M)+sucrose (0.4 M) for *X. bisporus* FRR 0025, glycerol (5.5 M)+NaCl (1.2 M) for *A. penicillioides* JH06THJ and glycerol (5.5 M)+NaCl (0.25 M)+sucrose (0.25 M)

---

[1]Pitt (1975); Gock et al. (2003); Williams and Hallsworth (2009); Chin et al. (2010); Alves et al. (2015); Leong et al. (2015); Stevenson et al. (2015a); Cray et al. (2016).

[2]Recently renamed as *Aspergillus halophilicus* (Hubka et al., 2013).

for *E. halophilicum* FRR 2471 and buffered to give pH values of media from 2.90 to 9.80 (see Table 2). Germination was characterized over a temperature range using the following media: glycerol (5.5 M)+NaCl (1.2 M) for *A. penicillioides* JH06THJ, glycerol (5.5 M)+sucrose (0.4 M) for *X. bisporus* FRR 0025 and glycerol (5.5 M)+NaCl (0.25 M)+sucrose (0.25 M) for *E. halophilicum* FRR 2471 over the range 2–50°C (Table 3).

Quantification of pH and water activity

The pH values for pre-autoclaved media were determined using a Mettler Toledo Seven Easy pH-probe (Mettler Toledo, Greifensee, Switzerland); values for solid media (post-autoclaved) were determined prior to inoculation using Fisherbrand colour-fixed pH indicator strips (Fisher Scientific Ltd, Leicestershire, UK). The water activity of all media was determined empirically using a Novasina Humidat-IC-II water-activity machine fitted with an alcohol-resistant humidity sensor and eVALC alcohol filter (Novasina, Pfäffikon, Switzerland). Water-activity measurements were taken at the same temperature at which cultures were to be incubated and several precautions were employed to ensure accuracy of readings, as described previously (Hallsworth and Nomura, 1999; Stevenson *et al.*, 2015a). The instrument was calibrated between each measurement using saturated salt solutions of known water activity (Winston and Bates, 1960). The water activity of each medium type was determined three times, and variation was within ± 0.001. Media chao-/kosmotropicity values were determined using the agar-gelation method described by Cray *et al.* (2013b). Extra-pure reagent-grade agar (Nacalai Tesque, Kyoto, Japan), at 1.5% w/v and supplemented with stressors at the concentrations used in the medium, was used to determine chao-/kosmotropicity values for added solutes (see Hallsworth *et al.*, 2003b; Cray *et al.*, 2013b). A Cecil E2501 spectrophotometer fitted with a thermoelectrically controlled heating block was used to determine the wavelength and absorbance values at which to assay compounds, and values for chao-/kosmotropic activity were calculated relative to those of the control (no added solute) as described by Cray *et al.* (2013b).

Replication; analysis and presentation of data

All measurements were carried out in triplicate. The values for maximal rates of percentage germination for the three model strains over a range of water activity (0.995–0.640), pH (2.80–9.80) and temperature (10–44°C) (Figs 1, 3 and 4) were determined according to the exponential part of curves plotted for these parameters over time (data not shown). The data obtained are presented as the maximum rate of progress of germination (% of total spores h^{-1}) versus water activity, pH and temperature (Figs 1, 3 and 4 respectively). Minimum water-activity values at which germination occurred, within a 30-day incubation period, were compared with those reported previously (Fig. 2). Optimum temperatures and pH values, and minimum water activity ranges, on which future studies to elucidate the potential role of glycerol as a determinant for the water-activity limit for life should focus were identified and indicated (Figs 1, 3 and 4). A further study involving biophysically diverse types of culture media (see also Williams and Hallsworth, 2009; Stevenson *et al.*, 2015a) was carried out to establish whether glycerol can enhance catalyse germination at the water-activity limit for life (Stevenson *et al.*, in press).

Conflict of interest

None declared.

References

Alves, F., Stevenson, A., Baxter, E., Gillion, J.L., Hejazi, F., Hayes, S., *et al.* (2015) Concomitant osmotic and chaotropicity-induced stresses in Aspergillus wentii: compatible solutes determine the biotic window. *Curr Genet* **61:** 457–477.

Ball, P., and Hallsworth, J.E. (2015) Water structure and chaotropicity: their uses and abuses. *Phys Chem Chem Phys* **17:** 8297–8305.

Bell, A.N.W., Magill, E., Hallsworth, J.E., and Timson, D.T. (2013) Effects of alcohols and compatible solutes on the activity of β-galactosidase. *Appl Biochem Biotech* **169:** 786–796.

Chin, J.P., Megaw, J., Magill, C.L., Nowotarski, K., Williams, J.P., Bhaganna, P., *et al.* (2010) Solutes determine the temperature windows for microbial survival and growth. *Proc Natl Acad Sci USA* **107:** 7835–7840.

Chitarra, G.S., Dijksterhuis, J., Breeuwer, P., Rombouts, F.M., and Abee, T. (2005) Differentiation inside multicelled macroconidia of *Fusarium culmorum* during early germination. *Fungal Genet Biol* **42:** 694–703.

Clark, N.G. (1977) *The Shapes of Organic Molecules.* London: John Murray.

Cray, J.A., Bell, A.N.W., Bhaganna, P., Mswaka, A.Y., Timson, D.J., and Hallsworth, J.E. (2013a) The biology of habitat dominance; can microbes behave as weeds? *Microbiol Biotechnol* **6:** 453–492.

Cray, J.A., Russell, J.T., Timson, D.J., Singhal, R.S., and Hallsworth, J.E. (2013b) A universal measure of chaotropicity and kosmotropicity. *Environ Microbiol* **15:** 287–296.

Cray, J.A., Stevenson, A., Ball, P., Bankar, S.B., Eleutherio, E.C., Ezeji, T.C., *et al.* (2015) Chaotropicity: a key factor in product tolerance of biofuel-producing microorganisms. *Curr Opin Biotechnol* **33:** 228–259.

Cray, J.A., Connor, M.C., Stevenson, A., Houghton, J.D.R., Rangel, D.E.N., Cooke, L.R., *et al.* (2016) Biocontrol agents promote growth of potato pathogens, depending on environmental conditions. *Microbial Biotechnol* **9:** 330–354.

Dagnas, S., Gougouli, M., Onno, B., Koutsoumanis, K.P., and Membré, J.M. (2015) Modeling red cabbage seed extract effect on *Penicillium corylophilum*: relationship between germination time, individual and population lag time. *Int J Food Microbiol* **211:** 86–94.

Dijksterhuis, J., Rodriquez de Massaguer, P., Silva, D., and Dantigny, P. (2013) Primary models for inactivation of fungal spores. In *Predictive Mycology*. Zwieterink, M., and Dantigny, P. (eds). New York: Nova Science Publishers, pp. 131–152.

Gock, M.A., Hocking, A.D., Pitt, J.I., and Poulos, P.G. (2003) Influence of temperature, water activity and pH on growth of some xerophilic fungi. *Int J Food Microbiol* **81:** 11–19.

Hallsworth, J.E., and Magan, N. (1994) Effects of KCl concentration on accumulation of acyclic sugar alcohols and trehalose in conidia of three entomopathogenic fungi. *Lett Appl Microbiol* **18:** 8–11.

Hallsworth, J.E., and Magan, N. (1995) Manipulation of intracellular glycerol and erythritol enhances germination of conidia at low water availability. *Microbiology* **141:** 1109–1115.

Hallsworth, J.E., and Nomura, Y. (1999) A simple method to determine the water activity of ethanol-containing samples. *Biotechnol Bioeng* **62:** 242–245.

Hallsworth, J.E., Nomura, Y., and Iwahara, M. (1998) Ethanol-induced water stress and fungal growth. *J Ferment Bioeng* **86:** 451–456.

Hallsworth, J.E., Prior, B.A., Nomura, Y., Iwahara, M., and Timmis, K.N. (2003a) Compatible solutes protect against chaotrope (ethanol)-induced, nonosmotic water stress. *Appl Environ Microbiol* **69:** 7032–7034.

Hallsworth, J.E., Heim, S., and Timmis, K.N. (2003b) Chaotropic solutes cause water stress in *Pseudomonas putida*. *Environ Microbiol* **5:** 1270–1280.

Hallsworth, J.E., Yakimov, M.M., Golyshin, P.N., Gillion, J.L.M., D'Auria, G., Alves, F.L., *et al.* (2007) Limits of life in MgCl$_2$-containing environments: chaotropicity defines the window. *Environ Microbiol* **9:** 803–813.

Hill, T.A. (1977) *The Biology of Weeds*. London, UK: E. Arnold.

Hubka, V., Kolarik, M., Kubátová, A., and Peterson, S.W. (2013) Taxonomic revision of the genus *Eurotium* and transfer of species to *Aspergillus*. *Mycologia* **105:** 912–937.

Lahouar, A., Marin, S., Crespo-Sempere, A., Saïd, S., and Sanchis, V. (2016) Effects of temperature, water activity and incubation time on fungal growth and aflatoxin B1 production by toxinogenic *Aspergillus flavus* isolates on sorghum seeds. *Rev Argent Microbiol* **48:** 78–85.

van Leeuwen, M.R., Wyatt, T.T., Golovina, E.A., Stam, H., Menke, H., Dekker, A., *et al.* (2013) The effect of natamycin on the transcriptome of conidia of *Aspergillus niger*. *Stud Mycol* **74:** 71–85.

Leong, S.L.L., Lantz, H., Pettersson, O.V., Frisvad, J.C., Thrane, U., Heipieper, H.J., *et al.* (2015) Genome and physiology of the ascomycete filamentous fungus *Xeromyces bisporus* the most xerophilic organism isolated to date. *Environ Microbiol* **17:** 496–513.

Lievens, B., Hallsworth, J.E., Belgacem, Z.B., Pozo, M.I., Stevenson, A., Willems, K.A., *et al.* (2015) Microbiology of sugar-rich environments: diversity, ecology, and system constraints. *Environ Microbiol* **17:** 278–298.

McCammick, E.M., Gomase, V.S., Timson, D.J., McGenity, T.J., and Hallsworth, J.E. (2010) Water-hydrophobic compound interactions with the microbial cell. In *Handbook of Hydrocarbon and Lipid Microbiology – Hydrocarbons, Oils and Lipids: Diversity, Properties and Formation*, Vol. **2**. Timmis, K.N. (ed). New York: Springer, pp. 1451–1466.

Neville, A.C. (1976) *Animal Asymmetry*. London: Edward Arnold.

Oren, A., and Hallsworth, J.E. (2014) Microbial weeds in hypersaline habitats: the enigma of the weed-like *Haloferax mediterranei*. *FEMS Microbiol Lett* **359:** 134–142.

Pitt, J.I. (1975) Xerophilic fungi and the spoilage of foods of plant origin. In *Water Relations of Foods*. Duckworth, R.B. (ed). London, UK: Academic Press, pp. 273–307.

Pitt, J.I., and Christian, J.H.B. (1968) Water relations of xerophilic fungi isolated from prunes. *Appl Environ Microbiol* **16:** 1853–1858.

Rummel, J.D., Beaty, D.W., Jones, M.A., Bakermans, C., Barlow, N.G., Boston, P., et al. (2014) A new analysis of Mars 'Special Regions', findings of the second MEPAG Special Regions Science Analysis Group (SR-SAG2). *Astrobiology* **14:** 887–968.

Samson, R.A., and Lustgraaf, B.V.D. (1978) *Aspergillus penicilloides* and *Eurotium halophilicum* in association with house-dust mites. *Mycopathologia* **64:** 13–16.

Santos, R., de Carvalho, C.C.R., Stevenson, A., Grant, I.R., and Hallsworth, J.E. (2015) Extraordinary solute-stress tolerance contributes to the environmental tenacity of mycobacteria. *Environ Microbial Rep* **7:** 746–764.

Stevenson, A., and Hallsworth, J.E. (2014) Water and temperature relations of soil Actinobacteria. *Environ Microbiol Rep* **6:** 744–755.

Stevenson, A., Cray, J.A., Williams, J.P., Santos, R., Sahay, R., Neuenkirchen, N., *et al.* (2015a) Is there a common water-activity limit for the three domains of life? *ISME J* **9:** 1333–1351.

Stevenson, A., Burkhardt, J., Cockell, C.S., Cray, J.A., Dijksterhuis, J., Fox-Powell, M., *et al.* (2015b) Multiplication of microbes below 0.690 water activity: implications for terrestrial and extraterrestrial life. *Environ Microbiol* **2:** 257–277.

Stevenson, A., Hamill, P.G., Medina, A., Kminek, G., Rummel, J.D., Dijksterhuis, J., *et al.* (in press) Glycerol enhances fungal germination at the water-activity limit for life. *Environ Microbiol*.

Touzain, F., Denamur, E., Medigne, C., Barbe, V., El Karoui, M., and Petit, M.A. (2010) Small variable segments constitute a major type of diversity of bacterial genomes at the species level. *Genome Biol* **11:** R45.

Williams, J.P., and Hallsworth, J.E. (2009) Limits of life in hostile environments; no limits to biosphere function? *Environ Microbiol* **11:** 3292–3308.

Winston, P.W., and Bates, P.S. (1960) Saturated salt solutions for the control of humidity in biological research. *Ecology* **41:** 232–237.

Wyatt, T.T., van Leeuwen, M.R., Gerwig, G.J., Golovina, E.A., Hoekstra, F.A., Kuenstner, E.J., *et al.* (2015a)

Functionality and prevalence of trehalose-based oligosaccharides as novel compatible solutes in ascospores of *Neosartorya fischeri* (*Aspergillus fischeri*) and other fungi. *Environ Microbiol* **17**: 395–411.

Wyatt, T.T., Golovina, E.A., Leeuwen, R., Hallsworth, J.E., Wösten, H.A., and Dijksterhuis, J. (2015b) A decrease in bulk water and mannitol and accumulation of trehalose and trehalose-based oligosaccharides define a two-stage maturation process towards extreme stress resistance in ascospores of *Neosartorya fischeri* (*Aspergillus fischeri*). *Environ Microbiol* **17**: 383–394.

Yakimov, M.M., Lo Cono, V., La Spada, G., Bortoluzzi, G., Messina, E., Smedile, F., *et al.* (2015) Microbial community of seawater-brine interface of the deep-sea brine Lake Kryos as revealed by recovery of mRNA are active below the chaotropicity limit of life. *Environ Microbiol* **17**: 364–382.

Engineering a bzd cassette for the anaerobic bioconversion of aromatic compounds

María Teresa Zamarro, María J. L. Barragán,[†] Manuel Carmona, José Luis García and Eduardo Díaz*
Centro de Investigaciones Biológicas, CSIC, C/Ramiro de Maeztu, 9, 28040 Madrid, Spain.

Summary

Microorganisms able to degrade aromatic contaminants constitute potential valuable biocatalysts to deal with a significant reusable carbon fraction suitable for eco-efficient valorization processes. Metabolic engineering of anaerobic pathways for degradation and recycling of aromatic compounds is an almost unexplored field. In this work, we present the construction of a functional bzd cassette encoding the benzoyl-CoA central pathway for the anaerobic degradation of benzoate. The bzd cassette has been used to expand the ability of some denitrifying bacteria to use benzoate as sole carbon source under anaerobic conditions, and it paves the way for future pathway engineering of efficient anaerobic biodegraders of aromatic compounds whose degradation generates benzoyl-CoA as central intermediate. Moreover, a recombinant *Azoarcus* sp. CIB strain harbouring the bzd cassette was shown to behave as a valuable biocatalyst for anaerobic toluene valorization towards the synthesis of poly-3-hydroxybutyrate (PHB), a biodegradable and biocompatible polyester of increasing biotechnological interest as a sustainable alternative to classical oil-derived polymers.

*For correspondence. E-mail ediaz@cib.csic.es

[†]Present address: U.S. Food and Drug Administration, Center for Drug Evaluation and Research (CDER), 10903 New Hampshire Avenue, Silver Spring, MD 20993, USA.

Founding Information
This work was supported by grants BIO2012-39501, BIO2016-79736-R and PCIN-2014-113 from the Ministry of Economy and Competitiveness of Spain; European Union FP7 Grant 311815; and by a grant of Fundación Ramón-Areces XVII CN.

Introduction

Aromatic compounds are the second most abundant class of organic compounds in nature after carbohydrates. Due to the thermodynamic stability of the aromatic ring, aromatic compounds are difficult to degrade and they tend to persist in the environment for long periods of time. Many of these compounds are toxic and/or carcinogenic thus representing major persistent environmental pollutants. Therefore, removal of aromatic compounds is very important both for a balanced global carbon budget and to protect wildlife and human health. Some specialized microorganisms (bacteria, archaea and fungi) have adapted to use aromatic compounds as sole carbon and energy source (mineralization) or, at least, partially degrade these molecules to less-toxic and persistent compounds (Carmona *et al.*, 2009; Fuchs *et al.*, 2011). These microorganisms constitute, thus, potential valuable biocatalysts to deal with a significant reusable carbon fraction suitable for eco-efficient valorization processes within the framework of a sustainable knowledge-based bio-economy.

There are two major strategies to degrade aromatic compounds depending on the presence or absence of oxygen. In the aerobic catabolism of aromatic compounds, oxygen is not only the final electron acceptor but also a cosubstrate for two key processes, i.e. the hydroxylation and oxygenolytic ring cleavage of the aromatic ring, carried out by oxygenases. In the absence of oxygen (anaerobic catabolism), the aromatic ring is dearomatized by reductive reactions (Carmona *et al.*, 2009; Fuchs *et al.*, 2011; Boll *et al.*, 2014; Rabus *et al.*, 2016). A wide variety of bacteria, pathways and associated gene clusters responsible for the aerobic catabolism of aromatic compounds have been studied and characterized (Díaz *et al.*, 2013). Moreover, genetic and metabolic engineering approaches have been applied to develop more efficient recombinant biocatalysts for the aerobic conversion of aromatic compounds to added value products, e.g. biopolymers, biofuels, commodity chemicals (Kosa and Ragauskas, 2012; Bugg and Rahmanpour, 2015; Wierckx *et al.*, 2015; Beckham *et al.*, 2016; Johnson *et al.*, 2016). In contrast, the anaerobic degradation of aromatic compounds has been much less well-studied than the aerobic degradation, especially regarding the genetic determinants that encode the anaerobic pathways, and there are only a few examples of recombinant anaerobic biodegraders (Coschigano *et al.*, 1994; Darley *et al.*, 2007; Zamarro *et al.*, 2016). However, anaerobic processes may offer significant benefits compared to aerobic bioprocesses, e.g. higher yields, less heat and oxidative stress generation, reduced biomass production and lower mechanical energy input, as they do not require aeration, which can significantly reduce production costs (Cueto-Rojas *et al.*, 2015). Thus, metabolic engineering of anaerobic pathways for degradation and recycling of waste aromatic compounds is still an almost unexplored field of great biotechnological potential.

Similar to the very well-known aerobic degradation strategies, the anaerobic degradation of aromatic compounds channels a wide variety of compounds into a few central intermediates through devoted peripheral degradation pathways (catabolic funnel). The different peripheral pathways converge into a few central pathways that carry out the reductive de-aromatization and further conversion of the central intermediates to compounds of the central metabolism of the cell. Most monocyclic aromatic compounds are channelled and activated to benzoyl-CoA. The catabolic genes encoding the benzoyl-CoA central pathway enzymes are usually arranged in large chromosomal clusters that also contain the specific transcriptional regulators (Egland *et al.*, 1997; Breese *et al.*, 1998; López-Barragán *et al.*, 2004; Rabus *et al.*, 2005, 2016; Wischgoll *et al.*, 2005; Carmona *et al.*, 2009; Holmes *et al.*, 2012; Carlström *et al.*, 2015; Hirakawa *et al.*, 2015).

Azoarcus sp. CIB is a facultative anaerobic beta-proteobacterium capable of degrading either aerobically and/or anaerobically (using nitrate as terminal electron acceptor) a wide range of aromatic compounds including some toxic hydrocarbons such as toluene (López-Barragán *et al.*, 2004; Martín-Moldes *et al.*, 2015; Zamarro *et al.*, 2016). Several pathways involved in the anaerobic degradation of aromatic compounds, including the benzoyl-CoA central pathway (*bzd* genes), have been characterized at the molecular level in strain CIB (López-Barragán *et al.*, 2004; Carmona *et al.*, 2009; Juárez *et al.*, 2013). In addition to this free-living lifestyle, *Azoarcus* sp. CIB also shows an endophytic lifestyle (Fernández *et al.*, 2014) and is able to resist some metals and metalloids, e.g. selenium oxyanions, producing nanoparticles of biotechnological interest (Fernández-Llamosas *et al.*, 2016). All these properties, together with the fact that the genome of strain CIB has been sequenced and annotated and different tools are available for its genetic manipulation (Martín-Moldes *et al.*, 2015), make *Azoarcus* sp. CIB a promising host for approaching metabolic engineering strategies to improve the anaerobic bioconversion of aromatic compounds.

In this work, we present the construction of a functional bzd cassette for anaerobic benzoate degradation and its application to the development of recombinant *Azoarcus* sp. CIB biocatalysts for toluene valorization towards the synthesis of poly-3-hydroxybutyrate (PHB), a biodegradable and biocompatible polyester of increasing biotechnological interest as a sustainable alternative to classical oil-derived polymers (Rehm, 2010; Nikodinovic-Runic *et al.*, 2013; Madbouly *et al.*, 2014).

Results and discussion

Construction of a functional bzd catabolic cassette for anaerobic degradation of benzoate

As indicated above, the benzoyl-CoA central pathway involved in the anaerobic degradation of benzoate and many other aromatic compounds whose peripheral pathways converge into benzoyl-CoA has been previously characterized in the denitrifying *Azoarcus* sp. CIB strain (López-Barragán *et al.*, 2004; Carmona *et al.*, 2009). Benzoate becomes initially activated to benzoyl-CoA by the benzoate-CoA ligase (BzdA) and is then de-aromatized by the action of a benzoyl-CoA reductase (BzdNOPQ), the only oxygen-sensitive enzyme within the benzoyl-CoA pathway, that uses a low-potential ferredoxin (BzdM) as electron donor and generates cyclohexa-1,5-diene-1-carbonyl-CoA (1,5-dienoyl-CoA) (Fig. 1A). Further degradation of 1,5-dienoyl-CoA resembles a modified β-oxidation pathway with addition of water to a double bond (BzdW dienoyl-CoA hydratase), a dehydrogenation reaction (BzdX hydroxyacyl-CoA dehydrogenase) and hydrolytic ring fission (BzdY oxoacyl-CoA hydrolase), generating finally 3-hydroxy-pimelyl-CoA that feeds into a lower pathway (Fig. 1A). The *bzd* genes encoding the bzd pathway enzymes are clustered together in a large operon driven by the P_N promoter (Fig. 1B). The specific transcriptional regulation of the *bzd* operon is conducted by the BzdR repressor that is encoded immediately upstream of the catabolic operon (Fig. 1B). Induction of the *bzd* genes requires the binding of the effector molecule, benzoyl-CoA, to the BzdR repressor (Durante-Rodríguez *et al.*, 2010). The *bzdNOPQMSTUVWXYZA* catabolic genes and the cognate *bzdR* regulatory gene have been engineered as a 19.6 kb DNA cassette into a broad-host range vector, giving rise to plasmid pLB1 (Fig. 1B). To construct the bzd cassette, the right end of the *bzd* cluster (genes *bzdXYZA*) from plasmid pECOR8 (Table 1) was first cloned as an EcoRI fragment into the EcoRI-digested broad-host range pIZ1016* vector, giving rise to plasmid pIZECO (Table 1). Then, the left end of the *bzd* cluster (genes *bzdRbzdNOPQMSTUVWX*) was cloned as a XbaI/NcoI-double digested fragment from the recombinant λBzd1 phage (López-Barragán *et al.*, 2004) into the

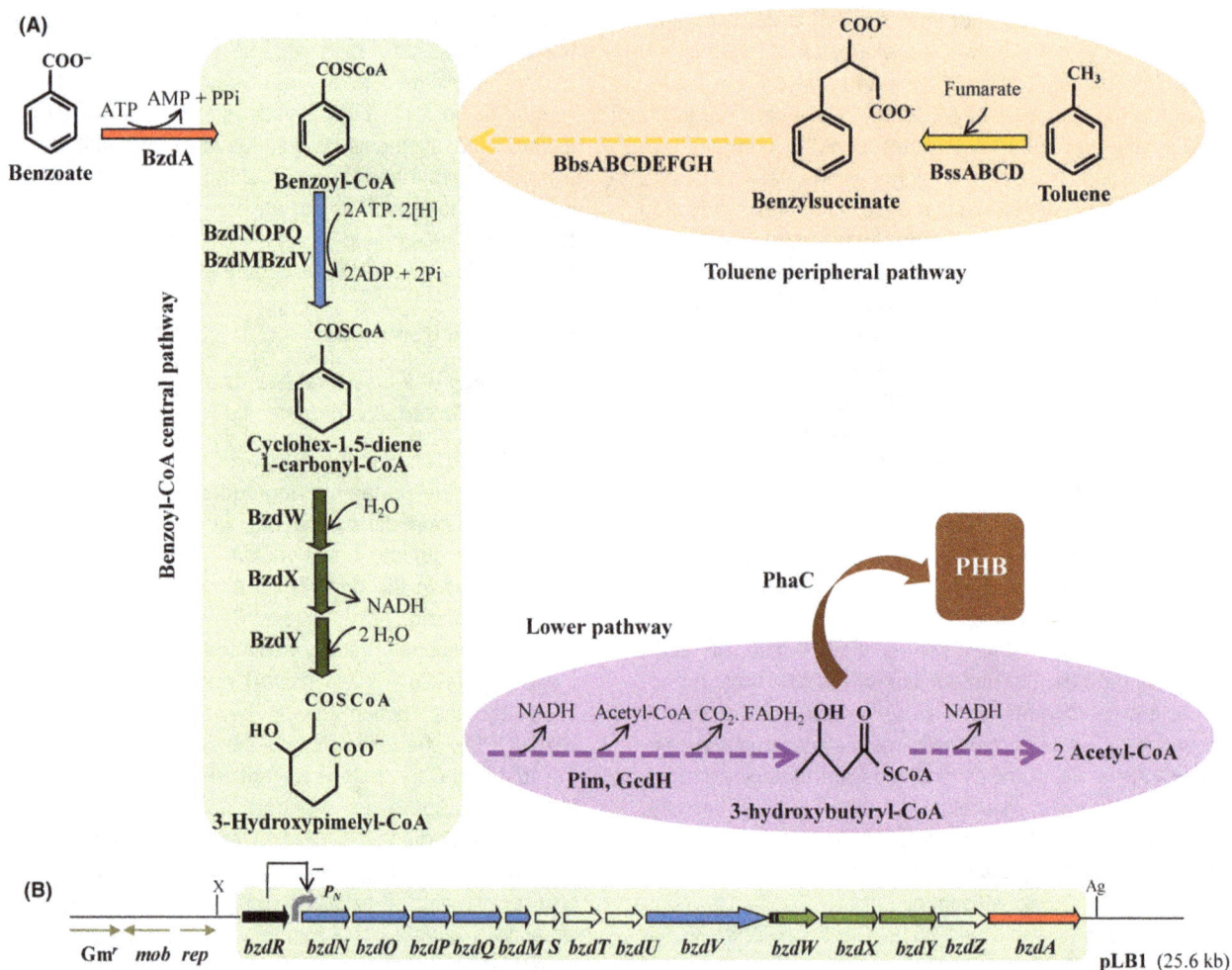

Fig. 1. Scheme of the anaerobic metabolism of benzoate and toluene, and gene organization of the bzd cassette in *Azoarcus* sp. CIB.
A. Scheme of peripheral pathway for the anaerobic conversion of toluene into benzoyl-CoA (orange), the activation of benzoate to benzoyl-CoA (red), the benzoyl-CoA central pathway (green), the lower pathway (violet) and the polymerization of 3-hydroxybutyryl-CoA to PHB (brown). Discontinuous arrows indicate that more than one enzymatic step is involved. Enzyme abbreviations: BssABCD, benzylsuccinate synthase; Bbs, enzymes involved in the modified β-oxidation of benzylsuccinate to benzoyl-CoA; BzdA, benzoate-CoA ligase; BzdNOPQ, benzoyl-CoA reductase; BzdM, ferredoxin; BzdV, putative NADPH:ferredoxin oxidoreductase; BzdW, cyclohex-1,5-diene-1-carbonyl-CoA hydratase; BzdX, 6-hydroxycyclohex-1-ene-1-carbonyl-CoA dehydrogenase; BzdY, 2-ketocyclohexane-1-carbonyl-CoA hydrolase; Pim, enzymes involved in β-oxidation of dicarboxylic acids; GcdH, glutaryl-CoA dehydrogenase; PhaC, PHB synthase.
B. Schematic representation of the *bzd* genes for anaerobic benzoate degradation engineered as a mobile DNA cassette in plasmid pLB1. Genes are indicated in the same colour code than the corresponding enzymes in panel A; i.e., genes encoding enzymes involved in the initial activation, de-aromatization and modified β-oxidation are indicated in red, blue and green colour respectively. The *bzdR* gene encoding a transcriptional repressor of the catabolic P_N promoter is shown in black. X and Ag, XbaI and AgeI restriction sites flanking the bzd cassette respectively. The gentamicin resistance gene (Gm^r), mobilization (*mob*) and replication (*rep*) functions are also indicated.

Xbal/Ncol-double digested pIZECO plasmid, giving rise to plasmid pLB1 (25.6 Kb) (Table 1). To check whether the bzd cassette was functional, it was transferred to a mutant strain, *Azoarcus* sp. CIBd*bzdN,* unable to use benzoate anaerobically because it contains a disruption insertion in the first gene of the *bzd* catabolic operon with avoids the expression of the rest of *bzd* genes (Table 1) (López-Barragán *et al.*, 2004). Plasmid pLB1 was transferred by biparental filter mating from *E. coli* S17-1λ*pir* (donor strain) to *Azoarcus* sp. CIBd*bzdN* (recipient strain) as previously described (López-Barragán

et al., 2004). Exconjugants harbouring the pLB1 plasmid, *Azoarcus* sp. CIBd*bzdN* (pLB1) (Table 1), were isolated aerobically on gentamicin (7.5 μg ml^{-1})-containing MC medium with 10 mM glutarate as sole carbon source for counterselection of donor cells. The presence of plasmid pLB1 in *Azoarcus* sp. CIBd*bzdN* cells restored their anaerobic growth on benzoate and caused the consumption of this carbon source, as in the case of the wild-type CIB strain containing plasmid pIZ1016 as control (Fig. 2). This result strongly suggested that the recombinant bzd cassette in plasmid pLB1 was functional. To

Table 1. Bacteria and plasmids used in this study.

Strain or plasmid	Relevant genotype and main characteristics	Reference or source
E. coli strains		
DH5α	*endA1 hsdR17 supE44 this-1 recA1 gyrA(Nar^r) relA1 Δ/argF-lac)* *U169 depRΦ80dlacd(lacZ)M15*	Sambrook and Russell, (2001)
S17-1λpir	Tp^r Sm^r *recA thi hsdRM*^+ RP42:::Tc::Mu::Km Tn7 *λpir* phage lysogen	de Lorenzo and Timmis, (1994)
Azoarcus strains		
Azoarcus sp. CIB	Wild-type strain	López-Barragán *et al.*, (2004)
Azoarcus sp. CIBd*bzdN*	Km^r, *Azoarcus* sp. CIB with a disruption in the *bzdN* gene	López-Barragán *et al.*, (2004)
Azoarcus communis SWub3	Wild- type strain (LMG22127)	Reinhold-Hurek *et al.*, (1993)
Azoarcus sp. CIB (pLB1)	CIB strain containing plasmid pLB1	This work
Azoarcus sp. CIBd*bzdN* (pLB1)	CIBd*bzdN* strain containing plasmid pLB1	This work
A. communis SWub3 (pLB1)	SWub3 strain containing plasmid pLB1	This work
Plasmids		
pECOR8	Ap^r, pUC19 harbouring a 5.4 Kb EcoRI DNA fragment from *Azoarcus* sp. CIB that contains the right end of the *bzd* cluster	López-Barragán *et al.*, (2004)
pIZ1016	Gm^r, pBBR1MCS-5 broad-host range cloning vector	Moreno-Ruiz *et al.*, (2003)
pIZ1016*	Gm^r, pIZ1016 derivative without the NcoI restriction site in the polylinker	This work
pIZECO	Gm^r, 11.4 Kb-pIZ1016* derivative harbouring the right end of the *bzd* cluster	This work
pLB1	Gm^r, 25.6 Kb-pIZ1016* derivative harbouring the complete *bzd* cassette	This work

confirm this, plasmid pLB1 was transferred to a closely related species, *Azoarcus communis* SWub3 (Table 1), that is an endophyte unable to degrade aromatic compounds under anaerobic conditions (Reinhold-Hurek *et al.*, 1993) but that can use aliphatic dicarboxylic acids, e.g. glutarate, that feed to the lower benzoyl-CoA pathway (Fig. 1A). Remarkably, the recombinant *A. communis* SWub3 (pLB1) strain was able to grow anaerobically using benzoate as sole carbon and energy source (doubling time of about 15 h), confirming that the bzd cassette was functional in heterologous hosts and conferred the ability to degrade benzoate in anoxic conditions (Fig. 2).

As far as we know, this is the first report on the construction of a DNA cassette encoding a transferable benzoyl-CoA central pathway that allows the expansion of the catabolic potential of certain facultative anaerobes towards the use of aromatic compounds under anaerobic conditions. Moreover, this result reinforces the recent thought that questions the previous idea that the *Azoarcus* genus comprises bacteria that fit into one of two major eco-physiological groups, i.e. either the free-living anaerobic biodegraders of aromatic compounds or the obligate endophytes unable to degrade aromatics under anaerobic conditions (Fernández *et al.*, 2014). Thus, here we show that a member of the subgroup of *Azoarcus* strains that are obligate endophytes unable to degrade aromatics under anaerobic conditions can evolve towards the use of these carbon sources when acquiring the cognate genetic determinants.

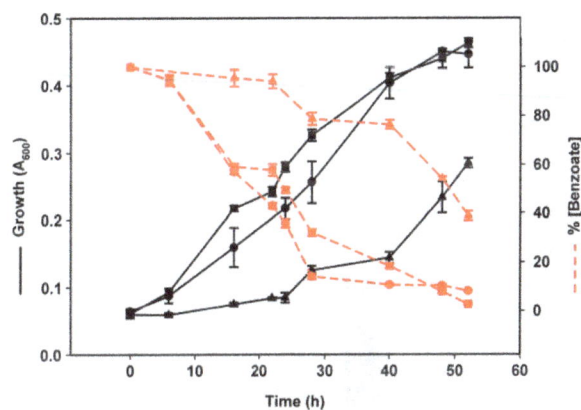

Fig. 2. Anaerobic growth curves and benzoate consumption in different *Azoarcus* strains. *Azoarcus* sp. CIB (pIZ1016) (circles), *Azoarcus* sp. CIBd*bzdN* (pLB1) (squares) and *Azoarcus communis* (pLB1) (triangles) were grown anaerobically at 30°C in minimal MC medium containing 3 mM benzoate and 10 mM nitrate as sole donor and electron acceptors respectively, as previously detailed (López-Barragán *et al.*, 2004). Gentamicin (7.5 μg ml⁻¹) was added to the medium to assure plasmid maintenance. Bacterial growth (black lines) was monitored by measuring the absorbance at 600 nm (A_{600}). The concentration of benzoate in the culture medium (red lines) was monitored spectrophotometrically at 273 nm (López-Barragán et al., 2004) and is indicated as a percentage of the initial concentration. Values are the mean of three different experiments. Error bars indicate standard deviation.

Azoarcus sp. CIB accumulates PHB when grown anaerobically in toluene

It has been reported that some anaerobic biodegraders of aromatic hydrocarbons, such as the bacterium *Aromatoleum aromaticum* EbN1, are able to accumulate poly

(3-hydroxybutyrate) (PHB) up to 5.2% of the cell dry weight (CDW) during anaerobic growth on toluene using nitrate as final electron acceptor (Trautwein *et al.*, 2008). PHB formation was predicted to contribute at two different levels, (i) enhancing consumption of surplus reducing equivalents generated during the anaerobic catabolism of aromatic hydrocarbons, and (ii) alleviating the cytotoxic effect of aromatic hydrocarbons by trapping them into the hydrophobic PHB granules (Trautwein *et al.*, 2008; Rabus *et al.*, 2014). To check whether *Azoarcus* sp. CIB was also able to accumulate PHB when grown anaerobically on toluene, PHB monomer composition and cellular PHB content of lyophilized cells were determined by gas chromatography-tandem mass spectrometry (GC-MS) of the methanolysed polyester as previously described (de Eugenio *et al.*, 2010). The chromatographic profile obtained by GC-MS of the products released showed a single peak corresponding to the 3-hydroxy-butyrate monomer of the PHB polymer. Figure 3A shows that *Azoarcus* sp. CIB cells grown anaerobically on toluene were able to accumulate PHB at a level of 13% CDW. Moreover, granules of PHB were

clearly observed in *Azoarcus* sp. CIB cells by transmission electron microscopy (Fig. 3B).

Taken together, these results show that *Azoarcus* sp. CIB is able to degrade toluene anaerobically and eventually reroute some of the metabolic flux towards the synthesis of a valuable biopolymer such as PHB. According to the known peripheral pathway for anaerobic toluene degradation in denitrifying bacteria, this aromatic hydrocarbon becomes activated to benzylsuccinate by the *bssABCD*-encoded benzylsuccinate synthase enzyme, and then the later is converted to benzoyl-CoA via a modified β-oxidation pathway encoded by the *bbsABC-DEFJGH* genes (Fig. 1A) (Heider *et al.*, 2016; Zamarro *et al.*, 2016). Benzoyl-CoA is then converted to 3-hydroxypimelyl-CoA via the upper benzoyl-CoA central pathway (Fig. 1A), and the later is further metabolized to the central metabolism by a dicarboxylic acid β-oxidation pathway (lower benzoyl-CoA pathway) (Fig. 1A) (López-Barragán *et al.*, 2004; Carmona *et al.*, 2009). As one of the final metabolites of the lower benzoyl-CoA pathway, i.e. 3-hydroxybutyryl-CoA, is the monomer used by the PHB synthase to produce the PHB polymer (Fig. 1A)

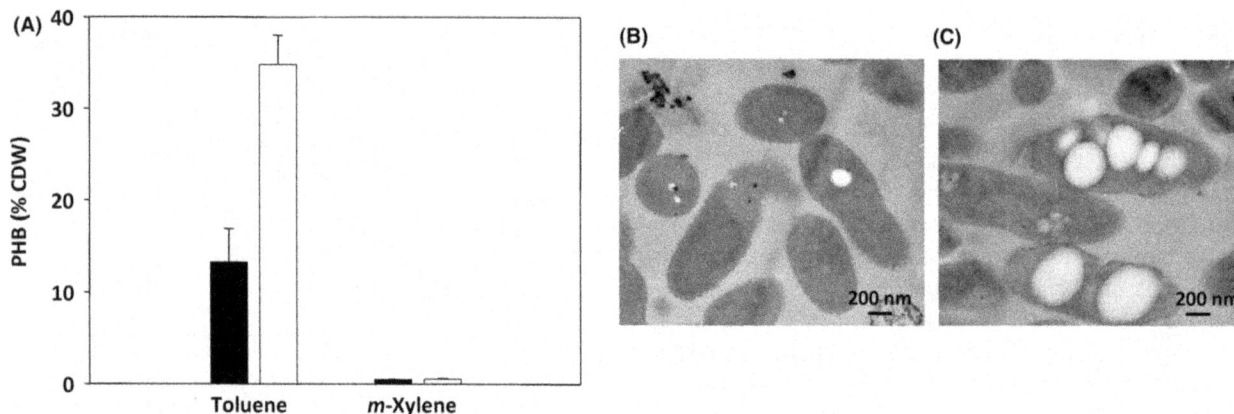

Fig. 3. Analysis of *Azoarcus* sp. CIB cells that contain plasmid pIZ1016 (control) or the pLB1 plasmid grown anaerobically on aromatic hydrocarbons.
A. PHB monomer content of *Azoarcus* sp. CIB (pIZ1016) (black bars) and *Azoarcus* sp. CIB (pLB1) (white bars) cells at mid-exponential phase of anaerobic growth at 30°C in minimal MC medium with 400 mM toluene or 275 mM *m*-xylene, supplied in 2,2,4,4,6,8,8-heptamethylnonane as an inert carrier phase, and 10 mM nitrate. PHB production was quantified by GC-MS of the methanolysed polyester and shown as the percentage of PHB monomer with respect to the total cell dry weight (CDW). Error bars represent standard derivation found in four different experiments. Methanolysis procedure was carried out by suspending 5–10 mg of lyophilized cells in 0.5 ml of chloroform and 2 ml of methanol containing 15% sulfuric acid and 0.5 mg ml^{-1} of 3-methylbenzoic acid (internal standard), and then incubated at 80°C for 7 h. After cooling, 1 ml of de-mineralized water and 1 ml of chloroform were added and the organic phase containing the resulting methyl esters of monomers was analysed by GC-MS (de Eugenio *et al.*, 2010). An Agilent series 7890A coupled with 5975C MS detector (EI, 70 eV) and a split–splitless injector were used for analysis. An aliquot (1 μl) of organic phase was injected into the gas chromatograph at a split ratio 1:20. Separation of compounds was achieved using an HP-5 MS capillary column (5% phenyl-95% methyl siloxane, 30 m × 0.25 mm film thickness). Helium was used as carrier gas at a flow rate of 1 ml min^{-1}. The injector and transfer line temperature were set at 275 and 300°C respectively. The oven temperature programme was as follows: initial temperature 80°C for 2 min, then from 80°C up 150°C at a rate of 5°C min^{-1} and held for 1 min. The mass spectra were recorded in full scan mode (m/z 40–550). 3-hydroxybutyric acid methyl ester was resolved using selected ion monitoring mode (SIM).
B and C. Representative view of *Azoarcus* sp. CIB (pIZ1016) (B) and *Azoarcus* sp. CIB (pLB1) (C) cells grown anaerobically on toluene. Cells were harvested, washed twice in PBS and fixed in 5% (w/v) glutaraldehyde. Afterwards, cells were suspended in 2.5% (w/v) OsO$_4$ for 1 h, gradually dehydrated in ethanol solutions [30%, 50%, 70%, 90% and 100% (v/v); 30 min each] and propylene oxide (1 h), and embedded in Epon 812 resin. Ultrathin sections were cut with a microtome using a Diatome diamond knife. The sections were picked up with 400-mesh cupper grids coated with a layer of carbon and observed using a Jeol-1230 transmission electron microscope. PHB granules can be observed as white spheres inside the cells.

(Rehm, 2010), the anaerobic metabolism of toluene might stimulate PHB production by increasing the 3-hydroxybutyryl-CoA levels in the cell. Therefore, it is tempting to speculate that increasing the flux through the benzoyl-CoA pathway could further enhance the anaerobic conversion of toluene into PHB. In *Azoarcus* sp. CIB, the *bss-bbs* genes are clustered together within an integrative and conjugative element ICE_{XTD} that enhances their copy dosage in the cell (Martín-Moldes *et al.*, 2015; Zamarro *et al.*, 2016), thus likely favouring the initial activation of toluene by benzylsuccinate synthase, a reaction that was shown to be a rate-limiting step in toluene catabolism. As several gene clusters have been predicted to encode dicarboxylic acid β-oxidation pathways in strain CIB (Martín-Moldes *et al.*, 2015), it appears that the single gene dosage *bzd* genes could represent a bottleneck when trying to enhance the metabolic flux through the benzoyl-CoA pathway for achieving a higher conversion of toluene into PHB. Thus, the use of the pLB1 multicopy plasmid could be a rational strategy to increase the *bzd* gene dosage and, eventually, enhance the anaerobic production of PHB in *Azoarcus* sp. CIB.

The bzd cassette enhances PHB accumulation in a recombinant Azoarcus *sp. CIB (pLB1) strain*

To check whether the expression of the *bzd* genes in the multicopy plasmid pLB1 could lead to a higher anaerobic benzoate metabolism, growth and benzoate consumption in *Azoarcus* sp. CIB (pLB1) cells were compared with those in the wild-type *Azoarcus* sp. CIB

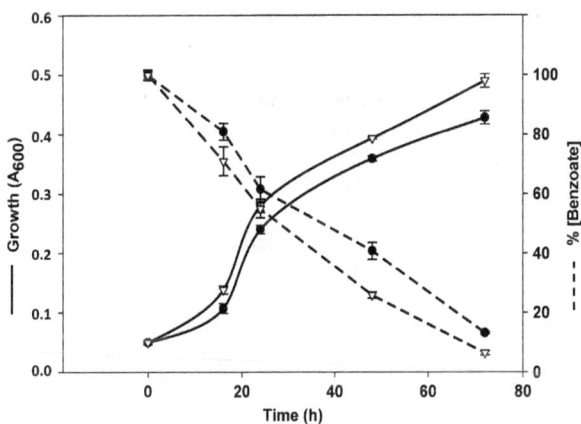

Fig. 4. Growth and benzoate consumption of *Azoarcus* sp. CIB harbouring plasmid pIZ1016 (circles) or the pLB1 plasmid (triangles). Bacteria were cultivated anaerobically at 30°C in minimal MC medium containing 3 mM benzoate and 10 mM nitrate. Bacterial growth, monitored by measuring the absorbance at 600 nm (A_{600}), is indicated with a continuous line. The concentration of benzoate in the culture medium was monitored spectrophotometrically at 273 nm, and the percentage of benzoate remaining in the culture medium is indicated with a dashed line. Values are the mean of three different experiments. Error bars indicate standard deviation.

cells. As shown in Fig. 4, a slightly increase in growth and benzoate removal was observed in cells containing the pLB1 plasmid, suggesting that increasing the *bzd* gene dosage enhances the anaerobic degradation of benzoate in *Azoarcus* sp. CIB. Then, we monitored the accumulation of PHB in *Azoarcus* sp. CIB (pLB1) cells grown anaerobically on 400 mM toluene supplied in 2,2,4,4,6,8,8-heptamethylnonane carrier phase. PHB production was quantified from cells harvested at mid-exponential phase, as previously done with the wild-type CIB strain. Interestingly, the bzd cassette caused accumulation of PHB up to 35% CDW in *Azoarcus* sp. CIB (pLB1) (Fig. 3A). Thus, these results show that when recombinant *Azoarcus* sp. CIB (pLB1) cells grow anaerobically on toluene, the bzd cassette enhances 2.7-fold the bio-production of PHB with respect to that observed in wild-type cells. The higher conversion of toluene into PHB was also confirmed by transmission electron microscopy, which revealed an increase in number and size of PHB granules in recombinant cells (Fig. 3C) with respect to wild-type cells (Fig. 3B). The level of PHB accumulation from the toxic hydrocarbon toluene was the highest described so far under anaerobic conditions, and it was even higher than that reported for the aerobic bioconversion of aromatic hydrocarbons to medium-chain-length polyhydroxyalkanoates (Ward *et al.*, 2005; Nikodinovic *et al.*, 2008; Ni *et al.*, 2010; Narancic *et al.*, 2012) or to PHB (Keum *et al.*, 2008; Hori *et al.*, 2009). As toluene is a major contaminant of high-volume waste streams at places where it is produced or used (e.g. petrochemical industry, solvents and painting markets, biopolymers (e.g. poly(ethylene terephthalate), polyurethanes), its valorization becomes a sustainable strategy for the recycling industry (Wierckx *et al.*, 2015).

Azoarcus sp. CIB is also able to grow anaerobically on *m*-xylene as sole carbon and energy source (Juárez *et al.*, 2013; Zamarro *et al.*, 2016). Although the peripheral pathway for anaerobic *m*-xylene degradation is likely encoded by the same *bss-bbs* genes responsible for the toluene peripheral pathway, the central intermediate of *m*-xylene degradation is 3-methylbenzoyl-CoA rather than benzoyl-CoA (as in the case of toluene) (Juárez *et al.*, 2013; Zamarro *et al.*, 2016). The 3-methylbenzoyl-CoA is further degraded via a specific mbd anaerobic central pathway, which is different to the common bzd pathway, and a lower pathway that does not generate 3-hydroxybutyryl-CoA (Juárez *et al.*, 2013; Zamarro *et al.*, 2016). Thus, anaerobic *m*-xylene degradation might not generate as much PHB as the anaerobic catabolism of toluene. In fact, when *Azoarcus* sp. CIB cells were grown in 275 mM *m*-xylene, supplied in 2,2,4,4,6,8,8-heptamethylnonane carrier phase, no significant PHB accumulation was detected (< 1% CDW) (Fig. 3A). As expected, no further increase of PHB accumulation was observed in *Azoarcus* sp. CIB (pLB1) cells grown on *m*-xylene (Fig. 3A). Therefore, these results strongly suggest that the observed effect of the multicopy bzd cassette enhancing PHB production from toluene might be due to an enhanced metabolic flux through the anaerobic benzoyl-CoA pathway.

Conclusion

This work is a proof of concept of the potential of metabolic engineering for the anaerobic bioconversion of aromatic compounds towards the production of valuable products. We have engineered the first broad-host range metabolic bzd cassette for the anaerobic degradation of benzoate. The bzd cassette has been used to expand the ability of some denitrifying bacteria to use benzoate as sole carbon source under anaerobic conditions, and it paves the way for future pathway engineering of efficient anaerobic biodegraders of aromatic compounds. Moreover, we have developed a recombinant *Azoarcus* sp. CIB strain harbouring the bzd cassette as a valuable environmentally friendly strategy for the efficient anaerobic conversion of some petrochemical waste into added value products, such as bioplastics, relevant to the circular economy.

Acknowledgements

We thank A. Valencia, Fernando de la Peña and Fernando Escolar for technical assistance, Secugen S.L. for DNA sequencing and M.A. Prieto for inspiring discussions. This work was supported by grants BIO2012-39501, BIO2016-79736-R and PCIN-2014-113 from the Ministry of Economy and Competitiveness of Spain; European Union FP7 Grant 311815; and by a grant of Fundación Ramón-Areces XVII CN.

Conflict of Interest

The authors declare no conflict of interest.

References

Beckham, GT, Johnson, CW, Karp, EM, Salvachúa, D, and Vardon, DR (2016) Opportunities and challenges in biological lignin valorization. *Curr Opin Biotechnol* **42:** 40–53.

Boll, M, Löffler, C, Morris, BE, and Kung, JW (2014) Anaerobic degradation of homocyclic aromatic compounds via arylcarboxyl-coenzyme A esters: organisms, strategies and key enzymes. *Environ Microbiol* **16:** 612–627.

Breese, K, Boll, M, Alt-Mörbe, J, Schägger, H, and Fuchs, G (1998) Genes coding for the benzoyl-CoA pathway of anaerobic aromatic metabolism in the bacterium *Thauera aromatica. Eur J Biochem* **256:** 148–154.

Bugg, TDH, and Rahmanpour, R (2015) Enzymatic conversion of lignin into renewable chemicals. *Curr Opin Chem Biol* **29:** 10–17.

Carlström, CI, Loutey, D, Bauer, S, Clark, IC, Rohde, RA, Iavarone, AT, *et al.* (2015) (Per)chlorate-reducing bacteria can utilize aerobic and anaerobic pathways of aromatic degradation with (per)chlorate as an electron acceptor. *MBio* **6:** e02287–14.

Carmona, M, Zamarro, MT, Blázquez, B, Durante-Rodríguez, G, Juárez, JF, Valderrama, JA, *et al.* (2009) Anaerobic catabolism of aromatic compound: a genetic and genomic view. *Microbiol Mol Biol Rev* **73:** 71–133.

Coschigano, PW, Häggblom, MM, and Young, LY (1994) Metabolism of both 4-chlorobenzoate and toluene under denitrifying conditions by a constructed bacterial strain. *Appl Environ Microbiol* **60:** 989–995.

Cueto-Rojas, HF, van Maris, AJ, Wahl, SA, and Heijnen, JJ (2015) Thermodynamics-based design of microbial cell factories for anaerobic product formation. *Trends Biotechnol* **33:** 534–546.

Darley, PI, Hellstern, JA, Medina-Bellver, JI, Marqués, S, Schink, B, and Philipp, B (2007) Heterologous expression and identification of the genes involved in anaerobic degradation of 1,3-dihydroxybenzene (resorcinol) in *Azoarcus anaerobius. J Bacteriol* **189:** 3824–3833.

Díaz, E, Jiménez, JI, and Nogales, J (2013) Aerobic degradation of aromatic compounds. *Curr Opin Biotechnol* **24:** 431–442.

Durante-Rodríguez, G, Valderrama, JA, Mancheño, JM, Rivas, G, Alfonso, C, Arias-Palomo, E, *et al.* (2010) Biochemical characterization of the transcriptional regulator BzdR from *Azoarcus* sp. CIB. *J Biol Chem* **285:** 35694–35705.

Egland, PG, Pelletier, DA, Dispensa, M, Gibson, J, and Harwood, CS (1997) A cluster of bacterial genes for anaerobic benzene ring biodegradation. *Proc Natl Acad Sci USA* **94:** 6484–6489.

de Eugenio, LI, Escapa, IF, Morales, V, Dinjaski, N, Galán, B, García, JL, and Prieto, MA (2010) The turnover of medium-chain-length polyhydroxyalkanoates in *Pseudomonas putida* KT2442 and the fundamental role of PhaZ depolymerase for the metabolic balance. *Environ Microbiol* **12:** 207–221.

Fernández, H, Prandoni, N, Fernández-Pascual, M, Fajardo, S, Morcillo, C, Díaz, E, and Carmona, M (2014) *Azoarcus* sp. CIB, an anaerobic biodegrader of aromatic compounds shows an endophytic lifestyle. *PLoS ONE* **9:** e110771.

Fernández-Llamosas, H, Castro, L, Blázquez, ML, Díaz, E, and Carmona, M (2016) Biosynthesis of selenium nanoparticles by *Azoarcus* sp CIB. *Microb Cell Fact* **15:** 109.

Fuchs, G, Boll, M, and Heider, J (2011) Microbial degradation of aromatic compounds – from one strategy to four. *Nat Rev Microbiol* **9:** 803–816.

Heider, J, Szaleniec, M, Martins, BM, Seyhan, D, Buckel, W, and Golding, BT (2016) Structure and function of benzylsuccinate synthase and related fumarate-adding glycyl radical enzymes. *J Mol Microbiol Biotechnol* **26:** 29–44.

Hirakawa, H, Hirakawa, Y, Greenberg, EP, and Harwood, CS (2015) BadR and BadM proteins transcriptionally regulate two operons needed for anaerobic benzoate degradation by *Rhodopseudomonas palustris*. *Appl Environ Microbiol* **81**: 4253–4262.

Holmes, DE, Risso, C, Smith, JA, and Lovley, DR (2012) Genome-scale analysis of anaerobic benzoate and phenol metabolism in the hyperthermophilic archaeon *Ferroglobus placidus*. *ISME J* **6**: 146–157.

Hori, K, Kobayashi, A, Ikeda, H, and Unno, H (2009) *Rhodococcus aeterivorans* IAR1, a new bacterial strain synthesizing poly (3-hydroxybutyrate-co-3-hydroxyvalerate) from toluene. *J Biosci Bioeng* **107**: 145–150.

Johnson, CW, Salvachúa, D, Khanna, P, Smith, H, Peterson, DJ, and Beckham, GT (2016) Enhancing muconic acid production from glucose and lignin-derived aromatic compounds via increased protocatechuate decarboxylase activity. *Metab Eng Commun* **3**: 111–119.

Juárez, JF, Zamarro, MT, Eberlein, C, Boll, M, Carmona, M, and Díaz, E (2013) Characterization of the *mbd* cluster encoding the anaerobic 3-methylbenzoyl-CoA central pathway. *Environ Microbiol* **15**: 148–166.

Keum, YS, Seo, JS, Li, QX, and Kim, JH (2008) Comparative metabolomic analysis of *Sinorhizobium* sp. C4 during the degradation of phenanthrene. *Appl Microbiol Biotechnol* **80**: 863–872.

Kosa, M, and Ragauskas, AJ (2012) Bioconversion of lignin model compounds with oleaginous *Rhodococci*. *Appl Microbiol Biotechnol* **93**: 891–900.

López-Barragán, MJ, Carmona, M, Zamarro, MT, Thiele, B, Boll, M, Fuchs, G, *et al.* (2004) The *bzd* gene cluster, coding for anaerobic benzoate catabolism, in *Azoarcus* sp. strain CIB. *J Bacteriol* **186**: 5762–5774.

de Lorenzo, V, and Timmis, KN (1994) Analysis and construction of stable phenotypes in gram-negative bacteria with Tn5- and Tn10-derived minitransposons. *Methods Enzymol* **235**: 386–405.

Madbouly, SA, Schrader, JA, Srinivasan, G, Liu, K, McCabe, KG, Grewell, D, *et al.* (2014) Biodegradation behavior of bacterial-based polyhydroxyalkanoate (PHA) and DDGS composites. *Green Chem* **16**: 1911–1920.

Martín-Moldes, Z, Zamarro, MT, del Cerro, C, Valencia, A, Gómez, MJ, Arcas, A, *et al.* (2015) Whole-genome analysis of *Azoarcus* sp. strain CIB provides genetic insights to its different lifestyles and predicts novel metabolic features. *Syst Appl Microbiol* **38**: 462–471.

Moreno-Ruiz, E, Hernáez, MJ, Martínez-Pérez, O, and Santero, E (2003) Identification and functional characterization of *Sphingomonas macrogolitabida* strain TFA genes involved in the first two steps of the tetralin catabolic pathway. *J Bacteriol* **185**: 2026–2030.

Narancic, T, Kenny, ST, Djokic, L, Vasiljevic, B, O'Connor, KE and Nikodinovic-Runic, J (2012) Medium-chain-length polyhydroxyalkanoate production by newly isolated *Pseudomonas* sp. TN301 from a wide range of polyaromatic and monoaromatic hydrocarbons. *J Appl Microbiol* **113**, 508–520.

Ni, YY, Kim, DY, Chung, MG, Lee, SH, Park, HY, and Rhee, YH (2010) Biosynthesis of medium-chain-length poly(3-hydroxyalkanoates) by volatile aromatic hydrocarbons-degrading *Pseudomonas fulva* TY16. *Bioresour Technol* **101**: 8485–8488.

Nikodinovic, J, Kenny, ST, Babu, RP, Woods, T, Blau, WJ and O'Connor, KE (2008) The conversion of BTEX compounds by single and defined mixed cultures to medium-chain-length polyhydroxyalkanoate. *Appl Microbiol Biotechnol* **80**, 665–673.

Nikodinovic-Runic, J, Guzik, M, Kenny, ST, Babu, R, Werker, A and O'Connor, KE (2013) Carbon-rich wastes as feedstocks for biodegradable polymer (polyhydroxyalkanoate) production using bacteria. *Adv Appl Microbiol* **84**, 139–200.

Rabus, R, Kube, M, Heider, J, Beck, A, Heitmann, K, Widdel, F, and Reinhardt, R (2005) The genome sequence of an anaerobic aromatic-degrading denitrifying bacterium, strain EbN1. *Arch Microbiol* **183**: 27–36.

Rabus, R, Trautwein, K, and Wöhlbrand, L (2014) Towards habitat-oriented systems biology of "*Aromatoleum aromaticum*" EbN1: chemical sensing, catabolic network modulation and growth control in anaerobic aromatic compound degradation. *Appl Microbiol Biotechnol* **98**: 3371–3388.

Rabus, R, Boll, M, Heider, J, Meckenstock, RU, Buckel, W, Einsle, O, *et al.* (2016) Anaerobic microbial degradation of hydrocarbons: from enzymatic reactions to the environment. *J Mol Microbiol Biotechnol* **26**: 5–28.

Rehm, BH (2010) Bacterial polymers: biosynthesis, modification and applications. *Nat Rev Microbiol* **8**: 578–592.

Reinhold-Hurek, B, Hurek, T, Gillis, M, Hoste, B, Vancanneyt, M, Kersters, K, and de Ley, J (1993) *Azoarcus* gen. nov., nitrogen-fixing proteobacteria associated with roots of kallar grass (*Leptochloa fusca* (L.) Kunth), and description of two species, *Azoarcus indigens* sp. nov. and *Azoarcus communis* sp. nov. *Int J Syst Bacteriol* **43**: 574–584.

Sambrook, JW, and Russell, DW (2001) *Molecular Cloning: A Laboratory Manual*, 3rd edn. Cold Spring Harbor, NY: Cold Spring Harbor Laboratory.

Trautwein, KS, Kühner, S, Wöhlbrand, L, Halder, T, Kutcha, K, Steinbüchel, A, and Rabus, R (2008) Solvent stress response of the denitrifying bacterium "*Aromatoleum aromaticum*" strain EbN1. *Appl Environ Microbiol* **74**: 2267–2274.

Ward, PG, Roo, G and O'Connor, KE (2005) Accumulation of polyhydroxyalkanoate from styrene and phenylacetic acid by *Pseudomonas putida* CA-3. *Appl Environ Microbiol* **71**, 2046–2052.

Wierckx, N, Prieto, MA, Pomposiello, P, de Lorenzo, V, O'Connor, K, and Blank, LM (2015) Plastic waste as a novel substrate for industrial biotechnology. *Microb Biotechnol* **8**: 900–903.

Wischgoll, S, Heintz, D, Peters, F, Erxleben, A, Sarnighausen, E, Reski, R, *et al.* (2005) Gene clusters involved in anaerobic benzoate degradation of *Geobacter metallireducens*. *Mol Microbiol* **58**: 1238–1252.

Zamarro, MT, Martín-Moldes, Z, and Díaz, E (2016) The ICE$_{XTD}$ of *Azoarcus* sp. CIB, an integrative and conjugative element with aerobic and anaerobic catabolic properties. *Environ Microbiol* **18**: 5018–5031.

Revealing the combined effects of lactulose and probiotic enterococci on the swine faecal microbiota using 454 pyrosequencing

Jong Pyo Chae,[1,#] Edward Alain B. Pajarillo,[1,#]
Ju Kyoung Oh,[1] Heebal Kim[2] and Dae-Kyung Kang[1,*]
[1]Department of Animal Resources Science, Dankook
University, Cheonan 330-714, Korea.
[2]Department of Agricultural Biotechnology, Seoul
National University, Seoul 151–921, Korea.

Summary

Demand for the development of non-antibiotic growth promoters in animal production has increased in recent years. This report compared the faecal microbiota of weaned piglets under the administration of a basal diet (CON) or that containing prebiotic lactulose (LAC), probiotic *Enterococcus faecium* NCIMB 11181 (PRO) or their synbiotic combination (SYN). At the phylum level, the *Firmicutes* to *Bacteroidetes* ratio increased in the treatment groups compared with the CON group, and the lowest proportion of *Proteobacteria* was observed in the LAC group. At the family level, *Enterobacteriaceae* decreased in all treatments; more than a 10-fold reduction was observed in the LAC (0.99%) group compared with the CON group. At the genus level, the highest *Oscillibacter* proportion was detected in PRO, the highest *Clostridium* in LAC and the highest *Lactobacillus* in SYN; the abundance of *Escherichia* was lowest in the LAC group. Clustering in the discriminant analysis of principal components revealed distinct separation of the feeding groups (CON, LAC, PRO and SYN), showing different microbial compositions according to different feed additives or their combination. These results suggest that individual materials and their combination have unique actions

*For corerspondence. E-mail dkkang@dankook.ac.kr

#Both authors contributed equally to this work.

Funding Information
This work was supported by a grant from the Next-Generation BioGreen 21 Program (PJ01115903), Rural Development Administration, Korea, and by the National Research Foundation of Korea grant funded by the Korean Government (NRF-2013R1A1A2012365).

and independent mechanisms for changes in the distal gut microbiota.

Introduction

In the animal industry, much-improved farming systems and cost-effective techniques for the production of pathogen-free high-quality meat are important goals of current research (Lallès *et al.*, 2007; Lee *et al.*, 2012; Bomba *et al.*, 2014). In recent years, the gastrointestinal (GI) tract and its resident microbiota have been recognized as influential host genetic elements and environmental factors in improving overall animal health, growth and performance (Kim and Isaacson, 2015). The microbial ecosystem of the GI tract plays an important role in preserving a stable and thriving gut environment through its impact on host physiology and functionality (Richards *et al.*, 2011), modulation of metabolic activities and immunological responses (Hemarajata and Versalovic, 2012), and provision of a natural defence system against pathogenic invasion (O'Connor *et al.*, 2014). Several factors that affect intestinal microbiota in pigs have recently been investigated, including age (Kim *et al.*, 2011), diet (Yan *et al.*, 2013), weaning (Pajarillo *et al.*, 2014a) and antibiotic growth promoters (AGP) (Unno *et al.*, 2014). These studies have revealed important concepts in animal farming and management, as well as in the development of non-AGP.

Pigs are constantly subjected to harsh and stressful conditions during their growth (Lallès *et al.*, 2007; Pluske *et al.*, 2009; Bomba *et al.*, 2014). Previous studies have shown that extreme physiological and morphological changes, including reshaping of the microbiota, occurs in the GI tract of piglets (Lallès *et al.*, 2007; Pajarillo *et al.*, 2014a). The ban of AGP in feeds in some countries caused serious bacterial and viral infections and increased mortality (Unno *et al.*, 2014). Hence, the development of AGP-alternatives, particularly eubiotics (e.g. probiotics, prebiotics and synbiotics), to boost animal health and enhance growth performance is important. Probiotics and prebiotics can improve the health of their host by balancing the gut microbiota (Hemarajata and Versalovic, 2012; Kim and Isaacson, 2015). Recent studies have shown that performance levels and growth indicators in weaned piglets were significantly increased by probiotic (i.e. lactobacilli) or prebiotic (i.e. inulin, lactulose) administration through increased butyrate supply, heightened villus height for intestinal integrity and better immunomodulation in the gut (Konstantinov *et al.*, 2004; Krause *et al.*, 2010; Lee *et al.*, 2012; Guerra-Ordaz *et al.*, 2014; Sattler *et al.*, 2014). Furthermore, the combination of probiotics and prebiotics, called synbiotics, has shown promising results as a non-AGP (Guerra-Ordaz *et al.*, 2014; Sattler *et al.*, 2014). The effects of probiotics and prebiotics in the swine gut microbiota were investigated using both culture-dependent and -independent approaches, such as quantitative PCR and denaturing gradient gel electrophoresis (Lee *et al.*, 2012; Martinez *et al.*, 2012; Guerra-Ordaz *et al.*, 2014; Sattler *et al.*, 2014; Pajarillo *et al.*, 2015a). However, previous studies have focused on a handful of microbial groups, typically fewer than 10 bacterial taxa. Using 16S rRNA gene pyrosequencing technology in combination with bioinformatics tools enables more efficient and informative quantification and statistical comparison of microbial diversity and composition across samples.

Recently, independent administration of the probiotic *E. faecium* NCIMB 11181 (Pajarillo *et al.*, 2015a) and the prebiotic lactulose (Chae *et al.*, 2015) in weaned piglets showed differences in faecal microbial diversity and bacterial community composition. The aim of this study was to examine the synbiotic effect of lactulose and the probiotic bacterium *Enterococcus faecium* NCIMB 11181 on the microbial diversity of weaned piglets and to compare the unexplored synergistic effect against the individual effects of the prebiotic and probiotic on the structure and composition of faecal microbiota, using pyrosequencing of the 16S rRNA genes.

Results

DNA sequence data and quality control

Seventy-nine piglets were divided into four groups: control (CON; $n = 15$), LAC ($n = 15$), PRO ($n = 20$) and SYN ($n = 29$). The pyrosequencing data were generated and pooled for each group. In total, 80 251, 100 172, 98 156 and 208 660 high-quality sequence reads were obtained in the CON, LAC, PRO and SYN groups respectively. The average numbers of sequence reads generated per pig were 5350, 6678, 4907 and 7195 in the CON, LAC, PRO and SYN groups respectively (Table S1).

Microbial diversity

α-diversity measurements comparing the microbial communities of these groups revealed significant differences in piglets among the LAC, PRO and SYN groups compared with the CON group (Fig. 1, Table S1). Rarefaction curves of pooled samples were determined at an operational taxonomic unit (OTU) definition of 97% identity (Fig. S1). The richness estimates, including abundance coverage estimate (ACE) and Chao1 for the LAC, PRO and SYN groups were significantly higher than those for the CON group ($P < 0.05$). The median values

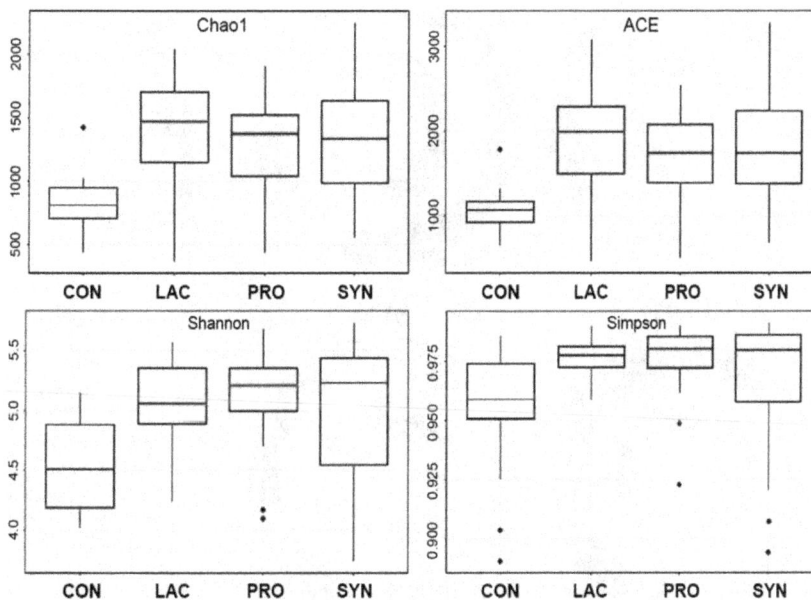

Fig. 1. α-Diversity measurements of pig faecal microbiota according to treatment. Microbial richness estimates (Chao1 and ACE) and diversity indices (Shannon and Simpson) provide measures of diversity within each community at an OTU identity cut-off of 97%. Each group is labelled accordingly: CON, control; LAC, prebiotic lactulose; PRO, probiotic *Enterococcus faecium* NCIMB 11181; SYN, synbiotic. The Kruskal–Wallis non-parametric test for significance was performed to assess differences among pig groups; $P_{Chao1} = 0.001$, $P_{ACE} = 0.001$, $P_{Shannon} = 0.002$, $P_{Simpson} = 0.03$.

for both ACE (1992) and Chao1 (1472) were highest in the LAC group (Table S1), indicating that the microbial communities in piglets that ingested the administered prebiotics exhibited increased numbers of unique species. The Shannon and Simpsonsl (1-D) diversity indices were also increased by feeding prebiotic lactulose and/or probiotic *E. faecium* NCIMB 11181 (Fig. 1, Table S1). The highest median Shannon value was observed in the SYN group (5.23), and the highest Simpson value was detected in the PRO group (0.980) (Table S1); by contrast, the lowest diversity index was found in the CON group. These diversity indices indicate the number of different bacterial OTUs and populations of microorganisms present in a sample; higher values denote greater diversity. Although the inclusion of prebiotics and/or probiotics in the diet significantly increased α-diversity compared with CON values, no differences were observed in richness or diversity values among the three treatment groups (LAC, PRO and SYN).

Comparison of faecal microbial shifts in response to the administration of prebiotics, probiotics and synbiotics: Taxon-based analysis

A taxon-based approach was performed using the EzTaxon database to investigate changes in the composition of the faecal microbiota of weaned piglets after administration of prebiotics, probiotics and synbiotics.

The relative abundances at the phylum and family levels are shown in Fig. 2. At the phylum level, the majority of sequences (> 90%) belonged to the *Firmicutes* and *Bacteroidetes*, regardless of the feed additive types. The ratio of *Firmicutes* to *Bacteroidetes* increased in the LAC group (Fig. 2 and Table S2). The abundance of *Proteobacteria* was highest in the CON group, whereas *Proteobacteria* abundances in the LAC, PRO and SYN groups were decreased. The lowest proportion of *Proteobacteria* was found in the LAC group.

At the family level (Fig. 2), the most abundant bacterial groups were *Ruminococcaceae* and *Prevotellaceae* in all pig groups (Table S2), regardless of the treatment. In addition, *Enterobacteriaceae*, *Veillonellaceae*, *Lactobacillaceae* and *Lachnospiraceae* were also detected as major bacterial groups (Fig. 2). After administration, the average population of *Lactobacillaceae* was increased by the LAC (9.20%), PRO (7.97%) and SYN (13.8%) treatments compared with the CON (5.67%) group. Additionally, the highest proportion of *Lachnospiraceae* was detected in LAC (7.07%), followed by SYN (6.26%) (Table S2). Furthermore, large decreases in the proportions of *Enterobacteriaceae* were found in all treatment groups; in particular, more than a 10-fold reduction in LAC (0.99%) was detected compared with the CON group (Fig. 2).

A total of 99 bacterial genera were identified from at least one faecal microbiota sample in this experiment,

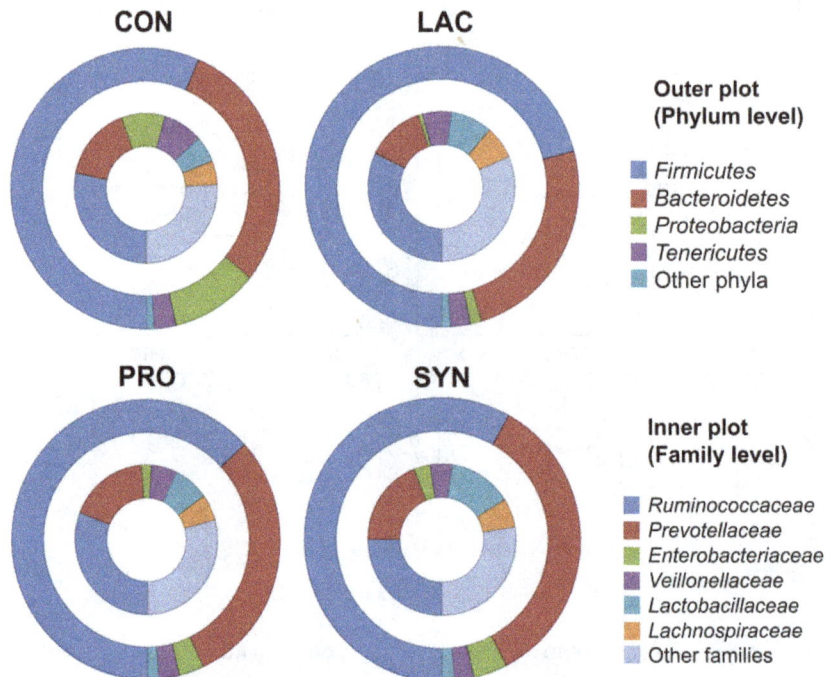

Fig. 2. Doughnut plots of the relative abundances of sequences at the phylum and family levels. The EzTaxon database was used to classify the taxon groups. Mean relative abundances were calculated from all samples in each group; outer and inner plots depict selected taxa at the phylum and family levels respectively.

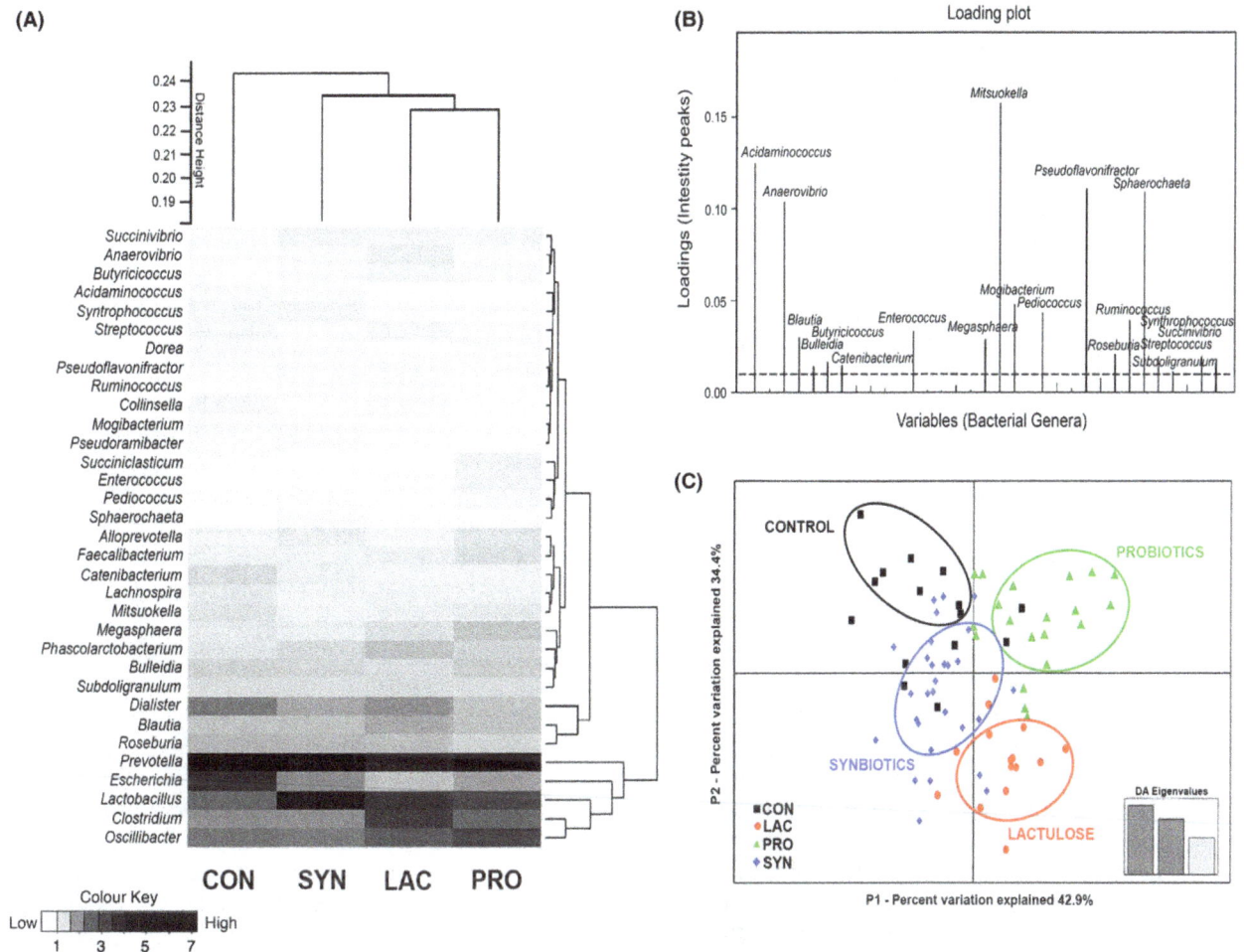

Fig. 3. Differentially abundant bacterial genera among the CON, LAC, PRO and SYN groups. Piglets in the control group received a basal diet during the entire experimental period, whereas piglets receiving treatments were fed the basal diet plus the assigned feed additive.
A. The heatmap shows the 33 abundant genera (> 0.1% mean relative abundance) after normalization. The normalized levels of abundance are depicted in the colour key, where white represents the lowest (min=0) and black (max=7) shows the highest level of abundance. Columns represent treatment groups, and rows indicate the bacterial genera.
B. The canonical loading plot shows peaks for the bacterial genera that had strong influences on the differentiation of the control from the treatment groups.
C. Clustering of the faecal microbiota according to treatment was performed by DAPC plot using the 33 differentially abundant bacterial genera as variables.

including 33 differentially abundant genera (> 0.1% of total sequences) (Fig. 3A). *Prevotella, Lactobacillus, Oscillibacter, Clostridium* and *Escherichia* genera were considered more abundant ($x > 1.0\%$ mean abundance) compared with the remaining 28 bacterial genera ($1.0\% > x > 0.1\%$ mean abundance) (Fig. 3A). Moreover, these five highly abundant bacterial genera were also identified as part of the core microbiota of the swine distal gut in a previous report (Pajarillo *et al.*, 2015b). Differential levels of abundance were detected among the feeding groups; the highest proportion of *Oscillibacter* was detected in the PRO group, the highest *Clostridium* in the LAC group, and the highest *Lactobacillus* in the SYN group. The administration of feed additives

decreased the number of *Escherichia*, especially in the LAC group.

A dendrogram was constructed using the Bray–Curtis dissimilarity matrix to assess the similarity of the bacterial communities among treatment groups (Fig. 3A). The distinguishing variables (bacterial genera) were plotted as discriminant peaks to determine which of the differentially abundant genera had the greatest influence on the dissimilarity among the CON, LAC, PRO and SYN groups (Fig. 3B). The canonical loading plot displayed the five most influential bacterial genera, namely, *Mitsuokella, Acidaminococcus, Pseudoflavonifractor, Sphaerochaeta* and *Anaerovibrio*. These discriminant peaks for each variable were directly proportional to the strength of influence on differences among groups, with higher peaks depicting stronger influence on the variation and *vice versa*. Next, the discriminant analysis of principal components (DAPC) showed the separation of individual microbial communities in a two-dimensional plot (Fig. 3C). This plot revealed the significantly separated clustering of pig faecal microbiotas according to treatment group (CON, LAC, PRO and SYN).

Comparison of faecal microbial shifts in response to the administration of prebiotics, probiotics and synbiotics: OTU-based analysis

An OTU-based approach was conducted for an all-inclusive membership analysis and in-depth ecological investigation of the bacterial communities under the influence of prebiotics, probiotics and synbiotics. The OTUs used in this analysis were defined at 95% sequence identity. In all, 397 bacterial OTUs were identified in at least one pig; 253, 287, 335 and 338 OTUs were detected in the CON, LAC, PRO and SYN groups respectively. Next, a Venn diagram was created to describe the core (shared) and unique (distinct) bacterial OTUs among pig groups (Fig. 4A). The overlap of two or more ellipses denotes the shared bacterial OTUs between two or more pig groups. Of the 397 OTUs, 21 were considered to be core (shared) OTUs. In terms of abundance, the core microbiota accounted for more than 50% of the total bacterial population (percent abundance) in the swine distal gut; however, the core only accounted for 5.2% of all bacterial phylotypes (21 of 397 bacterial OTUs) detected in at least one pig sample.

The canonical loading plot was applied to the 397 bacterial OTUs to determine the most influential bacterial phylotypes after the administration of feed additives. According to the EzTaxon database, the three bacterial OTUs that had the highest peaks were DQ905455_g, AM406061_g and EU259447_g; the two former bacterial phylotypes were also found in the core microbiota, whereas the last (EU259447_g) was present at a higher abundance in the faecal microbiota of pigs belonging to the LAC group (Fig. S2). However, other bacterial OTUs showed relatively higher peaks that may help discriminate among the pig treatment groups (Fig. 4B). The DAPC clustering of the faecal microbiotas according to treatment group illustrates the distinct separation of the feeding groups (CON, LAC, PRO and SYN), showing that microbial communities have different compositions according to the type of feed additives (Fig. 4C).

Discussion

Prebiotics, probiotics and synbiotics have beneficial effects on animal health and nutrition (Krause *et al.*,

2010; Martinez *et al.*, 2012; Guerra-Ordaz *et al.*, 2014; Sattler *et al.*, 2014; Umu *et al.*, 2015). The administration of the probiotic *E. faecium* NCIMB 11181 or the prebiotic lactulose, which are used to improve animal health and performance, showed significant shifts in the swine faecal microbiota (Chae *et al.*, 2015; Pajarillo *et al.*, 2015a). In this study, the synbiotic effects of the lactulose and *E. faecium* NCIMB 11181 combination were compared with the effects of probiotic or prebiotic administration on the overall microbial diversity and bacterial composition of swine faeces using a high-throughput pyrosequencing method.

First, significant shifts in the structure and proportion of specific bacterial phyla, families, genera and OTUs were detected in the CON, LAC, PRO and SYN piglets. Although *Firmicutes* and *Bacteroidetes* remained the most dominant bacterial groups regardless of treatment, the ratio of *Firmicutes* to *Bacteroidetes* significantly increased in the LAC compared with the CON group. The increase in the *Firmicutes* to *Bacteroidetes* ratio in the LAC group suggests that lactulose promotes the proliferation of some groups in the *Firmicutes* phylum, which may promote various metabolic activities and fermentation of complex plant-based diets (Hooda *et al.*, 2012; Sattler *et al.*, 2014; Umu *et al.*, 2015). Several studies show that lactulose supplementation improves growth performance, short-chain fatty acid composition and microbial populations in swine and poultry (Fleige *et al.*, 2007; Cho and Kim, 2013; Zheng *et al.*, 2014a). The effect of lactulose in the gut increases bacterial diversity and stimulates the growth of many bacteria belonging to *Firmicutes*, including lactobacilli and clostridia (Konstantinov *et al.*, 2004; Mao *et al.*, 2014). Lactulose-utilizing bacteria contribute to short chain fatty acids (SCFA) and equol production, which may induce anti-inflammatory and antioxidant properties in swine intestines (Ito *et al.*, 1997; Zheng *et al.*, 2014b; Ziar *et al.*, 2014). In addition, the higher ratio of *Firmicutes* to *Bacteroidetes* in younger piglets may be advantageous for increasing intestinal SCFA and reducing infection (Molist *et al.*, 2012), which is also correlated with increasing body weight (Guo *et al.*, 2008).

At the family level, the SYN group had the highest abundance of *Lactobacillaceae*, which suggests complementary effects between the prebiotic oligosaccharide and probiotic bacterium. The *Lactobacillaceae* family comprises well-known probiotic bacteria that are generally recognized as safe and are highly adapted to the GI environment (Etzold *et al.*, 2014), which improve overall GI integrity and functionality. This family is described as the energy-generating machinery in humans and animals by increasing the levels of short-chain fatty acids, particularly acetate, propionate and butyrate, in the gut (Hooda *et al.*, 2012; Guerra-Ordaz *et al.*, 2014). In a previous study, the synbiotic mixture of *L. plantarum* and

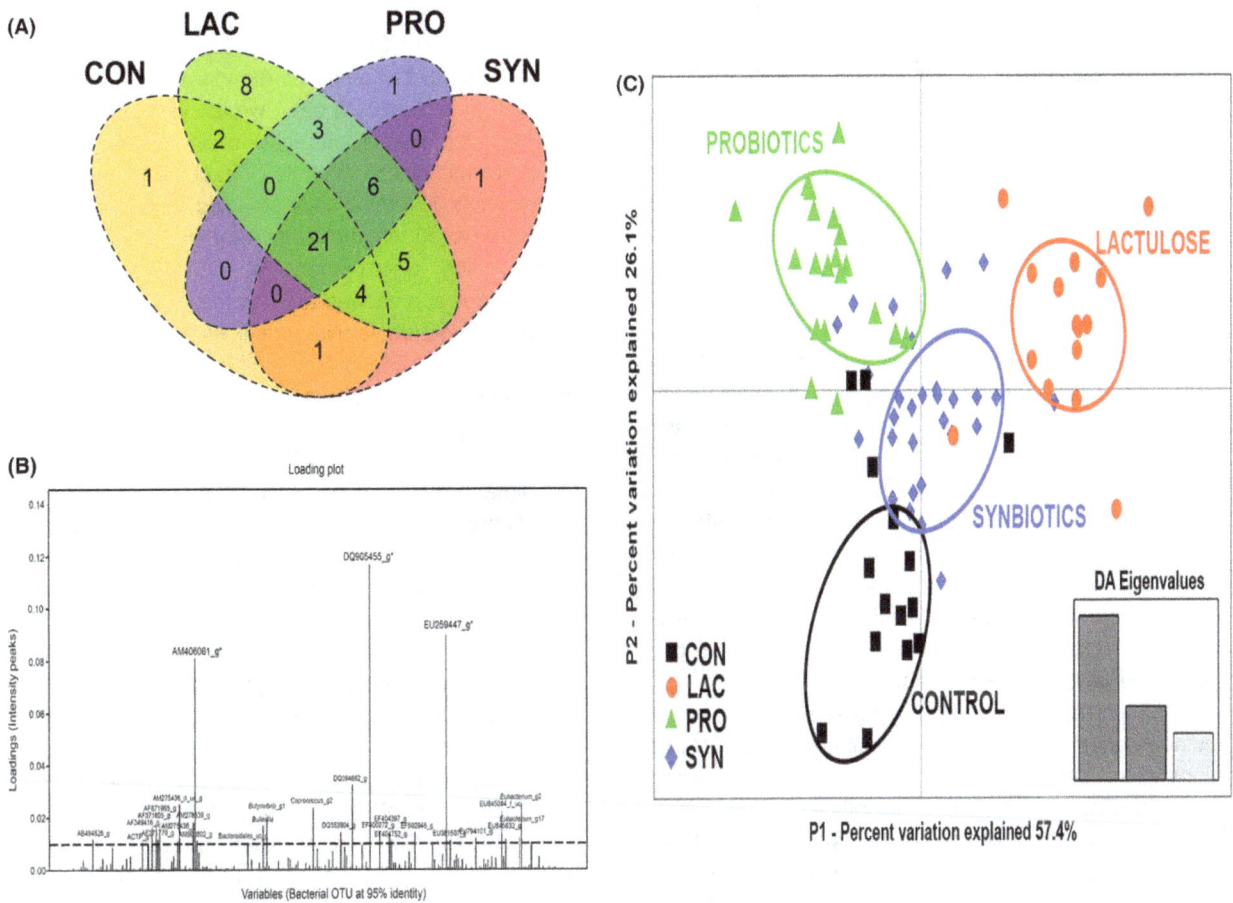

Fig. 4. OTU-based community structures and compositions in the faecal microbiota after treatment.
A. Venn diagram showing the distribution of OTUs for the CON, LAC, PRO and SYN groups. Numbers indicate the number of OTUs that were unique and the number shared (core) by two or more groups, as depicted by non-intersecting and intersecting ellipses respectively.
B. The canonical loading plot shows the peaks of all bacterial OTUs that had strong influences on the separation of the control and treatment groups.
C. Clustering of the faecal microbiota according to treatment was performed by DAPC plot using all bacterial OTUs at a 95% identity cut-off as variables.

lactulose produced different effects on the microbial populations in pigs (Guerra-Ordaz et al., 2014). Consequently, the combination of *E. faecium* NCIMB 11181 and lactulose in this study suggests the formation of unique metabolites or compounds that trigger the growth of *Lactobacillaceae*. Decreased abundances of *Enterobacteriaceae* and *Veillonellaceae* families were also observed in all treatment groups. Low populations of *Enterobacteriaceae* are favourable for animal production, because increases in this family are likely associated with high mortality in pigs, caused by bacterial infection (Pluske et al., 2009; Krause et al., 2010). Reduction in the abundance of *Veillonellaceae* may also have positive effects on pig health, due to its association with nasopharyngeal infections and GI-associated diseases, as well as with cirrhosis and extreme levels of bile acids in the gut (Bajaj et al., 2011; Gevers et al., 2014).

At the genus level, the abundances of many bacterial genera were affected. First, the genus *Lactobacillus* increased in abundance in all treatment groups, which was notably highest in the SYN group. Lactobacilli are responsible for higher levels of anti-inflammatory and systemic responses and for out-competing and exclusively displacing pathogenic bacteria along the mucosal surfaces of the host (Etzold et al., 2014; Johnson and Klaenhammer, 2014). However, the synbiotic effect on the increased lactobacilli population may be dependent on the specific probiotics and prebiotics used, as well as their dosages, because the synbiotic effect of *L. plantarum* and lactulose in the *Lactobacillus* population did not exceed the individual effects of singular administration of the prebiotic or probiotic (Guerra-Ordaz et al., 2014). On the other hand, the number of *Escherichia* decreased sharply in all treatment groups, most remarkably in the LAC group. Previous reports showed that the lactulose effect was most distinguishable on the population of *Escherichia*, specifically enterotoxigenic and enteropathogenic *E. coli* K88 at post-weaning

(Konstantinov *et al.*, 2004; Krause *et al.*, 2010; Guerra-Ordaz *et al.*, 2014).

Furthermore, the identified discriminating bacterial genera may have contributions to gut functions in relation to the feed additives with which they are associated. Investigating the influence of probiotic and prebiotic interventions not only in a handful of bacterial groups, which was common in previous studies (Guerra-Ordaz *et al.*, 2014; Zheng *et al.*, 2014a), but on the total gut microbiota revealed unique bacterial genera that may be beneficial or harmful to pigs; specifically *Mitsuokella*, *Acidaminococcus*, *Pseudoflavonifractor*, *Sphaerochaeta*, and *Anaerovibrio* genera were highly influential in the separation of pig groups. The most discriminating bacterium, *Mitsuokella*, was most abundant in the PRO group (Fig. S2). This have significant implications on the functional properties in vivo of these discriminating bacterial genera; however, genomic and biochemical information remains limited for many of these discriminating genera. Future isolation and detailed characterization of these bacteria will increase our understanding of their potential roles in pig health.

A taxon-independent (OTU-based) analysis is a robust and appropriate method for the overall assessment of variations in the swine faecal microbiota comprising numerous unclassified bacterial phylotypes. The linear discriminant analysis of the variables (bacterial OTUs) suggested that three unclassified bacteria were the most influential in the overall separation of the groups. Specifically, the bacterial phylotype EU259447_g may be associated with lactulose in feed in the post-weaning diet (Fig. S2); the OTU was later identified as the closest relative to *Eubacterium coprostanoligenes* based on the neighbour-joining tree of closely related organisms (Fig. S4). This bacterium is a member of *Clostridium* cluster IV, which may be highly associated with fermentable carbohydrates (e.g. inulin, lactulose) (Sattler *et al.*, 2014). Establishing the relationship of these species with the scope of the prebiotic, probiotic or synbiotic effects in swine physiology will increase our understanding of the functional and metabolic benefits of these feed additives.

Despite a previous in vitro study showing that lactulose can promote the growth of *E. faecium* (Mao *et al.*, 2014), synbiotic administration did not result in significant proliferation of probiotic enterococci in the faecal microbiota of piglets. However, the synergistic action of the probiotic *E. faecium* and prebiotic lactulose is shown in the DAPC plot, which revealed distinct and separate clusters of microbial communities among treatments. It is possible that the effects of the synbiotic combination are complementary; however, this study detected the cumulative effects on the microbial composition (i.e. increased α–diversity, decreased pathogenic bacteria and increased lactobacilli population).

In conclusion, the synbiotic combination of lactulose and *E. faecium* NCIMB 11181 generated differences in the gut microbiota compared the individual effects of the prebiotic or probiotic. In other words, individual materials and their combination can lead to different results in the distal gut microbiota through independent or synergistic mechanisms. Further understanding of gut microbiota changes based on the administration of eubiotics will lead to the development of AGP-alternatives and improve our approaches and strategies for environment- and animal-friendly farming practices.

Experimental procedures

Animal and sample collection

All animal protocols used in this study were approved by the Dankook University Animal Care Committee. Seventy-nine healthy piglets raised on a farm (Cheonan, Korea) were selected randomly and allocated into control and treatment groups. All piglets were born from different sows on the same day and were weaned at 4 weeks of age. Following weaning, all piglets were given the same basal feed for the next 2 weeks (Table S3) without the administration of antibiotics or feed additives. At 6 weeks, piglets were grouped into the following feeding treatments: control (CON, $n = 15$), prebiotic lactulose (LAC, $n = 15$) (Chae *et al.*, 2015), probiotic *E. faecium* NCIMB 11181 (PRO, $n = 20$) (Pajarillo *et al.*, 2015a) and their synbiotic combination (SYN, $n = 29$). Pigs in the CON group continued consuming the basal diet for 2-weeks. The probiotic *E. faecium* NCIMB 11181 (Lactiferm®; Chr. Hansen, Nienburg, Germany) was given at a concentration of 1.0×10^9 colony forming units (CFU) kg^{-1} feed, and the prebiotic lactulose was given at a concentration of 5 g kg^{-1} feed. The daily feed allotment was provided as two meals at 12-h intervals. Animals were sheltered in an environmentally controlled room with a slatted plastic floor. Each pen was equipped with a one-sided self-feeder and a nipple water-feeder for ad libitum access to feed and water throughout the experiment. The housing conditions were room temperature (25°C), 60% humidity, a mechanical ventilation system, and artificial light for 12 h using fluorescent lights.

Fresh faecal samples were collected individually from the rectum of each piglet after 2-weeks of daily administration of prebiotic lactulose, probiotic *E. faecium* NCIMB 11181 or their synbiotic combination. Purified DNA extracts were obtained using the UltraClean Faecal DNA isolation kit (MO BIO Laboratories, Carlsbad, CA, USA) from rectal faecal grabs of individual piglets, as described previously (Pajarillo *et al.*, 2014a,b). The quantity and concentration of DNA extracts were checked using the Optizen UV/Vis spectrophotometer (Mecasys, Daejeon, Korea) and sorted according to treatment group.

454-pyrosequencing

PCR amplification of the DNA extracts was performed according to parameters and conditions described previously (Pajarillo et al., 2014a,b). PCR primers targeting the V1–V3 hypervariable regions of the bacterial 16S rRNA gene were used in this study. The PCR conditions consisted of an initial denaturing phase at 94°C for 3 min; 35 cycles of 94°C for 1 min, annealing at 55°C for 45 s, and extension at 72°C for 1 min, with a final extension at 72°C for 8 min. The visualization of PCR amplicons was performed in a 1.5% (w/v) agarose gel stained with ethidium bromide. Clear DNA amplicons visualized in agarose gels without primer dimers or contaminant bands were used in subsequent experiments. Pyrosequencing was performed using the Roche 454 GS-FLX titanium system (454 Life Sciences, Branford, CT, USA). Raw sequence reads were processed and analysed from each faecal sample, as described previously (Jeon et al., 2013).

Processing of 16S rRNA gene sequences

Downstream analysis of sequences was performed based on a previous study (Pajarillo et al., 2015b). The GS-FLX pig faecal dataset for the SYN group was analysed and compared with the previous pig faecal datasets for the CON, LAC and PRO groups (Chae et al., 2015; Pajarillo et al., 2015a). Briefly, raw sequence reads generated by the 454-pyrosequencer were demultiplexed (barcodes were removed and sequences sorted into categorical groups). Sequence reads with fewer than 300 bases were eliminated. The valid pyrosequencing reads from each pig sample was summarized in Table S4. Next, chimeras were checked and removed from the sequence data using the Bellerophon method, and sequence data were then denoized in Mothur (Schloss et al., 2009). The average length of high-quality sequences without primers was 477 bp, and these sequences were used for further analysis. Using the CD-HIT program (Li and Godzik, 2006), OTUs were assigned at a > 97% identity level. Taxonomic ranking and classification were performed using the EzTaxon database (Chun et al., 2007). During classification, when sequences could not be assigned into a sublevel, 'uc' was added to the end of the name (e.g. Ruminococcaceae_uc for OTUs that could be classified only at the family level). If the taxon was still unknown, the genus name was written first, and the initial letter of each unknown taxon level was written at the end of the name (e.g. if the genus name was unknown, a 'g' was written after the name, e.g., Prevotella_g; the same pattern was used for the species (s), genus (g), family (f) and accession numbers of unidentified phylotypes). The following cut-off values were used for taxonomic assignment: species ($x \geq 97\%$), genus ($97\% > x \geq 94\%$), family ($94\% > x \geq 90\%$), order ($90\% > x \geq 85\%$), class ($85\% > x \geq 80\%$) and phylum ($80\% > x \geq 75\%$), where x corresponds to the sequence identity between sequences within a certain OTU (Chun et al., 2007).

Statistical analyses

The summaries of the percent abundances of the classified taxon groups were generated using CLCommunity software (ChunLab Inc., Seoul, Korea). Microbial richness estimates and diversity indices, including Chao1, ACE, Shannon, and Simpson (1-D), were calculated using Mothur (version 1.32.1), with OTUs defined at the 97% identity level (Schloss et al., 2009). Both individual and pooled diversity indices and richness estimators are shown in the boxplot illustration (boxplot {graphics}). Differences in α-diversity values among the CON, LAC, PRO and SYN groups were calculated at $P < 0.05$ using the Kruskal–Wallis test (kruskal.test {stats}).

The R software (v. 3.1.0; R Core Team, Auckland, New Zealand) was used for the following statistical and multivariate analyses. The pooled percent abundance data were imported (read.table {utils}) from CLCommunity to R software data. Differences in relative abundance among the CON, LAC, PRO and SYN groups were calculated using one-way analysis of variance (aov {car}) for multiple independent groups and Tukey's test (TukeyHSD {car}) for the subsequent post-hoc analysis. For taxon-dependent analysis, we used 99 bacterial genera detected in at least one pig faecal sample. Metastats was employed to sort the differentially abundant genera from samples (Paulson et al., 2011). After removal of bacterial genera with less than 0.1% relative mean abundance, 33 differentially abundant genera with greater than 0.1% relative mean abundance remained. Next, we applied a square root (sqrt {base}) transformation to the abundance data of 33 differentially abundant bacterial genera. A heatmap (heatmap {vegan}) was generated from the square-root-transformed data of 33 differentially abundant genera generated above. For sample clustering analysis, we based our methods on two distance metrics: the Bray–Curtis dissimilarity matrix and the Euclidean distance, both of which were calculated in R (vegdist {vegan}). The stable algorithm used was the 'average' method in hierarchical clustering (hclust {stats}). The distances were calculated from 33 differentially abundant bacterial genera in each group.

For multivariate analysis of bacterial genera and bacterial OTUs (95% identity cut-off), the adegenet package in R was used to reduce multi-dimensionality in the multivariate framework of microbial community studies (Jombart and Ahmed, 2011). A DAPC (dapc {adegenet}) plot was constructed using a square root-transformed data table for individual pigs from each pig group. Here, clustering of pigs was defined prior to construction of the plot based on the independent categorical variable, that is, by treatment group. Individual pig samples containing either differentially abundant bacterial genera or all bacterial OTUs (at a 95% identity cut-off) were used to create two different DAPC plots. The principal components were selected to correspond to ≥80% cumulative variance, explained by the Eigen values of the plot, which were then subjected to linear discriminant analysis. The graphical output from the DAPC plots and canonical loading plots was then created using scatter plots (scatter {ade4}) and (loadingplot {adegenet}) respectively. The canonical loading plots were used to identify bacterial genera capable of differentiating the microbial communities according to the defined clustering groups using the user-defined threshold (0.05) (Pajarillo *et al.*, 2014b). The normalized abundance of the discriminating variables (i.e. bacterial genera, OTU) were compared among groups using a boxplot. For the phylogenetic tree reconstruction of the three most discriminating bacterial OTUs, ClustalX (version 2.1) was used to align the 16S rRNA gene sequences with other known bacterial phylotypes from the Ribosomal Database Project (RDP) Naïve Bayesian Classifier (version 2.10) using default parameters. Next, phylogenetic trees were constructed using the neighbour-joining method in MEGA5 (Tamura *et al.*, 2011; Pajarillo *et al.*, 2014a). The stability of the nodes was tested by bootstrap analysis using the adjusted values of 1000 replicates.

References

Bajaj, J.S., Ridlon, J.M., Hylemon, P.B., Thacker, L.R., Heuman, D.M., Smith, S., *et al.* (2011) Linkage of gut microbiome with cognition in hepatic encephalopathy. *Am J Physiol Gastrointest Liver Physiol* 302: G168–G175.

Bomba, L., Minuti, A., Moisá, S.J., Trevisi, E., Eufemi, E., Lizier, M., *et al.* (2014) Gut response induced by weaning in piglet features marked changes in immune and inflammatory response. *Funct Integr Genomics* 14: 657–671.

Chae, J.P., Pajarillo, E.A.B., Park, C.-S., and Kang, D.-K. (2015) Lactulose increases bacterial diveresity and modulates the swine faecal microbiota as revealed by 454-pyrosequencing. *Anim Feed Sci Tech* 209: 157–166.

Cho, J.H., and Kim, I.H. (2013) Effects of lactulose supplementation on performance, blood profiles, excreta microbial shedding of *Lactobacillus* and *Escherichia coli*, relative organ weight and excreta noxious contents in broilers. *J Anim Physiol Anim Nutr (Berl)* 98: 424–430.

Chun, J., Lee, J.H., Jung, Y., Kim, M., Kim, S., Kim, B.K., and Lim, Y.W. (2007) EzTaxon: a web-based tool for the identification of prokaryotes based on 16S ribosomal RNA gene sequences. *Int J Syst Evol Microbiol* 57: 2259–2261.

Etzold, S., Kober, O.I., MacKenzie, D.A., Tailford, L.E., Gunning, A.P., Walshaw, J., *et al.* (2014) Structural basis for adaptation of lactobacilli to gastrointestinal mucus. *Environ Microbiol* 16: 888–903.

Fleige, S., Preiβinger, W., Meyer, H.H.D., and Pfaffl, M.W. (2007) Effect of lactulose on growth performance and intestinal morphology of pre-ruminant calves using a milk replacer containing *Enterococcus faecium*. *Animal* 1: 367–373.

Gevers, D., Kugathasan, S., Denson, L.A., Vázquez-Baeza, Y., Van Treuren, W., Ren, B., *et al.* (2014) The treatment-naive microbiome in new-onset Crohn's disease. *Cell Host Microbe* 15: 382–392.

Guerra-Ordaz, A.A., González-Ortiz, G., La Ragione, R.M., Woodward, M.J., Collins, J.W., Pérez, J.F., and Martín-Orúe, S.M. (2014) Lactulose and *Lactobacillus plantarum*, a potential complementary synbiotic to control postweaning colibacillosis in piglets. *Appl Environ Microbiol* 80: 4879–4886.

Guo, X., Xia, X., Tang, R., Zhou, J., Zhao, H., and Wang, K. (2008) Development of a real-time PCR method for *Firmicutes* and *Bacteroidetes* in faeces and its application to quantify intestinal population of obese and lean pigs. *Lett Appl Microbiol* 47: 367–373.

Hemarajata, P., and Versalovic, J. (2012) Effects of probiotics on gut microbiota: mechanisms of intestinal immunomodulation and neuromodulation. *Therap Adv Gastroenterol* 6: 39–51.

Hooda, S., Boler, B.M.V., Serao, M.C.R., Brulc, J.M., Staeger, M.A., Boileau, T.W., *et al.* (2012) 454 pyrosequencing reveals a shift in fecal microbiota of healthy adult men consuming polydextrose or soluble corn fiber. *J Nutr* 142: 1259–1265.

Ito, Y., Moriwaki, H., Muto, Y., Kato, N., Watanabe, K., and Ueno, K. (1997) Effect of lactulose on short-chain fatty acids and lactate production and on the growth of faecal flora, with special reference to *Clostridium difficile*. *J Med Microbiol* 46: 80–84.

Jeon, Y.-S., Chun, J., and Kim, B.-S. (2013) Identification of household bacterial community and analysis of species shared with human microbiome. *Curr Microbiol* 67: 557–563.

Johnson, B.R., and Klaenhammer, T.R. (2014) Impact of genomics on the field of probiotic research: historical perspectives to modern paradigms. *Antonie Van Leeuwenhoek* 106: 141–156.

Jombart, T., and Ahmed, I. (2011) adegenet 1.3-1: new tools for the analysis of genome-wide SNP data. *Bioinformatics* 27: 3070–3071.

Kim, H.B., and Isaacson, R.E. (2015) The pig gut microbial diversity: understanding the pig gut microbial ecology through the next generation high throughput sequencing. *Vet Microbiol* 177: 242–251.

Kim, H.B., Borewicz, K., White, B.A., Singer, R.S., Sreevatsan, S., Tu, Z.J., and Isaacson, R.E. (2011) Longitudinal investigation of the age-related bacterial diversity in the feces of commercial pigs. *Vet Microbiol* 153: 124–133.

Konstantinov, S.R., Awati, A., Smidt, H., Williams, B.A., Akkermans, A.D.L., and de Vos, W.M. (2004) Specific response of a novel and abundant *Lactobacillus amylovorus*-like phylotype to dietary prebiotics in the guts of weaning piglets. *Appl Environ Microbiol* **70**: 3821–3830.

Krause, D.O., Bhandari, S.K., House, J.D., and Nyachoti, C.M. (2010) Response of nursery pigs to a synbiotic preparation of starch and an anti-*Escherichia coli* K88 probiotic. *Appl Environ Microbiol* **76**: 8192–8200.

Lallès, J.-P., Bosi, P., Smidt, H., and Stokes, C.R. (2007) Nutritional management of gut health in pigs around weaning. *Proc Nutr Soc* **66**: 260–268.

Lee, J.S., Awji, E.G., Lee, S.J., Tassew, D.D., Park, Y.B., Park, K.S., *et al.* (2012) Effect of *Lactobacillus plantarum* CJLP243 on the growth performance and cytokine response of weaning pigs challenged with enterotoxigenic *Escherichia coli*. *J Anim Sci* **90**: 3709–3717.

Li, W., and Godzik, A. (2006) Cd-hit: a fast program for clustering and comparing large sets of protein or nucleotide sequences. *Bioinformatics* **22**: 1658–1659.

Mao, B., Li, D., Zhao, J., Liu, X., Gu, Z., Chen, Y.Q., *et al.* (2014) In vitro fermentation of lactulose by human gut bacteria. *J Agric Food Chem* **62**: 10970–10977.

Martinez, R.C.R., Cardarelli, H.R., Borst, W., Albrecht, S., Schols, H., Gutiérrez, O.P., *et al.* (2012) Effect of galactooligosaccharides and *Bifidobacterium animalis* Bb-12 on growth of *Lactobacillus amylovorus* DSM 16698, microbial community structure, and metabolite production in an in vitro colonic model set up with human or pig microbiota. *FEMS Microbiol Ecol* **84**: 110–123.

Molist, F., Manzanilla, E.G., Pérez, J.F., and Nyachoti, C.M. (2012) Coarse, but not finely ground, dietary fibre increases intestinal Firmicutes:Bacteroidetes ratio and reduces diarrhoea induced by experimental infection in piglets. *Br J Nutr* **108**: 9–15.

O'Connor, E.M., O'Herlihy, E.A., and O'Toole, P.W. (2014) Gut microbiota in older subjects: variation, health consequences and dietary intervention prospects. *Proc Nutr Soc* **73**: 441–451.

Pajarillo, E.A.B., Chae, J.P., Balolong, M.P., Kim, H.B., and Kang, D.-K. (2014a) Assessment of fecal bacterial diversity among healthy piglets during the weaning transition. *J Gen Appl Microbiol* **60**: 140–146.

Pajarillo, E.A.B., Chae, J.P., Balolong, M.P., Kim, H.B., Seo, K.-S., and Kang, D.-K. (2014b) Pyrosequencing-based analysis of fecal microbial communities in three purebred pig lines. *J Microbiol* **52**: 646–651.

Pajarillo, E.A.B., Chae, J.P., Balolong, M.P., Kim, H.B., Park, C.-S., and Kang, D.-K. (2015a) Effects of probiotic *Enterococcus faecium* NCIMB 11181 administration on swine faecal microbiota diversity and composition using barcoded pyrosequencing. *Anim Feed Sci Tech* **201**: 80–88.

Pajarillo, E.A.B., Chae, J.P., Kim, H.B., Kim, I.H., and Kang, D.-K. (2015b) Barcoded pyrosequencing-based metagenomic analysis of the faecal microbiome of three purebred

pig lines after cohabitation. *Appl Microbiol Biotechnol* **99**: 5647–5656.

Paulson, J.N., Pop, M., and Bravo, H.C. (2011) Metastats: an improved statistical method for analysis of metagenomic data. *Genome Biol* **12(Suppl** 1): P17.

Pluske, J.R., Pethick, D.W., Hopwood, D.E., and Hampson, D.J. (2009) Nutritional influences on some major enteric bacterial diseases of pig. *Nutr Res Rev* **15**: 333–371.

Richards, J.D., Gong, J., and de Lange, C.F.M. (2011) The gastrointestinal microbiota and its role in monogastric nutrition and health with an emphasis on pigs: current understanding, possible modulations, and new technologies for ecological studies. *Can J Anim Sci* **85**: 421–435.

Sattler, V.A., Bayer, K., Schatzmayr, G., and Klose, V. (2014) Impact of a probiotic, inulin, or their combination on the piglets' microbiota at different intestinal locations. *Benef Microbes* **1**: 1–11.

Schloss, P.D., Westcott, S.L., Ryabin, T., Hall, J.R., Hartmann, M., Hollister, E.B., *et al.* (2009) Introducing mothur: open-source, platform-independent, community-supported software for describing and comparing microbial communities. *Appl Environ Microbiol* **75**: 7537–7541.

Tamura, K., Peterson, D., Peterson, N., Stecher, G., Nei, M., and Kumar, S. (2011) MEGA5: molecular evolutionary genetics analysis using maximum likelihood, evolutionary distance, and maximum parsimony methods. *Mol Biol Evol* **28**: 2731–2739.

Umu, Ö.C.O., Frank, J.A., Fangel, J.U., Oostindjer, M., da Silva, C.S., Bolhuis, E.J., *et al.* (2015) Resistant starch diet induces change in the swine microbiome and a predominance of beneficial bacterial populations. *Microbiome* **3**: 16.

Unno, T., Kim, J., Guevarra, R.B., and Nguyen, S.G. (2014) Effects of antibiotic growth promoter and characterization of ecological succession in swine gut microbiota. *J Microbiol Biotechnol* **25**: 431–438.

Yan, H., Potu, R., Lu, H., Vezzoni de Almeida, V., Stewart, T., Ragland, D., *et al.* (2013) Dietary fat content and fiber type modulate hind gut microbial community and metabolic markers in the pig. *PLoS ONE* **8**: e59581.

Zheng, W., Hou, Y., Su, Y., and Yao, W. (2014a) Lactulose promotes equol production and changes the microbial community during in vitro fermentation of daidzein by fecal inocula of sows. *Anaerobe* **25**: 47–52.

Zheng, W., Hou, Y., and Yao, W. (2014b) Lactulose increases equol production and improves liver antioxidant status in barrows treated with daidzein. *PLoS ONE* **9**: e93163.

Ziar, H., Gérard, P., and Riazi, A. (2014) Effect of prebiotic carbohydrates on growth bile survival and cholesterol uptake abilities of dairy-related bacteria. *J Sci Food Agric* **94**: 1184–1190.

16

Molecular optimization of rabies virus glycoprotein expression in *Pichia pastoris*

Safa Ben Azoun,[1] Aicha Eya Belhaj,[1] Rebecca Göngrich,[2] Brigitte Gasser[2] and Héla Kallel[1,*]

[1]*Laboratory of Molecular Microbiology, Vaccinology and Biotechnology Development, Biofermentation Unit, Institut Pasteur de Tunis, 13, place Pasteur. BP. 74, Tunis, 1002, Tunisia.*
[2]*Department of Biotechnology, BOKU - University of Natural Resources and Life Sciences Vienna, Muthgasse 18, Vienna, 1190, Austria.*

Summary

In this work, different approaches were investigated to enhance the expression rabies virus glycoprotein (RABV-G) in the yeast *Pichia pastoris*; this membrane protein is responsible for the synthesis of rabies neutralizing antibodies. First, the impact of synonymous codon usage bias was examined and an optimized RABV-G gene was synthesized. Nevertheless, data showed that the secretion of the optimized RABV-G gene was not tremendously increased as compared with the non-optimized one. In addition, similar levels of RABV-G were obtained when α-factor mating factor from *Saccharomyces cerevisiae* or the acid phosphatase PHO1 was used as a secretion signal. Therefore, sequence optimization and secretion signal were not the major bottlenecks for high-level expression of RABV-G in *P. pastoris*. Unfolded protein response (UPR) was induced in clones containing high copy number of RABV-G expression cassette indicating that folding was the limiting step for RABV-G secretion. To circumvent this limitation, co-overexpression of five factors involved in oxidative protein folding was investigated. Among these factors only *PDI1*, *ERO1* and *GPX1* proved their benefit to enhance the expression. The highest expression level of RABV-G reached 1230 ng ml^{-1}. Competitive neutralizing assay confirmed that the recombinant protein was produced in the correct conformational form in this host.

*For correspondence. E-mail: hela.kallel@pasteur.rns.tn

Introduction

The methylotrophic yeast *Pichia pastoris* (*Komagataella* sp.) has become a substantial workhorse for biotechnology, especially for heterologous protein production (Kurtzman, 2009; Ahmad *et al.*, 2014). Using this system, a variety of proteins of different origins (human, animal, plant, fungal, bacterial and viral) has been produced with varying degrees of success (Sreekrishna *et al.*, 1997; Leonardo *et al.*, 2012).

However, not all recombinant proteins are efficiently secreted. High-level expression in *P. pastoris* could face some potential bottlenecks, such as limitations in gene dosage, mRNA transcription, protein processing and folding in the endoplasmic reticulum (ER) (Agaphonov *et al.*, 2002) and translocation which is depending on the secretion signal peptide (Koganesawa *et al.*, 2001).

Genetic modification and molecular biotechnology proved to be most useful and effective tools for high-level expression (Bai *et al.*, 2011). Several genetic factors can be modified to enhance protein expression such as sequence optimization (Bai *et al.*, 2011), gene copy number (Norden *et al.*, 2011; Shen *et al.*, 2012), promoter selection (Shen *et al.*, 1998; Hohenblum *et al.*, 2004), secretion signal (Koganesawa *et al.*, 2001) and coexpression of folding-assistant proteins (Li *et al.*, 2010; Shen *et al.*, 2012).

Previous investigations have demonstrated that the limiting steps are dependent on several factors such the protein to be expressed, the promoter and the host strain employed (Cereghino and Cregg, 2000; Li *et al.*, 2007; Ashe and Bill, 2011). Generally, sequence optimization to adapt the codon usage of the gene of interest to the preferred host codon usage has been identified as one of the important factors influencing heterologous expression in *P. pastoris* (Cereghino and Cregg, 2000). For instance production of *Aspergillus niger* lipase (Yang and Liu, 2010), yeast multidrug resistance protein MDR1 (Bai *et al.*, 2011) and *Streptomyces rimosus* GDS(L)-lipase (Vujaklija *et al.*, 2002) in *P. pastoris* were increased through codon optimization by 5.3-fold, three-fold and 22-fold respectively. In addition, numerous studies have shown that gene dosage of the foreign protein has a high impact on recombinant protein production (Yu *et al.*, 2009; Zhu *et al.*, 2009). Yu *et al.* (2010) have demonstrated that increasing the gene copy number of lip2 from *Yarrowia lipolytica* enhanced the protein expression level by twofold. In many cases, increasing the target gene copy number dramatically enhances the production of foreign protein in *P. pastoris*. Although in some cases, such as human trypsinogen (Hohenblum *et al.*, 2004) and Na-ASPI (Inan *et al.*, 2006) opposite results were obtained; increased gene dosage led to a reduction of the expression level. This was attributed to limitations in folding and secretion or a saturation of the secretory pathway. Secretory protein production usually requires the presence of a secretion signal sequence at the N-terminus of the foreign protein to target it to the secretory pathway; different secretion signals are available to target protein secretion in *P. pastoris*, such as PHO1 secretion signal (Chang *et al.*, 1986) and *Saccharomyces cerevisiae* α-mating factor secretion signal (Payne *et al.*, 1995). Each signal has its particular advantage, and there is no common rule which allows the identification of the most effective sequence (Hashimoto *et al.*, 1998; Damasceno *et al.*, 2012; Gasser *et al.*, 2013).

During the journey of a protein through different cellular compartments, namely the ER, the Golgi apparatus, and finally, vesicular transport to the extracellular environment several post-translational modifications occur (Vanz *et al.*, 2014). However, not all recombinant proteins are efficiently secreted and ER retention during high-level production can be a problem. In particular, aberrant folding properties of the target protein and/or high-level production can lead to the accumulation of unfolded or even aggregated proteins in the ER (Inan *et al.*, 2006; Hesketh *et al.*, 2013) which can initiate the unfolded protein response (UPR) (Hohenblum *et al.*, 2004; Whyteside *et al.*, 2011; Zhu *et al.*, 2011) and ER-associated degradation (ERAD) (Whyteside *et al.*, 2011; Vanz *et al.*, 2012).

The ER contains several chaperones and foldases such as protein disulfide isomerase (PDI), the hsp70 member Kar2/BiP, calreticulin and calnexin, which are mainly involved in protein folding processes. Their co-overexpression with the product gene has significantly improved the productivity of several secreted proteins (Inan *et al.*, 2006; Damasceno *et al.*, 2007; Li *et al.*, 2010). Protein folding and ER stress have a severe impact on the redox state of the ER as well as on the cytosolic redox balance which can affect the process of folding its self as described in Delic *et al.* (2012). By altering the levels of redox active enzymes such as glutathione peroxidase (GPX) or the antioxidant transcription factor *YAP1*, the amount of a secreted heterologous protein was improved (Delic *et al.*, 2012, 2014).

In this work, we report our efforts to enhance the heterologous production of rabies virus glycoprotein (RABV-G) in *P. pastoris*. This protein which is exposed on the surface of rabies virus has been identified as the major antigen that induces protective immunity and thereby affords complete protection against rabies virus challenge (Cox *et al.*, 1977). To identify the limiting steps in RABV-G expression in this yeast we investigated the effect of different factors that can have a significant impact on heterologous gene expression; these factors are: gene optimization, secretion signal sequence, gene copy number and the coexpression of different proteins. In particular, we studied the effect of the coexpression of proteins which are involved in the oxidative protein folding process (*PDI1* and *ERO1*), those related to glutathione (*GPX1* and *GLR1*) and the transcriptional regulator factor *YAP1*, which responds to oxidative stress and was shown to be induced during methanol growth (Yano *et al.*, 2009). *YAP1* activates the expression of the glutathione redox system including glutathione reductase (*GLR1*).

To explore the importance of codon bias and sequence optimization on the expression of RABV-G in *P. pastoris*, the sequence of wild-type (wt)-RABV-G gene was optimized for *P. pastoris* and the expression of optimized (opt)-RABV-G in the selected transformants was monitored by Western blot and enzyme linked immunosorbent assay (ELISA). To understand the effect of RABV-G gene dosage, the copy number of the expression cassette inserted into host genome was determined by RT-qPCR in the transformed clones and correlated with the expression level. Two secretion signals (α factor and PHO1) were also studied. Finally the effect of coexpression of five intracellular proteins (*PDI1*, *ERO1*, *GPX1*, *GLR1* and *YAP1*) on RABV-G production level was investigated. The results of these explorations are presented in this study.

Results

Effect of gene optimization, signal sequence and gene dosage on RABV-G production

Rabies virus glycoprotein is a type I membrane glycoprotein containing 505 amino acids in the native form, and it is the mediator of binding to cellular receptors and entry to host cells (Anilionis *et al.*, 1981). It is composed of a cytoplasmatic domain, a transmembrane domain, and an ectodomain; the protein forms homotrimers and is anchored on the membrane envelope of the virion (Gaudin *et al.*, 1992). RABV-G has seven disulfide bonds and three potential N-glycosylation sites (Gaudin *et al.*, 1992; Walker and Kongsuwan, 1999). The important immunogenic property of RABV-G makes it as an attractive alternative that can be used as a vaccine or as diagnostic antigen in ELISA for detecting anti-glycoprotein antibodies in immunized host.

The expression of native RABV-G gene in *P. pastoris* under the control of AOX1 promoter and using α-factor as a secretion signal, has led a low expression level around 60 ng ml^{-1} (Ben Azoun et al., in press). It is worth to note that the expression of this membrane protein in this host was not reported so far in the literature.

In an effort to improve the expression level of RABV-G by increasing translational efficiency, an optimized gene of RABV-G (opt-RABV-G) supplied by GeneCust was used to transform P. pastoris cells. The sequence of wt-RABV-G was optimized with an appropriate algorithm that replaces rare codons by preferred ones in P. pastoris and optimizes other aspects of mRNA structure for optimal expression in this host. Some codons used in RABV-G were converted to the high frequency codon preferences in P. pastoris summarized in Table S1. A total of 188 codons in the wt gene were changed to codon preferred by P. pastoris, 317 codons remained unchanged. To design the codon-optimized gene, two restriction endonuclease sites were introduced at 5′ and 3′ ends of the coding sequence for an easy cloning.

To determine the effect of sequence optimization on the yield of RABV-G two expression plasmids were constructed, the optimized sequence of RABV-G gene was cloned in pPICZαA containing the α-MF-prepro secretion signal of S. cerevisiae and in pHIL-S1 containing the acid phosphatase PHO1 signal sequence of P. pastoris. The expression plasmids containing the opt-RABV-G gene were linearized and transformed into the P. pastoris strains KM71H (pPICZαA) or GS115His$^-$ (pHIL-S1). The expression cassette was integrated into the P.pastoris genome at the 5′AOX1 locus via homologous recombination, giving rise to (His$^+$; Mut^S) and (His$^+$; Mut$^+$) phenotype in the transformants of KM71H and GS115 His$^-$ respectively.

For each construction, we selected six recombinant strains and determined the copy number of the integrated expression cassette by real-time q-PCR (Fig. 1A). For clones transformed with pPICZαA-opt-RABV-G, two clones named α-8, harbouring eight copies of RABV-G gene, and α-7 containing seven copies were isolated. Clones bearing intermediate number of the expression cassette were also identified, and named α-3 and α-4. Clones with low copy were isolated, and designated as α-1 and α-2.

A similar approach was applied for clones transformed with pHIL-S1-opt-RABV-G and resulted in the isolation of clones containing different copies of the expression cassette. p-7 clone contains seven copies, whereas p-4 and p-5 harbour intermediate copy number of the expression cassettes: four and five copies respectively. Clones with low copy number were isolated, and named p-1, p-2 and p-3. These clones contain one copy, two copies and three copies of the expression cassette respectively (Fig. 1B).

The selected clones were cultivated in deep well plates, after induction of RABV-G by methanol during 72 h; RABV-G levels in culture medium were determined by Western blot using anti-rabies polyclonal antibodies as shown in Fig. 1C and D. Only a protein band of 66 kDa corresponding to RABV-G was clearly found in all culture supernatants of the recombinant strains. No band was detected in the transformed clones with empty vectors (pPICZαA and pHIL-S1), suggesting that the RABV-G protein was successfully secreted into the culture medium after methanol induction. The expression level of RABV-G by the different clones was measured by ELISA. Figure 1E and F shows that RABV-G level depends on the number of integrated expression cassette, but not on the type of signal secretion. Clones α-7 and α-8 bearing seven and eight copies, respectively, showed an expression level of 128.5 ng ml^{-1}. For clones containing lower RABV-G gene copies, we observed a correlation between the inserted copies of the expression cassette and the expression level; although no linear correlation was observed between these factors. Therefore, these data show that there is an optimal copy number of the inserted expression cassette beyond which the efficiency of folding is limited (Fig. 1E). p-7 clone showed an expression level of 129 ng ml^{-1} which was similar to that obtained with α-7 clone, harbouring the same copy number of the expression cassette but using PHO1 as a signal secretion.

These data indicate that sequence optimization had increased the expression level only by 2.1-fold; nevertheless the use of different secretion signals did not show a marked impact on product yield. Therefore, these factors do not seem to be major bottlenecks for high-level expression of RABV-G. On the contrary, the improvement of the RABV-G level seems to be due to the enhancement of RABV-G transcription; the correlation between mRNA of RABV-G transcript level, gene dosage and protein level is shown in Fig. S1.

The transcription level of the RABV-G gene increased with the increase of gene dosage in both constructs (Fig. S1). The transcription levels of RABV-G in clones containing equal copy numbers of α-factor-opt-RABV-G and PHO1-opt-RABV-G were compared with respective clones harbouring one copy of each expression cassette. The transcription level of RABV-G increased in clones containing two gene copies (α-2 and p-2) by 1.12-fold when compared with the single copy clones (α-1 and p-1). For clones harbouring three and four copies of RABV-G gene, namely α-3, α-4, p-3 and p-4, the enhancement of RABV-G transcript level was higher and was in the range of 1.5-fold for clones with three copies to 2.3 for clones with four gene copies.

Clones containing high copy number of each expression cassette showed a remarkable increase of mRNA

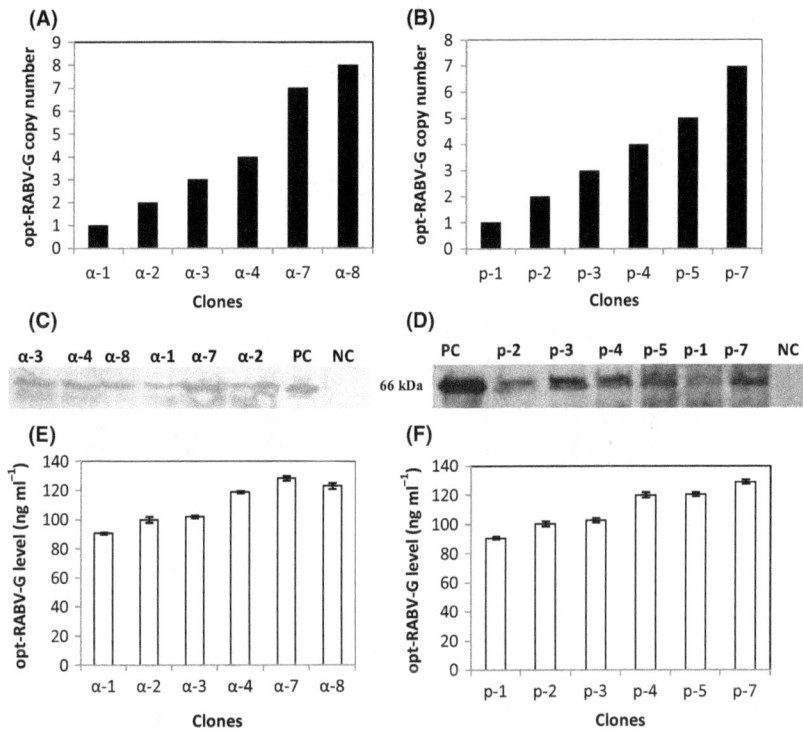

Fig. 1. RABV-G protein expression in the selected recombinant *Pichia pastoris* strains. Copy number of selected recombinant *P. pastoris* strains of (A) pPICZαA-RABV-G and (B) pHIL-S1-RABV-G transformants. Western blot analysis of 15 µl of culture supernatants (concentrated 10-fold by centricon) of (C) pPICZαA-RABV-G, and (D) pHIL-S1-RABV-G transformants. PC (positive control): inactivated and purified rabies virus. NC (negative control): KM71H and GS115His⁻ strains transformed with empty pPICZαA and pHIL-S1 vectors respectively. The amount of RABV-G produced by the different clones with α-factor (E) or PHO1 (F) signal sequence at 72 h of induction as determined by ELISA.

level of RABV-G gene. α-7 and p-7 clones which integrated seven copies of RABV-G gene exhibited an enhancement factor of 4.7-fold and 4.3-fold respectively. Clone α-8 which contains eight copies of the expression cassette reached the highest increase of RABV-G transcript (5.5-fold). These data show that there is a linear correlation between the mRNA level of the gene of interest and the expression level. The increase of gene dosage resulted in higher mRNA levels but did not significantly enhance the secreted RABV-G protein level. This can be due to depletion of precursors and energy (Baumann *et al.*, 2011) or to an accumulation of the protein in the cell caused by limitations in folding and/or secretion. These data also indicate that mRNA levels of RABV-G gene were not the limiting factor for high-level expression of RABV-G, thus we thought that RABV-G could be retained in the cell.

Determination of intracellular RABV-G protein in different recombinant clones

To check if RABV-G protein was retained within the cell, cell extracts were divided into soluble and membrane-associated fractions (including the secretory organelles)

according to Hohenblum *et al.* (2004) and analysed by Western blot (Fig. 2). The intensity of bands obtained by different clones was more pronounced in the soluble fraction than in the membrane-associated fraction for clones utilizing either α-factor or PHO1 as a secretion signal. Figure 2 shows that the gene copy number impacts the number and the intensity of the intracellular product bands. High copy strains such as α-8 and α-7 showed more intense bands than intermediate copy strains (α-4 and α-3). By contrast, for low copy strains (α-1 and α-2), minor level of RABV-G was seen in the soluble fraction (Fig. 2). In addition, for high copy strains (as α-8 and α-7), a band with a molecular weight lower than 66 kDa was observed. This band could be due to degradation of the protein of interest. For clones where the secretion of RABV-G protein was directed by PHO1 signal, the accumulation of full length RABV-G protein in the soluble fraction was lower than that seen for α-factor clones (Fig. 2). Nevertheless, the degradation of RABV-G protein was more prominent even for the low copy strains. However for α-factor clones, the proportion of RABV-G retained in the membrane-associated fraction was more prominent than that obtained for PHO1-strains. This effect was prominent for the high copy strains (Fig. 2C and D).

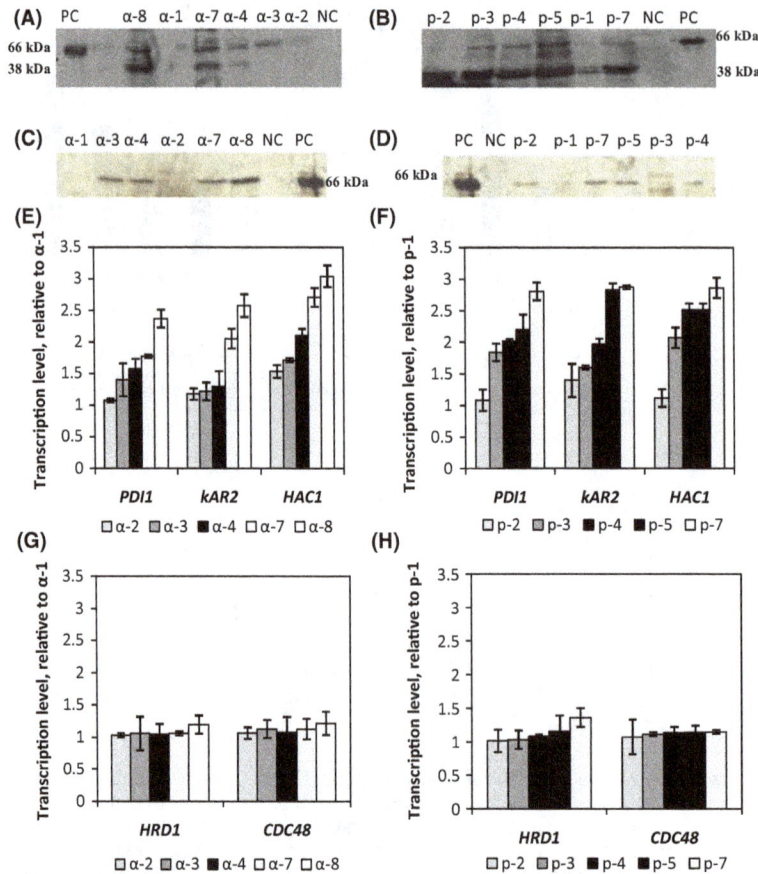

Fig. 2. Soluble intracellular RABV-G protein accumulation in clones with (A) α-factor and (B) clones with PHO1 signal sequence. Membrane-associated RABV-G protein level in clones with (C) α-factor and (D) in clones with PHO1 signal secretion. Transcription levels of selected UPR-related genes (*PDI1*, *KAR2*, *HAC1*) in clones with (E) α-factor and (F) PHO1 as a signal secretion. Transcription levels of ERAD-related genes (*HRD1* and *CDC48*) in the different *Pichia pastoris* recombinant strains with α-factor (G) and (H) PHO1 signal secretion. Gene transcript levels were normalized relative to α-1 or p-1 strains containing single copy of RABV-G gene. α-2, α-3, α-4, α-7 and α-8 correspond to selected clones of recombinant *P. pastoris* strains harbouring different copies of RABV-G gene and where RABV-G secretion was driven by the α-factor. p-2, p-3, p-4, p-5 and p-7: selected clones where RABV-G was directed by PHO1 signal secretion and containing different copies of RABV-G gene.

Transcription study of key genes involved in UPR and ERAD pathway

To determine if the retention of RABV-G inside the cell was due to folding limitations in the ER, the transcription levels of three typical UPR signal proteins (*HAC1*, *PDI1* and *KAR2*) were determined. In addition, the transcript levels of two key proteins involved in ERAD pathway (*HRD1* and *CDC48*) were measured, to determine if the UPR response had triggered an ERAD response.

The transcription levels of *HAC1* and the two chaperones located in the ER, *KAR2* and *PDI1* determined in all clones after methanol induction for 72 h are displayed in Fig. 2E and F. As compared with the strain containing one copy of RABV-G gene, the transcription level of *HAC1* was 2.7-fold and 2.8-fold higher in α-7 and p-7, respectively, but lower in the low and medium copy number strains. On the other side *HAC1* transcript level

elevated to threefold in the α-8 strain which contains eight copies of the RABV-G gene.

Therefore, increasing the RABV-G copy number resulted in an expected increase in the *HAC1* mRNA level. These results suggest that RABV-G over expression had induced the UPR in all strains. The transcription level of the other two UPR related genes confirmed this hypothesis. *PDI1* and *KAR2* transcripts showed a similar trend, with an increase of 1.7-fold in α-7 and 2.8-fold in p-7 for *PDI1* mRNA level, and 2.05-fold in α-7 and 2.8-fold in p-7 for *KAR2* transcript (Fig. 2E and F).

To determine the destiny of unfolded RABV-G, the transcription level of two key genes involved in ERAD pathway were determined in clones with either the α-factor or PHO1 secretion signal. Figure 2G and H shows the relative transcription levels of *HRD1* and *CDC48* in the different strains. No significant difference in transcription of these two genes in all clones was seen as compared with α-1 or p-1; this demonstrates that the ERAD pathway was not activated.

These data clearly suggest a folding limitation of the expressed protein. Therefore, we attempted to coexpress five proteins to improve the folding of RABV-G protein. These proteins are (i) *PDI1* and *ERO1* which are the main players of the oxidative protein folding machinery in the ER; (ii) Glutathione-related genes such as *GLR1* which is responsible for converting oxidized glutathione to reduced glutathione and *GPX1* involved in the detoxification of reactive oxygen species (ROS) at the expense of reduced glutathione, and (iii) Yap1which is the transcription factor of the oxidative stress response.

Coexpression of the oxidative protein folding (PDI1 and ERO1)

Protein disulfide isomerase (*PDI1*) is an ER-resident foldase which plays a crucial role in the formation, isomerization and reduction of disulfide bonds. During disulfide bond formation, *PDI1* accepts electrons from cysteine residues in nascent proteins, leading to the reduction of *PDI1* (Delic *et al.*, 2012). ER oxido-reduction (*ERO1*), on the other hand, is an enzyme which interplays with *PDI1* during this process and transfers the electrons to molecular oxygen or other electron acceptors to restore the oxidized form of *PDI1* (Tu *et al.*, 2000).

Plasmids overexpressing *PDI1*or *ERO1* under control of glyceraldehyde-3-phosphate dehydrogenase (GAP) promoter were transformed by electroporation into α-7/KM71H and p-7/GS115⁻ strains that contained seven copies of RABV-G gene.

Six clones from each construction were selected and their copy number was estimated by real-time q-PCR (Table S2). α-7 clones containing low, medium and high copy numbers of *PDI1* gene, namely α-7/P1, α-7/P3 and α-7/P6 were selected. p-7/P1, p-7/P6 and p-7/P3 bearing different copy numbers of *PDI1* were also isolated. In addition, α-7 and p-7 clones harbouring different copy numbers of *ERO1* were selected. The production of RABV-G by the selected clones was studied and compared with α-7 and p-7 strains which contain only seven copies of RABV-G gene. The expression level was monitored by Western blot and ELISA of culture supernatants after 72 h of methanol induction.

Western blot analysis of culture supernatants of the selected recombinant strains expressing *PDI1*or *ERO1* shows the presence of a single band at the expected size of RABV-G (Fig. 3A-1). The intensity of the band depends on the chaperone coexpressed. Coexpression of *PDI1* dramatically improved the level of secreted RABV-G, independent of the secretion signal used; this effect increased with higher gene copy number of *PDI1*. Insertion of six copies of *PDI1* resulted in 9.6-fold enhancement of secreted RABV-G when compared with α-7 and p-7 strains. Insertion of three or one copy of *PDI1* increased the production level 8.6-fold and 7.9-fold respectively (Fig. 3A-2).

ERO1 coexpression improved RABV-G expression in all clones, as seen in Western blot analysis of culture supernatants. However, when compared with *PDI1* expression, the effect was lower. Here, we observe an enhancement factor around three independent of the *ERO1* gene copy number (Fig. 3A-1 and A2).

These results clearly demonstrate that *PDI1* overexpression led to a significantly higher level of RABV-G when compared with *ERO1* strains.

Coexpression of glutathione-related genes (GPX1 and GLR1)

Glutathione reductase is a key enzyme in the conversion of oxidized GSSG to its reduced form GSH and is crucial for the maintenance of the cellular glutathione redox potential (Toledano *et al.*, 2013). *GPX1*, on the other hand, is an enzyme involved in the detoxification of ROS in particular H_2O_2, at the expense of reduced glutathione, thereby generating GSSG (Toledano *et al.*, 2013). The effect of these two enzymes on the RABV-G production was evaluated, and the selection of α-7 and p-7 clones harbouring low, medium and high copy number of *GPX1* and *GLR1* genes was performed (Table S3).

The expression of the RABV-G protein by the selected clones was analysed by Western blot and ELISA (Fig. 3B-1 and B2). We showed that *GPX1*coexpression had a positive effect on the expression level of RABV-G protein. A single protein band with the molecular weight of 66 kDa was revealed in all culture supernatants; its intensity increased with *GPX1* gene copy number (Fig. 3B-1). This result was also confirmed by ELISA (Fig. 3B-2); the insertion of six copies of *GPX1* gene in α-7 and p-7 strains enhanced the expression of RABV-G protein by 8.2-fold. The insertion of one or three copies of *GPX1* gene improved the expression level by six- and 6.8-fold respectively. Coexpression of *GLR1* with RABV-G had only a minor effect on the expression level of RABV-G protein (Fig. 3B-1 and B-2). Gene dosage of *GLR1* does not seem to have a significant impact on the expression level for α-7 and p-7 strains, as *GLR1* coexpression enhanced RABV-G production by 1.1- to 1.28-fold for all clones. These data indicate that the coexpression of *GPX1* has a large impact on RABV-G when compared with *GLR1*.

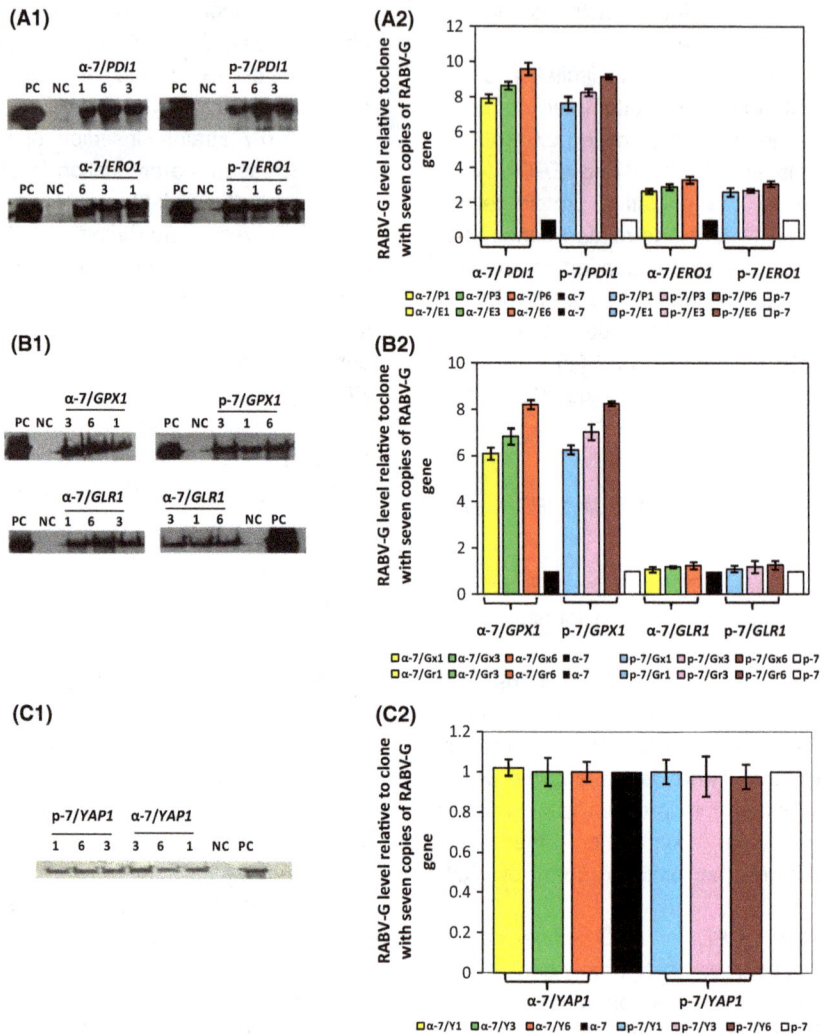

Fig. 3. Effect of coexpression of five factors involved in oxidative protein folding on the expression level of RABV-G. Western blot analysis of 15 μl of culture supernatants (concentrated 10-fold by centricon) of yeast clones with α-factor or PHO1 sequence leader to direct RABV-G protein secretion coexpressing (A-1), *PDI1* or *ERO1* genes, (B-1) *GPX1* or *GLR1* and (C-1) *YAP1* genes. PC (positive control): inactivated and purified rabies virus. NC (negative control): KM71H and GS115His⁻ transformed with an empty pPICZαA and pHIL-S1 respectively. The numbers 1, 3 and 6 correspond to copy number of different genes contained in each clone. The amount of RABV-G coexpressed with (A-2) *PDI1* or *ERO1*, (B-2) *GPX1* or *GLR1* and (C-2) *YAP1* produced by the different clones with α-factor or PHO1 as a signal sequence at 72 h of methanol induction as determined by ELISA. Clone abbreviations are explained in Tables S2, S3 and S4.

Coexpression of the transcription factor in stress response (YAP1)

The transcription factor *YAP1* is a major oxidative stress regulator. During growth on methanol, *YAP1* activates the expression of the glutathione redox system and upregulates *GLR1* (Yano *et al.*, 2009). The impact of the coexpression of this protein on RABV-G production was evaluated in strains with low, medium and high copy number of *YAP1*, named as α-7/Y1, α-7/Y3, α-7/Y6 and p-7/Y1, p-7/Y3, p-7/Y6. The expression of RABV-G was compared with α-7 and p-7 containing seven copies of RABV-G gene (Table S4).

Overexpression of *YAP1* had no effect on the level of RABV-G when compared with their respective controls (α-7 and p-7), as determined by Western bolt and ELISA (Fig. 3C-1 and C-2). RABV-G expression level reached after 72 h of methanol induction remained unchanged for all clones.

Competition of neutralization activity of rabies-immune serum by RFFIT

To prove that the RABV-G expressed by clone α-7/P6 was secreted in its native form and can compete with rabies virus to react with anti-rabies neutralizing serum and to block cell infection, the RFFIT test was used with the modifications introduced by Li *et al.* (2010). The neutralizing activity of rabies-immune serum was evaluated in the presence of different levels of recombinant RABV-G varying from 0.8 to 49 μg ml^{-1}. As shown in Fig. 4, the neutralizing titre of the rabies-immune serum was significantly reduced in the presence of RABV-G compared with that seen in the absence of recombinant RABV-G. These results suggest that the RABV-G produced by *P. pastoris* was recognized by rabies virus neutralizing antibodies.

Discussion

High-level expression of a given protein in a heterologous system can be affected by several factors; a practical solution is to identify major bottlenecks which are in general, host- and product-dependent. Codon optimization, signal sequence and gene dosage are the main factors studied to improve the expression of a target protein. Several examples have been mentioned in literature that demonstrate that the use of an optimized sequence, the increase of gene copy number and the choice of leader sequence can significantly enhance protein productivity (Vujaklija *et al.*, 2002; Zhao *et al.*, 2009; Shen *et al.*, 2012).

In this work, we first studied the effect of gene optimization on the expression of RABV-G in *P. pastoris*. We showed that the expression of opt-RABV-G led to approximately 2.1-fold increase in RABV-G production compared with wt-RABV-G (S. Ben Azoun, unpublished). These data indicate that sequence optimization was not a critical parameter for RABV-G expression in *P. pastoris*, as gene optimization sequence did not result in a tremendous increase of the expression level.

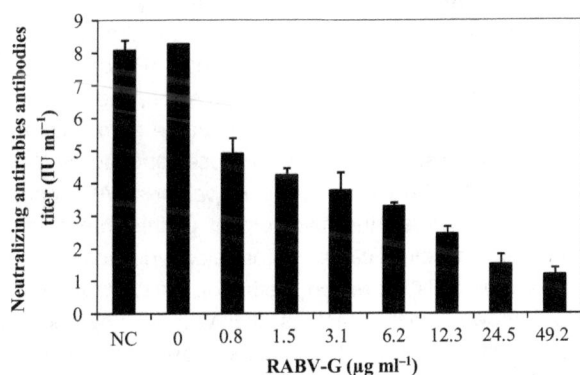

Fig. 4. Competition of the neutralizing activity of the rabies-immune serum by recombinant RABV-G. The neutralizing titres (IU ml^{-1}) were determined in the presence of varying amounts of RABV-G expressed by α-7/P6 recombinant clone, NC (negative control): culture supernatant of KM71H strain transformed with empty pPICZαA vector.

Numerous studies have demonstrated that the choice of secretion signal had a profound impact on the expression level of heterologous proteins (Ide *et al.*, 2007; Ghosalkar *et al.*, 2008). It controls the entry of proteins into the oxidative protein folding compartments such the ER in eukaryotes and the periplasm in prokaryotes (Killian *et al.*, 1990; Von Heijne, 1990; Rapoport, 1992). For *P. pastoris*, few secretion leader sequences are known and applied for the secretion of heterologous proteins (Damasceno *et al.*, 2012; Gasser *et al.*, 2013). The most widely used secretion signal is the *S. cerevisiae* α-mating factor leader sequence, whereas the *P. pastoris* acid phosphatase PHO1 signal sequence has been used in few cases (Heimo *et al.*, 1997; Romero *et al.*, 1997). The *S. cerevisiae* SUC2 gene signal sequence was used occasionally (Paifer *et al.*, 1994). It was reported that the expression level of hIFN-α2b in *P. pastoris* was higher with α-factor signal sequence than with its native signal sequence (Ghosalkar *et al.*, 2008). Expression of ocanase in *P. pastoris* under the direction of the PHO1 secretion signal was less effective than α-factor (Zhao *et al.*, 2009). While Koganesawa *et al.* (2001) demonstrated that the silkworm lysozyme expression using α-factor leader was so unstable that it could be easily attacked by proteases, they suggested that the expression level and the stability of secreted heterologous protein were greatly affected by the selection of the appropriate secretion signal sequence.

PHO1 and α-factor secretion leader were evaluated in this work for the secretory expression of RABV-G in *P. pastoris*. We found that the expression level of secreted RABV-G was similar in clones containing equal copies of the product gene from both constructs. However, intracellular RABV-G retention profile was different; a band with a molecular weight lower than 66 kDa was observed. This band could be due to degradation of the protein of interest; this band was more important in clones where the secretion of RABV-G protein was directed by PHO1 signal. These clones are derived from the strain *P. pastoris* GS115, they have the phenotype Mut$^+$. Hence, they have faster cellular machinery on methanol compared with KM71H recombinant clones which are Mut^s (clones with α-factor sequence leader).

The optimization of the gene copy number is another factor that can be modulated to improve the expression level in *P. pastoris*. Several studies have shown that this parameter is critical for heterologous protein expression in *P. pastoris* (Shen *et al.*, 2012; Yang *et al.*, 2013). We demonstrated that increasing RABV-G copy number resulted in an increased RABV-G transcript level as well as secreted RABV-G level. This positive correlation between the copy number and the expression was only observed in clones bearing less than eight copies of RABV-G gene, despite the high transcriptional rate observed in this clone. This can be due to the saturation of the folding capacity in this high copy number clone, and to the activation of UPR mechanism as described for secreted heterologous proteins in *S. cerevisiae* (Kauffman *et al.*, 2002) or in *P. pastoris* (Inan *et al.*, 2006).

The retention of RABV-G in the *P. pastoris* strains was investigated, we have found that RABV-G protein was partially accumulated in intracellular compartments; this accumulation increased with the gene copy number, implying that the secretory protein pathway might be saturated in high copy strains.

One can assume that the product is retained in the ER due to limitations in folding and/or disulfide bond formation and the set up of ER stress as a consequence. This triggers the activation UPR pathway which aims at reducing ER stress conditions by induction of genes involved in protein folding (Hoseki *et al.*, 2010; Kohno, 2010). To test whether the UPR was induced, the transcript levels of three key genes (*HAC1*, *PDI1*, *KAR2*) involved in UPR pathway were analysed. Seventy-two hours after induction of RABV-G expression lead to a significant increase of *HAC1*, *PDI1* and *KAR2* transcripts and an accumulation of intracellular RABV-G (Fig. 2). These data clearly indicate that UPR was induced upon induction of RABV-G expression. Often the UPR is followed by ERAD (Lünsdorf *et al.*, 2011; Vanz *et al.*, 2012; Lin *et al.*, 2013). The ERAD pathway has been investigated through the monitoring of the transcripts of two key genes (*HRD1* and *CDC48*). There was no significant difference between the recombinant strains and the negative control indicating that ERAD was not triggered in RABV-G expressing *P. pastoris* (Fig. 2).

Protein folding in the ER is a critical step, both, in mammals and yeasts and is a prerequisite for secretion (Helenius *et al.*, 1992; Inan *et al.*, 2006). The newly synthesized heterologous protein enters the lumen of the ER; it encounters a change in the redox environment that ultimately promotes the formation of intra-chain and/or inter-chain disulfide bonds (Inan *et al.*, 2006). Chaperone proteins play a critical role in protein folding in the ER and efficient processing is necessary to obtain high levels of proteins (Ngiam *et al.*, 2000).

Coexpression of the oxidative protein folding factors *PDI1* or *ERO1* has been successful in increasing the amount of some heterologous proteins in *P. pastoris* (Lodi *et al.*, 2005; Baumann *et al.*, 2011; Vad *et al.*, 2005; Inan *et al.*, 2012) although in some cases this approach was not effective to improve heterologous protein expression (Damasceno *et al.*, 2007).

In the current study, coexpression of the two folding factors *PDI1* or *ERO1* remarkably increased the expression level of RABV-G by 9.5-fold and 3.3-fold, respectively, in the high copy of RABV-G gene strains (Fig. 3).

This indicates that restriction in folding or misfolding of RABV-G are the major bottlenecks for optimal expression of this protein. Recently, Delic *et al.* (2012) proved that protein folding and ER stress have a severe impact on the cytosolic redox balance which may be a major factor during folding.

In *P. pastoris*, several genes of gluthatione redox system, such as the gluthatione reductase gene (*GLR1*), the *GPX1*, are induced by the transcriptional regulator *YAP1* which responds to oxidative stress (Yano *et al.*, 2009).

In this study, we coexpressed the glutathione-related genes *GPX1* and *GLR1* and the transcription factor involved in stress response *YAP1*. It has been reported by Delic *et al.* (2012) that coexpression of *GPX1* has improved the level of trypsinogen 1.6-fold in glucose-driven protein production conditions compared with *GLR1* which failed to increase the production of this protein in *P. pastoris*. *GPX1* overexpression correlated with more oxidizing redox conditions in the ER, as did *PDI1* overexpression but not *ERO1* or *GLR1* overexpression (Delic *et al.*, 2012). On the other hand, *YAP1* overexpression increased trypsinogen secretion by twofold and re-established cytosolic redox state to the state of glucose-grown wt strains (Delic *et al.*, 2014).

The coexpression of *GPX1*, *GLR1* or *YAP1* genes in high RABV-G copy strains yielded different results. The highest relative improvement was observed with clones that overexpressed *GPX1*, which showed 8.2-fold increase of RABV-G expression level, whereas the coexpression of *GLR1* slightly increased the RABV-G expression (1.2-fold). These results clearly show that cell engineering towards conditions previously shown to be beneficial for protein secretion by enhancing ER oxidation in glucose-grown *P. pastoris* is also improving production of secreted product in methanol grown cells.

Surprisingly, the coexpression of YAP1 has failed to enhance RABV-G expression; conversely the yield of RABV-G protein was slightly decreased in the clone α-7/Y6 (Fig. 3). Hence, overexpression of YAP1 gene is not required to enhance RABV-G expression, the available YAP1 level allows the cells to manage with oxidative stress. This result is in line with those reported by Yano *et al.* (2009) who identified the *P. pastoris* YAP1 homologue and showed the involvement of this transcription factor in the detoxification of formaldehyde and ROS in *P. pastoris* cells grown on methanol. Delic *et al.* (2014) also showed that YAP1 is involved in physiological detoxification of ROS formed upon oxidative folding in the ER. YAP1 is required to activate the antioxidant enzymes, which are quenching ROS. Cells with significantly lowered YAP1 level react to increased secretory folding load with accumulation of ROS and strong flocculation.

Therefore, YAP1 level is already high in methanol grown cells (Yano *et al.*, 2009); this might explain why the increase of YAP1 gene copies does not elicit any further antioxidant response.

However, it still remains to be analysed in future if the reduction of the cytosolic redox state upon production of secretory proteins and ER stress observed in glucose conditions (Delic *et al.*, 2012, 2014) is also occurring in methanol grown *P. pastoris*, or if the constant high YAP1 level on methanol prevent or diminish this response.

The RABV-G protein produced by *P. pastoris* was able to significantly reduce the neutralizing activity of the human immune rabies serum, thus demonstrating that the recombinant protein was produced in the correct conformational form.

In conclusion, in the current study we investigated different approaches to enhance RABV-G expression in *P. pastoris*. The data obtained showed that using an optimized gene sequence, different signal sequence or increasing gene copy number were not the major bottlenecks to improve the amount of recombinant RABV-G. Coexpression of oxidative folding proteins such as *PDI1* or *ERO1* and the glutathione-related genes (*GPX1* or *GLR1*) improved the secretion of RABV-G by 9.5-fold, 3.3-fold, 8.2-fold and 1.2-fold respectively. The secreted RABV-G protein was also able to react with neutralizing antirabies serum, demonstrating therefore a correct folding of the recombinant protein. Overall, these results demonstrate through combined engineering of the expression construct and the cellular oxidative protein folding machinery, *P. pastoris* can produce sufficient amounts of functional recombinant RABV-G which can be used as diagnostic antigen for detecting rabies virus neutralizing antibodies in immunized hosts or as a vaccine.

Experimental procedures

Strains, plasmids and media

Pichia pastoris KM71H and GS115His$^-$ (Invitrogen, Carlsbad, CA, USA) were used as host strains for protein expression. *Escherichia coli* DHB10 (Invitrogen) was used for the propagation of recombinant vectors.

The plasmids pPICZαA with alpha factor secretion signal (Invitrogen) and pHIL-S1 with PHO1 signal sequence were used for cloning and expression of opt-RABV-G protein (GenBank accession number KT878717). The plasmid Ppuzzle (Delic *et al.*, 2012) was used for cloning and expression of five folding-assisting factors: *PDI1* (GenBank accession number EU_805807.1), *ERO1* (GenBank accession number XM_002489600.1), *GPX1* (GenBank accession number AB_472088.1), *GLR1* (GenBank accession number AB_472087.1) and *YAP1* (GenBank accession number AB_472084.1). Gene

expression was under control of the glycolytic glyceraldehyde-3-phosphate dehydrogenase promoter P$_{GAP}$.

Yeast peptone dextrose (YPD) agar (2% peptone, 1% yeast extract, 2% glucose and 20 g l^{-1} agar) with zeocin (100 mg ml^{-1}) was used to select the positive clones which contain the recombinant plasmid pPICZαA-opt-RABV-G.

RDB (Regeneration Dextrose Medium + His) agar lacking histidine (1 M sorbitol, 1% glucose, 1.34% yeast nitrogen base, 4×10^{-5}% biotin, 0.005% of L-glutamic acid, L-methionine, L-lysine, L-leucine and L-isoleucine, and 20 g l^{-1} agar) was employed to select the His-prototrophic clones transformed with the recombinant plasmid pHIL-S1-opt-RABV-G.

BMGY (Buffered Glycerol-complex Medium) (1 mM potassium phosphate pH 6, 2% peptone, 1% yeast extract, 1.34% yeast nitrogen base, 1% glycerol and 4×10^{-5}% biotin) was used for cell growth before the induction of recombinant clones of *P. pastoris* (KM71H, GS115His$^-$) strains.

BMMY (Buffered Methanol-complex medium) (1 mM potassium phosphate pH 6, 2% peptone, 1% yeast extract, 1.34% yeast nitrogen base, 1% methanol and 4×10^{-5}% biotin) containing methanol as a carbon source was used to induce the expression of RABV-G protein.

Construction of expression vectors containing the optimized rabies virus glycoprotein sequence

Codon-opt-RABV-G was obtained from GeneCust (Dudelange, Luxembourg). The opt-RABV-G lacking its native signal peptide sequence was cloned in two plasmids containing two different secretion signal sequences: pPICZαA with *S. cerevisiae* α-mating factor as secretion signal and pHIL-S1 with *P. pastoris* PHO1 as signal sequence.

The opt-RABV-G was amplified by PCR from the recombinant plasmid PUC 19 (GeneCust) using two different primer pairs listed in Table S5. αG-F/αG-R primers were used to clone RABV-G gene into pPICZαA, the second pair pG-F/pG-R was employed for the cloning of RABV-G gene into pHIL-S1.

After amplification, PCR products were purified and ligated into pPICZαA or pHIL-S1 and transformed into *E. coli*; positive clones from the ligation reactions were analysed by restriction digestion and sequencing of the insert fragment.

Construction of expression vectors containing PDI1, ERO1, GPX1, GLR1 and YAP1 sequences

P. pastoris genes *PDI1*, *ERO1*, *GPX1*, *GLR1* and *YAP1* were isolated from the recombinant pPuzzle vectors

(Delic et al., 2012) which contain a zeocin selection marker. The genes were cloned into a pPuzzle vector under control of the GAP promoter and with KanMX as selection marker. The vectors were integrated into the 5'AOX1 region of P. pastoris recombinant clones harbouring seven copies of RABV-G gene, after linearization of the respective sequences. After transformation by electroporation, positive transformants of P. pastoris were selected on YPD plates with zeocin and G418 or YPD plates with G418.

Expression of recombinant Opt-RABV-G protein in deep 24-well cell culture plates

Recombinant P. pastoris clones were grown in YPD agar plates or RDB agar plates at 30°C. For expression studies, 2 ml of BMGY in culture plate (Dominique Dutscher, Brumath, France) were grown overnight at 30°C and 250 r.p.m. After 14–16 h, optical density was measured at 600 nm, and cells were resuspended in 2 ml of fresh BMMY medium to an initial OD_{600} of 1. Cultures were performed at 250 r.p.m., 30°C up to 72 h with methanol added to 1% every 24 h. Aliquots were taken every 24 h, cells were pelleted; supernatants and cells were stored at -20°C for further analysis.

Lysis of cells and detection of proteins

Cell pellets (the equivalent of 1 ml at an OD_{600} of ~50) were washed twice in phosphate buffered saline, then resuspended in 500 µl of yeast breaking buffer (50 mM sodium phosphate, 1 mM PMSF (Phenylmethylsulfonyl floride), 1 mM EDTA (Ethylenediaminetetraacetic acid) and 5% (v/v) glycerol), as described by Shen et al. (2012) and mixed with an equal volume of acid-washed glass beads (Sigma Aldrich, St Louis, MO, USA). The yeast cell wall was broken by vortexing 10 times for 1 min with 1 min chilling on ice. The lysate was centrifuged at 16 000 g for 20 min at 4°C. The supernatants containing the cytosolic proteins were collected. The pellet containing the membrane proteins was further treated with 400 µl yeast breaking buffer plus 2% (w/v) SDS. After centrifugation at 4000 g for 5 min at 4°C, the supernatants containing the membrane proteins were collected and analysed by Western blot.

Western blot analysis

Secreted proteins, soluble proteins and membrane proteins were analysed on a 12% SDS-polyacrylamide gel using Bio-Rad cell system. Proteins were transferred to a nitrocellulose membrane (GE Healthcare, Uppsala, Sweden). The membrane was incubated overnight at 4°C and then incubated with horse anti-rabies virus polyclonal antibody for 1 h. After washing, the membrane was incubated with HRP-conjugated polyclonal anti-horse IgG (antibodies online, Germany). The membrane was finally incubated for two minutes with ECL solution (GE Healthcare).

Enzyme-linked immunosorbent assay test

An indirect ELISA was performed to quantify of RABV-G expressed by the different clones. About 100 µl per well of either the sample or the standard (inactivated and purified rabies virus) were incubated for 2 h at 37°C. Thereafter, monoclonal antibody anti-glycoprotein TW1 (NIBSC, Hertfordshire, UK) was added to the wells and incubated for 1 h at 37°C. Finally, anti-human antibody coupled to peroxidase (Sigma Aldrich) were added and incubated 30 min at 37°C. After tetramethylbenzidine addition, the reaction intensity was measured at 450 nm. OD values higher than 0.150 were considered as positive.

Competitive neutralization assay

The neutralizing titre of rabies-immune serum was determined by RFFIT according to the standard methodology (Smith et al., 1973; Li et al., 2010). To determine whether RABV-G expressed in P. pastoris reacted with neutralizing antibodies present in the human immune rabies serum, serial dilutions of recombinant RABV-G were mixed with 50 µl rabies human immune serum (European Pharmacopeia, Strasbourg, France) and incubated at 37°C for 1 h. The serum and RABV-G mixtures were then incubated at 37°C for 1 h with a constant dose of challenge rabies virus that causes infection in 80% of BHK-21 cells. BHK-21 cells were then added to the samples and incubated for 20 h at 37°C, 5% CO_2. BHK-21cells were fixed in 80% acetone stained with fluorescein-labelled anti-rabies nucleocapsid immunoglobulins (Sanofi Diagnostic Pasteur, Marnes la Coquette, France) to detect the presence of non-neutralized virus (fluorescent foci).

RNA extraction

For transcript quantification, frozen cells were resuspended in Trizol reagent (Invitrogen) and disrupted with glass beads in FastPrep™ cell homogenizer (Thermo Scientific, Waltham, MA, USA). Total RNA was then extracted using the RNeasy Kit from Qiagen (Hilden, Germany), following the manufacturer's instructions. RNA was tested in 1% agarose gel, and was quantified by measuring OD 230/260/280 using a NanoDrop (Thermo Scientific).

Quantitative real-time PCR to determine transgene copy number and gene transcriptional level

Individual reactions were carried out in 10 µl containing 5 µl Maxima™ SYBR® Green qPCR Master Mix (Roche, Mannheim, Germany), 0.25 µl fw- and rev-primers (listed in Table S5) and 4 ng template genomic DNA or 2.5 µg cDNA. RT-qPCR was performed using a thermal cycler (Bio-Rad, Hercules, CA, USA). PCR conditions were as followed: pre-incubation at 95°C for 10 min, the thermal cycler was programmed to perform 45 cycles of: 15 s at 95°C; 20 s at 60°C; 15 s at 72°C. A melting curve was carried out to ensure that only a specific amplification product was obtained. RT-qPCR was run in duplicate with biological replicates (independent experiments) to allow for the statistical confidence in differential gene expression.

Data analysis

The relative gene expression was calculated for each sample; biological replicates (independent experiments) were conducted for all studies. As amplification efficiencies of the target and reference genes were not the same, Pfaffl method (Pfaffl, 2005) was chosen for the relative quantification of RT-qPCR results.

Acknowledgements

This work is supported by the Tunisian Ministry of Scientific Research and Higher Education (grants LR11IPT01 and "Valorisation des Résultats de la Recherche").

Conflict of Interest

The authors declare that they have no conflict of interest.

References

Agaphonov, M.O., Romanova, N.V., Trushkina, P.M., Smirnov, V.N., and Ter-Avanesyan, M.D. (2002) Aggregation and retention of human urokinase type plasminogen activator in the yeast endoplasmic reticulum. *BMC Mol Biol*, **3:** 15–23.

Ahmad, M., Hirz, M., Pichler, H., and Schhwab, H. (2014) Protein expression in *Pichia pastoris*: recent achievements and perspectives for heterologous protein production. *Appl Microbiol Biotechnol*, **12:** 5301–5317.

Anilionis, A., Wunner, W.H., and Curtis, P.J. (1981) Structure of the glycoprotein gene in rabies virus. *Nature*, **294:** 275–278.

Ashe, M.P., and Bill, R.M. (2011) Mapping the yeast host cell response to recombinant membrane protein production relieving the biological bottlenecks. *Biotechnol J*, **6:** 707–714.

Bai, J., Swartz, D.J., Protasevich, I., Brouillette, C., Harrell, P., Gasser, B., *et al.* (2011) A gene optimization strategy that enhances production of fully functional P-glycoprotein in *Pichia pastoris*. *PlosOne*, **6:** 1–15.

Baumann, K., Adelantado, N., Lang, C., Mattanovich, D., and Ferrer, P. (2011) Protein trafficking, ergosterol biosynthesis and membrane physics impact recombinant protein secretion in *Pichia pastoris*. *Microb Cell Fact*, **10:** 93–108.

Ben Azoun, S., Ben Zakour, M., Sghaeir, S. and Kallel, H. (2016) Expression of rabies virus glycoprotein in the methylotrophic yeast *Pichia pastoris*. *Biotechnol Appl Biochem*. DOI: 10.1002/bab.1471 (in press)

Cereghino, J.L., and Cregg, J.M. (2000) Heterologous protein expression in the methylotrophic yeast *Pichia pastoris*. *FEMS Microbiol Rev*, **24:** 45–66.

Chang, C.N., Matteucci, M., Perry, J., Wulf, J.J., Chen, C.Y., and Hitzeman, R.A. (1986) *Saccharomyces cerevisiae* secretes and correctly processes human interferon hybrid proteins containing yeast invertase signal peptides. *Mol Cell Biol*, **6:** 1812–1819.

Cox, J.H., Dietzschold, B., and Schneider, L.G. (1977) Rabies virus glycoprotein. II. Biological and serological characterization. *Infect Immun*, **16:** 754–759.

Damasceno, L.M., Anderson, K.A., Ritter, G., Cregg, J.M., Old, L.J., and Batt, C.A. (2007) Cooverexpression of chaperones for enhanced secretion of a single-chain antibody fragment in *Pichia pastoris*. *Appl Microbiol Biotechnol*, **74:** 381–389.

Damasceno, L.M., Huang, C.J., and Batt, C.A. (2012) Protein secretion in *Pichia pastoris* and advances in protein production. *Appl Microbiol Biotechnol*, **93:** 31–39.

Delic, M., Rebnegger, C., Wanka, F., Puxbaum, V., Troyer, C., Hann, S., *et al.* (2012) Oxidative protein folding and unfolded protein response elicit differing redox regulation in endoplasmic reticulum and cytosol of yeast. *Free Radic Biol Med*, **52:** 2000–2012.

Delic, M., Graf, A.B., Koellensperger, G., Haberhauer-Troyer, C., Hann, S., Mattanovich, D., and Gasser, B. (2014) Overexpression of the transcription factor Yap1 modifies intracellular redox conditions and enhances recombinant protein secretion. *Microb Cell.*, **1:** 376–386.

Gasser, B., Prielhofer, R., Marx, H., Maurer, M., Nocon, J., Steiger, M., *et al.* (2013) *Pichia pastoris* - protein production host and model organism for biomedical research. *Future Microbiol*, **8:** 191–208.

Gaudin, Y., Ruigrok, R., Tuffereau, C., Knossow, M., and Flamand, A. (1992) Rabies glycoprotein is a trimer. *Virology*, **187:** 627–632.

Ghosalkar, A., Sahai, V., and Srivastava, A. (2008) Secretory expression of interferon-alpha 2b in recombinant *Pichia pastoris* using three different secretion signals. *Protein Exp Purif*, **60:** 103–109.

Hashimoto, Y., Koyabu, N., and Imoto, T. (1998) Effects of signal sequences on the secretion of hen lysozyme by yeast: construction of four secretion cassette vectors. *Protein Eng*, **11:** 75–77.

Heimo, H., Palmu, K., and Suominen, I. (1997) Expression in *Pichia pastoris* and purification of *Aspergillus awamori* glucoamylasecatalytic domain. *Protein Exp Purif*, **10:** 70–79.

Helenius, A., Marquardt, T., and Braakman, I. (1992) The endoplasmic reticulum as a protein-folding compartment. *Trends Cell Biol*, **2**: 227–231.

Hesketh, A.R., Castrillo, J.I., Sawyer, T., Archer, D.B., and Oliver, S.G. (2013) Investigating the physiological response of *Pichia* (*Komagataella*) *pastoris* GS115 to the heterologous expression of misfolded proteins using chemostat cultures. *Appl Microbiol Biotechnol*, **97**: 9747–9762.

Hohenblum, H., Gasser, B., Maurer, M., Borth, N., and Mattanovich, D. (2004) Effects of gene dosage, promoters, and substrates on unfolded proteinstress of recombinant *Pichia pastoris*. *Biotechnol Bioeng*, **85**: 367–375.

Hoseki, J., Ushioda, R., and Nagata, K. (2010) Mechanism and components of endoplasmic reticulum-associated degradation. *J Biochem*, **147**: 19–25.

Ide, N., Masuda, T., and Kitabatake, N. (2007) Effects of pre- and pro-sequence of thaumatin on the secretion by *Pichia pastoris*. *Biochem Biophys Res Commun*, **363**: 708–714.

Inan, M., Aryasomayajula, M., Sinha, J.D., and Meagher, M.M. (2006) Enhancement of protein secretion in *Pichia pastoris* by overexpression of protein disulfide isomerase. *Biotechnol Bioeng*, **93**: 771–778.

Kauffman, K.J., Pridgen, E.M., Doyle, F.J., Dhurjati, P.S., and Robinson, A.S. (2002) Decreased protein expression and intermittent recoveries in BiP levels result from cellular stress during heterologous protein expression in *Saccharomyces cerevisiae*. *Biotechnol Prog*, **18**: 942–950.

Killian, J.A., de Jong, A.M., Bijvelt, J., Verklei, A.J., and de Kruijfff, B. (1990) Induction of non-bilayer lipid structures by functional signal peptides. *EMBO J*, **9**: 815–819.

Koganesawa, N., Aizawa, T., Masaki, K., Matsuura, A., Nimori, T., Bando, H., *et al.* (2001) Construction of an expression system of insect lysozyme lacking thermal stability: the effect of selection of signal sequence on level of expression in the *Pichia pastoris* expression system. *Prot Eng*, **14**: 705–710.

Kohno, K. (2010) Stress-sensing mechanisms in the unfolded protein response: similarities and differences between yeast and mammals. *J Biochem*, **147**: 27–33.

Kurtzman, C.P. (2009) Biotechnological strains of *Komagataella* (*Pichia*) *pastoris* are Komagataella phaffii as determined from multigene sequence analysis. *J Ind Microbio lBiotechnol*, **36**: 1435–1438.

Leonardo, M., Damasceno, C., and Batt, C. (2012) Protein secretion in *Pichia pastoris* and advances in protein production. *Appl Microbiol Biotechnol*, **93**: 31–39.

Li, P., Anumanthan, A., Gao, X., Ilangovan, K., Suzara, V., Duzgunes, N., and Renugopalakrishnan, V. (2007) Expression of recombinant proteins in *Pichia pastoris*. *Appl Biochem Biotechnol*, **142**: 105–124.

Li, X., Luo, J., Wang, S., Shen, Y., Qiu, Y., Wang, X., *et al.* (2010) Engineering, expression, and immune-characterization of recombinant protein comprising multi-neutralization sites of rabies virus glycoprotein. *Prot Exp Purif*, **70**: 183.

Lin, X.Q., Liang, S.L., Han, S.Y., Zheng, S.P., Ye, Y.R., and Lin, Y. (2013) Quantitative iTRAQ LC-MS/MS proteomic reveals the cellular response to heterologous protein overexpression and the regulation of HAC1 in *Pichia pastoris*. *J Proteomics*, **91**: 58–72.

Lodi, T., Neglia, B., and Donnini, C. (2005) Secretion of human serumalbumin by *Kluyveromyces lactis* overexpressing KlPDI1 and KlERO1. *Appl Environ Microbiol*, **71**: 4359–4363.

Lünsdorf, H., Gurramkonda, C., Adnan, A., Khanna, N., and Rinas, U. (2011) Virus-like particle production with yeast: ultrastructural and immunocytochemical insights into *Pichia pastoris* producing high levels of the Hepatitis B surfaceantigen. *Microb Cell Fact*, **10**: 48–58.

Ngiam, C., Jeenes, D.J., Punt, P.J., Hondel, C.A., and Archer, D.B. (2000) Characterisation of a foldase, protein disulphide isomerase A, in the proteinsecretory pathway of *Aspergillus Niger*. *Appl Environ Microbiol*, **66**: 775–782.

Norden, K., Agemark, M., Danielson, J., Alexandresson, E., Kjelbom, P., and Johanson, U. (2011) Increasing gene dosage greatly enhances recombinant expression of aquaporins in *Pichia pastoris*. *BMC Biotechnol*, **11**: 47–59.

Paifer, E., Margolles, E., Cremata, J., Montesino, R., Herrera, L., and Delgado, J.M. (1994) Efficient expression and secretion of recombinant alpha amylase in *Pichia pastoris* using two different signal sequences. *Yeast*, **10**: 1415–1419.

Payne, W.E., Gannon, P.M., and Kaiser, C.A. (1995) An inducible acidphosphatase from the yeast *Pichia pastoris*: characterization of the gene and its product. *Gene*, **163**: 19–26.

Pfaffl, M. (2005) A new mathematical model for relative quantification in real-time RT-PCR. *Nucleic Acids Res*, **29**: 2003–2007.

Rapoport, T.A. (1992) Transport of proteins across the endoplasmic reticulum membrane. *Science*, **258**: 931–936.

Romero, P.A., Lussier, M., Sdicu, A.M., Bussey, H., and Herscovics, A. (1997) Ktr1p is an a-1, 2-mannosyltransferase of *Saccharomyces cerevisiae*: comparison of the enzymatic properties of soluble recombinant Ktr1p and Kre2p/Mnt1p produced in *Pichia pastoris*. *Biochem J*, **321**: 289–295.

Shen, S., Sulter, G., and Cregg, J. (1998) A strong nitrogen source-regulated promoter for controlled expression of foreign genes in the yeast *Pichia pastoris*. *Gene*, **216**: 93–102.

Shen, Q., Ming, W., Hai-Bin, W., Hua, N., and Shu-Qing, C. (2012) The effect of gene copy number and co-expression of chaperone on production of albumin fusion proteins in *Pichia pastoris*. *Appl Gen Mol Biotechnol*, **96**: 763–772.

Smith, J.S., Yager, P.A. and Baer, G.M. (1973) A rapid tissue culture test for determining rabies neutralizing antibody. In: *Laboratory Techniques in Rabies*. Kaplan, M.M. and Koprowski, H. (eds). Geneva: WHO, pp. 354–357.

Sreekrishna, K., Brankamp, R.G., Kropp, K.E., Blankenship, D.T., Tsay, J.T., Smith, P.L., *et al.* (1997) Strategies for optimal synthesis and secretion of heterologous proteins in the methylotrophic yeast *Pichia pastoris*. *Gene*, **190**: 55–62.

Toledano, M.B., Delaunay-Moisan, A., Outten, C.E., and Igbaria, A. (2013) Functions and cellular compartmentation of the thioredoxin and glutathione pathways in yeast. *Antioxid Redox Signal*, **18**: 1699–1711.

Tu, B.P., Ho-Schleyer, S.C., Traves, K.J. and Weissman, J.S. (2000) Biochemical basis of oxidative protein folding in the endoplasmic reticulum. *Science*, **290**: 1571–1574.

Vad, R., Nafstad, E., Dahl, L., and Gabrielsen, O. (2005) Engineering of a *Pichia pastoris* expression system for secretion of high amounts of intact human parathyroid hormone. *J Biotechnol*, **116**: 251–260.

Vanz, A.L., Lunsdorf, H., Adnan, A., Nimtz, M., Gurramkonda, C., Khanna, N., and Rinas, U. (2012) Physiological response of *Pichia pastoris* GS115 to methanol-induced high level production of the Hepatitis B surface antigen: catabolic adaptation, stress responses, and autophagic processes. *Microb Cell Factories*, **11**: 103–113.

Vanz, A.L., Nimtz, M., and Rinas, U. (2014) Decrease of UPR- and ERAD-related proteins in *Pichia pastoris* during methanol-induced secretory insulin precursor production in controlled fed-batch cultures. *Microb Cell Factories*, **13**: 23–33.

Von Heijne, G. (1990) The signal peptide. *J Memb Biol*, **115**: 195–201.

Vujaklija, D., Schro, W., and Abramic, M. (2002) A novel streptomycete lipase: cloning, sequencing and high-level expression of the *Streptomyces rimosus* GDS(L)-lipase gene. *Arch Microbiol*, **178**: 124–130.

Walker, P.J., and Kongsuwan, K. (1999) Deduced structural model for animal rhabdovirus glycoproteins. *J Gen Virol*, **80**: 1211–1220.

Whyteside, G., Nor, R.M., Alcocer, M.J., and Archer, D.B. (2011) Activation of the unfolded protein response in *Pichia pastoris* requires splicing of a HAC1 mRNA intron and retention of the C-terminal tail of Hac1p. *FEBS Lett*, **585**: 1037–1041.

Yang, J.K., and Liu, L.Y. (2010) Codon optimization through a two-step gene synthesis leads to a high-level expression of *Aspergillus niger* lip 2 gene in *Pichia pastoris*. *J Mol Cat B Enzymatic*, **63**: 164–169.

Yang, S., Kuang, Y., Li, H., Liu, Y., Hui, X., Li, P., *et al.* (2013) Enhanced production of recombinant secretory proteins in *Pichia pastoris* by optimizing Kex2 P1site. *PLoS ONE*, **8**: 75347–75358.

Yano, T., Takigami, E., Yurimoto, H., and Sakai, Y. (2009) Yap1-regulated glutathione redox system curtails accumulation of formaldehyde and reactive oxygen species in methanol metabolism of *Pichia pastoris*. *Eukaryot Cell*, **8**: 540–549.

Yu, X., Wang, L., and Xu, Y. (2009) *Rhizopuschinensis* lipase: gene cloning, expression in *Pichia pastoris* and-properties. *J Mol Cat B Enzymatic*, **57**: 304–311.

Yu, M., Wen, S., and Tan, T. (2010) Enhancing production of *Yarrowia lipolytica* lipase Lip2 in *Pichia pastoris*. *Eng Life Sci*, **10**: 458–464.

Zhao, H., He, Q., Xue, C., Sun, B., Yao, X., and Liu, Z. (2009) Secretory expression of glycosylated and aglycosylated mutein of onconase from *Pichia pastoris* using different secretion signals and their purification and characterization. *FEMS Yeast Res*, **9**: 591–599.

Zhu, T., Guo, M., Tang, Z., Zhang, M., Zhuang, Y., Chu, J., and Zhang, S. (2009) Efficient generation of multi-copy strains for optimizing secretory expression of porcine insulin precursor in yeast *Pichia pastoris*. *J Appl Microbiol*, **107**: 954–963.

Zhu, T., Guo, M., Zhuang, Y., Chu, J., and Zhang, S. (2011) Understanding the effect of foreign gene dosage on the physiology of *Pichia pastoris* by transcriptional analysis of key genes. *Appl Microbiol Biotechnol*, **89**: 1127–1135.

17

Synergistic chemo-enzymatic hydrolysis of poly (ethylene terephthalate) from textile waste

Felice Quartinello,[1] Simona Vajnhandl,[2] Julija Volmajer Valh,[2] Thomas J. Farmer,[3] Bojana Vončina,[2] Alexandra Lobnik,[2] Enrique Herrero Acero,[3] Alessandro Pellis[1,*] and Georg M. Guebitz[1,4]

[1]Department of Agrobiotechnology IFA-Tulln, University of Natural Resources and Life Sciences Vienna, Inst. of Environ. Biotech., Konrad Lorenz Strasse 20, 3430, Tulln a. d. Donau, Austria.

[2]Laboratory for Chemistry and Environmental Protection, Institute of Engineering Materials and Design, Faculty of Mechanical Engineering, University of Maribor, Smetanova ulica 17, 2000, Maribor, Slovenia.

[3]Green Chemistry Centre of Excellence, Department of Chemistry, University of York, Heslington, York YO10 5DD, UK.

[4]Austrian Centre of Industrial Biotechnology, Division Polymers & Enzymes, Konrad Lorenz Strasse 20, 3430, Tulln a. d. Donau, Austria.

Summary

Due to the rising global environment protection awareness, recycling strategies that comply with the circular economy principles are needed. Polyesters are among the most used materials in the textile industry; therefore, achieving a complete poly(ethylene terephthalate) (PET) hydrolysis in an environmentally friendly way is a current challenge. In this work, a chemo-enzymatic treatment was developed to recover the PET building blocks, namely terephthalic acid (TA) and ethylene glycol. To monitor the monomer and oligomer content in solid samples, a Fourier-transformed Raman method was successfully developed. A shift of the free carboxylic groups (1632 cm^{-1}) of TA into the deprotonated state (1604 and 1398 cm^{-1}) was observed and bands at 1728 and 1398 cm^{-1} were used to assess purity of TA after the chemo-enzymatic PET hydrolysis. The chemical treatment, performed under neutral conditions (T = 250 °C, P = 40 bar), led to conversion of PET into 85% TA and small oligomers. The latter were hydrolysed in a second step using the *Humicola insolens* cutinase (HiC) yielding 97% pure TA, therefore comparable with the commercial synthesis-grade TA (98%).

*For correspondence. E-mail alessandro.pellis@boku.ac.at;

Funding Information
No funding information provided.

Introduction

Global population and rising living standards are directly correlated to the continuous increase of textile waste (Allwood *et al.*, 2006). Overproduction of fabrics since 2010 is driven by the increased rate of replacement of products. In 2008, around 14 M tons of textile waste was generated in Europe, but only 5 M tons was recovered (Zamani, 2014). About 75% of the recovered fabrics were reused or recycled mainly in industrial applications (Jorgensen *et al.*, 2008). The remaining collected waste textiles are either landfilled or incinerated. However, the recycling of used textiles would lead to several environmental benefits such as energy saving, as this process requires less energy than the production of the same products from virgin materials and reduction of the carbon footprint of the overall process (Clark *et al.*, 2016).

The textile and clothing industry is a heterogeneous business which covers different types of fibres, with a consistent 54% that is represented by man-made synthetic materials. The consumption of these synthetic fibres increased by 77% between 2000 and 2012 (Harder *et al.*, 2017). On the other hand, the growth share of synthetic fibres in global consumption results in the rising demand for petroleum-based chemicals (Pellis *et al.*, 2016a). Among the synthetic textiles, poly(ethylene terephthalate) (PET) is one of the most widely used polymers in the global textile industry (Herrero Acero *et al.*, 2011). PET is a semi-crystalline thermoplastic polymer which shows excellent tensile strength, chemical resistance and high thermal stability. Two PET grades are dominating the global market: the fibre grade PET, with a M_w of 15–20 Kg mol^{-1} and intrinsic viscosity between 0.4 and 0.75 dl g^{-1}, and the bottle grade PET, which refers to a higher M_w polymer (> 20 Kg mol^{-1}) with an intrinsic viscosity above 0.95 dl g^{-1} (Tasca *et al.*, 2010; Al-Sabagh *et al.*, 2016).

Due to its wide production and utilization (Cavaco-Paulo and Guebitz, 2008), PET represents a broad disposal inert textile. The non-toxic nature, durability and crystal clear transparency of PET during use are the principal advantages of this polyester, while its rather slow biodegradability is the major cause of concern to the environmentalists. Recycling the textile waste-derived polyesters can significantly cut down the energy usage, resource depletion and greenhouse gas emissions. Unfortunately, different factors such as colouring dyes (Giannotta *et al.*, 1994) and other chemicals such as detergents, fuels and pesticides (Demertzis *et al.*, 1997) reduce the quality of recycled PET reducing the number of the possible applications. When compared to mechanical recycling and incineration, chemical hydrolysis could lead to higher value products (Paszun and Spychaj, 1997). The most common chemical-based PET hydrolysis processes are alkaline hydrolysis using 4–20% NaOH/KOH solutions (Karayannidis *et al.*, 2002), phase transfer catalysts and acidic hydrolysis using concentrated sulphuric acid or other mineral acids (Yoshioka *et al.*, 2001). All these processes are very costly and toxic because the chemicals required and laborious purification steps are needed. In recent years, more environmentally friendly PET recycling strategies based on neutral hydrolysis (carried out using water or steam at 1–4 MPa and temperatures of 200–300 °C) were reported (Launay *et al.*, 1994). In last decade, the interest of biotechnologies towards polyesters biodegradation and recycling is gaining a key role. Yoshida *et al.* (2016) showed a novel bacterium, *Ideonella sakaiensis* 201-F6, able to break down the plastic using two enzymes to hydrolyse PET and assimilate its building block for growth. Earlier, various studies demonstrated that a class of enzymes belonging to the α/β hydrolase family, namely cutinases, are able to hydrolyse the ester bonds of PET and several other polyesters (Pellis *et al.*, 2015, 2016b; Barth *et al.*, 2016; Wei *et al.*, 2016). Among them, cutinase are currently under investigation for the bioprocessing of PET textiles on an industrial scale (Silva *et al.*, 2005). Earlier, it was reported that cutinases from *Thermobifida fusca* and *Humicola insolens* were able to hydrolyse low-crystallinity PET while complete hydrolysis by enzymes only seems to be difficult if not impossible for PET with higher crystallinity (Mueller, 2006; Nimchua *et al.*, 2007; Ronkvist *et al.*, 2009). In this work, we propose an innovative synergistic chemo-enzymatic hydrolysis of PET for the production of high-purity TA (97%) avoiding harsh chemical treatments.

Results and discussion

FT-Raman analysis

In the past, several methods, mainly HPLC-related, have been established to follow enzymatic hydrolysis of PET in aqueous solutions (Herrero Acero *et al.*, 2011; Pellis

et al., 2016c). However, water-insoluble oligomers would obviously escape quantification of TA in the presence of insoluble oligomers; hence, a novel method based on FT-Raman and triethylamine was adapted for this study. TA, p-$C_6H_4(COOH)_2$, has two carboxylic groups which are detectable using Fourier transform Raman spectroscopy. The analysis of benzene dicarboxylic acids such as isophthalate and terephthalate using FT-Raman was previously described by Arenas and Tellez (Arenas, 1979; Téllez *et al.*, 2001). According to these reports, the -COOH group of solid TA reveals a typic-centred band at 1631 cm^{-1}, mainly given by the asymmetric stretching (ν_{as}) of C=O. As described by Tellez, a coupled vibrational mode is characteristic of the single band at 1286 cm^{-1}. The discrimination of monomeric TA from esterified species was performed converting these functional groups into the corresponding anions. The simplest reaction of carboxylic acid is salification by a base. This reaction causes the shift of (ν_{as}) of the C=O group and appearance of new bands due to the in-phase and out of phase -COO$^{(-)}$ stretching vibration (Socrates, 2004). The shift of the acid peak of the carboxylic group from 1720 cm^{-1} to 1580 cm^{-1} after alkaline treatment was previously reported for the grafting of cotton with cyclodextrins (Voncina and Majcen, 2005) or for detection of end groups of fluoropolymers (Pianca *et al.*, 1999).

The conversion of the -COOH group into the carboxylate species was carried out using triethylamine (TEA) in chloroform solution (Di Robert *et al.*, 2015). The deprotonated anion obtained *via* incubation with tertiary aliphatic amine caused the shift of ν_{as}C=O and the increase of the single band of the symmetric stretching (ν_s) of C=O at 1400 cm^{-1} and an ammonium-related band in the 2700 cm^{-1} region.

Figure 1 shows the shift of the acid peak of solid TA from 1631 to 1604 cm^{-1} after deprotonation with TEA. Furthermore, the clear appearance of a peak at 1398 cm^{-1} was observed due the symmetric stretching mode of the carboxylate moiety. On the other hand, the peak at 1286 cm^{-1} was strongly reduced. The 1:5 TA/TEA ratio showed the clearest signal for the monomers due to a complete conversion of the desired groups (Fig. S2).

To assess the suitability of this method to quantify TA in the presence of oligomers, BHET, DMT and PET were incubated with the tertiary aliphatic amine (1:5 ratio). Expectedly, the shift of 1632 and 1395 cm^{-1} typical for TA did not occur as all carboxylic groups of these oligomers are esterified (Figs S3–S5).

Water-based PET hydrolysis

Preliminary experiments showed that water-based PET depolymerization was not achieved with ratio 1:4 PET/

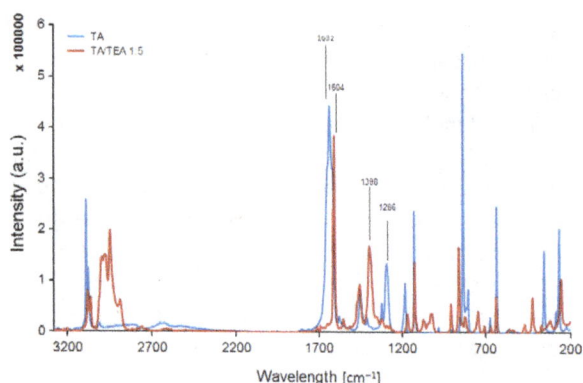

Fig. 1. FT-Raman analysis showing the deprotonation of the TA's carboxylic acid moieties (blue) *via* incubation with a 1:5 TA/TEA ratio (red). Spectra were normalized between the region 2000 and 2200 cm^{-1}.

H_2O and incubation for 90 min. Therefore, a series of experiments were performed, to define the appropriate depolymerization conditions (Table 1).

The incubation of PET fibre at 180 °C and 12 bar did not lead to any depolymerization of the sample, and hence, there was no difference of spectra. In Fig. S6, it is possible to observe how the spectra of raw PET fibre and Sample 2 are very similar with no remarkable differences that could be spotted, showing that reaction temperature and pressure were not optimal for the hydrolysis of the polymer. Finally, when temperature and pressure were increased to 250 °C and 39 bar, respectively, the polymer was completely reduced in a whitish powder (Fig. S7). In Fig. S8, it is possible to see that spectrum of PET depolymerization product for Sample 4 (blue) was similar to the spectrum of commercial TA. Using these conditions, a powder consisting of 85% TA was obtained from the water-based PET depolymerization according to HPLC and ^{1}H-NMR analysis. Unfortunately, further variation of the reaction conditions

including addition of zinc acetate did not lead to a higher yield of TA. Hence, enzymatic hydrolysis of the remaining oligomers was assessed in the next step. To note that experiments with various steady-state lengths were performed but did not lead to an improvement of the TA yield, therefore the enzymatic finishing step was critical in order to obtain high-purity TA.

Enzymatic PET hydrolysis

The kinetics of the formation of degradation products *via* enzymatic treatment depends on various factors, including the chemical–physical structure of polyester substrate (Eberl *et al.*, 2008). Already, various authors have previously demonstrated that cutinase preferentially hydrolyze the amorphous region on PET (Donelli *et al.*, 2009; Gamerith *et al.*, 2017). On the other hand, it was shown that enzymatic hydrolysis of PET oligomers is way faster than of a long-chain polymer (Ribitsch *et al.*, 2011). Therefore, due to the high crystallinity of the PET fibres used in the textile industry, a chemical pre-hydrolysis was established in this work.

The sample from chemical pre-treatment with the highest degree of hydrolysis (Sample 4, 85% TA) was further incubated with different concentrations of *Humicola insolens* cutinase (HiC) to hydrolyse remaining oligomers (Fig. 2). The highest amount of soluble TA (6.5 mM) was obtained after 6 h of incubation both when 1 or 2 mg ml^{-1} of HiC was applied without any further increase until 24 h of incubation. The lower concentrations of enzyme of 0.1 and 0.5 mg ml^{-1} led to the release of 0.53 and 1.9 mM of TA respectively.

Table 1. Design of experiments strategy used for performing the water-based hydrolysis of PET.

Experiment	Initial MW (IV) range of PET	T [°C]	PET/H_2O ratio[a]	Steady-state time [min][b]
1	0.62	180	1/4	0
2			1/10	30
3		250	1/4	
4			1/10	0
5		180	1/4	
6			1/10	30
7		250	1/4	
8			1/10	0

All reported conditions were also tested; zinc acetate was added to the reaction mixture.
a. 25 g of PET.
b. After the transient period (to reach the desired T).

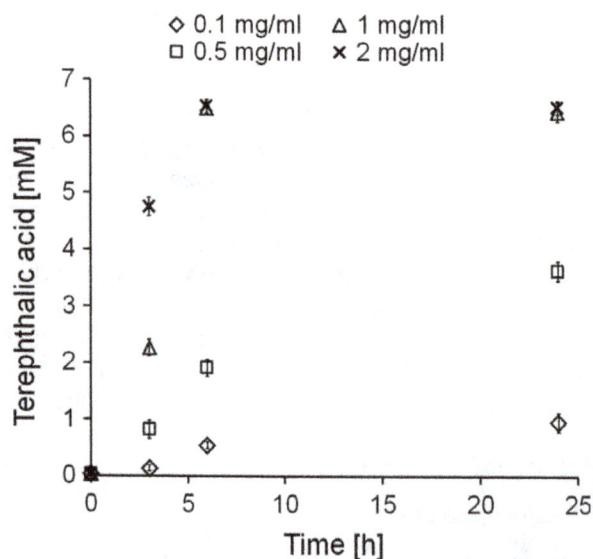

Fig. 2. Enzymatic hydrolysis of chemically pre-treated PET with different concentrations of *Humicola insolens* cutinase (HiC).

Terephthalic acid with 97% purity was obtained after enzymatic treatment of the chemical-treated PET which is comparable to synthesis-grade TA (98% pure). When chemical pre-hydrolysis of PET was performed in the presence of zinc acetate as a catalyst, a negative influence on enzymatic hydrolysis was observed (Fig. 3). In fact, there was no further increase in the amount of TA seen (Fig. 3).

Zinc and other metal ions are well known to be potential inhibitors of many enzymes, including cutinases as recently reported by Chen *et al.* that biochemically characterized *Thermobifida fusca* cutinases (Chen *et al.*, 2010).

FT-Raman detection of depolymerized PET

While in the chemically pre-treated sample oligomers were present, the peak at 1728 cm^{-1} indicative of ester bond was considerably reduced after enzymatic hydrolysis (Fig. 4). In parallel, the increase in the band at 1398 cm^{-1} indicated formation of TA. Finally, the signal at 1286 cm^{-1}, as expected, had the opposite trend of the asymmetric stretching of CO, further confirming the reduction in the oligomers to TA (Fig. 5).

In addition to FT-Raman, ^{1}H-NMR analysis was performed. Likewise, ^{1}H spectra (recorded in DMSO-d_6)

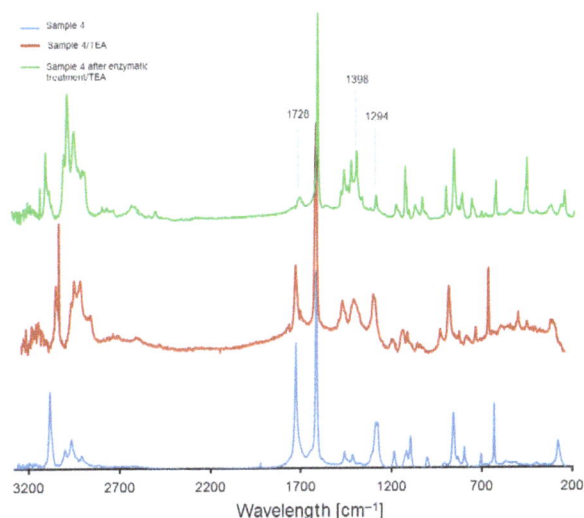

Fig. 4. FT-Raman analysis showing the incubation of Sample 4 with a 1:5 TEA ratio before (red) and after (green) enzymatic hydrolysis. Spectra were normalized between the 2000 and 2200 cm^{-1} region.

indicate the presence of some longer oligomers in the chemically pre-treated sample while the spectra of the reaction product after subsequent enzymatic hydrolysis suggest their conversion into TA. Based on the ^{1}H-NMR-performed calculations (see ESI for details), the oligomers are mostly comprised of less than four (but nearer two) constitutional repeat units, and TA end groups dominate over EG suggesting the oligomerization is more effective at releasing free EG than free TA, or that the purification steps involved preferentially remove EG.

Materials and Methods

Chemicals, substrates and enzymes

Buffer components, bovine serum albumin (BSA), *para*-nitrophenyl-butyrate (*p*-NPB), methanol, zinc acetate, TA and formic acid were purchased from Sigma-Aldrich (USA). All other chemicals and reagents used in this work were of analytical grade and used without further purification if not otherwise specified. *Humicola insolens* cutinase (HiC) was a gift from Novozymes (Beijing, China). All hydrolyses were performed using Wellman PET fibre with a viscosity of 0.62 dl g^{-1}.

Water-based PET hydrolysis

Water-based PET hydrolysis was performed in 1 l stainless steel high-pressure and high-temperature reactor at different temperatures (180 and 250 °C) and with and without the addition of zinc acetate as catalyst. All experiments were carried out with 25 g of virgin PET fibre in 250 ml of deionized water. The reactions were stopped after 60 and 90 min, respectively (30 min after reaching

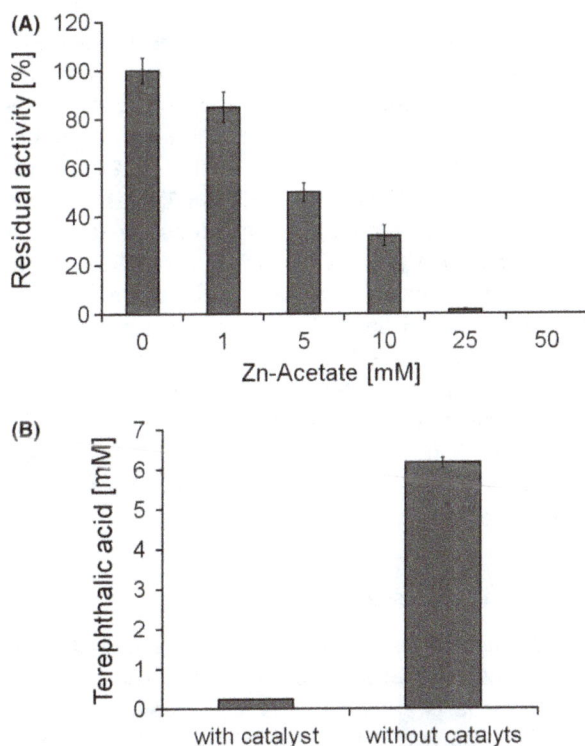

Fig. 3. Residual activity of HiC in the presence of Zn acetate (A) and enzymatic hydrolysis of PET chemically pre-treated in the presence of Zn acetate (B).

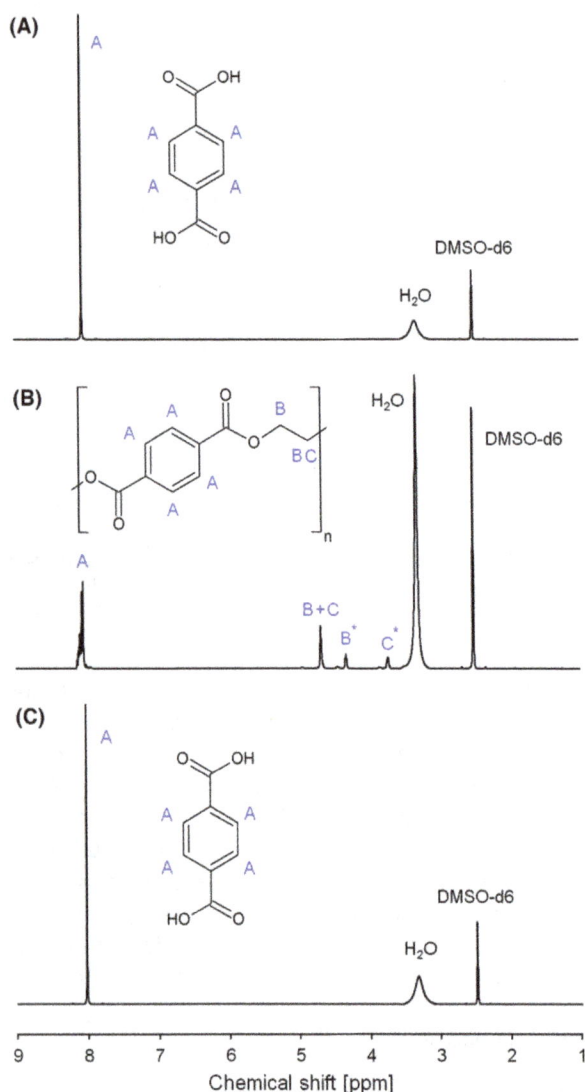

Fig. 5. ¹H-NMR of pure TA (A), PET degradation after the chemical treatment (Sample 4, B) and PET degradation after enzymatic finishing (C). All spectra were recorded in DMSO-d₆. All samples were fully soluble in the selected solvent. For detailed proton assignments, please see ESI.

the steady state), and two different ratios PET/H₂O were assessed. A temperature and pressure profile of a typical reaction is shown in Fig. S1. At the end of the process, H₂O and EG were removed and a white powder was obtained (Fig. S7B). The EG stream can be further processed anaerobically due to the current low cost of the compound.

ATR FT-IR analysis

ATR FT-IR spectra were recorded on a Perkin-Elmer Spectrum GX spectrometer. The ATR accessory (supplied by Specac Ltd., UK) contained a diamond crystal. A total of 16 scans for each sample were taken with a resolution

of 4 cm⁻¹. All spectra were recorded at 21 °C over the wavelength interval between 4000 and 650 cm⁻¹.

FT-Raman analysis

To remove the water, terephthalic acid (TA), bis(2-hydroxyethyl) terephthalate (BHET), dimethyl terephthalate (DMT) and poly(ethylene terephthalate) PET as standards were lyophilized for 24 h; 50 mg of samples was incubated with different volumes of pure triethylamine in 10 ml of chloroform for 6 h. Afterwards, chloroform was removed by evaporation at 21 °C for 24 h. The virgin material and the powders obtained from the chemical process and after the enzymatic treatment were also incubated as described above. The FT-Raman spectra were recorded using a Perkin-Elmer Raman station 400, coupled with a 785 nm laser. Spectra were collected at a resolution of 2 cm⁻¹ for 25 scans and normalized in the region 2200–2400 cm⁻¹ before any data processing. The bands were assigned as follows: 3000–2700 cm⁻¹ ν(Et-NH₄⁺), 1728 cm⁻¹ ν(C=O, bended), 1632–1604 cm⁻¹ νas(C=O), single band 1398 cm⁻¹ δ(C-O), single band 1286 cm⁻¹ coupled mode ν(C=O)+δ(COH).

Protein quantification and SDS-PAGE analysis

Protein concentration was measured using the Bio-Rad Protein Assay Kit (Bio-Rad, Vienna, Austria). BSA was used as protein standard; 10 μl of the sample was added into the wells of a 96-well plate. Afterwards, 200 μl of the prepared Bio-Rad reagent solution was added (Bio-Rad Reagent diluted 1:5 with MQ-water). The plate was incubated for 5 min at 21 °C and 400 rpm. Buffer (100 mM Tris-HCl pH 7) was used as blank. The absorption after 5 min was measured at λ = 595 nm in a plate reader (Tecan INFINITE M200) and the concentration calculated from the average of triplicates. Sodium dodecyl sulphate–polyacrylamide gel electrophoresis (SDS-PAGE) was performed according to Laemmli using 4% stacking gels and 15% separating gels and run at 150 V (Laemmli, 1970). Pre-stained protein marker IV (Peqlab, Germany) was used as a molecular mass marker. Proteins were stained with Coomassie method.

Esterase activity assay

Esterase activity was measured at 25 °C using p-nitrophenyl-butyrate (p-NPB) as a substrate according to Biundo et al. (Biundo et al., 2015). The final assay was carried out by mixing 200 μl of the substrate stock solution, in 100 mM Tris–HCl buffer pH 7, with 20 μl of enzyme solution. The increase in the absorbance at 405 nm due to the release of p-nitrophenol (ε 405 nm = 10.27 ml (μmol

$cm)^{-1}$) was measured for 5 min, every 18 s with a plate reader. A blank was included using 20 µl of buffer instead of enzyme solution. The activity was calculated in units (U), where 1 unit is defined as the amount of enzyme required to hydrolyse 1 µmol of substrate per minute under the given assay conditions. The activity assay was also performed in the presence of 1, 5, 10, 25 or 50 mM of the chemical catalyst (zinc acetate). All the experiments were performed in triplicates.

Enzymatic hydrolysis of PET oligomers

The enzymatic hydrolysis of the PET powder resulting from the chemical treatments was performed as previously described with some modifications (Herrero Acero et al., 2011; Pellis et al., 2016a). Briefly, 10 mg of pretreated PET powder and 2.0 ml of enzyme solution (diluted in 100 mM Tris-HCl buffer pH 7) were added in a test tube. The mixture was incubated for 24 h at 50 °C and 600 rpm in an orbital shaker. Different concentrations 0.1, 0.5, 1 and 2 mg ml^{-1} of enzyme were used to understand the optimal concentration of HiC. All the experiments were performed in triplicates.

High-Performance Liquid Chromatography

Proteins were removed using a 1:1 (v v^{-1}) ice-cold methanol precipitation. Therefore, samples were centrifuged at 14 000 rpm (Centrifuge Beckman JU-MI) at 0 °C for 15 min. The supernatant was acidified by adding 8 µl of 6 M HCl and then transferred into HPLC vials. The released products were analysed by high-performance liquid chromatography (HPLC) (Agilent Technologies, Palo Alto, CA) coupled with a UV detector, at 241 nm, using a water/methanol linear gradient (Table S1). To determine the TA percentage in the mixture, 10 mg of PET powder resulting from before and after the enzymatic hydrolysis was diluted in 100 ml of 100 mM Tris-HCl buffer pH 7 and analysed as described before.

^{1}H-NMR analysis

^{1}H nuclear magnetic resonance was performed on a Bruker Avance II 400 spectrometer (resonance frequency of 400.13 MHz for ^{1}H) equipped with a 5 mm observe broadband probe head (BBFO) with z-gradients. DMSO-d_6 was used as NMR solvent for all samples.

Conclusions

In this study, the synergism of chemical and enzymatic hydrolysis of PET was demonstrated on a multigram scale. To monitor the PET hydrolysis and formation of

TA, a FT-Raman method based on the deprotonation of the –COOH group was established. The chemical pretreatment performed in an environmentally friendly way under neutral conditions, therefore avoiding harsh chemicals, leads to depolymerization of the polyester-composed waste textiles yielding about 85% TA. The enzymatic hydrolysis performed in a second reaction step leads to further hydrolysis of the remaining oligomers yielding TA with a purity of 97%. Future studies should consider the chemo-enzymatic treatment of different PET containing textiles wastes as well as studies on synthesis of PET based on the recovered TA.

Acknowledgements

This project received funding from the European Union's Horizon 2020 research and innovation programme under the grant agreement 641942 (Resyntex project).

Author contributions

The manuscript was written through contributions of all authors. All authors have given approval to the final version of the manuscript.

Conflict of interest

None declared.

References

Allwood, J.M., Laursen, S.E., deRodriguez, C.M. and Bocken, N.M.P. (2006) Well dressed? The present and future sustainability of clothing and textiles in the United Kingdom. Technical annex. University of Cambridge.

Al-Sabagh, A.M., Yehia, F.Z., Eshaq, Gh., Rabie, A.M. and ElMetwally, A.E. (2016) Greener routes for recycling of polyethylene terephthalate. Egypt J Petrol 25: 53–64.

Arenas, J.F. (1979) Infrared and Raman spectra of phthalate, isophthalate and terephthalate ions. Spectrochim Acta, Part A 35: 355–363.

Barth, M., Honak, A., Oeser, T., Wei, R., et al. (2016) A dual enzyme system composed of a polyester hydrolase and a carboxylesterase enhances the biocatalytic degradation of polyethylene terephthalate films. Biotechnol J 11: 1082–1087.

Biundo, A., Hromic, A., Pavkov-Keller, T., Gruber, K., et al. (2015) Characterization of a poly(butylene adipate-co-terephthalate)-hydrolyzing lipase from Pelosinus fermentans. Appl Microbiol Biotechnol 100: 1753–1764.

Cavaco-Paulo, A., and Guebitz, G.M. (2008) Enzymes go big: surface hydrolysis and functionalisation of synthetic polymers. Trends Biotechnol 26: 32–38.

Chen, S., Su, L., Billig, S., Zimmermann, W., Chen, J., and Wu, J. (2010) Biochemical characterization of the cutinases from Thermobifida fusca. J Mol Catal B: Enzym 63: 121–127.

Clark, J.H., Farmer, T.J., Herrero-Davila, L., and Sherwood, J. (2016) Circular economy design considerations for research and process development in the chemical sciences. *Green Chem* **14**: 3914–3934.

Demertzis, P.G., Johansson, F., Lievens, C., and Franz, R. (1997) Studies on the development of a quick inertness test procedure for multi-use PET containers-sorption behaviour of bottle wall strips. *Packag Technol Sci* **10**: 45–58.

Di Robert, M., Silverstein, F.W., David, J. and Bryce, D.L. (2015). Spectrometric Identification of Organic Compounds.

Donelli, I., Taddei, P., Smet, P.F., Poelman, D., Nierstrasz, V.A., *et al.* (2009) Enzymatic surface modification and functionalization of PET: a water contact angle, FTIR, and fluorescence spectroscopy study. *Biotechnol Bioeng* **103**: 845–856.

Eberl, A., Heumann, S., Kotek, R., Kaufmann, F., *et al.* (2008) Enzymatic hydrolysis of PTT polymers and oligomers. *J Biotechnol* **135**: 45–51.

Gamerith, C., Zartl, B., Pellis, A., Guillamot, F., *et al.* (2017) Enzymatic recovery of polyester building blocks from polymer blends. *Proces Biochem* [In press] https://doi.org/10.1016/j.procbio.2017.01.004.

Giannotta, G., Po, R., Cardi, N., Tampellini, E., *et al.* (1994) Processing effects on poly(ethylene terephthalate) from bottle scraps. *Polym Eng Sci* **34**: 1219–1223.

Harder, R., Kalmykova, Y., Morrison, G.M., Feng, F., *et al.* (2017) Quantification of goods purchases and waste generation at the level of individual households. *J Ind Ecol* **18**: 227–241.

Herrero Acero, E., Ribitsch, D., Steinkellner, G., Gruber, K., *et al.* (2011) Enzymatic surface hydrolysis of pet: effect of structural diversity on kinetic properties of cutinases from thermobifida. *Macromolecules* **44**: 4640.

Jorgensen, A., Le Bocq, A., Nazarkina, L., *et al.* (2008) Methodologies for social life cycle assessment. *Int J LCA* **13**: 96–103.

Karayannidis, G.P., Chatziavgoustis, F., Achilias, A.P., *et al.* (2002) Poly(ethylene terephthalate) recycling and recovery of pure terephthalic acid by alkaline hydrolysis. *Adv Polym Technol* **21**: 250–259.

Laemmli, U.K. (1970) Cleavage of structural proteins during the assembly of the head of bacteriophage T4. *Nature* **227**: 680–685.

Launay, A., Thominette, F., and Verdu, J. (1994) Hydrolysis of poly(ethylene terephthalate): a kinetic study. *Polym Degrad Stabil* **46**: 319–324.

Mueller, R.-J. (2006) Biological degradation of synthetic polyesters-Enzymes as potential catalysts for polyester recycling. *Process Biochem* **41**: 2124–2128.

Nimchua, T., Punnapayak, H., and Zimmermann, W. (2007) Comparison of the hydrolysis of polyethylene terephthalate fibers by a hydrolase from *Fusarium oxysporum* LCH I and *Fusarium solani* f. sp. pisi. *Biotechnol J* **2**: 361–364.

Paszun, D., and Spychaj, T. (1997) Chemical recycling of Poly(ethylene terephthalate). *Ind Eng Chem Res* **36**: 1373–1383.

Pellis, A., Herrero Acero, E., Weber, H., and Obersriebnig, M. (2015) Biocatalyzed approach for the surface functionalization of poly(L-lactic acid) films using hydrolytic enzyme. *Biotechnol J* **10**: 1739–1749.

Pellis, A., Herrero Acero, E., Ferrario, V., Ribitsch, D., *et al.* (2016a) The closure of the cycle: enzymatic synthesis and functionalization of bio-based polyesters. *Trends Biotechnol* **34**: 316–328.

Pellis, A., Gamerith, C., Ghazaryan, G., Ortner, A., *et al.* (2016b) Ultrasound-enhanced enzymatic hydrolysis of poly (ethylene terephthalate). *Biores Technol* **218**: 1298–1302.

Pellis, A., Haernvall, K., Pichler, C.M., Ghazaryan, G., *et al.* (2016c) Enzymatic hydrolysis of poly(ethylene furanoate). *J Biotechnol* **235**: 47–53.

Pianca, M., Esposto, G., and Radice, S. (1999) End groups in fluoropolymers. *J Fluorine Chem* **95**: 71–84.

Ribitsch, D., Heumann, S., Trotscha, E., Herrero Acero, E., *et al.* (2011) Hydrolysis of polyethyleneterephthalate by p-nitrobenzylesterase from Bacillus subtilis. *Biotechnol Prog* **27**: 951–960.

Ronkvist, A.M., Xie, W., Lu, W., and Gross, R.A. (2009) Cutinase-catalyzed hydrolysis of Poly(ethylene terephthalate). *Macromolecules* **42**: 5128–5138.

Silva, C., Carneiro, F., O'Neill, A., Fonseca, L.P., Cabral, J.M.S., Guebitz, G. and Cavaco-Paulo, A. (2005) Cutinase- a new tool for biomodification of synthetic fibers. *J Polym Chem* **43**, 2448–2450.

Socrates, G. (2004) *Infrared characteristic group frequencies tables and charts*, 3rd edn. John Wiley & Sons.

Tasca, F., Gorton, L., Kujawa, M., Patel, I., *et al.* (2010) Increasing the coulombic efficiency of glucose biofuel cell anodes by combination of redox enzymes. *Biosens Bioelectron* **25**: 1710–1716.

Téllez, A.C., Hollauer, E., Mondragon, M.A. and Castaño, V.M. (2001) Fourier transform infrared and Raman spectra, vibrational assignment and ab initio calculations of terephthalic acid and related compounds. *Spectrochim Acta Part A: Mol Biomol Spectrosc* **57**, 993–1007.

Voncina, B. and Majcen, L. M.A. (2005) Grafting of cotton with β-cyclodextrin via poly(carboxylic acid). *J Appl Polym Sci* **96**, 1323–1328.

Wei, R., Oeser, T., Schmidt, J., Meier, R., *et al.* (2016) Engineered bacterial polyester hydrolases efficiently degrade polyethylene terephthalate due to relieved product inhibition. *Biotechnol Bioeng* **113**: 1658–1665.

Yoshida, S., Hiraga, K., Takehana, T., Taniguch, I., Yamaji, H., Maeda, Y., *et al.* (2016) A bacterium that degrades and assimilates poly(ethylene terephthalate). *Science* **351**: 1196–1199.

Yoshioka, T., Motoki, T., and Okuwaki, A. (2001) Kinetics of hydrolysis of Poly(ethylene terephthalate) powder in sulfuric acid by a modified shrinking-core model. *Ind Eng Chem Res* **40**: 75–79.

Zamani, B. (2014) Towards Understanding Sustainable Textile Waste Management: Environmental impacts and social indicators. Thesis for the degree of licentiate of engineering: Chalmers University of Technology.

18

Statistical test for tolerability of effects of an antifungal biocontrol strain on fungal communities in three arable soils

Kai Antweiler,[1] Susanne Schreiter,[2,†] Jens Keilwagen,[3] Petr Baldrian,[4] Siegfried Kropf,[1,*] Kornelia Smalla,[2] Rita Grosch[5] and Holger Heuer[2]

[1]*Department for Biometry and Medical Informatics, Otto-von-Guericke University Magdeburg, Magdeburg, Germany*
[2]*Department of Epidemiology and Pathogen Diagnostics, Julius Kühn-Institut – Federal Research Centre for Cultivated Plants, Braunschweig, Germany*
[3]*Department of Biosafety in Plant Biotechnology, Julius Kühn-Institut – Federal Research Centre for Cultivated Plants, Quedlinburg, Germany*
[4]*Laboratory of Environmental Microbiology, Institute of Microbiology of the CAS, Prague, Czech Republic.*
[5]*Leibniz Institute of Vegetable and Ornamental Crops, Grossbeeren, Germany.*

Summary

A statistical method was developed to test for equivalence of microbial communities analysed by next-generation sequencing of amplicons. The test uses Bray–Curtis distances between the microbial community structures and is based on a two-sample jackknife procedure. This approach was applied to investigate putative effects of the antifungal biocontrol strain RU47 on fungal communities in three arable soils which were analysed by high-throughput ITS amplicon sequencing. Two contrasting workflows to produce abundance tables of operational taxonomic units from sequence data were applied. For both, the developed test indicated highly significant equivalence of the fungal communities with or without previous exposure to RU47 for all soil types,

*For Correspondence. E-mail siegfried.kropf@med ovgu de

†Present address: Department of AgroEcology, Rothamsted Research, West Common, Harpenden, Hertfordshire, AL5 2JQ, UK.

Funding Information
Bundesministerium für Bildung und Forschung (Grant / Award Number: 031B0025B, 03MS642A, 03MS642H); Akademie Věd České Republiky (Grant / Award Number: RVO61388971); Deutsche Forschungsgemeinschaft (Grant / Award Number: KR 2231/6-1).

with reference to fungal community differences in conjunction with field site or cropping history. However, minor effects of RU47 on fungal communities were statistically significant using highly sensitive multivariate tests. Nearly all fungal taxa responding to RU47 increased in relative abundance indicating the absence of ecotoxicological effects. Use of the developed equivalence test is not restricted to evaluate effects on soil microbial communities by inoculants for biocontrol, bioremediation or other purposes, but could also be applied for biosafety assessment of compounds like pesticides, or genetically engineered plants.

Introduction

The emergence of high-throughput sequencing techniques now allows a detailed analysis of how microbial communities are influenced by the environmental application of microbial inoculants (Trabelsi and Mhamdi, 2013), pesticides (Jacobsen and Hjelmsø, 2014), transgenic crops (Verbruggen et al., 2012) or other human activities with a potential risk for microbial ecosystem services. The effect on the microbial community structure has to be assessed on the basis of high-dimensional abundance data, typically considering several hundreds or thousands of different operational taxonomic units (OTUs), while the number of samples in typical studies is small. Statistical methods for such high-dimensional data are available (DeSantis et al., 2007; Kropf et al., 2007; Kropf and Adolf, 2009; Ding et al., 2012) but are usually directed to detect differences between groups representing treatments, soil types, cultivars, etc.. In contrast, statistical methods to show that differences among microbial communities are negligible still have to be established for ecological risk assessment of human activities to support decision making (Heuer et al., 2002; Suter, 2006; Weinert et al., 2009). This inversed problem is investigated in statistical equivalence tests. In univariate equivalence tests, a tolerance threshold for the dependent (target) variable is defined that is just acceptable as difference for the mean expected outcome of the two groups to be considered as sufficiently similar. Then, modifications of the classical tests for difference are used to show that the real differences are smaller than this limit with probability of at least $1 - \alpha$ (α is the significance level of the test). That method can be extended to the case of several target variables (low-dimensional multivariate data). One available method would be the so-called intersection–union principle, where univariate tests are performed for each dependent variable at the unadjusted alpha level, and multivariate equivalence is accepted if equivalence could be proven in each of the univariate tests. In high-dimensional data, it would, however, be difficult to define appropriate tolerance thresholds for each variable. Moreover, the claim to prove equivalence in each dependent variable is hard to meet in the high-dimensional case. Therefore, we utilize multivariate distance measures between the high-dimensional sample vectors (e.g. Euclidean distances). Chervoneva et al. (2007) have carried out this before in a different way. Our approach is more versatile as it allows using non-Euclidean distances and even dissimilarity measures that do not satisfy the metric axioms of distances. Therefore, we use the term dissimilarity measure in the rest of the article. An ecologically justified limit has to be fixed that defines which distance can be tolerated. The derivation of such an appropriate limit is an essential part of the statistical procedure proposed here.

The application of microbial inoculants is an important component of an environmentally sustainable crop production system. There is an increasing demand for healthy food without chemical residues. In the last years, the market for products based on microbial inoculants, including biofertilizers, plant strengtheners and biocontrol agents, is growing by 10% per year (Berg, 2009). Biocontrol, as part of an integrated pest management, is well suited to partially replace synthetic pesticides which have led to increasing problems with pesticide resistance and which often affect human health and environmental quality (Hajek, 2004; Pérez-García et al., 2011). It is well documented that the treatment of plants with microbial inoculants originated from plant-associated microenvironments (e.g. soil, rhizosphere, phyllosphere) can efficiently protect plants from pathogens or pests (Berg, 2009; Hallmann et al., 2009; Lugtenberg and Kamilova, 2009; Andrews et al., 2010; Pérez-García et al., 2011; Kupferschmied et al., 2013; Adam et al., 2014). However, the application of microbial inoculants to agricultural soils can lead to changes in the indigenous microbial communities, which raises concerns regarding their biosafety (Trabelsi and Mhamdi, 2013). The biocontrol strain Pseudomonas jessenii RU47 was isolated from a disease-suppressive soil and showed antagonistic activity against different phytopathogenic strains of the fungal species Rhizoctonia solani and Fusarium oxysporum (Adesina et al., 2007, 2009). Efficient biocontrol of the important pathogen R. solani AG1-IB was shown in three different soil types, which makes this strain a promising biocontrol

agent (Schreiter et al., 2014a). Inoculation experiments with strain RU47 gave evidence for at least temporary effects on indigenous bacterial communities in soil (Schreiter et al., 2014b). As the control targets of strain RU47 are fungal pathogens, and the observed antibiosis of RU47 against two species of fungi makes non-target effects likely, it should be evaluated to what extent fungal communities in the agroecosystem are affected. Deep sequencing of barcoded fungal rDNA-ITS regions amplified from soil DNA (Voříšková et al., 2014) provides an excellent opportunity to approach the ecological risk assessment of microbes introduced into agroecosystems to promote cultivated plants. Microorganisms selected for biological control of phytopathogens typically have the potential to produce biocidal compounds like siderophores, antibiotics, biocidal volatiles or lytic enzymes (Saraf et al., 2014), which raises concerns in the approval procedure (EC regulation 1107/2009 concerning the placing of plant protection products on the market). However, this physiological potential for non-target effects might not be ecologically relevant in the environment so that an ecological risk should rather be experimentally assessed in situ. Such an experimental approach is impeded by the lack of experimental designs and statistical methods to test for tolerable effects of biocontrol agents on microbial communities.

The objective of this study was the development of a statistical method to test for equivalence of fungal community structures in soil with and without application of a biocontrol agent. This approach was applied to investigate putative effects of the antifungal biocontrol strain RU47 on fungal communities in three arable soils. The effects of RU47 were statistically evaluated with reference to fungal community differences in conjunction with field site or cropping history. Application of such an equivalence test is not restricted to environmental risk assessment of biocontrol agents, but could be widely applied to evaluate effects of strains released for bioremediation or other purposes, of pesticides applied on agricultural fields, or effects of genetically engineered plants on soil microbial communities.

Results

Analysis of fungal community structures in three arable soils

To evaluate the influence of the biocontrol strain P. jessenii RU47 on the fungal soil community, bulk soil samples were taken from three soils in separated plots that were treated with strain RU47 in the previous season and from soil untreated with RU47 (experimental station IGZ in Großbeeren, GB), as indicated in Fig. 1. Soil types were diluvial sand (DS), alluvial loam (AL) and loess loam (LL). The three soils have been translocated

40 years ago. Soil LL was also sampled from the original field near Klein Wanzleben (KW; Germany), 150 km apart from GB. The difference in the fungal community structure in soil LL between the two sites reflected an acceptable deviation caused by different weather conditions, crop rotation and agricultural practice. Two blocks in GB were sampled to determine the deviation of the fungal community structure of each soil caused by slightly different cropping histories or random drift due to spatial separation. The differences between these two blocks are considered as alternative approach for defining a threshold for acceptable deviations here.

Fungal ITS regions were amplified and analysed by barcoded high-throughput pyrosequencing. The amplicon sequencing data were processed by two contrasting strategies to reduce the risk of missing putative effects due to biased assembly of OTUs. The first approach aimed to reliably assign as many sequences as possible to a minimal number of OTUs by a database-dependent strategy (DBDS). For that, all ITS sequences were assigned to the most similar species hypothesis (SH) in the UNITE database (Kõljalg et al., 2013). If a sequence had the same similarity to more than one SH, then it was assigned to the more frequent SH in the dataset. OTUs with low similarity to any fungal ITS were discarded. Thereby, 97.9% of 407 239 sequences were assigned to 1607 OTUs. The 585 OTUs with sequences from at least five samples were further analysed, representing 95.5% of all sequences. The second strategy applied a strict quality control of the sequences and a database-independent assignment of sequences to OTUs using the pipeline SEED (Větrovský and Baldrian, 2013). In the final SEED dataset, 61.4% of all sequences were retained that were assigned to 2688 OTUs. The 828 OTUs with sequences from at least three samples were further analysed, representing 59.1% of all sequences.

The method to generate the OTU-abundance table, either by DBDS or by SEED, did hardly affect the representation of the fungal community structure (Figs 2 and 3). The fungal communities in the three soils in site GB and in soil LL in sites GB and KW were clearly separated in principal component analysis (Fig. 2). The fungal communities from the loamy soils LL and AL in GB were highly similar on the first and second principal components, but well separated on the third principal component with the exception of a single sample from LL which clustered with the AL samples. Communities of soils DS (GB) and LL (KW) were least similar, as these were best separated on the first principal component which explained more of the variance than the second and third principal component. The first three principal components explained slightly less of the variance for the SEED dataset compared to DBDS. The long-term spatial separation of the two blocks in GB (15 m apart) is reflected by differences in fungal community structure (Fig. 3, where site KW is omitted from the analysis). The block effect is most evident in the third principal component. The soil type has a much stronger influence on the fungal community structure, as this effect is reflected by the first and second principal components (Fig. 3). The block effect is stronger when phylogenetic information is used as additional source of information. Then, it is visible also with inclusion of soils from both sites even though this spatial effect is not evident for the field site LL (KW) where the soil is mixed by tillage (Fig. 4).

Most of the fungal ITS in all three soils analysed belonged to the Ascomycota, which comprised about 72% in the loamy soils AL and LL and considerably less (60%) in the sandy soil DS (Table 1). Basidiomycota and Zygomycota were also major phyla in these soils with on average 14% or 11% respectively. Their relative abundances significantly differed between soils with specifically low abundance of Basidiomycota in soil LL and Zygomycota in soil AL. Chytridiomycota, Glomeromycota and Rozellomycota were rather minor components of the fungal communities (Table 1). The most abundant families in all three soils were Nectriaceae and Mortierellaceae. Their relative abundance significantly differed

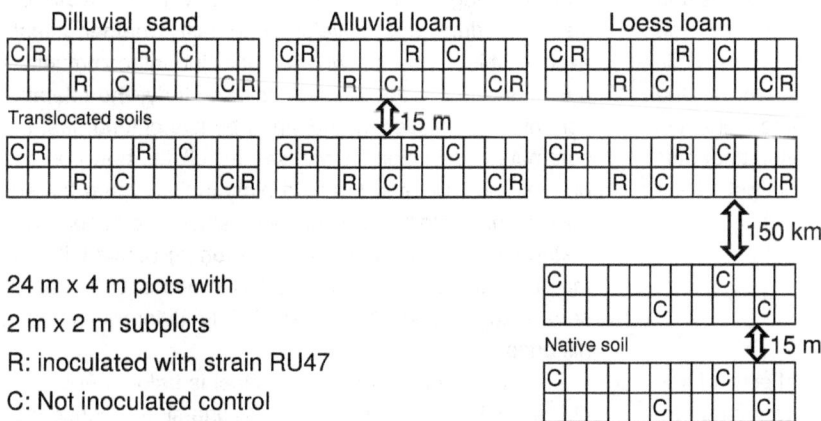

Fig. 1. Scheme of the experimental plot systems in Großbeeren (Germany) with three soil types in two blocks, and the field near Klein Wanzleben (Germany), where soils from inoculated and control plots were sampled.

24 m x 4 m plots with

2 m x 2 m subplots

R: inoculated with strain RU47

C: Not inoculated control

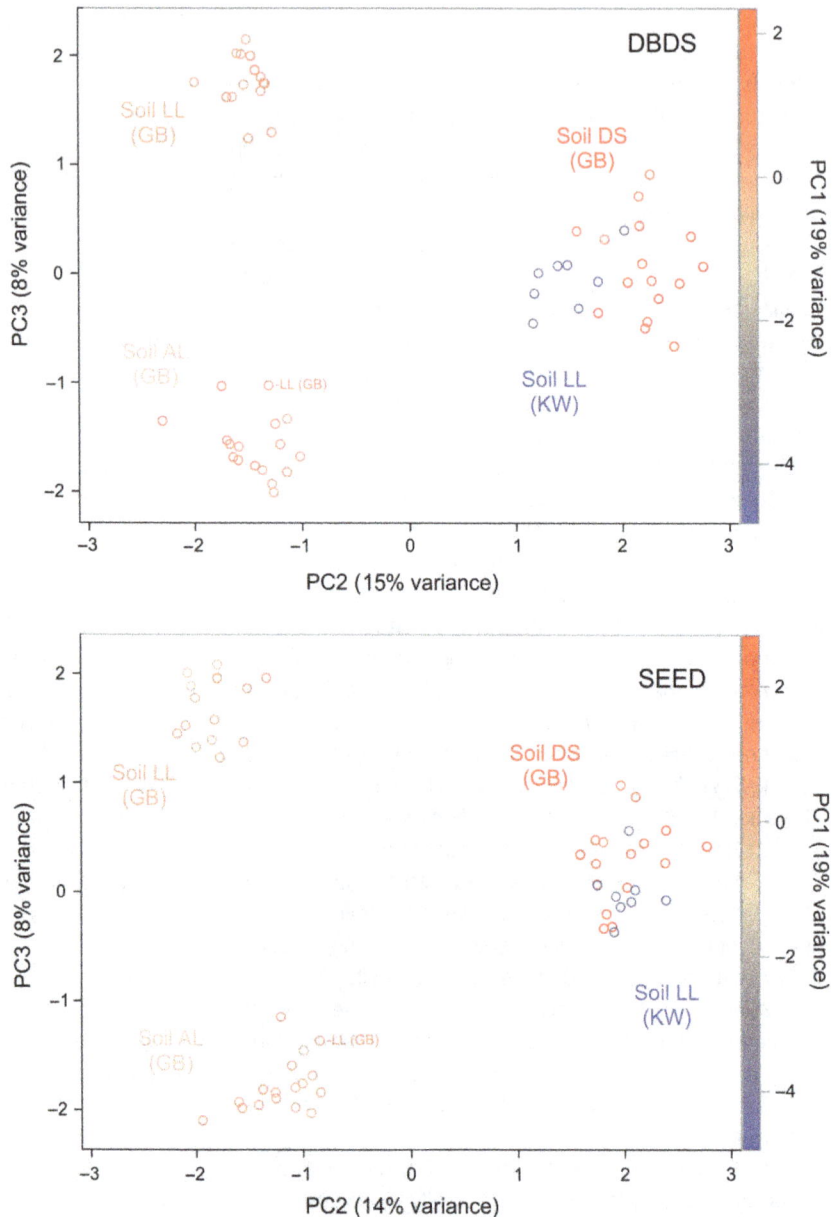

Fig. 2. Principal component analysis of fungal communities at two sites, Klein Wanzleben (KW) and Großbeeren (GB), and in three soils (LL for both sites, AL and DS for site GB). Fungal OTU tables were retrieved by two contrasting strategies for sequence assignment, DBDS (upper plot) or SEED (lower plot), as explained in the text. The first principal component (PC1) is indicated by a colour gradient.

among soils and was negatively correlated ($R^2 = 0.38$). On genus level, fungal ITS assigned to *Cryptococcus* and *Mortierella* were most frequently detected in all three soils, but with significant differences between soils (Table 2). *Mortierella* was especially abundant in the loamy soils AL and LL, while *Cryptococcus* was highest in the sandy soil DS. The abundance of the phytopathogenic species *Rhizoctonia solani*, which was the target of the biocontrol strain RU47, was below detection limit.

Equivalence of fungal communities in soils with and without exposure to RU47

To objectively evaluate whether putative effects of the inoculated biocontrol strain RU47 on non-target fungi are

acceptable in a risk evaluation, a tolerable fungal community change was biologically defined and a statistical test procedure was developed to test whether effects exceed this range or whether the fungal communities are equivalent in this respect. In this study, the tolerable community deviation was defined by two criteria, first the community deviation in the same soil LL between the original field KW and the field plots in GB where the soil was translocated. The second stricter criterion was defined by the deviation between fungal communities in the same soil at the same site GB in two equally treated separated blocks which had slightly different cropping histories.

To statistically prove that an effect is below the threshold at 5% significance level, a one-sided confidence

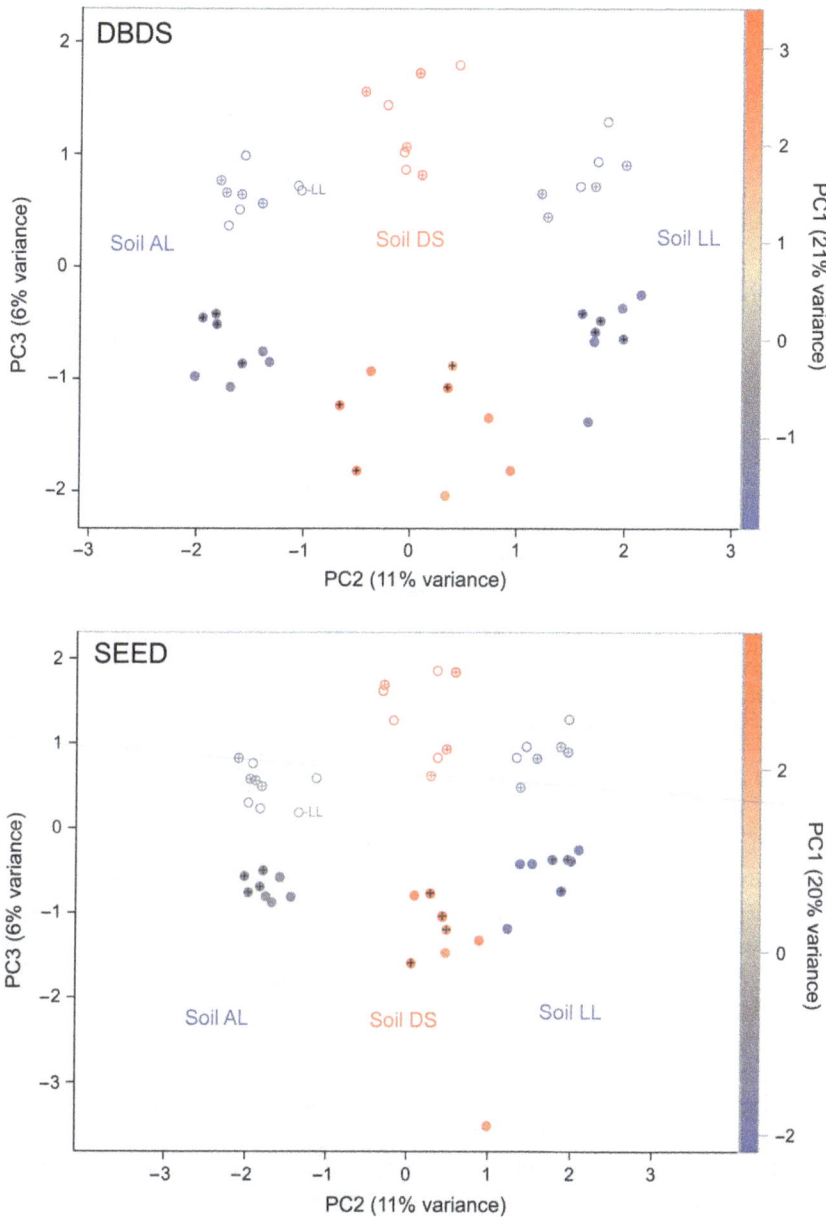

Fig. 3. Principal component analysis of fungal communities in three soils sampled at the experimental plot systems in Großbeeren (Fig. 1).
Samples from plots treated with the bacterial inoculant strain RU47 are indicated by a cross. Full or open circles indicate samples from adjacent blocks. The first principal component (PC1) is indicated by a colour gradient. The underlying OTU tables were generated by two contrasting strategies for sequence assignment, DBDS (upper plot) or SEED (lower plot), as explained in the text.

interval for a meaningful statistic S expressing the dissimilarity between treated and untreated soils can be constructed. The interval must lie completely below that threshold. For each soil type, a separate test was carried out as it would not be acceptable to have very similar samples in one soil type overrule dissimilar samples in another type of soil. Given a boundary, the test used OTU counts from soil GB only. Counts from site KW were just used to compute this boundary for the acceptable region that the statistic S has to be guaranteed to fall into with a given probability. To calculate that probability a normal distribution-based approach was used because bootstrap methods did perform too liberal in simulation studies with the given sample sizes. This put

a constraint on the choice of dissimilarity measure that can be used in the test procedure as the resulting distribution of the test statistic has to match well enough. The relative Bray–Curtis distance was chosen as dissimilarity measure. The details of the procedure are given in the Experimental procedures section.

Application of the statistical test for equivalence showed with high significance that fungal communities in RU47 treated and in control plots had smaller dissimilarities than the reference thresholds (Table 3). This equivalence was significant for both boundary criteria, the site differences in community structure for untreated LL soils and the stricter block differences. For the latter, the boundary is on average half as high as for the site

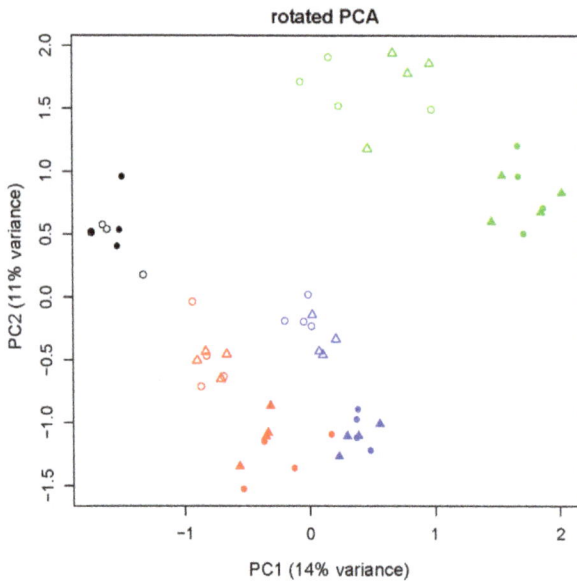

Fig. 4. Rotated principal component analysis of fungal communities at two sites, Klein Wanzleben and Großbeeren, and in three soils (LL red–black in Klein Wanzleben, AL blue and DS green). Triangles indicate RU47-treated plots and circles controls. Full or open symbols indicate samples from adjacent blocks. Fungal OTU tables were retrieved by sequence assignment strategy DBDS, as explained in the text.

difference, but still the measured relative Bray–Curtis distance between RU47 treated and control plots is only half of the boundary on average as well. Equivalence of the fungal communities was shown with both datasets, SEED and DBDS.

To validate the method further, we reversed the roles of the block factor and the control/RU47 factor for another analysis. This has not been carried out for biological reasons but to demonstrate the sensitivity of the procedure on the choice of the boundary. Now it is tested if the different blocks of the same kind of soil and the same treatment (control/RU47) can be proven

equivalent. The boundary criterion that is constructed from differences among sites remains unchanged. The second boundary criterion now is constructed from the differences between control and RU47 plots. The tests are significant for the loam soils LL und AL when the site differences are used as criterion. All other tests are not significant (Table 4).

Even though the blocks of each of the loam soils have to be considered equivalent if site differences are considered a negligible difference, they are different enough that RU47 treated soils and their controls have to be considered equivalent in comparison to them.

Testing for (tolerable) effects of strain RU47 on fungal communities in the three soils

Although the equivalence tests above have shown that the putative effects of RU47 on the fungal communities are tolerable, that does not mean that they do not exist at all. Therefore, we looked with multivariate methods for differences between fungal communities at location GB with and without previous exposure to RU47. Principal component analysis suggested small differences between the fungal composition of inoculated and non-inoculated soils, albeit much smaller than those between different soils or blocks (Fig. 3).

To test these putative differences for significance, we used two different recently developed tests for high-dimensional data. The first one (called PCuniRot) uses principal components in a very condensed test statistic (Ding *et al.*, 2012). Significance is assessed in repeated computations of the test statistic in rotated samples. The second test version (called Pearson test here, Kropf and Adolf, 2009) uses a multivariate similarity measure, in this case the Pearson correlation coefficients. Both tests have been applied to the DBDS version of the OTU table as well as to the SEED version, always using the log-transformed abundances and

Table 1. Structure of the fungal communities on phylum level in soils treated with the biocontrol strain RU47 or controls (C) based on the assignment of ITS sequences by DBDS.

Phylum	Percentages of sequences assigned to phylum ± SE ($n = 8$)						
	Soil DS (GB)		Soil AL (GB)		Soil LL (GB)		Soil LL (KW)
	C	RU47	C	RU47	C	RU47	C
Ascomycota	59 ± 5	65 ± 3	73 ± 2	70 ± 1	75 ± 2	75 ± 1	73 ± 1
Basidiomycota	23 ± 5	13 ± 2	17 ± 2	18 ± 2	10 ± 1	8 ± 1	13 ± 2
Zygomycota	14 ± 2	16 ± 2	6.5 ± 0.4	7.9 ± 0.6	11 ± 1	12 ± 1	12 ± 1
Chytridiomycota*	2.6 ± 0.4	3.4 ± 0.6	2.3 ± 0.3	3.2 ± 0.8	2.4 ± 0.3	3.4 ± 0.6	1.2 ± 0.1
Glomeromycota	1.0 ± 0.8	0.1 ± 0.0	0.2 ± 0.1	0.2 ± 0.1	0.7 ± 0.4	0.5 ± 0.3	0.1 ± 0.0
Rozellomycota*	0.3 ± 0.1	1.3 ± 0.2	0.1 ± 0.0	0.2 ± 0.0	0.1 ± 0.0	0.3 ± 0.1	0.1 ± 0.1
Unidentified	1.3 ± 0.1	1.9 ± 0.5	1.0 ± 0.5	0.7 ± 0.1	0.8 ± 0.1	1.2 ± 0.3	0.4 ± 0.1

*. Significantly different abundance in RU47-treated samples compared to controls (univariate stratified permutation tests, unadjusted $P < 0.05$).

Table 2. Relative abundances of the most frequent genera of the fungal communities in the analysed soils based on the assignment of ITS sequences by DBDS.

| | Percentages of sequences assigned to the genus ± SE (n = 8) | | | | | | |
| | Soil DS (GB) | | Soil AL (GB) | | Soil LL (GB) | | Soil LL (KW) |
Genus	C	RU47	C	RU47	C	RU47	C
Cryptococcus	9 ± 1	8 ± 1	15 ± 2	15 ± 2	6.2 ± 0.9	4.8 ± 0.4	3.3 ± 0.9
Mortierella	13 ± 2	15 ± 2	6.4 ± 0.4	7.7 ± 0.6	11 ± 1	12 ± 1	12 ± 1
Pseudeurotium	0.7 ± 0.3	0.4 ± 0.3	0.4 ± 0.1	0.4 ± 0.1	5 ± 1	5 ± 1	0.9 ± 0.2
Humicola	3.9 ± 0.5	4.5 ± 0.4	3.0 ± 0.6	2.4 ± 0.3	2.0 ± 0.4	1.8 ± 0.3	1.3 ± 0.1
Tetracladium	3.8 ± 0.5	3.7 ± 0.7	3.3 ± 0.5	3.6 ± 0.5	4.4 ± 0.6	3.8 ± 0.6	0.7 ± 0.1
Guehomyces	7 ± 4	1.5 ± 0.5	0.5 ± 0.1	0.4 ± 0.1	2.2 ± 0.7	1.6 ± 0.3	6 ± 1
Chaetomium	1.9 ± 0.3	1.7 ± 0.3	1.0 ± 0.1	0.7 ± 0.1	2.5 ± 0.3	3.1 ± 0.4	0.7 ± 0.1
Cladorrhinum	1.8 ± 0.6	1.4 ± 0.3	2.8 ± 0.6	2.0 ± 0.4	1.8 ± 0.3	1.6 ± 0.4	0.1 ± 0.0
Fusarium	0.6 ± 0.1	0.5 ± 0.1	2.2 ± 0.2	2.2 ± 0.3	2.5 ± 0.5	2.2 ± 0.3	2.5 ± 0.3
Ascobolus	0.8 ± 0.4	2.5 ± 2.1	0.0 ± 0.0	0.0 ± 0.0	0.1 ± 0.0	0.1 ± 0.1	0.0 ± 0.0
Stachybotrys	0.1 ± 0.0	0.4 ± 0.1	1.1 ± 0.1	1.2 ± 0.1	2.2 ± 0.2	2.2 ± 0.2	0.1 ± 0.0
*Rhizophlyctis**	0.08 ± 0.03	0.3 ± 0.1	0.04 ± 0.02	0.1 ± 0.0	0.2 ± 0.1	0.2 ± 0.1	0.1 ± 0.0

*. Significantly different abundance in RU47-treated samples compared to controls (univariate stratified permutation tests, unadjusted $P < 0.05$). This was also shown for the low abundant genera (< 0.05%) *Entoloma*, *Ascochyta*, *Candida*, *Scytinostroma*, *Amylocorticium*, *Pholiota*, *Serpula*, *Calonectria*, *Coccidioides*, *Hymenoscyphus* and *Dichotomomyces*.

Table 3. Equivalence of fungal communities in RU47-treated and non-treated soils.

Assessment of upper boundary	Soil	Dataset	Boundary	Bray–Curtis (SD)	Equivalence test (P-value)
Difference among sites (soil LL)	DS	DBDS	0.61	0.24 (0.03)	4.1E-08
		SEED	0.66	0.26 (0.03)	1.1E-08
	AL	DBDS	0.61	0.15 (0.02)	3.7E-12
		SEED	0.66	0.17 (0.02)	1.2E-12
	LL	DBDS	0.61	0.16 (0.03)	1.1E-09
		SEED	0.66	0.19 (0.03)	6.0E-10
Difference among blocks at site GB	DS	DBDS	0.38	0.24 (0.03)	4.8E-04
		SEED	0.40	0.26 (0.03)	3.9E-04
	AL	DBDS	0.22	0.15 (0.02)	9.8E-04
		SEED	0.22	0.17 (0.02)	7.4E-03
	LL	DBDS	0.25	0.16 (0.03)	4.9E-03
		SEED	0.26	0.19 (0.03)	1.2E-02

Table 4. Equivalence of fungal communities in different blocks.

Assessment of upper boundary	Soil	Dataset	Boundary	Bray–Curtis (SD)	Equivalence test (P-value)
Difference among sites (soil LL)	DS	DBDS	0.61	0.38 (0.03)	8.9E-01
		SEED	0.66	0.40 (0.03)	2.6E-01
	AL	DBDS	0.61	0.22 (0.03)	2.6E-05
		SEED	0.66	0.22 (0.03)	4.7E-04
	LL	DBDS	0.61	0.25 (0.03)	7.6E-04
		SEED	0.66	0.26 (0.03)	1.1E-03
Difference among control/RU47 at site GB	DS	DBDS	0.24	0.38 (0.03)	1.0E-00
		SEED	0.26	0.40 (0.03)	1.0E-00
	AL	DBDS	0.15	0.22 (0.03)	9.9E-01
		SEED	0.17	0.22 (0.03)	9.3E-01
	LL	DBDS	0.16	0.25 (0.03)	1.0E-00
		SEED	0.19	0.26 (0.03)	9.8E-01

including all three soils in a common analysis. We used factorial models with three factors for the effects of soil type, block and inoculation (Table 5). As can be seen from the P-values of all test versions and from the R^2 values for the Pearson versions, the soil type was the major influence on the fungal community. The block effects were also highly significant but distinctly smaller in the effect measure R^2. Inoculation effects of RU47 were by a magnitude smaller in the effect measure. Nevertheless, they were statistically significant in nearly all versions with exception of the SEED version of the PCuniRot test. Interactions between soil type and inoculation were significant only in the DBDS versions with only very small effect measures. The latter fact is also illustrated in Fig. 3, where for soil LL, the dots with crosses (RU47-treated plots) are shifted a bit compared to the dots without crosses in the third principal component, whereas this effect is not seen so clearly in the other soils.

To identify putative responders to RU47, multiple univariate stratified permutation tests on effects of RU47 on fungal groups on different taxonomic levels were carried out. The low abundant phyla *Chytridiomycota* and *Rozellomycota* tended to increase in RU47-treated plots, resulting in unadjusted P-values below 0.05 (Table 1). None among the abundant families and only *Rhizophlyctis* among the abundant genera gave some evidence for a response to RU47 albeit not a decrease (Table 2). Table 6 shows the OTUs with most probable responses to RU47 as revealed by stratified permutation tests. Responding OTUs from SEED were assigned to the genera *Mortierella*, *Cylindrocarpon*, *Cryptococcus*, *Myrothecium* and an unidentified SH of the *Ascomycota*. Responding OTUs from DBDS were assigned to *Cryptococcus terricola*, *Spizellomyces dolichospermus* and an unidentified SH of the *Rozellomycota*. Less abundant responders (< 0.1%) did not contribute much to the result of the multivariate tests. The P-values either decreased or hardly changed when they were removed from the SEED OTU table (0.023 or 0.004 for PCuniRot or Pearson test, respectively), while removing the high abundant responders resulted in higher P-values (0.33 or 0.09 for PCuniRot and Pearson test, respectively), indicating that changes in the relative abundance of these OTUs mostly contributed to the significance of the putative RU47 effects.

Table 5. Multivariate statistical tests on the effects of RU47, soil type or block on the fungal community structure.

Effect	P-value from PCuniRot test		P-value (R^2) from Pearson test	
	DBDS data	SEED data	DBDS data	SEED data
RU47	0.033	0.120	0.006 (0.014)	0.005 (0.013)
Soil	< 0.001	< 0.001	< 0.001 (0.433)	< 0.001 (0.723)
Block	< 0.001	< 0.001	< 0.001 (0.066)	< 0.001 (0.316)
RU47 × Soil	0.043	0.368	0.008 (0.001)	0.775 (0.002)

Discussion

In this study, we developed a statistical procedure to test whether an inoculated microbial strain has at most an acceptable effect on the indigenous microbial community. By the application of this test procedure, we showed that the antifungal inoculant RU47 that targeted the disease caused by *R. solani* had such a minor effect on the fungal communities in the three soils that the inoculation can be considered as practically equivalent. The method of how the high-throughput amplicon sequencing data were processed and assigned to OTUs did hardly affect the results, although two highly contrasting methods were chosen, SEED and DBDS. Probably the most delicate part of the equivalence test is the determination of the thresholds for a tolerable change in the microbial community. We determined that boundary experimentally. This presupposes the existence of an influence of other factors on the sample elements or a proper subset of the samples. The effect of this influence must be acceptable from an ecological point of view. We considered the deviation caused by having the same type of soil in different locations and under different agricultural practice as a starting point, as such influences on fungal communities are generally accepted and have never been associated with any risk. We were able to make this difference easier to accept by not only having the same type of soil but actually the same soil separated years ago in a region of comparable climatic conditions. We choose the difference between non-inoculated LL soils at these two sites as a boundary for the acceptable region. It might be reasonable to question the acceptability of the difference between the LL soil sites as boundary for the other types of soil. To back up the evidence provided by this boundary, we constructed a second type of boundary as well. The drift in fungal community structure due to separation of soils in two blocks at the site GB and slight differences in cropping history of these blocks provided us with a measurable boundary for each type of soil that should be even more generally acceptable.

Samples that are used to determine the boundaries have to be checked to exclude effects of unusual variation. With increasing number of replicates, this becomes less important. We found one replicate of soil LL in GB that was more similar to soil type AL than to the samples from soil LL in a principle component analysis (Figs 2 and 3). We decided to not exclude this replicate from the analysis since it equally influences the computation of the boundary as well as the difference between control and inoculated groups. As the latter has to be significantly smaller than the former, its effect would in the case of a failure of the first kind increase the distance between the test groups more than it would increase the boundary,

Table 6. OTUs showing a response to RU47 inoculation as indicated by univariate stratified permutation tests (unadjusted $P < 0.05$) for fungal OTUs with at least 0.1% abundance.

Data	SH (07FU)	Unadjusted P-value	Genus, Phylum / Species	BLASTN% identity	No. of sequences[a]	
					C	RU47
SEED	SH183635	0.002	*Mortierella, Ascomycota*	100.0	598	1231
	SH202969	0.038	*Cylindrocarpon, Ascomycota*	100.0	370	498
	SH183335	0.004	*Geomyces, Ascomycota*	98.6	311	346
	SH190017	0.016	*Cryptococcus, Basidiomycota*	100.0	51	279
	SH204317	0.002	Unidentified, *Ascomycota*	99.3	137	89
	SH174294	0.023	*Cylindrocarpon, Ascomycota*	100.0	82	160
	SH175276	0.039	*Myrothecium, Ascomycota*	99.5	78	155
DBDS	SH190017	0.007	*Cryptococcus terricola*	100.0	133	679
	SH183868	0.035	*Spizellomyces dolichospermus*	99.6	27	507
	SH180899	0.038	Unidentified, *Rozellomycota*	96.8	64	177

[a]Sum of fungal ITS sequences in all three soils (GB) adjusted according to the total numbers of sequences from RU47-treated and control plots.

thereby decreasing the probability that this failure might occur. All *P*-values for the test of equivalence stayed significant when that one LL replicate was excluded from boundary determination or from the whole analysis. Special care was taken to not unnecessarily increase the complexity of the computer program that was written to test the data, simulate the experiment to guide the choices in the construction of the algorithm and validate its non-liberality on bootstrapped samples from the real data.

The relative Bray–Curtis distance was chosen as dissimilarity measure because it was mostly conservative in our studies that used random simulated data and was completely conservative in our simulation studies that used bootstrap samples from the real dataset. Apart from its ecological explanation, the Bray–Curtis distance also has meaning as an information theoretical divergence and has the basic structure of some phylogenetically enhanced distances, as explained in the supplement. While some authors prefer to use model-based approaches (Warton *et al.*, 2012), others show that transformed data analysed with the usual methods work just as well (ter Braak and Šmilauer, 2015). We also believe that there is no theoretical framework that describes ecological data perfect so far, although we prefer to see the samples as empirical distributions of OTUs when they are assessed in its entirety (instead of assessed by their most characteristic OTU).

To estimate the standard deviation, a two-sample jackknife procedure avoiding systematic underestimation of the real standard deviation under given conditions (Karlin and Rinott, 1982) was used. The estimation of the quantile of the statistic S (that also is used in the calculation of the *P*-value) is the only step in the procedure that depends on normality. Asymptotically that assumption is met, but sample sizes are usually small in current ecological studies. As it is the statistic S that has to be

normal and not the single OTU, it is not straight forward to convince one selves of this property given a small sample set. Because normality is only a sufficient condition for non-liberality of the confidence interval, we used bootstrapped samples to test the coverage rate of that interval directly. The coverage rate was 100% in 10 000 runs of simulation. We observed the first confidence interval that missed its parameter once when we dropped to 30% confidence. The situation was different in the planning phase, where we used normal distributed OTUs and saw liberal estimations for some sets of low-dimensional simulation parameters although the Bray–Curtis distance behaved relatively well and never dropped below 90% coverage. Some statistically motivated dissimilarities that we constructed dropped below 20%. We designed the procedure to be applicable to as many meaningful measures as possible, including measures that utilize phylogenetical information (Fukuyama *et al.*, 2012).

Different equivalence tests can be combined without the need to adjust the level of the tests. For instance, if there were some OTUs known that should under no circumstances change in relative abundance beyond a known limit, standard univariate equivalence tests could be calculated for each of these OTUs additionally to the community-based approach described here. Only if each of the tests was significant with unadjusted *P*-value, equivalence is proven. Each test added to that procedure decreases the power of the procedure. Those OTUs that are checked separately do not need to be included in the community-based approach. This also holds true for OTUs that are the intended targets of the inoculated strain. The same principle can be used to show equivalence in multiple kinds of populations like bacteria or mesofauna depending on expected effects. The results of the equivalence tests presented in this paper would not change, but equivalence would only be assumed if the tests results for bacteria or other groups were significant, too. We did not consider this when we planned the experiment, because RU47 targets fungi and it is there where we suspect its strongest effects. Although we believe that is true, statistically everybody has the right to doubt it, because significant equivalence tests only prove equivalence (in respect to the boundary values) in those variables that actually are analysed. This trivial fact is much more important in tests for equivalence than it is in tests for difference, where a difference anywhere shows that groups differ. We do not intend to propose focussing on single populations as a standard for good agricultural research practices – especially given the dropping costs for analyses.

Logically related to the question what should be measured is the question when to measure it. The effect of a treatment often is maximal shortly after it is supplied. In terms of risk assessment, the interesting effects are those that last. We chose a time to gather the samples before a new crop is planted which might be affected by a modified fungal community. We believe that this is the most important point in time to know about any effects.

We tried to incorporate phylogenetic information in our analysis. This performed poorly with the methods that try to combine abundance and phylogenetic data directly as was to be expected regarding the low correlation ($r = 0.02$) between the phylogenetic similarity of pairs of OTUs and the correlation of their abundances in our data. We therefore switched to thinking about how to improve just the PCA plot that we already have. In factor analysis, a given PCA result is sometimes rotated to maximize the correlation of its components with the original variables to increase interpretability. There are far too many variables in this dataset for this strategy to be useful. We thought of rotating the PCA result such that it correlates maximally with phylogenetic information. We found out that this can be carried out with an ordinary canonical correlation analysis if advantage of the dual problem of the PCA is taken and the input matrices are reduced in their dimensions first. The result accentuates the block effect to some extend and shows block, soil and site effect in a two-dimensional plot (Fig. 4). The types of soil now lay ordered as we would have initially expected: the loamy soils are separated from DS, and the KW samples are next to samples from the same type of soil. We failed to produce a picture of equal quality by random rotations.

The RU47 treatment increased the abundance of the members of the Chytridiomycota and Rozellomycota. Chytridiomycota are typically saprobic fungi with flagellate gametes, degrading refractory materials such as chitin and can also act as mycoparasites (Barr, 1990; Kirk et al., 2008). Most of the diversity of the phylum

Rozellomycota is known only from environmental sequences (Hibbett and Taylor, 2013). In the SEED dataset, Nectria was the only genus that responded to RU47 by decreasing abundance in the treated soils. The members of this genus are typically saprotrophs or parasites of trees (Kirk et al., 2008). The increase in putative responders indicated that effects of the inoculum on fungal communities are rather due to the added nutrients than caused by the antifungal activity of RU47.

Compared to the DBDS, the SEED-based data processing resulted in a substantially lower number of sequence reads that passed the quality control steps. However, the representations of the fungal community structure from both datasets were surprisingly similar (Fig. 3). In multivariate statistical tests of RU47 effects on fungal communities, the smaller SEED dataset was less sensitive than DBDS (PCUniRot, Table 5). In both datasets, SH190017 was identified as a responder to RU47 which was based on 330 assigned sequences for SEED and 812 sequences in DBDS. So DBDS might allow for a better sensitivity using a higher percentage of the sequences and assignment to less OTUs, but with a higher risk of false assignments and thus might increase the noise in the dataset.

The disease suppression effects of beneficial microbial inoculants often are based on an antagonistic mode of action which microbes use to establish in competition with other microbes in a natural ecosystem. The majority of commercially available microbial inoculants belong to the genera Pseudomonas, Bacillus and Trichoderma which are dominant representatives of the natural soil and plant microbiome (Chet, 1987; Berg et al., 2005, 2006; Haas and Défago, 2005). Still, the application of microbial inoculants in the environment is an irreversible process and, if applied to plant-associated microenvironments such as the root zone in sufficient numbers, may perturb indigenous microbial populations and the ecological functions associated therewith (Bankhead et al., 2004; Winding et al., 2004). To date, only a few cultivation-independent studies have focused on the effects of beneficial microbial inoculants with disease-suppressive activity (Scherwinski et al., 2007) or commercialized plant stimulants (Chowdhury et al., 2013) on indigenous microbes. The inoculant RU47 investigated in this study showed in vitro weak antifungal activity against the target pathogen R. solani (Adesina et al., 2007) and did not produce known antibiotic substances. We found only minor effects on fungal communities in both soils which were tolerable based on our boundary criteria. However, dependent on the properties of the inoculant the non-target effects can be more severe. Hence, model studies which assess the impact of various biocontrol strains with different properties on non-target population are needed for environmental risk assessment. Also, such studies should be part of the development process of potential biocontrol strains and will support economically meaningful decisions in the beginning of the product development. Here, we provide the biometrical tools for the data analysis of such an environmental risk assessment of biocontrol strains, which could be analogously applied to other environmental applications like plant growth promoting microbes, biodegraders, genetically modified organisms or ecotoxicological studies.

Experimental procedures

Experimental design, sampling and sample processing

Bulk soil samples were taken from experimental plot systems in Großbeeren (Germany, 52.4° N, 13.3° E) and a field near Klein Wanzleben (Germany, 52.1° N, 11.4° E) as indicated in Fig. 1. The plot systems in Großbeeren contained three different soils which have been translocated there in the year 1972 (Ruehlmann, 2013). Two blocks of plots, 15 m apart from each other, were sampled. Each block consisted of three plots, each of which contained a different soil type and was divided into 2 m × 2 m subplots. Soil types were Arenic-Luvisol (diluvial sand, DS), Gleyic-Fluvisol (alluvial loam, AL) and Luvic-Phaeozem (loess loam, LL) (Rühlmann and Ruppel, 2005). To evaluate the influence of the biocontrol strain *P. jessenii* RU47 on the fungal soil community, samples were collected in spring from plots that were treated with strain RU47 in the previous season and from untreated control plots ($n = 4$ per treatment, soil and block). In the preceding season, each lettuce seedling was treated with 2×10^8 cells of RU47 one week before planting and with 3×10^9 cells two days after planting in the field (Schreiter *et al.*, 2014b).

On the same day, samples were taken in the field near Klein Wanzleben from where soil LL originated. The difference in the fungal community structure in soil LL between the two sites, Großbeeren and Klein Wanzleben, gives an estimate of acceptable deviation caused by different weather conditions and crop rotations. The two sites are 150 km apart. Deviation of the fungal community structures of each soil between the two blocks in Großbeeren, which were spatially separated for 40 years, gives an estimate of random drift of the fungal community or changes caused by slightly different cropping histories of the two blocks. Crops planted in the years 2000 to 2012 were pumpkin, nasturtium, nasturtium, phacelia, amaranth, wheat, pumpkin, nasturtium, wheat, broccoli, wheat, Teltow turnip, lettuce, lettuce for block 5, and pumpkin, nasturtium, pumpkin, amaranth, wheat, wheat, pumpkin, nasturtium, wheat, wheat, lettuce, lettuce for block 6.

Each subplot was sampled by mixing ten cores (30 cm of top soil; 2 cm core diameter) of bulk soil. DNA was extracted from 0.5 g of soil using the FastDNA SPIN Kit for Soil after two 30 s lysis steps with a FastPrep FP120 bead beating system, and further purified by the GENECLEAN SPIN Kit, as described by the manufacturer (MP Biomedicals, Heidelberg, Germany). The fungal ITS fragments were amplified using the primer pair ITS1F (CTTGGTCATTTAGAGGAAGTAA) / ITS4 (TCCTCCGCTTATTGATATGC) as previously described (Weinert *et al.*, 2009). The products were purified with a Minelute PCR purification kit (Qiagen, Hilden, Germany). Barcoded amplicon pyrosequencing was performed at the Biotechnology Innovation Center (BIOCANT, Cantanhede, Portugal) on a 454 Genome Sequencer FLX platform according to standard 454 protocols (Roche – 454 Life Sciences, Branford, CT, USA). Briefly, the purified PCR products were used as target to amplify the ITS1 region with fusion primers containing the Roche-454 A and B Titanium sequencing adapters, an eight-base barcode sequence in adaptor A, and specific sequences ITS1F / ITS2 (GCTGCGTTCTTCATCGATGC) targeting fungal ribosomal genes. The data were submitted to NCBI SRA with accession number SRP073893.

Generation of OTU-abundance tables from ITS sequences

The amplicon sequencing data were processed by two contrasting strategies. DBDS aimed to reliably assign as many sequences as possible to a minimal number of OTUs. For that, all ITS sequences were assigned to the most similar species hypothesis (SH) in the UNITE version 7.0 database (Koljalg *et al.*, 2013) using Megablast (Camacho *et al.*, 2009). The SH database is available at https://unite.ut.ee/repository.php. If a sequence had the same bit score to more than one SH, then it was assigned to the most abundant SH in the dataset. For processing the MEGABLAST results, the java tool BLASTPARSER was written and integrated into a Galaxy workflow (https://galaxyproject.org). It makes a unique assignment of the sequences to an OTU and generates the OTU-abundance table. OTUs were discarded when the assigned sequences were < 95% similar to all SH, or had < 100-bp alignment length or had highest similarity to non-fungal ITS. The other approach applied the pipeline SEED 1.2.1 (Větrovský and Baldrian, 2013) to achieve a strict quality control of the sequences and a database-independent assignment of sequences to an OTU. Briefly, pyrosequencing noise reduction was performed using the DENOISER 0.851 (Reeder and Knight, 2010). Chimeric sequences were detected using USEARCH 7.0.1090 (Edgar, 2010) and deleted. Only sequences longer than 310 bases were retained, and full ITS2 regions of these sequences were extracted using ITSX (Bengtsson-Palme *et al.*, 2013). Full ITS2 regions were clustered using

UPARSE implemented within USEARCH (Edgar, 2013) at a 97% similarity level. Consensus sequences were constructed for each OTU, and the closest hits at a genus or species level were identified using BLASTN against UNITE version 7 and GenBank for fungi. Ecology was assigned based on genus-level best hits to those taxa whose genera show consistency in this respect using published data (Tedersoo et al., 2014).

Statistical test for equivalence of fungal communities

The equivalence testing procedure is based on the relative Bray–Curtis distance between fungal communities of the control group and the RU47-treated group. The procedure computes a test statistic S which represents the dissimilarity between two groups of samples. For that statistic, a conservative approximation of its variance is constructed (Karlin and Rinott, 1982). Our samples consist of relative counts of OTUs which sum to one for each sample, i.e. discrete empirical probability densities on the space of all OTU types. We start by separating our samples corresponding to their stratum and group. We have two groups: treatment and control. In our case, we also have two strata: samples from the first block and samples from the second block. This should not be confused with the distinction of the different soil types, which was considered by completely separate analyses for each soil. All strata that are used in that procedure must be allowed to compensate each other's results. The assumption is that if equivalence is true for one stratum, it also is true for the others. Although it is hard to find a realistic case where variances differ between strata while that assumption is met, the variance estimator is guaranteed to be unbiased or conservative for any statistic, always, but it tends to be suboptimal in terms of power. For each stratum, two discrete empirical probability densities are computed by averaging all its samples that belong to the treatment group for the first distribution and all that belong to the control group for the second distribution. From both empirical distributions, we calculated the relative Bray–Curtis distance as dissimilarity measure, separately for all strata. This can be carried out by calculating the sum of absolute values of the group difference and scaling the result by 0.5, as shown in the supplements. The average of these values over all strata is our test statistic S. We used an unweighted average. Instead of averaging the samples, we could have averaged pairwise Bray–Curtis distances. Both are valid procedures, but the approach taken here has the advantage of mitigating the difference between group distance and average pairwise sample distance.

To estimate the variance of S, a stratified two-sample jackknife procedure was used. Technically, a variance for the test statistic with a reduced number of samples is estimated. At the start of the procedure, a value r is fixed that describes the proportion of sample elements used in each step of the jackknife procedure. For each group $j \in \{1, 2\}$ in each stratum $i \in \{1, \ldots, m\}$, a reduced sample size $q_{ij} = r \cdot n_{ij}$ is calculated where n_{ij} are the original sample sizes. We chose $r = 3/4$, which in our case means that in each combination of stratum and group 3 of the 4 available sample elements are used in a jackknife step. The procedure is repeated over all possible combinations of samples with exactly q_{ij} samples in stratum i and group j (yielding $N = \prod_{i,j} \binom{n_{ij}}{q_{ij}}$ runs). In each run, a test statistic $S_k^*(k = 1, \ldots, N)$ is calculated as described above, starting with averaging of the selected samples in each stratum-group-entity and ending with averaging the relative Bray–Curtis distances over the strata. No sample is allowed to switch group or stratum. These N values of the jackknife test statistics are averaged $\overline{S^*} = N^{-1} \sum_{k=1}^{N} S_k^*$, and the square of the difference between each value and that average is calculated and averaged again and scaled by the fraction of left out samples per entity. The final variance estimate is given by the formula: $\widehat{\sigma^2}(S) = (1 - r)^{-1} N^{-1} \sum_{k=1}^{N} (S_k^* - \overline{S^*})^2$

The value of the test statistic and its standard deviation can be used under assumption of normality to compute the upper limit UL of the one-sided confidence interval for S by $= S + t_{1-\alpha, n-2m} \cdot \sqrt{\widehat{\sigma^2}(S)}$. We used a t-distribution instead of a normal distribution to be on the conservative side. For degrees of freedom, we chose to use $n - 2m$, i.e. the total sample size minus number of groups times number of strata.

The boundary B is calculated in the same way as the test statistic S. In our case where we chose the block factor to compute the boundary, this was carried out by switching the variable group with strata. In the other case where we used the location (GB vs. KWL) for untreated samples of LL soil instead, the location was used in place of group in the procedure described above. Concerning the equivalence test, this also is the only place where samples from the second location where used. There is no need to estimate the variance of B. If S changes systematically with changing sample sizes, B must have the same structure of sample sizes as S has (or the resulting bias should be proved to be conservative, i.e., decrease B in respect to S).

The equivalence test can be finalized by comparing the upper limit UL of the one-sided 95% confidence interval of the test statistic S with the boundary B. The test is significant (i.e. equivalence is proven) if $\leq B$. The probability of the set of points right to the equivalence boundary B is the P-value of the test (using the t-distribution as described above).

We suggest using a balanced design, but the implemented program does work for all designs for which it is possible to define r as long as there is no empty resampled entity and the maximal sample size in an entity is below 64. Both latter limitations were introduced for computational convenience and are unrelated to the algorithm described. The first limitation is both for computational and theoretical reasons. This jackknife procedure is only guaranteed to be non-liberal if the ratio r is constant over all group-strata combinations. A small deviation from that ratio may be not too far away from the theoretical results. The program accepts an input parameter for a tolerance value to check if the ratio of the number of chosen samples and the sample size is approximately equal to r (e.g. if we would have excluded the one LL sample that clustered with AL soil from the analysis, there would be only 3 samples instead of 4). When three-fourth of those three samples, i.e. 2.25, are to be selected, only 2 would be selected and checked that $| \frac{2}{3} - \frac{3}{4} | <$ tolerance value.

The program code including a description ('README') and the raw data of this study can be downloaded from 'https://www.researchgate.net/publication/301770482_data_Statistical_test_for_tolerability_of_effects', doi: 10.13140/RG.2.1.3287.6407.

Statistical testing for effects of RU47 on fungal communities

Assignments of sequences to OTUs by DBDS and SEED resulted in two tables representing the OTU-abundance structure of the fungal communities of all samples. Relative abundances within each sample were log-transformed (log[relative abundance * 1000 + 1]) to ameliorate deviations from normal distribution. Samples from plot systems in Großbeeren were analysed by multivariate statistics to test for significant effects of the inoculated strain RU47 on fungal communities, while taking the additional factors soil and block into account. Two factorial multivariate statistical tests with three factors were applied. PCUniRot is based on principal component analysis combined with a modified ANOVA test statistic for the framework of a general linear model (Ding et al., 2012). This statistic uses a weighted combination of the sums of squares for the first q principal components (q determined by the Kaiser criterion). Rotation tests are then applied to derive the P-value. The other test (called Pearson test here) is based on Pearson correlation coefficients used as multivariate similarity measures for pairs of sample vectors (Kropf and Adolf, 2009). The test statistic describes the multiple correlations between the similarity measures for all pairs of sample vectors and the corresponding differences in the factor level of the factor of interest for the same pairs of sample vectors after eliminating the influence of all other factors. The P-value for the test is again derived in rotation tests. Additionally, the squared multiple correlation coefficient R^2 as effect measure can be interpreted as the proportion of variability in the observed similarity measures explained by the factor tested. These tests are more powerful in high-dimensional settings with small sample sizes than competing tests in many situations (Ding et al., 2012).

Individual OTUs which were likely influenced by the inoculation of RU47 were determined by a stratified permutation test (Good, 2000). In contrast to an ANOVA, the concept of this type of test remains valid even for groups with zero variance as was common in our data. The data were not log-transformed as this test procedure works correct on relative counts. As a test statistic, the sums of the absolute differences between replicates in each group and stratum were added. Stratification allows pooling evidence. Soil type and block were chosen as strata. Control/RU47 labels were randomly permutated 10 000 times. Permutations were restricted to stay inside the same stratum.

PCA rotation by canonical correlation to phylogenetic similarities

The phylogenetic similarities between all pairs of OTUs are given as a $p \times p$ matrix M. The abundance table X in this case is a $p \times n$ matrix of relative OTU abundancies transformed using the logarithm and centralized for each OTU afterwards. The motivation of the method suggests double-centred data, but this would be limited to PCAs on covariance matrices. The way described here also works on top of correlation matrices. We use a covariance matrix, but the results between double-centred and centring OTUs only were hardly visible. The n abundancy and p similarity columns are different kinds of measurements for each OTU. A canonical correlation between both matrices produces an orthonormal matrix as a result, which can be used as generalized rotation transformation. First, we have to reduce the dimensions of matrices to get meaningful results. The reduced abundance matrix used corresponds to the principle components that were used in our plots. The PCA was performed with the OTUs seen as variables resulting in a $n \times r$ matrix Y of principle components. This matrix cannot be used directly in our canonical correlation, because there the OTUs are seen mainly as sample elements which correspond to a transposed view of the problem. But for each PCA exists a dual PCA formulation that we can use, such that $(p - 1)Y = X'U\Lambda^{-1/2}$, where the eigenvector column matrix U and eigenvalue diagonal matrix Λ are solutions to the eigenvalue equation $XX'U = U\Lambda$. The scaling constant $(p - 1)$ and the column-scaling matrix $\Lambda^{-1/2}$ are not important for our current purposes, because we can standardize the components afterwards as we like. The matrix U, we will use as input for the canonical correlations, and the resulting $r \times r$ orthonormal matrix R will be used to define the 'rotated' PCAs $\tilde{Y} := X'UR$. The dimension of the phylogenetic matrix M has to be reduced, because it is so big that the columns of U already lie in a subspace of the column space of M, without any rotation. We use a PCA on M as well (i.e. an eigenvalue equation on $M'M$) to reduce its dimension and take the k first eigenvectors as input for the canonical correlations. We used Scree-plots to determine r and k.

Rotating the eigenvectors rearranges the variance that corresponds to each eigenvector (i.e. their eigenvalue). It can be necessary to change the order of those vectors after the procedure described above to choose the vectors that contain the maximum of variance. Because the columns of X' are centralized, and those of $X'U$ orthogonal (as those of $X'U\Lambda^{-1/2}$ are), the columns of $X'U$ are uncorrelated. $X'UR$ is a linear transformation of those columns; therefore, the variances of its columns can be calculated by adding the variances of the columns $X'U$ times the squares of the corresponding factors which are the squares of the entries of R. If λ is the vector of column variances corresponding to Y (i.e. the diagonal entries of Λ) and R^2 is the matrix R squared element-wise, the rotated variances are equal to $\lambda'R^2$. The proportion of total variance can be calculated by dividing that vector by the total variance.

Acknowledgements

This study was part of the project MÄQNU of the German Ministry of Education and Research (BMBF 03MS642A, 03MS642H). Development of tools for NGS data analysis was supported by the project BonaRes-ORDIAmur funded by the BMBF (031B0025B). PB was supported by the research concept of the Institute of Microbiology of the CAS (RVO61388971). The incorporation of phylogenetic information in the analysis was supported by a grant of the German Research Council (DFG KR 2231/6-1).

Conflict of interest

The authors have no conflict of interest to declare.

References

Adam, M., Heuer, H., and Hallmann, J. (2014) Bacterial antagonists of fungal pathogens also control root-knot nematodes by induced systemic resistance of tomato plants. *PLoS ONE* **9**: e90402.

Adesina, M.F., Lembke, A., Costa, R., Speksnijder, A., and Smalla, K. (2007) Screening of bacterial isolates from various European soils for in vitro antagonistic activity towards *Rhizoctonia solani* and *Fusarium oxysporum*: site-dependent composition and diversity revealed. *Soil Biol Biochem* **39**: 2818–2828.

Adesina, M.F., Grosch, R., Lembke, A., Vatchev, T.D., and Smalla, K. (2009) In vitro antagonists of *Rhizoctonia solani* tested on lettuce: rhizosphere competence, biocontrol efficiency and rhizosphere microbial community response. *FEMS Microbiol Ecol* **69**: 62–74.

Andrews, M., Hodge, S., and Raven, J.A. (2010) Positive plant microbial interactions. *Ann Appl Biol* **157**: 317–320.

Bankhead, S.B., Landa, B.B., Lutton, E., Weller, D.M., and Gardener, B.B.M. (2004) Minimal changes in rhizobacterial population structure following root colonization by wild type and transgenic biocontrol strains. *FEMS Microbiol Ecol* **49**: 307–318.

Barr, D.J.S. (1990) Phylum Chytridiomycota. In *Handbook of Protoctista*. Margulis, L., Corliss, J.O., Melkonian, M. and Chapman, D.J. (eds). Boston, MA: Jones & Bartlett, pp. 454–466.

Bengtsson-Palme, J., Ryberg, M., Hartmann, M., Branco, S., Wang, Z., Godhe, A., *et al.* (2013) Improved software detection and extraction of ITS1 and ITS2 from ribosomal ITS sequences of fungi and other eukaryotes for analysis of environmental sequencing data. *Methods Ecol Evol* **4**: 914–919.

Berg, G. (2009) Plant-microbe interactions promoting plant growth and health: perspectives for controlled use of microorganisms in agriculture. *Appl Microbiol Biot* **84**: 11–18.

Berg, G., Krechel, A., Ditz, M., Sikora, R.A., Ulrich, A., and Hallmann, J. (2005) Endophytic and ectophytic potato-associated bacterial communities differ in structure and antagonistic function against plant pathogenic fungi. *FEMS Microbiol Ecol* **51**: 215–229.

Berg, G., Opelt, K., Zachow, C., Lottmann, J., Gotz, M., Costa, R., and Smalla, K. (2006) The rhizosphere effect on bacteria antagonistic towards the pathogenic fungus Verticillium differs depending on plant species and site. *FEMS Microbiol Ecol* **56**: 250–261.

ter Braak, C.J., and Šmilauer, P. (2015) Topics in constrained and unconstrained ordination. *Plant Ecol* **216**: 683–696.

Camacho, C., Coulouris, G., Avagyan, V., Ma, N., Papadopoulos, J., Bealer, K., and Madden, T.L. (2009) BLAST plus: architecture and applications. *BMC Bioinform* **10**: 421.

Chervoneva, I., Hyslop, T., and Hauck, W.W. (2007) A multivariate test for population bioequivalence. *Stat Med* **26**: 1208–1223.

Chet, I. (1987) Trichoderma – application, mode of action, and potential as a biocontrol agent of soilborne plant pathogenic fungi. In *Innovative Approaches to Plant Disease Control*. Chet, I. (ed). New York: John Wiley & Sons, pp. 137–160.

Chowdhury, S.P., Dietel, K., Randler, M., Schmid, M., Junge, H., Borriss, R., *et al.* (2013) Effects of *Bacillus amyloliquefaciens* FZB42 on lettuce growth and health under pathogen pressure and its impact on the rhizosphere bacterial community. *PLoS ONE* **8:** e68818.

DeSantis, T.Z., Brodie, E.L., Moberg, J.P., Zubiela, I.X., Piceno, Y.M., and Andersen, G.L. (2007) High-density universal 16S rRNA microarray analysis reveals broader diversity than typical clone library when sampling the environment. *Microb Ecol* **53:** 371–383.

Ding, G.C., Smalla, K., Heuer, H., and Kropf, S. (2012) A new proposal for a principal component-based test for high-dimensional data applied to the analysis of Phylo-Chip data. *Biom J* **54:** 94–107.

Edgar, R.C. (2010) Search and clustering orders of magnitude faster than BLAST. *Bioinformatics* **26:** 2460–2461.

Edgar, R.C. (2013) UPARSE: highly accurate OTU sequences from microbial amplicon reads. *Nat Methods* **10:** 996–998.

Fukuyama, J., McMurdie, P.J., Les Dethlefsen, D.A.R. and Holmes, S. (2012) Comparisons of distance methods for combining covariates and abundances in microbiome studies. In *Pacific Symposium on Biocomputing. Pacific Symposium on Biocomputing* (p. 213). NIH Public Access.

Good, P. (2000) *Permutation Tests.* New York: Springer-Verlag.

Haas, D., and Défago, G. (2005) Biological control of soil-borne pathogens by fluorescent pseudomonads. *Nat Rev Microbiol* **3:** 307–319.

Hajek, A.E. (2004) *Natural Enemies: An Introduction to Biological Control.* Cambridge, UK: Cambridge University Press.

Hallmann, J., Davies, K.G., and Sikora, R.A. (2009) Biological control using microbial pathogens, endophytes and antagonists. In *Root-Knot Nematodes.* Perry, R.N., Moens, M., and Starr, J.L. (eds). Wallingford, GB: CAB International, pp. 380–411.

Heuer, H., Kroppenstedt, R.M., Lottmann, J., Berg, G., and Smalla, K. (2002) Effects of T4 lysozyme release from transgenic potato roots on bacterial rhizosphere communities are negligible relative to natural factors. *Appl Environ Microbiol* **68:** 1325–1335.

Hibbett, D.S., and Taylor, J.W. (2013) Fungal systematics: is a new age of enlightenment at hand? *Nat Rev Microbiol* **11:** 129–133.

Jacobsen, C.S., and Hjelmsø, M.H. (2014) Agricultural soils, pesticides and microbial diversity. *Curr Opin Biotechnol* **27:** 15–20.

Karlin, S., and Rinott, Y. (1982) Applications of ANOVA type decompositions for comparisons of conditional variance statistics including jackknife estimates. *Ann Stat* **10:** 485–501.

Kirk, P.M., Cannon, P.F., Minter, D.W., and Stalpers, J.A. (2008) *Dictionary of the Fungi.* Wallingford, UK: CAB International.

Koljalg, U., Nilsson, R.H., Abarenkov, K., Tedersoo, L., Taylor, A.F.S., Bahram, M., *et al.* (2013) Towards a unified paradigm for sequence-based identification of fungi. *Mol Ecol* **22:** 5271–5277.

Kropf, S., and Adolf, D. (2009) Rotation test with pairwise distance measures of sample vectors in a GLM. *J Stat Plan Inference* **139:** 3857–3864.

Kropf, S., Lux, A., Eszlinger, M., Heuer, H., and Smalla, K. (2007) Comparison of independent samples of high-dimensional data by pairwise distance measures. *Biom J* **49:** 230–241.

Kupferschmied, P., Maurhofer, M., and Keel, C. (2013) Promise for plant pest control: root-associated pseudomonads with insecticidal activities. *Front Plant Sci* **4:** 287.

Lugtenberg, B., and Kamilova, F. (2009) Plant-growth-promoting rhizobacteria. *Annu Rev Microbiol* **63:** 541–556.

Pérez-García, A., Romero, D., and de Vicente, A. (2011) Plant protection and growth stimulation by microorganisms: biotechnological applications of Bacilli in agriculture. *Curr Opin Biotechnol* **22:** 187–193.

Reeder, J., and Knight, R. (2010) Rapidly denoising pyrosequencing amplicon reads by exploiting rank-abundance distributions. *Nat Methods* **7:** 668–669.

Ruehlmann, J. (2013) The Box Plot Experiment in Grossbeeren after eight rotations: nitrogen, carbon and energy balances. *Arch Agron Soil Sci* **59:** 1159–1176.

Rühlmann, J., and Ruppel, S. (2005) Effects of organic amendments on soil carbon content and microbial biomass – results of the long-term box plot experiment in Grossbeeren. *Arch Agron Soil Sci* **51:** 163–170.

Saraf, M., Pandya, U., and Thakkar, A. (2014) Role of allelochemicals in plant growth promoting rhizobacteria for biocontrol of phytopathogens. *Microbiol Res* **169:** 18–29.

Scherwinski, K., Wolf, A., and Berg, G. (2007) Assessing the risk of biological control agents on the indigenous microbial communities: *Serratia plymuthica* HRO-C48 and *Streptomyces* sp. HRO-71 as model bacteria. *Biocontrol* **52:** 87–112.

Schreiter, S., Sandmann, M., Smalla, K., and Grosch, R. (2014a) Soil type dependent rhizosphere competence and biocontrol of two bacterial inoculant strains and their effects on the rhizosphere microbial community of field-grown lettuce. *PLoS ONE* **9:** e103726.

Schreiter, S., Ding, G.-C., Grosch, R., Kropf, S., Antweiler, K., and Smalla, K. (2014b) Soil type-dependent effects of a potential biocontrol inoculant on indigenous bacterial communities in the rhizosphere of field-grown lettuce. *FEMS Microbiol Ecol* **90:** 718–730.

Suter, G.W. (2006) *Ecological Risk Assessment.* Baca Raton: CRC Press.

Tedersoo, L., Bahram, M., Polme, S., Koljalg, U., Yorou, N.S., Wijesundera, R., *et al.* (2014) Global diversity and geography of soil fungi. *Science* **346:** 1078.

Trabelsi, D., and Mhamdi, R. (2013) Microbial inoculants and their impact on soil microbial communities: a review. *Biomed Res Int* **2013:** 863240.

Verbruggen, E., Kuramae, E.E., Hillekens, R., de Hollander, M., Kiers, E.T., Roling, W.F., *et al.* (2012) Testing potential effects of maize expressing the *Bacillus thuringiensis* Cry1Ab endotoxin (Bt maize) on mycorrhizal fungal communities via DNA- and RNA-based pyrosequencing and molecular fingerprinting. *Appl Environ Microbiol* **78:** 7384–7392.

Větrovský, T., and Baldrian, P. (2013) Analysis of soil fungal communities by amplicon pyrosequencing: current approaches to data analysis and the introduction of the pipeline SEED. *Biol Fertil Soils* **49:** 1027–1037.

Voříšková, J., Brabcová, V., Cajthaml, T., and Baldrian, P. (2014) Seasonal dynamics of fungal communities in a temperate oak forest soil. *New Phytol* **201:** 269–278.

Warton, D.I., Wright, S.T., and Wang, Y. (2012) Distance-based multivariate analyses confound location and dispersion effects. *Methods Ecol Evol* **3:** 89–101.

Weinert, N., Meincke, R., Gottwald, C., Heuer, H., Gomes, N.C., Schloter, M., *et al.* (2009) Rhizosphere communities of genetically modified zeaxanthin-accumulating potato plants and their parent cultivar differ less than those of different potato cultivars. *Appl Environ Microbiol* **75:** 3859–3865.

Winding, A., Binnerup, S.J., and Pritchard, H. (2004) Non-target effects of bacterial biological control agents suppressing root pathogenic fungi. *FEMS Microbiol Ecol* **47:** 129–141.

Single-cell genomics based on Raman sorting reveals novel carotenoid-containing bacteria in the Red Sea

Yizhi Song,[1] Anne-Kristin Kaster,[2] John Vollmers,[2] Yanqing Song,[3] Paul A. Davison,[4] Martinique Frentrup,[2] Gail M. Preston,[5] Ian P. Thompson,[1] J. Colin Murrell,[6] Huabing Yin,[3] C. Neil Hunter[4] and Wei E. Huang[1,*]

[1]Department of Engineering Science, University of Oxford, Parks Road, Oxford OX1 3PJ, UK.
[2]Leibniz Institute DSMZ, Deutsche Sammlung von Mikroorganismen und Zellkulturen GmbH, Inhoffenstrasse 7 B, 38124, Braunschweig, Germany.
[3]Division of Biomedical Engineering, School of Engineering, University of Glasgow, Glasgow G12 8QQ, UK.
[4]Department of Molecular Biology and Biotechnology, University of Sheffield, Sheffield S10 2TN, UK.
[5]Department of Plant Sciences, University of Oxford, South Parks Road, Oxford OX1 3RB, UK.
[6]School of Environmental Sciences, University of East Anglia, Norwich NR4 7TJ, UK.

Summary

Cell sorting coupled with single-cell genomics is a powerful tool to circumvent cultivation of microorganisms and reveal microbial 'dark matter'. Single-cell Raman spectra (SCRSs) are label-free biochemical 'fingerprints' of individual cells, which can link the sorted cells to their phenotypic information and ecological functions. We employed a novel Raman-activated cell ejection (RACE) approach to sort single bacterial cells from a water sample in the Red Sea based on SCRS. Carotenoids are highly diverse pigments and play an important role in phototrophic bacteria, giving strong and distinctive Raman spectra. Here, we showed that individual carotenoid-containing cells from a Red Sea sample were isolated based on the characteristic SCRS. RACE-based single-cell

genomics revealed putative novel functional genes related to carotenoid and isoprenoid biosynthesis, as well as previously unknown phototrophic microorganisms including an unculturable *Cyanobacteria* spp. The potential of Raman sorting coupled to single-cell genomics has been demonstrated.

Introduction

Microbes are the most diverse and abundant organisms on earth and play critical roles in biogeochemical carbon and nitrogen cycling (Paterson *et al.*, 1997; Whitman *et al.*, 1998; Schleifer, 2004; Huang *et al.*, 2009a,b). However, the vast majority of microbes in natural environments have not yet been grown in the laboratory using traditional cultivation methods (Amann *et al.*, 1995; Venter, 2003; Venter *et al.*, 2004; Daniel, 2005; Swan *et al.*, 2011; Rinke *et al.*, 2013; Hedlund *et al.*, 2014). Furthermore, to understand the function of microbes in a community, it is desirable to place even cultivable microorganisms in their ecological and environmental context instead of just studying pure cultures (Huang *et al.*, 2015).

Metagenomics is a powerful cultivation-independent approach for studying microbes. It can provide an overview of the diversity and metabolic blueprints of potential functions of microbes in environmental samples. Unfortunately, assembling individual discrete genomes from metagenomics data is quite difficult and rather costly, especially for samples with a high diversity of microorganisms. Although binning algorithms, which group contigs and assign DNA sequences to operational taxonomic units, have massively improved over the years, metagenomes assembled in this way may still be mosaics of DNA from different strains. Single-cell genomics can complement metagenomics approaches and be used to establish genetic linkages between DNA sequences within individual cells (Swan *et al.*, 2011; Lasken, 2012; Rinke *et al.*, 2013). Although fluorescent-activated cell sorting has been successfully used to sort single cells and subsequently perform single-cell genomics (Rinke *et al.*, 2013), Raman-activated cell sorting (RACS) can provide a valuable alternative to sort cells based on single-cell Raman spectra (SCRSs) which can reflect phenotypic and intrinsic biochemical fingerprints of cells (Huang *et al.*, 2004, 2015). An SCRS is able to provide a label-free biochemical profile of a single cell, which could contain information on nucleic acids, proteins, carbohydrates, lipids within the cell and specific

*For correspondence. E-mail wei.huang@eng.ox.ac.uk;

Funding Information
WEH acknowledges support from EPSRC (EP/M002403/1 and EP/M02833X/1) and NERC (NE/M002934/1) and CNH thanks BBSRC sLoLa (BB/M000265/1) in the UK for the financial support. JCM acknowledges funding from the Gordon and Betty Moore Foundation Marine Microbiology Initiative Grant #3303.

Raman-active compounds (e.g. carotenoids, isoprenoid, cytochrome *c*, poly-β-hydroxybutyrate and glycogen) (Li *et al.*, 2012a,b,c; Zhang *et al.*, 2015a,b). RACS-mediated single-cell genomics should, therefore, help to reveal phenotypic information and couple certain phenotypes to genotypes of cells.

Raman-activated cell sorting can be achieved by combining Raman single-cell detection with optical tweezers (Huang *et al.*, 2009a,b), microfluidic devices (Zhang *et al.*, 2015a,b), or the so-called Raman-activated cell ejection (RACE) (Wang *et al.*, 2013). Optical tweezers and microfluidic-based RACS methods are, however, quite labour intensive and difficult to apply to complex samples. Previously, RACE has been performed by using two separate instruments: Raman and laser microdissection microscopes (Wang *et al.*, 2013). In this study, we integrated Raman single-cell detection and cell ejection into one system. This new system is now able to characterize single cells based on SCRS and accurately isolate cells of interest on a slide surface by employing laser-induced forward

transfer (LIFT) (Hopp *et al.*, 2005). LIFT applies a laser to a thin-layer-coated surface which causes immediate and *local* evaporation of the coated layer, therefore pushing or ablating the selected cell on the layer into a collection device (Fig. 1).

Carotenoids are one of the most diverse chemicals in bacteria (Takaichi, 2008) and give strong and distinctive Raman spectra (Kochendoerfer *et al.*, 1999; Krebs *et al.*, 2003; Robert, 2009; Li *et al.*, 2012a,b,c). Due to the resonance Raman effect, the Raman spectra of carotenoids are sharp and strong, providing typical and unambiguous carotenoid bands (Kochendoerfer *et al.*, 1999; Krebs *et al.*, 2003; Robert, 2009) making them ideal targets for the RACE system. Carotenoids are biologically important molecules, present in nearly all photosynthetic cells (Li *et al.*, 2012a,b,c). For example, carotenoids are associated with both chlorophyll-based photochemical reaction centres and rhodopsin-based light-activated proton pumps (Bryant and Frigaard, 2006), which are two important light-harvesting (LH) systems of

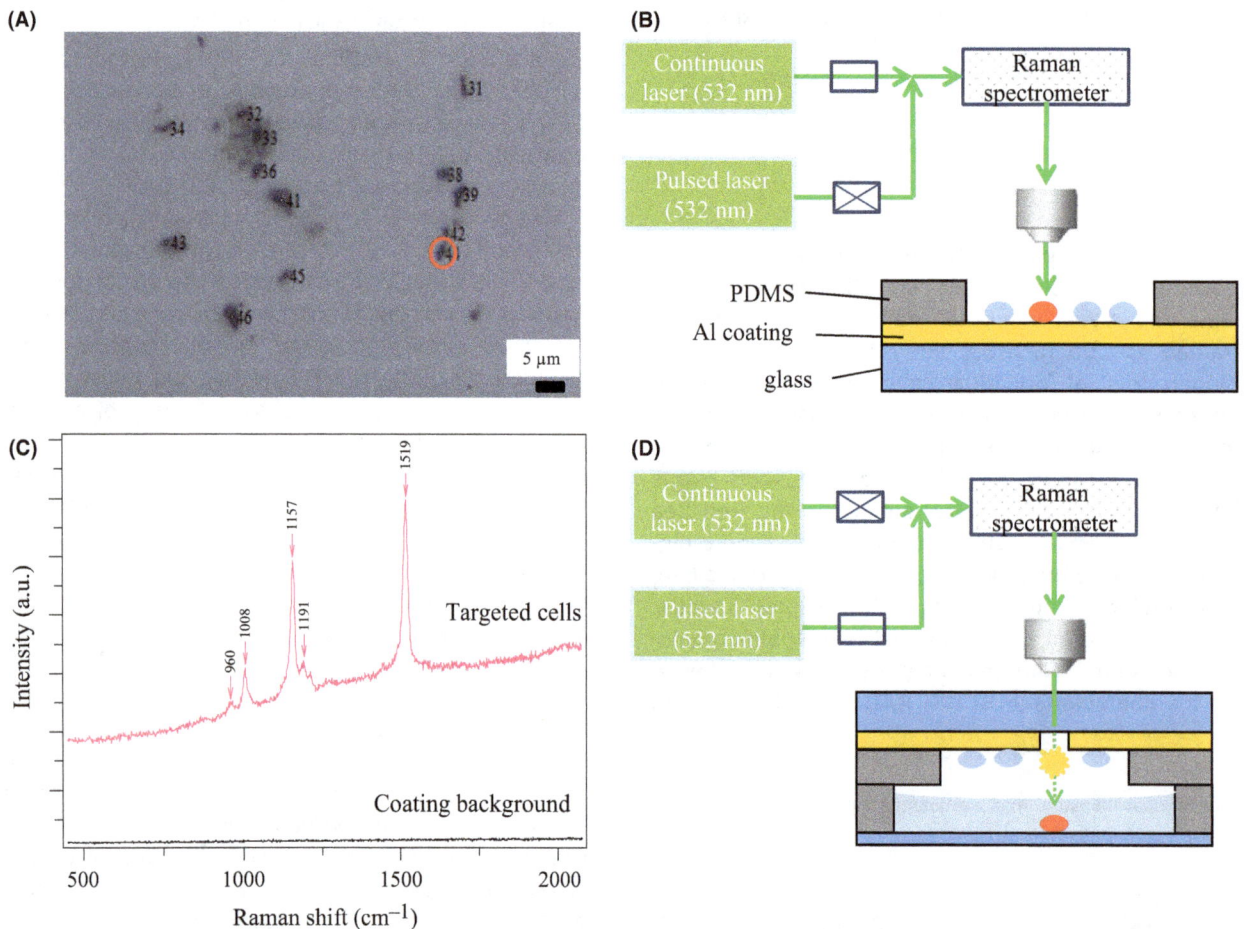

Fig. 1. Illustration of resonant Raman-activated single-cell ejection (RACE). (A) Microscopic image of cells on a RACE chip. (B) The continuous laser is used for acquiring single-cell Raman spectra. PDMS: Polydimethylsiloxane (C) Spectra of one single cell and the slide coating background. (D) The RACE chip is turned over and the target cell is ejected into the collector by a pulsed laser.

photosynthesis. In chlorophyll-based LH systems, carotenoids absorb solar energy, quench free radicals and are structurally important parts of chlorophyll-based LH complexes (Garcia-Asua *et al.*, 1998). In rhodopsin-based LH systems, a carotenoid, namely β-carotene, is a precursor of retinal, which is an essential component of bacteriorhodopsin and proteorhodopsin (Gonzalez *et al.*, 2011). Hence, carotenoids are important indicative molecules related to light harvesting. Sorting carotenoid-containing cells would, therefore, help to probe novel uncultured phototrophic bacteria. In this study, RACE was used to isolate carotenoid-containing bacteria from the Red Sea. Both new photosynthetic bacteria and putative novel functional genes were discovered.

Results

Establishing the RACE system by integrating a cell ejection system with a Raman micro-spectroscope

A 532 nm pulsed laser for single-cell ejection has been integrated into a Raman micro-spectroscope, which is designated as RACE (Fig. 1). Cells were added onto a RACE sampling chip (Fig. 1A and Fig. S1) and examined by the Raman micro-spectroscope using a continuous 532 nm laser (Fig. 1B). SCRSs were used as sorting criteria to distinguish the phenotypic 'profile' of cells (in this case, the presence of carotenoids), and the positions of targeted cells were recorded (Fig. 1C). The RACE sampling chip was turned over, and single cells of interest were then ejected and harvested in a RACE collection chip (Fig. 1D and Fig. S2).

Multiple displacement amplification from single cells isolated from RACE

Genomes of single or multiple (3–8) cells sorted with the RACE system were amplified by multiple displacement amplification (MDA; Blanco *et al.*, 1989). As an initial test, *Escherichia coli* JM109 cells containing the plasmid p18GFP were used to examine the performance of single-cell ejection. The green fluorescent protein gene *gfp* on the plasmid p18GFP and the universal stress protein A gene *uspA* (Chen and Griffiths, 1998) on the chromosome were used to assess gene recovery. The *gfp* on this plasmid has about 20–40 copies per cell, whereas *uspA* is a single copy gene on the *E. coli* chromosome. Figure S3A shows that the negative control did not contain amplified DNA, and three of seven samples contained MDA products from single *E. coli* cells. The other four samples did not yield visible MDA products (the products were too little to be seen in the gel), which may be due to the inefficient MDA of these cells. The PCR product of the *gfp* gene was recovered from all seven samples, while only one of the seven samples contained

the *uspA* PCR product (Fig. S3B and S3C). This is probably due to amplification bias in the MDA, since multiple-copy *gfp* genes have a better chance of being amplified than a single-copy gene such as *uspA* in the one cell. Although sometimes the MDA yield was not high enough to be clearly viewed in agarose gel, the amount of DNA obtained was sufficient to perform PCR amplification (Fig. S3B and S3C). Nonetheless, this shows that it is possible that a single-copy functional gene can be recovered by MDA from a single-cell isolated by RACE.

Sorting carotenoid-containing bacteria from a Red Sea water sample based on their resonant Raman spectra

Raman-activated cell ejection has been applied to sort bacterial cells sampled from surface seawater at Eilat on the Red Sea coast. SCRS of 5321 single cells have been obtained and analysed, and 33% of the analysed cells exhibited strong fluorescence and 67% had distinguishable SCRS for sorting. Figure 2 shows the principal component analysis (PCA) of those spectra, and the spots within the red line area are the cells containing carotenoids according to Raman spectra. The most significant loadings of PCA at axis 1 are 997–1007, 1145–1161 and 1503–1526 cm^{-1} (Fig. S4), corresponding to characteristic *v1*, *v2* and *v3* Raman bands of carotenoids (Robert, 2009; Li *et al.*, 2012a,b,c). This is in a good agreement with the fact that carotenoid-containing cells are distributed along axis 1 within the red line area (Fig. 2).

Figure 3A shows a few examples of SCRS, including a typical bacterial cell, individual cells showing

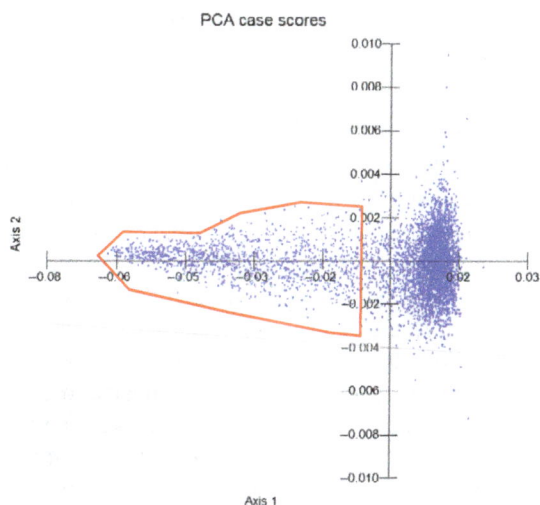

PCA case scores

Fig. 2. The PCA analysis of 5321 SCRS from the Red Sea sample. The group within red line area is cells containing various carotenoids, and the rest of cells have no carotenoids.

fluorescence, cells containing poly-β-hydroxybutyrate (PHB) and unidentified compounds (cell images are shown in Fig. S5A). Since SCRS can reflect the biochemical phenotypes of cells (Huang *et al.*, 2004, 2007a,b, 2009a,b; Li *et al.*, 2012a,b,c; Berry *et al.*, 2015), cells can be sorted based on the Raman biomarker bands of SCRS. For example, cells containing PHB have distinguishable Raman biomarkers (Majed and Gu, 2010) at 839, 1058, 1403 and 1123 cm^{-1} (Fig. 3A), which could be used to sort PHB containing cells. Interestingly, some SCRS also indicate that cells contain unidentified novel compounds, e.g. Fig. 3A shows a SCRS of a cell with a hydroxyisoquinoline-like compound (Fig. S5A panel v). Since RACS is a new technology, a database with well-characterized biomarkers is needed for the future analysis of cells.

In this study, the characteristic *v1, v2* and *v3* Raman bands of carotenoids were used as sorting criteria (Fig. 3). The *v1* band is the methyl rocking mode, and the *v2* and *v3* bands vary due to the different lengths of conjugated C=C bonds and stretching modes of C-C bonds (Robert, 2009) in different carotenoids. Since the Raman spectral resolution is about 1 cm^{-1}, variations in the positions of *v1, v2* and *v3* indicate the different structures of carotenoid presented in the cells. According to SCRS, 1223 (23%) cells contained carotenoids. Based on their carotenoid SCRS, 30 attempts of single-cell ejection and whole genome amplification (WGA) were performed. Seven of the 30 attempts were successful in which 16S rRNA was recovered from the genome amplification. The SCRS of these seven successful cases are shown in Fig. 3B and 3C, and their information are listed in Table 1. Collectively these seven sorted samples could be resolved into five different types of carotenoids. Optical images of the sorted cells show that the cells had size of 0.3–1 μm (Fig. S5B).

Genome sequencing of isolated cells

The estimated degree of genome completeness and putative contamination was based on universal marker genes analyses. The overall contamination based on CheckM analyses (Parks *et al.*, 2015) was estimated to be relatively low in all cases (Table S3). However, sequence analyses also indicated relatively low genome coverage – less than 20% in all cases. The genome sequence data revealed functional genes related to carotenoid biosynthesis, CO_2 fixation, haem/chlorophyll biosynthesis and fatty acid, alcohol and aldehyde metabolism. These data are summarized in Table 1. A detailed list of functional genes is given in Table S4. Despite the low estimated genome coverage, genes associated with carotenoid biosynthesis were identified in six of the seven sorted samples (P728-5, B728-3,

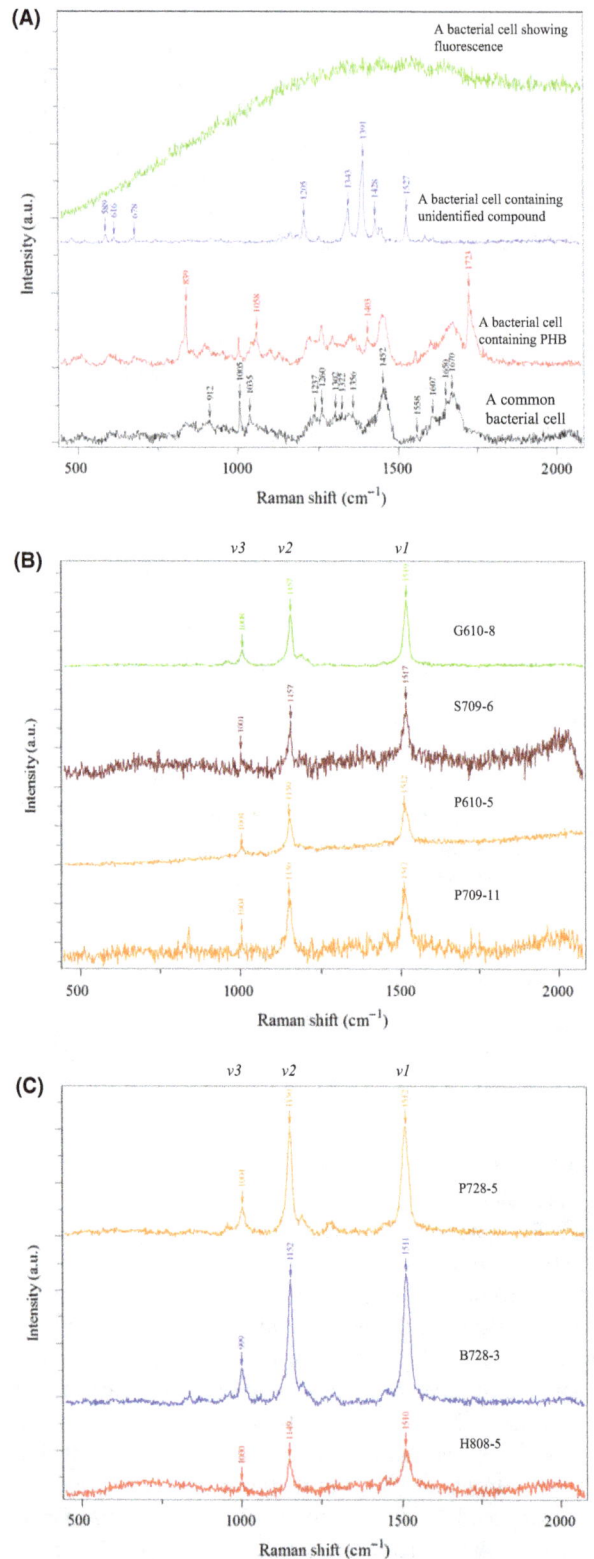

Fig. 3. (A) Single cell Raman spectra of some typical bacteria in a Red Sea sample. (B) Raman spectra of four groups of ejected cells. Each group contains 3–8 cells which had the same carotenoid spectra. (C) Raman spectra of three groups of ejected individual cells. The positions of the carotenoid *v1, v2* and *v3* bands are labelled.

Table 1. Summary of RACE-isolated cells and WGA results

Sample name	P728-5	B728-3	H808-5	S709-6	P709-11	G610-8	P610-5
Cells being ejected	1	1	1	3	5	8	8
Identification	*Pelomonas* spp.	*Bradyrhizobium* spp.	*Halomonas* spp.	*Shigella* spp.	*Pelomonas* spp.	*Cyanobacteria* spp.	*Pelomonas* spp.
Genome coverage (%)	8.18	6.65	4.17	10.74	6.9	8.95	19.29
Contamination	0	0	0	0	0	0	0
Putative genes for carotenoid biosynthesis	*crtI*	*ispA, crtE, crtI*	*shc*[a]	*crtI*	*idi, ispA, crtE, crtI*	*idi*	*crtI, crtC, crtJ*
Putative genes for CO$_2$ fixation	Aerobic-type *coxL/cutL*;				CO$_2$ concentrating mechanism/ carboxysome shell protein; *pepc*		CO$_2$-fixation; Calvin-Benson-cycle related gene
Putative genes for haem/ chlorophyll biosynthesis	*urod, cobN, chlD, ppox, fech*	*cobN*, haem biosynthesis	*cobN*	*cobN, uros, chl*	*cobN, chl, chlD, cpox, pbdg, ppox, fech*		*cobN, chlD, chl,* haem oxygenase, *fech, pbdg, cpox*
Putative genes for fatty acid, alcohol, aldehyde metabolism	*acc*: Acyl-CoA synthetases, short-chain alcohol dehydrogenases, *mcat*, Molybdenum cofactor biosynthesis enzyme	Short-chain alcohol dehydrogenases, short-chain dehydrogenases, Acyl-CoA synthetases, *fhl* subunit 4	Short-chain alcohol dehydrogenases, Short-chain dehydrogenases, Acyl-CoA synthetases; *acc*; uncharacterized *fdh*	Acyl-CoA synthetases; *hadh; acc*; short-chain alcohol dehydrogenases; Predicted acyltransferases; *fhl* subunit 3 and 4, *mnhD* subunit; uncharacterized protein for *fdh* activity	Acyl-CoA synthetases; Alcohol dehydrogenase, class IV; NAD+; uncharacterized protein for *fdh* activity	Alternatively TCA cycle; short-chain alcohol dehydrogenases; Acyl-CoA synthetases; Acyl-CoA dehydrogenases; NAD+; Predicted acyltransferases; *fhl* subunit 3 and 4, *mnhD* subunit	Acyl-CoA synthetases; *fhl* subunit 3 and 4, *mnhD* subunit

[a]The substrate of Shc (Squalene-hopene cyclase) is a carotenoid compound.

acc, acetyl-CoA carboxylase, carboxyltransferase component; *acp*, acyl carrier protein (used in FA synthesis); *chlD, chlI*: Mg-chelatase subunit; *cobN*, cobalamin biosynthesis protein CobN and related Mg-chelatases; *coxL/cutL* homologues (used in carbon fixation via reductive acetyl-CoA pathway); *cpox*, coproporphyrinogen III oxidase; *crtC*, hydroxyneurosporene synthase; *crtE*, geranylgeranyl pyrophosphate synthase; *crtI*, phytoene desaturase (dehydrogenase); *crtJ*, CrtJ protein which is involved in spheroidene biosynthesis; *fdh*, formate dehydrogenase; *fech*, Protoheme ferro-lyase (ferrochelatase); *fhl*, formate hydrogenase; *hadh*, 3-hydroxyacyl-CoA dehydrogenase; *idi*, isopentenyl diphosphate isomerase (or IPI, isopentenyl pyrophosphate isomerase); *ispA*, farnesyl diphosphate synthase; *mcat*, malonyl-CoA:acyl carrier protein transacylase; *mnhD*, Na+/H+ antiporter subunit D; *pbgd*, Porphobilinogen deaminase; *pepc*, Phosphoenolpyruvate carboxylase; *ppox*, Protoporphyrinogen oxidase; *shc*, Squalene-hopene cyclase; *urod*, Uroporphyrinogen-III decarboxylase; *uros*, Uroporphyrinogen-III synthase.

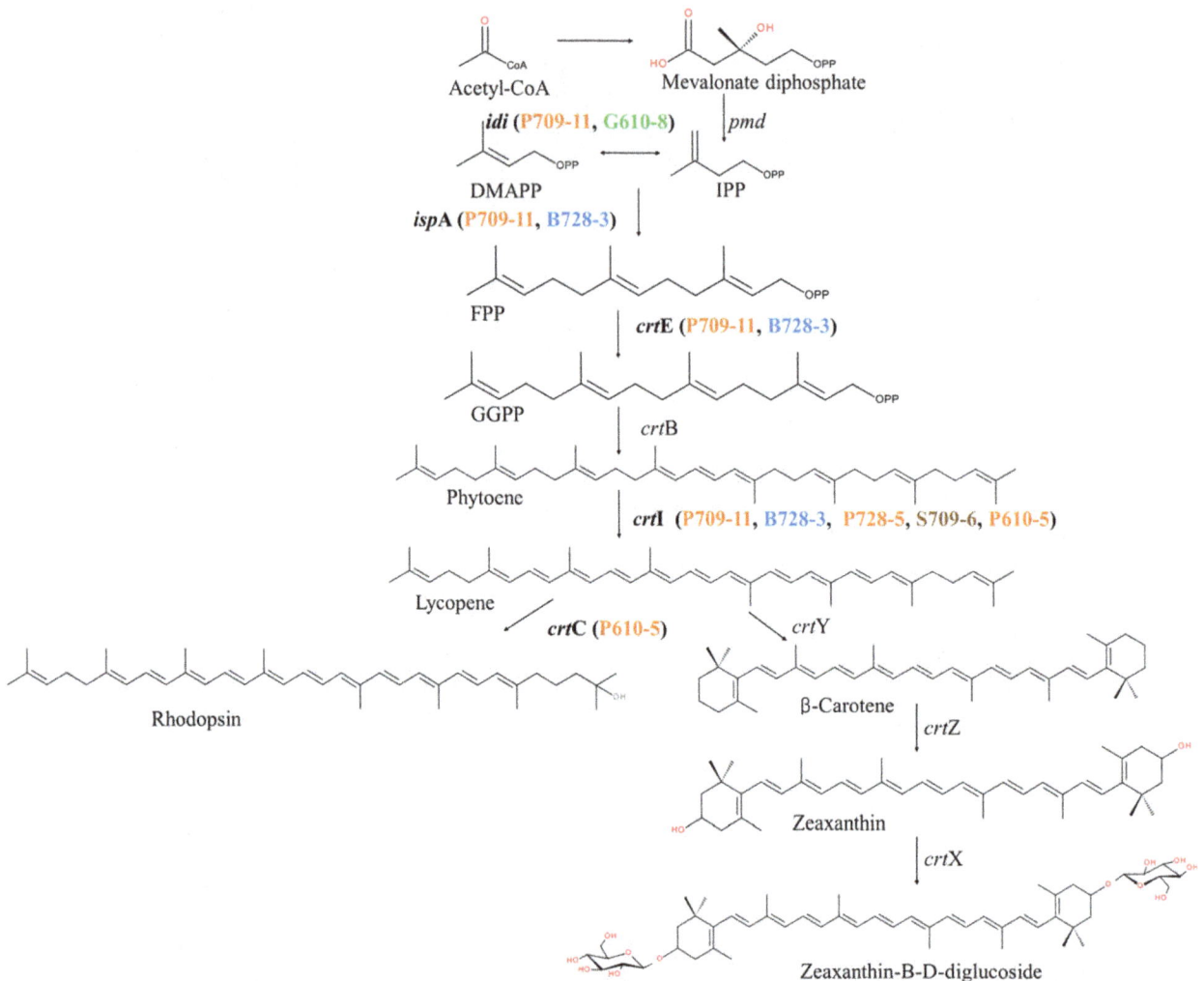

Fig. 4. The identified genes encoding putative proteins in bacterial carotenoid synthesis pathway from Red Sea single cell(s) MDA. The ejected single cell(s) which contain the genes are bracketed and coloured. *idi*: isopentenyl diphosphate isomerase; *isp*A: farnesyl diphosphate synthase; *crt*E: Geranylgeranyl pyrophosphate synthase; *crt*B: phytoene synthase; *crt*I: phytoene desaturase (dehydrogenase); *crt*C: hydroxyneurosporene synthase; *crt*D: methoxyneurosporene dehydrogenase; *crt*F: hydroxyneurosporene-O-methyltransferase; *crt*Y: lycopene cyclase; *crt*Z: carotene hydroxylase; *pmd*: phosphomevalonate decarboxylase.

P709-11, S709-6, G610-8 and P610-5). The recovered genes, which are directly involved in carotenoid biosynthesis pathways, are shown in Fig. 4. Specifically, isopentenyl diphosphate isomerase and geranylgeranyl pyrophosphate synthase (*isp*A) are involved in making the colourless substrate farnesyl pyrophosphate, and phytoene dehydrogenase (*crt*I) is responsible for synthesis of the red pigment lycopene (Fig. 4). H808-5 contained two types of putative *shc* (squalene-hopene cyclase) genes whose substrates are carotenoid compounds (Table 1), although no gene directly related to carotenoid synthesis was found in this sample (probably due to low genome coverage). These results validate the RACE method since the sorted cells were expected to contain carotenoids according to their SCRS.

A phylogenetic tree of the isolated samples is shown in Fig. 5. The carotenoid-containing bacteria sorted from the Red Sea sample were broadly distributed over several different groups. According to 16S rRNA sequencing and BLAST (Altschul *et al.*, 1990) against the NCBI bacterial 16S rRNA database, one of the sorted cells, G610-8, is a new cyanobacterial order (GenBank accession no. KU667126), which gives a best hit (84% homology) to an uncultivated *Melainabacteria* spp. (Fig. 5). The 1.2 kb 16S rRNA sequencing has been independently repeated four times by three Sanger sequencing runs and one Illumina MiSeq sequencing run. All sequencing runs gave identical results, which rules out sequencing errors. This microorganisms is described as a novel, as yet uncultivated order of cyanobacteria (Soo *et al.*, 2014).

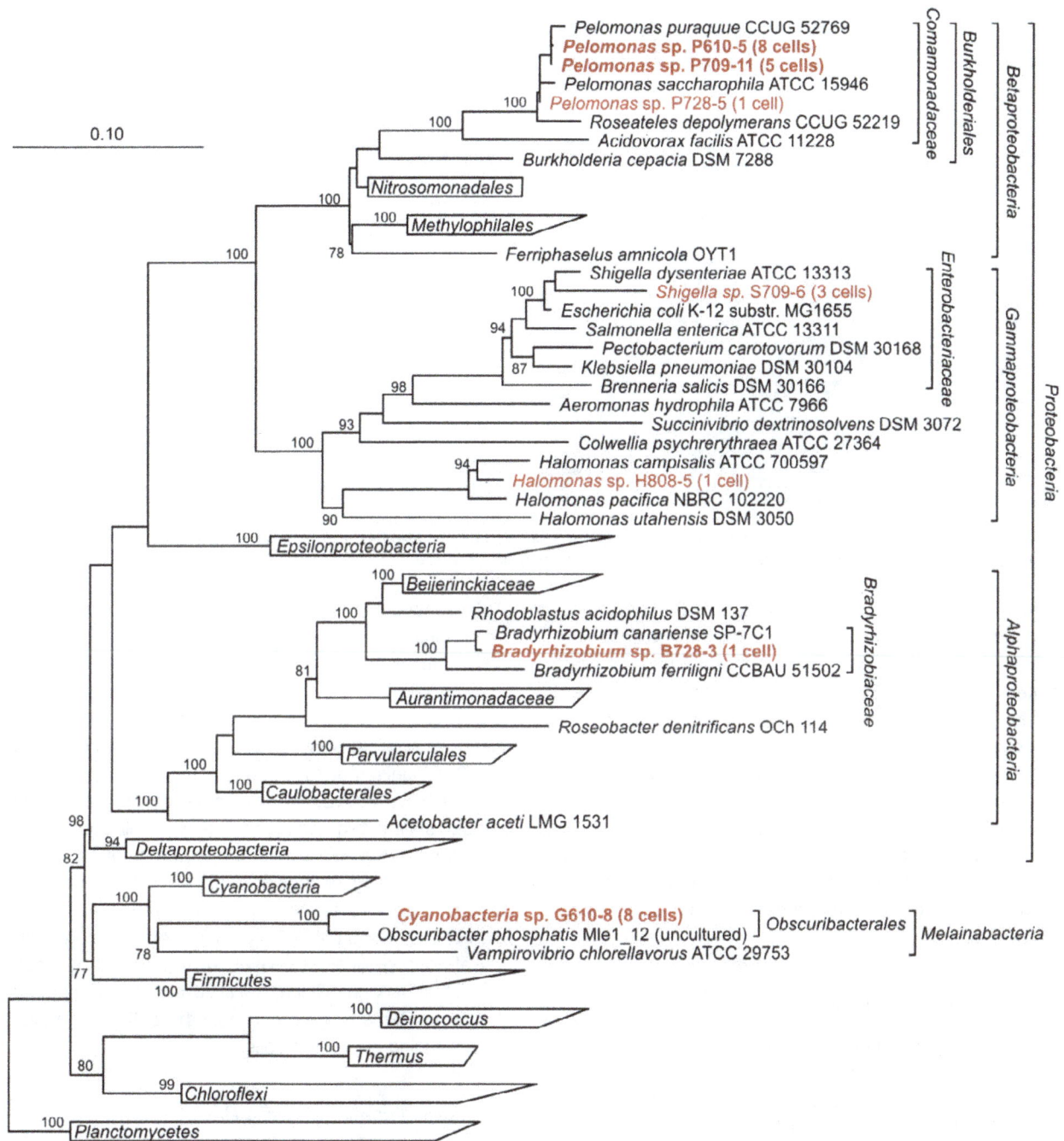

Fig. 5. Neighbor joining tree indicating the phylogenetic relationships of cells sorted by Raman-activated cell ejection and reference genomes. The tree is based on 16S rRNA gene sequences. Bootstrap support values above 75% are indicated at the respective nodes. Leaves referring to 16S rRNA gene sequences obtained from PCR products of MDA treated sorted cells, using universal bacterial primers, are marked in plain red lettering. Leaves referring to 16S rRNA gene sequences obtained from assembled single-cell genome sequencing data of sorted cells are marked in bold red lettering. Black lettering indicates reference sequences.

Discussion

The RACE technology has been applied to a Red Sea water sample to sort seven groups of cells and perform subsequent genome analyses (Table 1) and is a proof of concept for RACE applications in single-cell genomics.

MDA and genome recovery

There are further refinements needed to improve this technique in future studies, such as keeping the sorted cells viable, refining single-cell lysis and genome amplification to increase genome coverage, and reducing

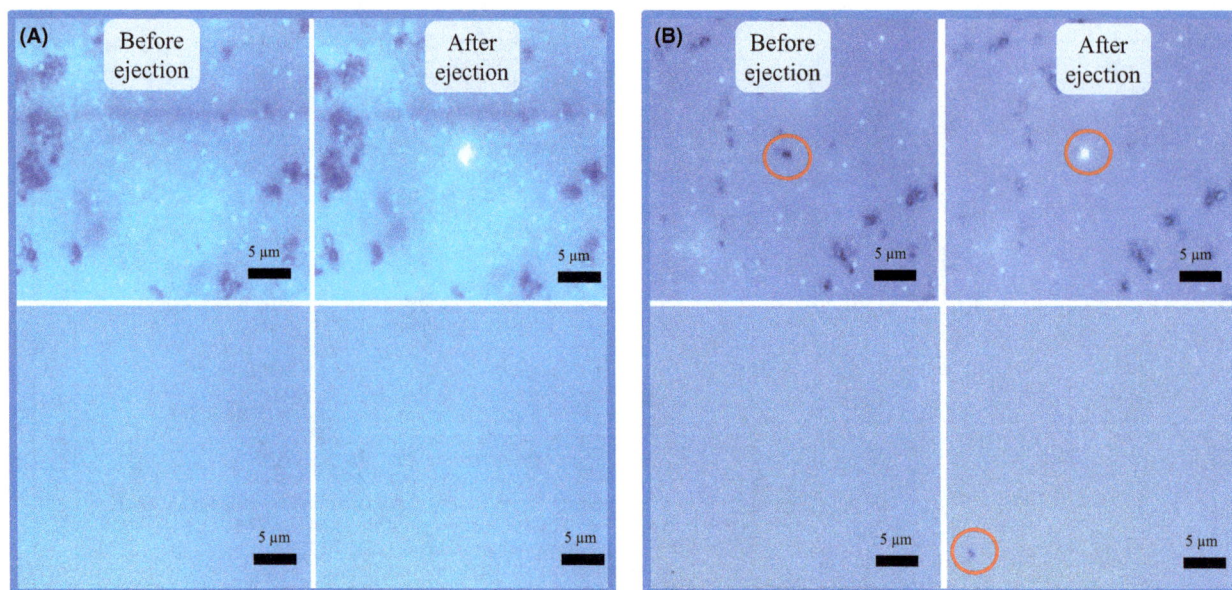

Fig. 6. Microscopy images of the RACE chip (top row) which holds a sample from the Red Sea and the collector (bottom row) before and after applying the pulsed laser. (A) Pulsed laser focused at a position without cells. (B) Pulsed laser focused on a cell.

contamination through better designed RACE chips. The low genome recovery is probably due to the following issues. (i) The cells may not be deposited into the lysis buffer in microwell (Fig. S1), the trajectory of cells can be further improved by adjusting the power of pulsed laser and exposure time. (ii) UV treatment of the Phi29 polymerase and inhibition residual hyperchloride used in the process, resulting in a loss of activity. Highly pure and DNA-free Phi29 polymerase is now available in commercial market. (iii) The cells were dried on the chip for several weeks, and genome DNA may be damaged or degraded during the time. It can be improved by using fresh cells.

Accurate single-cell ejection

The thin coating material was able to absorb the energy of the 532 nm pulsed laser and to eject a single cell sitting on it (Fig. 6 and Fig. S2). Figure 6A and Fig. S2A show that the coating material got ablated after laser ejection, leaving a ~1.5 μm mark on the coating slide. No debris could be found on the collection chip after ejection in the blank controls, suggesting that the coating material was completely vapourized. Figure 6B and Fig. S2B indicate that single cells could be accurately isolated and collected by the RACE collection chip.

For single-cell isolation using LIFT, the pulsed laser causes local vapourization of laser absorbing materials in the coating layer of the glass slide, which pushes the cells sitting on it to the collection well (Fig. 6 and Fig. S2). It has been shown that a 532 nm pulsed laser was able to vapourize metallic materials such as Cu and

Ag (Bohandy *et al.*, 1988). It has been previously demonstrated that cells can survive after isolation by LIFT, including bacterial cells (*E. coli*), various mammalian cells and sensitive mouse embryonic stem cells (Pique *et al.*, 1999; Fernandez-Pradas *et al.*, 2004; Ringeisen *et al.*, 2004; Hopp *et al.*, 2005; Kattamis *et al.*, 2007) (Ringeisen *et al.*, 2002).

To achieve RACE, the thin metal coating in the chip should satisfy two criteria: (i) give a minimal Raman background (Fig. 1C), which ensures high-quality SCRS for distinguishing phenotypes; (ii) be able to absorb a pulsed laser and cause *local* vapourization and expulsion of cells without thermal damage. After testing a variety of coating slides, a coating slide provided by Hesen Biotech (Shanghai, China) satisfied the above criteria as it has no Raman signal and absorbs a 532 nm pulsed laser causing ejection of cells.

Discovering new genes using RACE

Three sorted samples (P728-5, P709-11 and P610-5) with similar resonant Raman spectra (Fig. 3) were identified as closely related *Pelomonas* spp. according to phylogenetic analysis (Fig. S6A). Sequence identities in overlapping genome regions even indicate that these genomes may belong to the same species, but due to the low genome coverage resulting in only a few overlaps between the genomes, this could not be verified with confidence. *Pelomonas* is an understudied genus, and there are only a few reports concerning their metabolism, one suggesting that *Pelomonas* spp. could grow chemolithoautotrophically with hydrogen, and indicating the presence of CO_2 fixation genes (Gomila *et al.*, 2007). The sequences obtained from RACE-based genomes of three *Pelomonas* spp. all included carotenoid and haem/chlorophyll biosynthesis and CO_2 fixation genes, suggesting that these cells might be able to perform photosynthesis in their natural habitat, the Red Sea. Genes identified that are involved in CO_2 fixation include phosphoenolpyruvate carboxylase (from P709-11) which converts bicarbonate to oxaloacetate; a gene involved in making carboxysomes (for CO_2 concentrating (from P709-11); and a CO dehydrogenase that is involved in carbon fixation *via* the reductive acetyl-CoA pathway (from P728-5). It is possible that these cells are chemolithotrophs/oligotrophs and use these functional genes to survive in an oligotrophic system such as Red Sea.

A few genes encoding haem/chlorophyll biosynthesis enzymes were also recovered in the three *Pelomonas* spp., as well as in the *Bradyrhizobium* spp. B728-3, *Halomonas* spp. H808-5 and *Shigella* spp. S709-6. Those genes included magnesium chelatase subunits, ferrochelatase and several enzymes upstream involved in protoporphyrin IX synthesis (Table 1 and summary of genome sequences). After performing a best match search against the NCBI GenBank database (Altschul *et al.*, 1990), putative novel genes were found in the sorted samples. These genes have homology to well-characterized genes but contain novel sequences. For example: a novel gene fragment (GenBank accession no. KX246394) encoding geranylgeranyl pyrophosphate synthase (CrtE) was found in *Bradyrhizobium* spp. B728-3 (Fig. S7A); and a gene (GenBank accession no. KX246395) encoding CobN/magnesium chelatase from *Pelomonas* spp. P728-5 is 87% homologous to the cobalamin gene *cob*N in *Bradyrhizobium* sp. S23321 DNA and 71% homologous to a gene encoding a CobN/magnesium chelatase in *Rhodobacter sphaeroides* (Fig. S7B). Alignment of two putative *shc* genes found in H808-5 genomic sequence (Fig. S7C and D, GenBank accession no. KX246396 and KX246397) indicates that they are novel. SHC encoded by *shc* can catalyse very complex reactions, and any chemical mimicry of these reactions has so far been unsuccessful (Siedenburg and Jendrossek, 2011). Using single-cell genomics, a link between genes and the identities of their host cells can be established, which may reveal the function of these microorganisms in their natural environment. This approach illustrates how genomic studies at the single-cell level can provide information on unculturable and so far unknown microorganisms.

RACE is able to distinguish cells with different Raman spectra and fluorescence

Raman micro-spectroscopy is not only able to distinguish cells with different Raman spectra but also can be used

to detect fluorescence at the single-cell level which usually produces a large broad 'bump' spectrum dwarfing the Raman signal of SCRS (Fig. 3A). Fluorescence in some cells can be photo-bleached by extending laser exposure time (Huang *et al.*, 2007a,b). Since SCRS is independent of laser excitation, fluorescence interference in SCRS can be overcome by changing the incident lasers. SCRS are usually more informative than fluorescence. A RACE system able to distinguish and sort cells based on SCRS could, therefore, be more useful for the study of bacterial phenotypes.

Conclusion

To our knowledge, this is the first report of the use of Raman-activated cell sorting coupled to single-cell genomics. Single-cell genomics based on Raman sorting can not only sort cells with specific compounds (e.g. carotenoids in this case) but also isolate cells with carbon, nitrogen and general metabolic activity when it is coupled with stable isotope probing (Radajewski *et al.*, 2000), which would help link the specific metabolism of single cells (e.g. carbon, nitrogen substrate metabolism or general metabolic activity) (Huang *et al.*, 2004, 2007a,b; Li *et al.*, 2012a,b,c; Wang *et al.*, 2013; Berry *et al.*, 2015) and assist in defining the ecological functions of uncultivated bacteria in the environment.

Experimental procedures

Chemicals, microorganisms and growth conditions

All chemicals used in this study were purchased from Sigma-Aldrich (Dorset, UK) unless otherwise stated.

Escherichia coli JM109 (Promega, Southampton, UK) containing a p18GFP plasmid was incubated at 37°C in LB broth supplemented with 100 μg ml^{-1} ampicillin.

Seawater samples

The seawater sample was collected from the pier of the Inter-University Institute for Marine Sciences (IUI) in Eilat, Israel, on May 2015. The pier is 50 m long and the water depth is 5 m. The samples were concentrated 20 times using a Centricon Plus-70 Ultracel PL-100 (Merck Millipore, Tullagreen Carrigtwohill, Ireland) and then sent to the laboratory in Oxford on ice.

Confocal Raman micro-spectroscopy and spectral processing

Cells in concentrated seawater samples were washed before being analysed by Raman micro-spectroscopy to remove salts that interfere with cell observation. Each cellular suspension (1 μl) was mounted in the designed mini-wells of the RACE chip and allowed to air dry before Raman analysis (Fig. S1). The Raman spectra were acquired using a confocal Raman microscope (LabRAM HR Evolution; HORIBA Scientific, London, UK) equipped with an integrated microscope (BX41; Olympus, Essex, UK). A 50× long working distance objective (MPLFLN, NA 0.8; Olympus) was used to observe and acquire Raman signals from single cells and the overall magnification is 500. The laser beam position was calibrated and marked by the software Labspec6 (HORIBA Scientific). Cells were visualized by an integrated colour camera and a motorized XYZ stage (0.1 μm step). The Raman scattering was excited with a 532 nm Nd:YAG laser (Ventus, Laser Quantum, Manchester, UK). Raman measurements with grating 600 l mm^{-1} resulted in a spectral resolution of $\sim$1 cm^{-1} with 1019 data points. The laser power on a single cell was about 0.5 mW. The detector was a $-70°$C air-cooled charge coupled device detector (Andor, Belfast, UK). LabSpec6 software was also used to control the Raman system and acquire the Raman spectra. Acquisition times for Raman spectra were 5 s for single-cell measurements. The spatial location of each measured cell from the Red Sea sample were recorded in Labspec6 and used to identify cells for subsequent single-cell ejection.

Integration of single-cell ejection into a Raman microspectroscope

A 532 nm pulsed laser (ALPHALAS GmbH, Goettingen, Germany) was integrated into a confocal Raman microscope, sharing the same continuous laser (532 nm) light path and objective lens for SCRS measurement without changing mirrors and filters (Fig. 1).

The peak power was 1.1 kW, about two magnitudes lower than the traditional UV microdissection lasers (Table S1). The power of this 532 nm pulsed laser is maximally 3.4 mJ cm^{-2} (Table S1), which is much lower than the maximum resistible power (400 mJ cm^{-2}) of the mirror and edge filters in the Raman microscope. An exposure test showed that the 532 nm pulsed laser did indeed cause no damage to the mirror and edge filters in this LabRAM HR Evolution Raman microscope (data not shown).

Sterilization condition and reagents for single-cell genomics

DNA contamination is a major issue in single-cell genomics, due to the very small amount of template DNA derived from a single cell which has to be amplified by WGA. Standard sterilization practices are, therefore, not sufficient for single-cell WGA and a modified WGA method had to be developed (Rinke et al., 2013, 2014) to minimize the influence of contaminant DNA. All the equipment and working surfaces were wiped with 10% (v/v) domestic bleach (Domestos, Surrey, UK). All the consumables were autoclaved and exposed to a UV light bulb (254 nm) inside a UV sterilized Laminar Flow cabinet for 1 h. All reagents were carefully sterilized, and the detailed process is described in Supplementary Information.

Raman-activated single-cell ejection (RACE)

Supplementary information provides the details of reagent preparation and procedures. Briefly, 1 μl of the lysis buffer and 1 μl TE buffer were added into each well on the collection chip, which was then attached to the RACE chip. The enclosed chips were moved to the stage of the Raman microscope with the RACE chip facing down. The target cells were located by their coordinates and confirmed by comparing the bright-field image taken during Raman spectra acquisition. The laser spot was manually aligned to the coating layer under the target cell. The target cells were ejected by evaporation of the coating layer upon application of a <1 s exposure of the 532 nm pulsed laser, and the cells were harvested in the collector. For a 12-well RACE chip, at least two control wells were also set up: one negative control, which remained cell-free, and one positive control, to which cell suspensions were added before WGA (see Data S1).

Whole-genome amplification on-chip

The enclosed chips were carefully moved to a laminar flow chamber and the two chips were separated. A sterile coverslip was then placed on top of the collector chip. Wells were separated and independent with interference, and each well is a closed space. Cell lysis was carried out with lysis buffer (Qiagen, Manchester, UK) pre-added in the collection chip (see Data S1). Three freeze-thaw cycles were performed to enhance cell lysis. Subsequently, the chip was heated at 65°C for 10 min in a thermocycler (C1000; Bio-Rad, Hemel Hempstead, UK) to ensure a complete lysis of cells. After adding 1 μl of Stop solution to neutralize the lysis buffer, 12 μl of reaction master mix (containing phi29 polymerase from Epicenter, Cambridge, Camlab, UK) was added to each well (Data S1). The collector chip was then covered by a coverslip and kept in the thermocycler at 30°C with the hot lid activated and set at 70°C. After incubation for 8 h, the Phi29 DNA polymerase was deactivated by heating to 65°C for 10 min. The MDA product was then transferred into sterilized 200 μl PCR tubes for storage.

Library preparation and sequencing

Amplified DNA from single cells and cell consortia was quantified using a Qubit 1.0 fluorometer and a dsDNA

HR assay kit (Life Technologies, Darmstadt, Germany). Libraries were constructed using a Nextera XT DNA Library Preparation Kit (Illumina, Cambridge, UK) and 1 ng of amplified input DNA, as per the manufacturer's instructions. Sequencing was done on an Illumina MiSeq machine with paired end settings and 301 cycles per read.

Sequence processing and assembly

The raw sequences were subjected to adapter clipping and quality trimming using Trimmomatic (Bolger *et al.*, 2014) with the following arguments: 'LEADING:3 TRAILING:3 SLIDINGWINDOW:4:15 MINLEN:60'. Residual contaminant PhiX reads were removed from the datasets using FastQ Screen (http://www.bioinformatics.babraham.ac.uk/). In order to minimize subjective bias, no other potential contaminants from *E. coli* and human DNA were filtered from the datasets. To reduce MDA-induced bias of overrepresented genome regions, the standard three-step digital normalization protocol of the khmer suite was employed (Crusoe *et al.*, 2015). The normalized reads were then assembled using Spades v3.6 (Bankevich *et al.*, 2012).

Sequence analyses

For each assembly, the genome completeness and degree of contamination was estimated using CheckM (Parks *et al.*, 2015). Prokka v1.10 was used for gene prediction and annotation. Universal marker gene products were extracted using fetchMG (http://www.bork.embl.de/software/mOTU/fetchMG.html), compared against the NCBI nr database using BLAST and phylogenetically classified using MEGAN5 (Huson and Weber, 2013). For 16S rRNA gene sequences extracted from assembled genomes or generated from screening PCRs, the rough taxonomical placement was determined based on BLAST searches against the NCBI bacterial 16S RNA database, and suitable reference sequences originating from described type strains were selected accordingly. All 16S rRNA sequences were aligned against the SILVA database using Sina (Pruesse *et al.*, 2012) and incorporated into phylogenetic trees using Arb v5.5 (Ludwig *et al.*, 2004). Only sequences longer than 1200 bp were used for calculation of the basic tree backbones, while sequences shorter than 1200 bp were subsequently added using the parsimony function of Arb.

Conflict of interest

The authors declare no conflict of interest.

References

Altschul, S.F., Gish, W., Miller, W., Myers, E.W., and Lipman, D.J. (1990) Basic local alignment search tool. *J Mol Biol* **215**: 403–410.

Amann, R.I., Ludwig, W., and Schleifer, K.H. (1995) Phylogenetic identification and in-situ detection of individual microbial-cells without cultivation. *Microbiol Rev* **59**: 143–169.

Bankevich, A., Nurk, S., Antipov, D., Gurevich, A.A., Dvorkin, M., Kulikov, A.S., *et al.* (2012) SPAdes: a new genome assembly algorithm and its applications to single-cell sequencing. *J Comput Biol* **19**: 455–477.

Berry, D., Mader, E., Lee, T.K., Woebken, D., Wang, Y., Zhu, D., *et al.* (2015) Tracking heavy water (D$_2$O) incorporation for identifying and sorting active microbial cells. *Proc Natl Acad Sci USA* **112**: E194–E203.

Blanco, L., Bernad, A., Lazaro, J.M., Martin, G., Garmendia, C., and Salas, M. (1989) Highly efficient DNA synthesis by the phage phi 29 DNA polymerase. DNA replication. *J Biol Chem* **264**: 8935–8940.

Bohandy, J., Kim, B.F., Adrian, F.J., and Jette, A.N. (1988) Metal deposition at 532 nm using a laser transfer technique. *J Appl Phys* **63**: 1158–1162.

Bolger, A.M., Lohse, M., and Usadel, B. (2014) Trimmomatic: a flexible trimmer for Illumina sequence data. *Bioinformatics* **30**: 2114–2120.

Bryant, D.A., and Frigaard, N.-U. (2006) Prokaryotic photosynthesis and phototrophy illuminated. *Trends Microbiol* **14**: 488–496.

Chen, J., and Griffiths, M.W. (1998) PCR differentiation of *Escherichia coli* from other Gram-negative bacteria using primers derived from the nucleotide sequences flanking the gene encoding the universal stress protein. *Lett Appl Microbiol* **27**: 369–371.

Crusoe, M.R., Alameldin, H.F., Awad, S., Boucher, E., Caldwell, A., Cartwright, R., *et al.* (2015) The khmer software package: enabling efficient nucleotide sequence analysis. *F1000Research* **4**: 900–900.

Daniel, R. (2005) The metagenomics of soil. *Nat Rev Microbiol* **3**: 470–478.

Fernandez-Pradas, J.M., Colina, M., Serra, P., Dominguez, J., and Morenza, J.L. (2004) Laser-induced forward transfer of biomolecules. *Thin Solid Films* **453**: 27–30.

Garcia-Asua, G., Lang, H.P., Cogdell, R.J., and Hunter, C.N. (1998) Carotenoid diversity: a modular role for the phytoene desaturase step. *Trends Plant Sci* **3**: 445–449.

Gomila, M., Bowien, B., Falsen, E., Moore, E.R.B., and Lalucat, J. (2007) Description of *Pelomonas aquatica* sp nov and *Pelomonas puraquae* sp nov., isolated from industrial and haemodialysis water. *Int J Syst Evol Microbiol* **57**: 2629–2635.

Gonzalez, J.M., Pinhassi, J., Fernandez-Gomez, B., Coll-Llado, M., Gonzalez-Velazquez, M., Puigbo, P., *et al.* (2011) Genomics of the proteorhodopsin-containing marine flavobacterium Dokdonia sp strain MED134. *Appl Environ Microbiol* **77**: 8676–8686.

Hedlund, B.P., Dodsworth, J.A., Murugapiran, S.K., Rinke, C., and Woyke, T. (2014) Impact of single-cell genomics and metagenomics on the emerging view of extremophile "microbial dark matter". *Extremophiles* **18**: 865–875.

Hopp, B., Smausz, T., Kresz, N., Barna, N., Bor, Z., Kolozs-vari, L., *et al.* (2005) Survival and proliferative ability of various living cell types after laser-induced forward transfer. *Tissue Eng* **11**: 1817–1823.

Huang, W.E., Griffiths, R.I., Thompson, I.P., Bailey, M.J., and Whiteley, A.S. (2004) Raman microscopic analysis of single microbial cells. *Anal Chem* **76**: 4452–4458.

Huang, W.E., Bailey, M.J., Thompson, I.P., Whiteley, A.S., and Spiers, A.J. (2007a) Single-cell Raman spectral profiles of *Pseudomonas fluorescens* SBW25 reflects in vitro and in planta metabolic history. *Microb Ecol* **53**: 414–425.

Huang, W.E., Stoecker, K., Griffiths, R., Newbold, L., Daims, H., Whiteley, A.S., and Wagner, M. (2007b) Raman-FISH: combining stable-isotope Raman spectroscopy and fluorescence in situ hybridization for the single cell analysis of identity and function. *Environ Microbiol* **9**: 1878–1889.

Huang, W.E., Ferguson, A., Singer, A., Lawson, K., Thompson, I.P., Kalin, R.M., *et al.* (2009a) Resolving genetic functions within microbial populations: in situ analyses using rRNA and mRNA stable isotope probing coupled with single-cell Raman-fluorescence in situ hybridization. *Appl Environ Microbiol* **75**: 234–241.

Huang, W.E., Ward, A.D., and Whiteley, A.S. (2009b) Raman tweezers sorting of single microbial cells. *Environ Microbiol Rep* **1**: 44–49.

Huang, W.E., Song, Y., and Xu, J. (2015) Single cell biotechnology to shed a light on biological 'dark matter' in nature. *Microb Biotechnol* **8**: 15–16.

Huson, D.H. and Weber, N. (2013) Microbial community analysis using MEGAN. In *Microbial Metagenomics, Metatranscriptomics, and Metaproteomics*. DeLong, E.F. (ed.). Elsevier Inc. The Netherlands: Chapter 21, pp. 465–485.

Kattamis, N.T., Purnick, P.E., Weiss, R., and Arnold, C.B. (2007) Thick film laser induced forward transfer for deposition of thermally and mechanically sensitive materials. *Appl Phys Lett* **91**: 171120.

Kochendoerfer, G.G., Lin, S.W., Sakmar, T.P., and Mathies, R.A. (1999) How color visual pigments are tuned. *Trends Biochem Sci* **24**: 300–305.

Krebs, R.A., Dunmire, D., Partha, R., and Braiman, M.S. (2003) Resonance Raman characterization of proteorhodopsin's chromophore environment. *Journal of Physical Chemistry B* **107**: 7877–7883.

Lasken, R.S. (2012) Genomic sequencing of uncultured microorganisms from single cells. *Nat Rev Microbiol* **10**: 631–640.

Li, M., Xu, J., Romero-Gonzalez, M., Banwart, S.A., and Huang, W.E. (2012a) Single cell Raman spectroscopy for cell sorting and imaging. *Curr Opin Biotechnol* **23**: 56–63.

Li, M., Canniffe, D.P., Jackson, P.J., Davison, P.A., FitzGerald, S., Dickman, M.J., *et al.* (2012b) Rapid resonance Raman microspectroscopy to probe carbon dioxide fixation by single cells in microbial communities. *ISME J* **6**: 875–885.

Li, M., Ashok, P.C., Dholakia, K., and Huang, W.E. (2012c) Raman-activated cell counting for profiling carbon dioxide fixing microorganisms. *J Phys Chem A* **116**: 6560–6563.

Ludwig, W., Strunk, O., Westram, R., Richter, L., Meier, H., Kumar, Y., *et al.* (2004) ARB: a software environment for sequence data. *Nucleic Acids Res* **32**: 1363–1371.

Majed, N., and Gu, A.Z. (2010) Application of Raman microscopy for simultaneous and quantitative evaluation of multiple intracellular polymers dynamics functionally relevant to enhanced biological phosphorus removal processes. *Environ Sci Technol* **44**: 8601–8608.

Parks, D.H., Imelfort, M., Skennerton, C.T., Hugenholtz, P., and Tyson, G.W. (2015) CheckM: assessing the quality of microbial genomes recovered from isolates, single cells, and metagenomes. *Genome Res* **25**: 1043–1055.

Paterson, E., Hall, J.M., Rattray, E.A.S., Griffiths, B.S., Ritz, K., and Killham, K. (1997) Effect of elevated CO_2 on rhizosphere carbon flow and soil microbial processes. *Glob Change Biol* **3**: 363–377.

Pique, A., Chrisey, D.B., Auyeung, R.C.Y., Fitz-Gerald, J., Wu, H.D., McGill, R.A., *et al.* (1999) A novel laser transfer process for direct writing of electronic and sensor materials. *Applied Physics a-Materials Science & Processing* **69**: S279–S284.

Pruesse, E., Peplies, J., and Gloeckner, F.O. (2012) SINA: accurate high-throughput multiple sequence alignment of ribosomal RNA genes. *Bioinformatics* **28**: 1823–1829.

Radajewski, S., Ineson, P., Parekh, N.R., and Murrell, J.C. (2000) Stable-isotope probing as a tool in microbial ecology. *Nature* **403**: 646–649.

Ringeisen, B.R., Chrisey, D.B., Pique, A., Young, H.D., Modi, R., Bucaro, M., *et al.* (2002) Generation of mesoscopic patterns of viable Escherichia coli by ambient laser transfer. *Biomaterials* **23**: 161–166.

Ringeisen, B.R., Kim, H., Barron, J.A., Krizman, D.B., Chrisey, D.B., Jackman, S., *et al.* (2004) Laser printing of pluripotent embryonal carcinoma cells. *Tissue Eng* **10**: 483–491.

Rinke, C., Schwientek, P., Sczyrba, A., Ivanova, N.N., Anderson, I.J., Cheng, J.F., *et al.* (2013) Insights into the phylogeny and coding potential of microbial dark matter. *Nature* **499**: 431–437.

Rinke, C., Lee, J., Nath, N., Goudeau, D., Thompson, B., Poulton, N., *et al.* (2014) Obtaining genomes from uncultivated environmental microorganisms using FACS-based single-cell genomics. *Nat Protoc* **9**: 1038–1048.

Robert, B. (2009) Resonance Raman spectroscopy. *Photosynth Res* **101**: 147–155.

Schleifer, K.H. (2004) Microbial diversity: facts, problems and prospects. *Syst Appl Microbiol* **27**: 3–9.

Siedenburg, G., and Jendrossek, D. (2011) Squalene-hopene cyclases. *Appl Environ Microbiol* **77**: 3905–3915.

Soo, R.M., Skennerton, C.T., Sekiguchi, Y., Imelfort, M., Paech, S.J., Dennis, P.G., *et al.* (2014) An expanded genomic representation of the phylum cyanobacteria. *Genome Biology and Evolution* **6**: 1031–1045.

Swan, B.K., Martinez-Garcia, M., Preston, C.M., Sczyrba, A., Woyke, T., Lamy, D., *et al.* (2011) Potential for chemolithoautotrophy among ubiquitous bacteria lineages in the dark ocean. *Science* **333**: 1296–1300.

Takaichi, S. (2008) Distribution and biosynthesis carotenoids. In *The Purple Phototrophic Bacteria*. Hunter, C.N. (ed.). The Netherlands: Springer Science, pp. 97–117.

Venter, J.C. (2003) Unleashing the power of genomics: understanding the environment and biological diversity. *Scientist* **17**: 8–8.

Venter, J.C., Remington, K., Heidelberg, J.F., Halpern, A.L., Rusch, D., Eisen, J.A., *et al.* (2004) Environmental genome shotgun sequencing of the Sargasso Sea. *Science* **304:** 66–74.

Wang, Y., Ji, Y., Wharfe, E.S., Meadows, R.S., March, P., Goodacre, R., *et al.* (2013) Raman activated cell ejection for isolation of single cells. *Anal Chem* **85:** 10697–10701.

Whitman, W.B., Coleman, D.C., and Wiebe, W.J. (1998) Prokaryotes: the unseen majority. *Proc Natl Acad Sci USA* **95:** 6578–6583.

Zhang, Q., Zhang, P., Gou, H., Mou, C., Huang, W.E., Yang, M., *et al.* (2015a) Towards high-throughput microfluidic Raman-activated cell sorting. *Analyst* **140:** 6163–6174.

Zhang, P., Ren, L., Zhang, X., Shan, Y., Wang, Y., Ji, Y., *et al.* (2015b) Raman-activated cell sorting based on dielectrophoretic single-cell trap and release. *Anal Chem* **87:** 2282–2289.

PERMISSIONS

All chapters in this book were first published in MB, by John Wiley & Sons Ltd.; hereby published with permission under the Creative Commons Attribution License or equivalent. Every chapter published in this book has been scrutinized by our experts. Their significance has been extensively debated. The topics covered herein carry significant findings which will fuel the growth of the discipline. They may even be implemented as practical applications or may be referred to as a beginning point for another development.

The contributors of this book come from diverse backgrounds, making this book a truly international effort. This book will bring forth new frontiers with its revolutionizing research information and detailed analysis of the nascent developments around the world.

We would like to thank all the contributing authors for lending their expertise to make the book truly unique. They have played a crucial role in the development of this book. Without their invaluable contributions this book wouldn't have been possible. They have made vital efforts to compile up to date information on the varied aspects of this subject to make this book a valuable addition to the collection of many professionals and students.

This book was conceptualized with the vision of imparting up-to-date information and advanced data in this field. To ensure the same, a matchless editorial board was set up. Every individual on the board went through rigorous rounds of assessment to prove their worth. After which they invested a large part of their time researching and compiling the most relevant data for our readers.

The editorial board has been involved in producing this book since its inception. They have spent rigorous hours researching and exploring the diverse topics which have resulted in the successful publishing of this book. They have passed on their knowledge of decades through this book. To expedite this challenging task, the publisher supported the team at every step. A small team of assistant editors was also appointed to further simplify the editing procedure and attain best results for the readers.

Apart from the editorial board, the designing team has also invested a significant amount of their time in understanding the subject and creating the most relevant covers. They scrutinized every image to scout for the most suitable representation of the subject and create an appropriate cover for the book.

The publishing team has been an ardent support to the editorial, designing and production team. Their endless efforts to recruit the best for this project, has resulted in the accomplishment of this book. They are a veteran in the field of academics and their pool of knowledge is as vast as their experience in printing. Their expertise and guidance has proved useful at every step. Their uncompromising quality standards have made this book an exceptional effort. Their encouragement from time to time has been an inspiration for everyone.

The publisher and the editorial board hope that this book will prove to be a valuable piece of knowledge for researchers, students, practitioners and scholars across the globe.

LIST OF CONTRIBUTORS

Wolf Röther, Jakob Birke and Dieter Jendrossek
Institute of Microbiology, University of Stuttgart, Stuttgart, Germany

Stephanie Grond and Jose Manuel Beltran
Institute of Organic Chemistry, Eberhard Karls Universitüat Tübingen, Tübingen, Germany

So Young Choi, Won Jun Kim and Sang Yup Lee
Metabolic and Biomolecular Engineering National Research Laboratory, Department of Chemical and Biomolecular Engineering (BK21 Plus Program), BioProcess Engineering Research Center, and KAIST Institute (KI) for the BioCentury, Korea Advanced Institute of Science and Technology (KAIST), 291 Daehak-ro, Yuseong-gu, Daejeon 34141, Korea

Seung Jung Yu and Sung Gap Im
Department of Chemical and Biomolecular Engineering (BK21 Plus Program), KAIST,291 Daehak-ro, Yuseonggu, Daejeon 34141, Korea.

Si Jae Park
Department of Chemical Engineering and Materials Science, Ewha Womans University, 52 Ewhayeodae-gil, Seodaemun-gu, Seoul 03760, Korea

Alan J. Stephen and Sophie A. Archer
Schools of Chemical Engineering

Rafael L. Orozco and Lynne E. Macaskie
School of Biosciences, University of Birmingham, Edgbaston, Birmingham, B15 2TT, UK

Beatriz Galán, Esther García-Fernández, Igor Martínez, Lorena Fernández-Cabezón and José L. García
Department of Environmental Biology, Centro de Investigaciones Biológicas, Consejo Superior de Investigaciones Científicas, Ramiro de Maeztu 9, 28040 Madrid, Spain

Iria Uhía
Department of Environmental Biology, Centro de Investigaciones Biológicas, Consejo Superior de Investigaciones Científicas, Ramiro de Maeztu 9, 28040 Madrid, Spain

MRC Centre for Molecular Bacteriology and Infection, Department of Medicine, Imperial College London, London SW7 2AZ, UK.

Esther Bahíllo, Juan L. de la Fuente and José L. Barredo
Department of Biotechnology, Gadea Biopharma, Parque Tecnológico de León, Nicostrato Vela s/n, 24009 León, Spain

Young Hoon Jung
School of Food Science and Biotechnology, Kyungpook National University, Daegu 41566, South Korea.

Sooah Kim, Jungwoo Yang and Kyoung Heon Kim
Department of Biotechnology, Graduate School, Korea University, Seoul 02841, South Korea.

Jin-Ho Seo
Department of Agricultural Biotechnology and Center for Food and Bioconvergence, Seoul National University, Seoul 08826, South Korea.

Stephanie Karmann
Institute of Life Technologies, University of Applied Sciences and Arts Western Switzerland (HES-SO Valais), Sion, Switzerland.
Department of Biosystems Science and Engineering, ETH Zurich (ETHZ), Basel, Switzerland.

Stéphanie Follonier and Manfred Zinn
Institute of Life Technologies, University of Applied Sciences and Arts Western Switzerland (HES-SO Valais), Sion, Switzerland.

Daniel Egger and Dirk Hebel
Infors AG, Bottmingen, Switzerland.

Sven Panke
Department of Biosystems Science and Engineering, ETH Zurich (ETHZ), Basel, Switzerland.

Jun-hua Liu, Meng-ling Zhang, Rui-yang Zhang, Wei-yun Zhu and Sheng-yong Mao
Laboratory of Gastrointestinal Microbiology, College of Animal Science and Technology, Nanjing Agricultural University, Nanjing, Jiangsu Province, China.

Silvio Matassa
Center of Microbial Ecology and Technology
(CMET), Ghent University, Coupure Links 653,
9000 Gent, Belgium.
Avecom NV, Industrieweg 122P, 9032 Wondelgem,
Belgium.

Nico Boon
Center of Microbial Ecology and Technology
(CMET), Ghent University, Coupure Links 653,
9000 Gent, Belgium.

Ilje Pikaar
The School of Civil Engineering, The University of
Queensland, St. Lucia, QLD 4072, Australia.

Willy Verstraete
Center of Microbial Ecology and Technology
(CMET), Ghent University, Coupure Links 653,
9000 Gent, Belgium.
Avecom NV, Industrieweg 122P, 9032 Wondelgem,
Belgium.
KWR Watercycle Research Institute, Post Box 1072,
3430, BB Nieuwegein, The Netherlands

**Vincent Peck, Liliana Quiza, Jean-Philippe Buffet,
Mondher Khdhiri, Audrey-Anne Durand, Claude
Guertin and Philippe Constant**
INRS-Institut Armand-Frappier, 531 boulevard des
Prairies, Laval, Québec Canada H7V 1B7.

Alain Paquette
Centre d'étude de la forêt, Université du Québec
à Montréal, Case postale 8888, succursale Centre-
ville, Montréal, Québec Canada H3C 3P8.

Nelson Thiffault
Centre d'étude de la forêt, Université du Québec
à Montréal, Case postale 8888, succursale Centre-
ville, Montréal, Québec Canada H3C 3P8.
Direction de la recherche forestiére, Ministére des
Forêts, de la Faune et des Parcs, 2700 Einstein,
Québec, Québec Canada G1P 3W8.

Christian Messier
Centre d'étude de la forêt, Université du Québec
à Montréal, Case postale 8888, succursale Centre-
ville, Montréal, Québec Canada H3C 3P8.
Institut des Sciences de la Forêt Tempérée (ISFORT),
Université du Québec en Outaouais (UQO), 58 rue
Principale, Ripon, Québec Canada J0V 1V0.

Nadyre Beaulieu
Produits Forestiers Résolu, 2419 Route 155 sud, La
Tuque, Québec Canada G9X 3N8

**Magali Ranchou-Peyruse, Marion Guignard and
Anthony Ranchou-Peyruse**
Université de Pau et des Pays de l'Adour, Equipe
Environnement et Microbiologie, IPREM-CNRS
5254, F-64013 Pau, France.

Cyrielle Gasc and Pierre Peyret
Université d'Auvergne, EA 4678 CIDAM, 63001
Clermont-Ferrand, France.

Thomas Aüllo
TIGF – Transport et Infrastructures Gaz France, 40
Avenue de l'Europe, CS20522, 64000 Pau, France.

David Dequidt
STORENGY – Geosciences Department, Bois-
Colombes, France

**Yilin Ren, Dechuan Meng, Jinchun Chen and
Qiong Wu**
Center for Synthetic and Systems Biology, School
of Life Science, Tsinghua-Peking Center for Life
Sciences, Tsinghua University, Beijing 100084, China.

Linping Wu
Guangzhou Institutes of Biomedicine and Health,
Chinese Academy of Sciences, Guangzhou 510530,
People's Republic of China.

Guo-Qiang Chen
Center for Synthetic and Systems Biology, School
of Life Science, Tsinghua-Peking Center for Life
Sciences, Tsinghua University, Beijing 100084, China.
Center for Nano and Micro Mechanics, Tsinghua
University, Beijing 100084, China.
MOE Key Lab of Industrial Biocatalysis, Dept
Chemical Engineering, Tsinghua University, Beijing
100084, China

**Olga Revelles, J. Luis García and M. Auxiliadora
Prieto**
Centro de Investigaciones Biológicas, CSIC, C/
Ramiro de Maeztu, 9, 28040 Madrid, Spain.

**Daniel Beneroso, J. Angel Menéndez and Ana
Arenillas**
Instituto Nacional del Carbón, CSIC, Apartado 73,
33080 Oviedo, Spain

Andrew Stevenson, Philip G. Hamill and John E. Hallsworth
Institute for Global Food Security, School of Biological Sciences, MBC, Queen's University Belfast, Belfast BT9 7BL, UK.

Jan Dijksterhuis
CBS-KNAW Fungal Biodiversity Centre, Uppsalalaan 8, CT 3584, Utrecht, The Netherlands.

María Teresa Zamarro, María J. L. Barragán, Manuel Carmona, José Luis García and Eduardo Díaz
Centro de Investigaciones Biológicas, CSIC, C/ Ramiro de Maeztu, 9, 28040 Madrid, Spain

Jong Pyo Chae, Edward Alain B. Pajarillo, Ju Kyoung Oh and Dae-Kyung Kang
Department of Animal Resources Science, Dankook University, Cheonan 330-714, Korea.

Heebal Kim
Department of Agricultural Biotechnology, Seoul National University, Seoul 151–921, Korea

Safa Ben Azoun, Aicha Eya Belhaj and Héla Kallel
Laboratory of Molecular Microbiology, Vaccinology and Biotechnology Development, Biofermentation Unit, Institut Pasteur de Tunis, 13, place Pasteur. BP. 74, Tunis, 1002, Tunisia.

Rebecca Göngrich and Brigitte Gasser
Department of Biotechnology, BOKU - University of Natural Resources and Life Sciences Vienna, Muthgasse 18, Vienna, 1190, Austria

Felice Quartinello and Alessandro Pellis
Department of Agrobiotechnology IFA-Tulln, University of Natural Resources and Life Sciences Vienna, Inst. of Environ. Biotech., Konrad Lorenz Strasse 20, 3430, Tulln a. d. Donau, Austria

Simona Vajnhandl, Julija Volmajer Valh, Vončina and Alexandra Lobnik
Laboratory for Chemistry and Environmental Protection, Institute of Engineering Materials and Design, Faculty of Mechanical Engineering, University of Maribor, Smetanova ulica 17, 2000, Maribor, Slovenia

Thomas J. Farmer and Bojana Enrique Herrero Acero
Green Chemistry Centre of Excellence, Department of Chemistry, University of York, Heslington, York YO10 5DD, UK

Georg M. Guebitz
Austrian Centre of Industrial Biotechnology, Division Polymers & Enzymes, Konrad Lorenz Strasse 20, 3430, Tulln a. d. Donau, Austria

Kai Antweiler and Siegfried Kropf
Department for Biometry and Medical Informatics, Otto-von-Guericke University Magdeburg, Magdeburg, Germany

Susanne Schreiter, Kornelia Smalla and Holger Heuer
Department of Epidemiology and Pathogen Diagnostics, Julius Kühn-Institut – Federal Research Centre for Cultivated Plants, Braunschweig, Germany

Jens Keilwagen
Department of Biosafety in Plant Biotechnology, Julius Kühn-Institut – Federal Research Centre for Cultivated Plants, Quedlinburg, Germany

Petr Baldrian
Laboratory of Environmental Microbiology, Institute of Microbiology of the CAS, Prague, Czech Republic.

Rita Grosch
Leibniz Institute of Vegetable and Ornamental Crops, Grossbeeren, Germany

Wei E. Huang, Yizhi Song and Ian P. Thompson
Department of Engineering Science, University of Oxford, Parks Road, Oxford OX1 3PJ, UK

Anne-Kristin Kaster, John Vollmers and Martinique Frentrup
Leibniz Institute DSMZ, Deutsche Sammlung von Mikroorganismen und Zellkulturen GmbH, Inhoffenstrasse 7 B, 38124, Braunschweig, Germany

Yanqing Song and Huabing Yin
Division of Biomedical Engineering, School of Engineering, University of Glasgow, Glasgow G12 8QQ, UK

Index

A

Anaerobic Bioconversion, 134-135, 140
Androstadienedione, 29, 40
Androstenedione, 29, 40
Antifungal Biocontrol Strain, 173-174
Aromatic Compounds, 46, 134-135, 137, 140
Aspergillus Flavus, 132
Aspergillus Penicillioides, 123-126, 128-129

B

Bacterial Microbiota, 63-64, 67-69, 71, 73
Bio-based Plastic, 107
Bioattenuation, 96-97, 101, 104
Biocatalysts, 134-135
Biocompatibility, 9-10, 16-17, 19, 53
Biocompatibility Test, 19
Biohydrogen, 21-23, 26-28, 62, 77
Biological Interpretation, 44
Bioreactor, 18, 33-34, 53-56, 59, 61-62
Bioremediation, 103-104, 173
Biosynthesis, 3, 10, 13, 17, 19-20, 62, 70, 108, 114-116, 140-141, 163, 189, 192-193, 197, 200
Bottlenecks, 21-22, 152, 154, 159-160, 163

C

Chromatography, 2-3, 18-19, 39, 43, 49, 53, 60-61, 64, 113-114, 119-120, 138, 171
Chromosomal Genes, 18-19
Clostridium Propionicum, 17, 108-109
Copolymer, 9, 20, 115-116, 118
Cyclization, 6-7

D

Desulphurization, 6
Dna Extraction, 38, 64, 85-86, 91

E

Ecological Integrity, 82-83
Escherichia Coli, 9-12, 16, 18-20, 22, 37-38, 50-51, 62, 107, 111, 114-116, 150-151, 161, 191, 197, 199-200
Ethylene Glycol, 11-13, 15, 18, 166
Eurotium Halophilicum, 123-126, 128-129, 132

F

Fed-batch Fermentation, 55, 57-60
Feed Digestion, 63, 71
Fermentation, 14, 21-27, 36, 42-44, 50-55, 57-63, 75-79, 107, 117-122, 146, 151
Furfural, 42-51
Future Sustainable Food, 74

G

Gas Chromatography, 18, 43, 53, 61, 64, 113, 138
Genetic Diversity, 97
Global Warming, 107
Glutathione Reductase, 157
Glyoxylate, 15, 47

H

Hybridization, 96-97, 99, 101, 103, 105-106, 200

I

Intracellular Metabolite
Profiling, 42

L

Lactulose, 142-144, 146-148, 150-151
Lignocellulose, 42, 50-51

M

Mechanical Site Preparation (MSP), 82
Megasphaera Elsdenii, 108
Microbial Diversity, 64, 89, 94-95, 98, 104, 143, 146, 187, 200
Microbial Hydrogen Production, 21
Microbial Protein, 74-77, 79
Microorganisms, 22, 27, 34, 42, 63, 74-75, 77, 79-80, 83, 88, 90, 96, 99-102, 104-105, 113, 117-118, 131, 134, 144, 174, 186, 189, 197, 200
Microwave Pyrolysis, 117, 122
Molecular Optimization, 152
Mono-aromatic Hydrocarbons, 96-97, 99, 101-102
Multifunctional Molecular Indicators, 82, 87
Municipal Solid Waste (MSW), 52
Mycobacterium Neoaurum, 31, 40
Mycobacterium Smegmatis, 29-30, 37, 39-40

O

Oligotrophic Aquifer, 96

P

Peptococcaceae, 96-97, 99-103

Pharmaceutical Steroid Synthons, 29, 37

Phb, 27, 52-60, 78, 80, 108, 110, 112, 116, 118-121, 134-140, 192

Phosphotransacetylase, 11, 13

Photosynthetic Bacteria, 24, 27, 191

Pichia Pastoris, 152, 155-156, 163-165

Plga, 9-17, 19-20

Poly(d-la-co-ga-co-d-2hb), 10, 13, 16-17

Polyhydroxyalkanoates (PHA), 107, 114, 117

Polyisoprene Cleavage, 3-4

Polylactide (PLA), 107, 115

Prebiotics, 142-144, 146-147, 150

Probiotic Enterococci, 142, 148

Probiotics, 71, 142-144, 146-147

Proteobacteria, 63, 66, 68-70, 85-89, 104, 141-142, 144

R

Rabies Virus Glycoprotein, 152-153, 161, 163-164

Rhodospirillum Rubrum, 28, 52, 57, 59, 61-62, 117-118, 122

Ruminal Content (RC), 63

Ruminal Epithelium (RE), 63

S

Saccharomyces Cerevisiae, 42, 48, 50-51, 62, 164

Solution Hybrid Selection (SHS), 99

Solvent Extraction, 1

Swine Faecal Microbiota, 142, 148, 150

Synbiotics, 142-144, 146

Syngas, 52-59, 61-62, 117-122

T

Terpolyester, 10, 107-108, 110-112, 114

Tolerability, 173

X

Xeromyces Bisporus, 123-126, 128-129

Xerophilic Fungi, 123, 132

Xerophilicity, 123-124, 127

Xylose, 9-12, 17-18, 45-46, 50, 77, 108, 115

Y

Yeast, 18, 42-43, 47-48, 50, 59, 61, 74-78, 80, 113, 124-126, 128-130, 152-153, 158, 161-165

www.ingramcontent.com/pod-product-compliance
Lightning Source LLC
Chambersburg PA
CBHW082029190326

41458CB00010B/3315